Ecological Studies

Volumes published since 1989 are listed at the end of this book.

Ecological Studies, Vol. 105

Analysis and Synthesis

Jürgen Overbeck Ryszard J. Chróst
Editors

Microbial Ecology of Lake Plußsee

With 143 illustrations, 1 in color

Springer-Verlag
New York Berlin Heidelberg London Paris
Tokyo Hong Kong Barcelona Budapest

Jürgen Overbeck
Max-Planck Institute for Limnology
August-Thienemann Strasse 2
Postfach 165
D-24302 Plön
Germany

Ryszard J. Chróst
Microbial Ecology Department
University of Warsaw
ul. Karowa 18
PL-00-927 Warsaw 64
Poland

Cover illustration: Three-dimensional projection of Lake Plußsee: azimuth = 60°, depth 10 times natural scale. (Courtesy H.-J. Krambeck)

Library of Congress Cataloging-in-Publication Data
Microbial ecology of Lake Plußsee/Jürgen Overbeck, Ryszard J. Chróst, editors;
contributors, D. Albrecht . . . [et al.].
p. cm. — (Ecological studies; v. 105)
Includes bibliographical references and index.
ISBN 0-387-94120-7
1. Freshwater microbiology—Germany. 2. Lake ecology—Germany.
3. Plusssee (Germany). I. Overbeck, Jürgen. II. Chróst, Ryszard.
III. Albrecht, D. (Dieter). IV. Series.
QR105.5.M495 1993
576′.15′0943512—dc20 93-5258

Printed on acid-free paper.

Production coordinated by Princeton Editorial Associates and managed by Henry Krell;
manufacturing supervised by Jacqui Ashri.
Typeset by Princeton Editorial Associates, Princeton, NJ.
Color separated by Veriscan, New York, NY; printed by New England Book Components, Hingham, MA.
Printed and bound by Edwards Brothers, Inc., Ann Arbor, MI.
Printed in the United States of America.

9 8 7 6 5 4 3 2 1

ISBN 0-387-94120-7 Springer-Verlag New York Berlin Heidelberg
ISBN 3-540-94120-7 Springer-Verlag Berlin Heidelberg New York

This volume is dedicated to the memory of Professor Waldemar Ohle (1908–1987), who was one of the first to show that microbial processes and dynamics are of the utmost importance in the functioning of lake ecosystems.

Preface

Since the mid-1960s, when the Department of General Limnology (later named the Department of Microbial Ecology) of the Max-Planck Institute for Limnology (Plön, Germany) was created, a variety of research projects has been carried out in Lake Plußsee on microbial ecophysiology and ecology. The Plußsee is a small (14-ha) but deep (29-m) naturally eutrophic lake. The funnel-shaped lake basin, minimal exposure to wind, and the absence of rivers or creeks result in an extremely stable stratification from the end of May to the end of October. The Plußsee is very suitable for limnological examination and especially for studies of microbial ecophysiology and ecology of the lake ecosystem.

The primary aim of the investigations discussed in this volume was to study in situ the range of microbial metabolism (predominantly bacteria) present in this lake with special reference to physiological and biochemical mechanisms of microbial adaptation, competition for resources, reproduction, and survival in an aquatic environment. This book does not describe an ecosystem study; it discusses instead the results of in situ studies in which ecophysiological, biochemical, and molecular approaches were applied to yield a better understanding of microbial activity (with special reference to carbon metabolism) and basic microbial ecological processes at the molecular, cellular, community, and lake ecosystem levels. All nutrient cycles are biochemically based processes, and microorganisms play a crucial role in those events; therefore, their complexity can only be understood if the basic control mechanisms of the microbial metabolism are known.

The chapters in this book were written by different specialists who studied particular aspects of microbial activity in Lake Plußsee throughout the last decade. A few of the contributions present earlier studies, but their scientific results and conclusions are still relevant in aquatic microbial ecology. Frequent comments of reviewers of multiauthored books are that the various chapters are uneven in quality or incompletely integrated. Yet it is the individuality of approaches that is the most striking, intrinsic, and inspiring characteristic of books reviewing research that is in progress. A book such as this cannot be comprehensive and all-encompassing, and many areas of aquatic microbial ecology must be omitted or treated minimally. It is hoped that both the topics selected and their method of treatment will be of interest to all concerned with the ecology of microorganisms, whether freshwater or marine. As a long-term study, begun in the 1960s by the senior editor, the book also reflects the development of methods and approaches in aquatic microbial ecology.

We want to express our gratitude to the contributors for coming in from the field long enough to prepare their manuscripts. We are also very indebted to Mrs. Renate Koht (Max-Planck Institute for Limnology, Plön) and Mrs. Nancy Weider for their help during preparation of this volume. The financial support of the Max-Planck Gesellschaft zur Förderung der Wissenschaften (Germany) for preparation of this book is greatly acknowledged.

Jürgen Overbeck

Plön, April 1993 Ryszard J. Chróst

Contents

Preface vii
Contributors xi

1. **Limnology of the Plußsee** 1
   Hans-Jürgen Krambeck, Dieter Albrecht, Barbara Hickel, Wolfgang Hofmann, and Hans-Hermann Arzbach
2. **Dissolved Organic Matter: Analysis of Composition and Function by a Molecular-Biochemical Approach** 24
   Uwe Münster and Dieter Albrecht
3. **Primary Production, Photosynthesis, and Chlorophyll$_a$ in the Plußsee** 63
   Maria-Elisabeth Meffert and Jürgen Overbeck
4. **Phytoplankton-Excreted Organic Carbon** 81
   Anne F. Sell
5. **Bacterial Secondary Production** 92
   Ryszard J. Chróst and Hakumat Rai
6. **Microbial Enzymatic Degradation and Utilization of Organic Matter** 118
   Ryszard J. Chróst
7. **Saprophytic and Oligotrophic Bacteria in the Plußsee** 175
   Jürgen Overbeck

8. Heterotrophic Potential of Bacteria 192
Jürgen Overbeck

9. Phosphoenolpyruvate Carboxylase: Allosteric Regulation at the Ecosystem Level by Dissolved Organic Matter 201
Jürgen Overbeck

10. Relations between *Limnothrix redekei* (Cyanophyta/Cyanobacteria) and Bacteria: Mutual Control of Growth by Extracellular Algal Products 217
Maria-Elisabeth Meffert

11. Zooplankton Grazing on Bacteria 229
Andrea Bemmer and Jürgen Overbeck

12. Nanoflagellate and Ciliate Grazing on Bacteria 251
Bodo G. Meier and Edith Reck

13. Viruses 270
Karl-Paul Witzel, Jutta Demuth, and Christian Schütt

14. Phototrophic Bacteria in the Plußsee: Ecology of the Sulfuretum 287
Rita Oberhäuser-Nehls, Konstantinos Anagnostidis, and Jürgen Overbeck

15. Iron Bacteria 326
Wolf-Dieter Schmidt and Jürgen Overbeck

16. Metabolism of Methane and the Carbon Cycle in the Plußsee 337
Monir Naguib

17. Bacterial Life in the Plußsee: General Remarks on Aquatic Microbial Ecology 356
Ryszard J. Chróst and Jürgen Overbeck

Index 367

Contributors

Dieter Albrecht	Max-Planck Institute for Limnology Department of Ecophysiology D-24302 Plön, Germany
Konstantinos Anagnostidis	Institute of Systematic Botany and Botanical Garden, University of Athens Athens, Greece
Hans-Hermann Arzbach	Institut für Hydrobiologie und Fischereiwissenschaft D-22765 Hamburg, Germany
Andrea Bemmer	Max-Planck Institute for Limnology Department of Microbial Ecology D-24302 Plön, Germany
Ryszard J. Chróst	Microbial Ecology Department University of Warsaw PL-00–927 Warsaw, Poland
Jutta Demuth	Institut für Virusforschung D-23701 Eutin-Sielbeck, Germany

Barbara Hickel	Max-Planck Institute for Limnology Department of Ecophysiology D-24302 Plön, Germany
Wolfgang Hofmann	Max-Planck Institute for Limnology Department of Ecophysiology D-24302 Plön, Germany
Hans-Jürgen Krambeck	Max-Planck Institute for Limnology Computer Center D-24302 Plön, Germany
Maria-Elisabeth Meffert	Max-Planck Institute for Limnology Department of Microbial Ecology D-24302 Plön, Germany
Bodo G. Meier	Max-Planck Institute for Limnology Department of Microbial Ecology D-24302 Plön, Germany
Uwe Münster	Max-Planck Institute for Limnology Department of Ecophysiology D-24302 Plön, Germany
Monir Naguib	Max-Planck Institute for Limnology Department of Microbial Ecology D-24302 Plön, Germany
Rita Oberhäuser-Nehls	Max-Planck Institute for Limnology Department of Microbial Ecology D-24302 Plön, Germany
Jürgen Overbeck	Max-Planck Institute for Limnology Department of Microbial Ecology D-24302 Plön, Germany
Hakumat Rai	Max-Planck Institute for Limnology Department of Ecophysiology D-24302 Plön, Germany
Edith Reck	Kreisverwaltung D-24302 Plön, Germany

Wolf-Dieter Schmidt — Regierung von Unterfranken
D-97070 Würzburg, Germany

Christian Schütt — Biologische Anstalt Helgoland
Abteilung Meeresmikrobiologie
Meeresstation
D-27498 Helgoland, Germany

Anne F. Sell — Technische Universität Dresden
Institut für Hydrobiologie
D-01277 Dresden, Germany

Karl-Paul Witzel — Max Planck Institute for Limnology
Department of Ecophysiology
D-24302 Plön, Germany

Aerial view of the Plußsee. (Photograph by Waldemar Ohle.)

The Plußsee

Location	54°10′N, 10°23′E
Height above sea level (m)	24.5
Area (ha)	14.3
Catchment area (ha)	42
Ratio catchment area/lake area	2.9
Volume (m^3)	1,348,224
Diameter (m)	approx. 400
Maximum depth (m)	29.2
Mean depth (m)	9.4
Volume of epilimnion (0–5 m depth)	44%

1. Limnology of the Plußsee

Hans-Jürgen Krambeck, Dieter Albrecht, Barbara Hickel, Wolfgang Hofmann, and Hans-Hermann Arzbach

Morphometry, Thermal Stratification, and Global Radiation

The Plußsee is situated in Schleswig-Holstein in northern Germany at 10° 23′ east and 54° 10′ north. During the last glacial period, about 12,000 years ago, the lake was formed as a kettle lake and has a very regular shape. Figure 1.1 shows the depth isopleths from 0 to 28 m and the ratio of depth to lake volume. The lake is situated in a valley surrounded by beech trees, which create a very effective wind shelter for the lake surface. The lake has no inflow except for such rare events as thunderstorms, when there is a lot of sudden precipitation. The mean annual precipitation is 750 mm as calculated from meteorological data of the last 77 years (1925–1992).

The funnel-shaped lake basin combined with its minimal exposure to wind and lack of disturbance by rivers or creeks result in an extremely stable stratification with high stability and low exchange rates, which can be seen in the clearly defined strata of bacteria below the thermocline (Overbeck 1968; Chapter 7, this

---

The sections on Morphometry, Thermal Stratification, and Global Radiation and on Impact of Physical Conditions on Limnological Processes were written by Hans-Jürgen Krambeck. The section on Chemistry of Lake Water was written by Dieter Albrecht. The section on Phytoplankton was written by Barbara Hickel. The section on Macrozooplankton was written by Wolfgang Hofmann. The section on Fish Fauna was written by Hans-Hermann Arzbach.

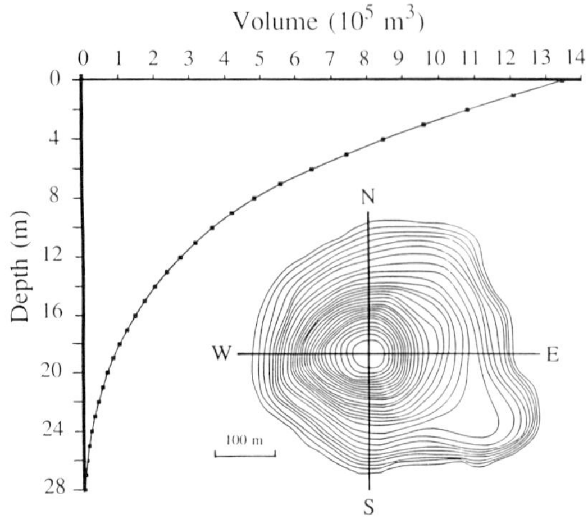

Figure 1.1. Depth isopleths (inset) and the relationship between the volume and depth of the Plußsee near Plön/Holstein (10° 23′ East, 54° 10′ North). The Plußsee area is 142,900 m^2; it has a volume of 1,345,000 m^3 and a mean depth of 9.41 m.

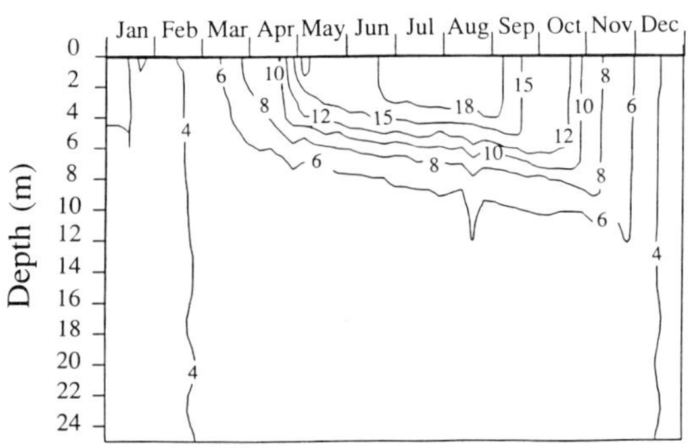

Figure 1.2. The representative annual pattern of the thermal stratification of the Plußsee in 1990. The lake is dimictic and has two water mixing periods in February–March and late November–December. The maximum temperature gradient throughout the thermocline is about 5°C/m during the summer stratification.

Table 1.1. Comparison of the Plußsee Exchange Coefficient *A* with Other Lakes

Lake	Area (km^2)	Maximum depth (m)	Depth considered (m)	Period considered	*A* (cm^2 sec^{-1} 10^{-2})
Plußsee	0.14	29.2	10	April–August	1.42
Mendota	39	25.6	12	15 June–15 August	7
Eastwhite water	1.0	16	12	July–August	3
Schleinsee	0.15	11.6	11	April–August	2

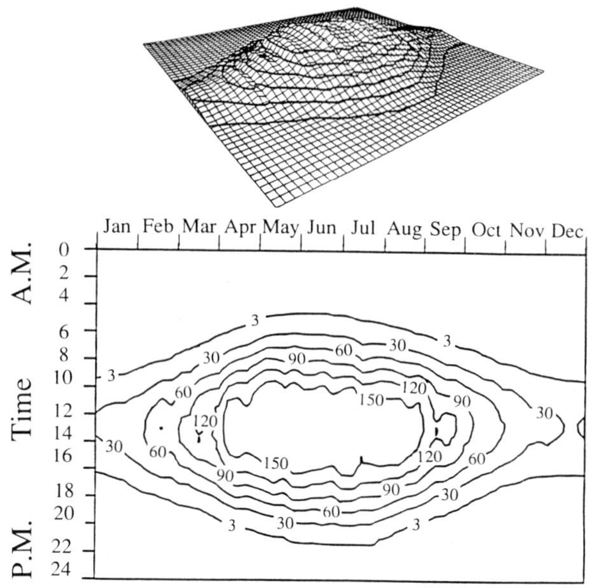

Figure 1.3. The annual global radiation at the surface of the Plußsee. Because of the lake's latitude of 54° North, the annual variation of global radiation is very high. The time between 3-J cm^{-2} h^{-1} isopleths shows the annual day length from sunrise to sunset.

volume). Figure 1.2 shows the regular annual pattern of thermal stratification of the Plußsee for 1990. Thermal stratification data can be used to calculate the stability of the lake as expressed by the mechanical work needed to make it homothermal and also to calculate the exchange rates that govern mass transport by means of eddy diffusion. The maximum stability of the lake is about 6×10^7 Nm (≈ 450 Nm/m^2), and the exchange coefficient A is 0.0014 ± 0.0003 cm^2/s (for the calculation method and further results, see Krambeck 1974). Table 1.1 presents the comparison of the Plußsee A coefficient with those of other lakes (Hutchinson 1957, Table 48).

Because the lake's latitude is quite high (about 54° N), the sun ascends to 59° in summer and to only 13° in winter. The light conditions thus cannot support primary production in the Plußsee during the winter months. Figure 1.3 shows a 10-year average of the global radiation at the lake surface. The 3-J cm^{-2} h^{-1} isoline describes approximately the time of sunrise and sunset.

Chemistry of Lake Water

This section presents examples of the results of chemical analyses of major nutrients and chemical parameters in the Plußsee. The physical and chemical conditions of the water were analyzed weekly for the past 25 years. The sampling and analytical methods are comparable and standardized for different years of the study. Lake water was sampled in the pelagial zone at depths of 0–15, 20, and 25 m. Water samples were taken with the use of a 2-L Ruttner sampler and collected in 500-ml polyethylene bottles. Samples taken for analysis of dissolved components were filtered through ash-free filters (Schleicher and Schuell, 589 Blue Ribbon).

Routine chemical analyses of water samples included fractions of phosphorus (dissolved reactive phosphate, dissolved total phosphorus, and particulate phosphorus), fractions of nitrogen (ammonia, nitrite, nitrate, dissolved organic nitrogen, and particulate nitrogen), and dissolved silicon dioxide. These compounds were analyzed by a Technicon continuous-flow system (Albrecht and Overbeck 1969, Albrecht 1973). Dissolved hydrogen sulfide was determined by photometric measurement of the buildup of methylene blue (Grasshoff et al. 1983).

Total water hardness (including calcium and magnesium) was determined by complexometry. Dissolved oxygen in the lake water was determined in situ with the use of an oxygen electrode. A glass electrode was used for pH measurement. Conductivity was determined with platinum electrodes placed 1 cm apart at 20°C.

Dissolved Oxygen

The Plußsee was very well stratified with respect to both temperature and oxygen. During the summer stratification, an epilimnion layer about 6 m deep developed. The concentration of oxygen depended on the amount of phytoplankton and their ability to photosynthesize. At the end of March or beginning of April, spring mixing of water masses and distinct phytoplankton development caused the whole water column to become very well oxygenated. The highest concentrations of dissolved oxygen (12–15 mg O_2/L) in the Plußsee were always determined during spring (Fig. 1.4). In the middle of June higher water temperature and an increase in the microbial heterotrophic activity involved in biological oxidation of organic matter strongly decreased the concentration of oxygen to less than 5 mg O_2/L at

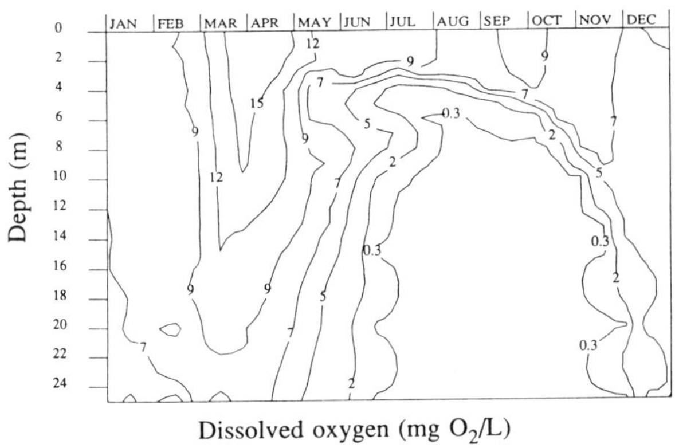

Figure 1.4. Spatial and temporal distribution of dissolved oxygen in the water column of the Plußsee in 1992.

depths below 10 m. Methanotrophic bacteria strongly influence the dynamics of oxygen in the lake. Up to 40–60% of the biological oxygen demand is used by methanotrophic bacteria for oxidation of methane (for further details see Chapter 16, this volume). From the beginning of July until the end of October, the lake water at depths below 10 m were anoxic. During the summer stratification in August and September, the anoxic part of the Plußsee extended up to the depth of 6 m (Fig. 1.4). During the autumn homothermy of the Plußsee, the mixing of water, which usually occurred at the end of November or beginning of December, again aerated the entire water column. During the winter, the concentration of dissolved oxygen in the water column varied from 5 to 9 mg O_2/L, and there were no anoxic conditions in the deep parts of the lake. For meromictic stages of the lake see Chapter 14, this volume.

pH of Water

Calcium hydrogen carbonate has buffering capacity that protects the water from rapid changes of pH. However, the carbonate–hydrogen carbonate buffer system is not as strong in the Plußsee as it is in other lakes of the Schleswig-Holstein area because of the high content of organic matter and relatively low concentrations of calcium and carbonate in the sediments. The annual variation of the pH of the Plußsee water samples is shown in Figure 1.5. During the spring algal bloom, as a result of the rapid utilization of carbon dioxide in the very active photosynthesis by algae, the pH increased to 9.6. From May to the beginning of October, the pH of epilimnetic water varies between 7.8 and 9.0. Characteristic pH values for water from the Plußsee at depths below 10–12 m do not fluctuate markedly and remain about pH 7.3 throughout the year.

Soluble Reactive Phosphorus

The concentration of soluble reactive phosphorus (SRP) in the euphotic zone of the Plußsee depended on the season and the activity of the microplankton. Algae

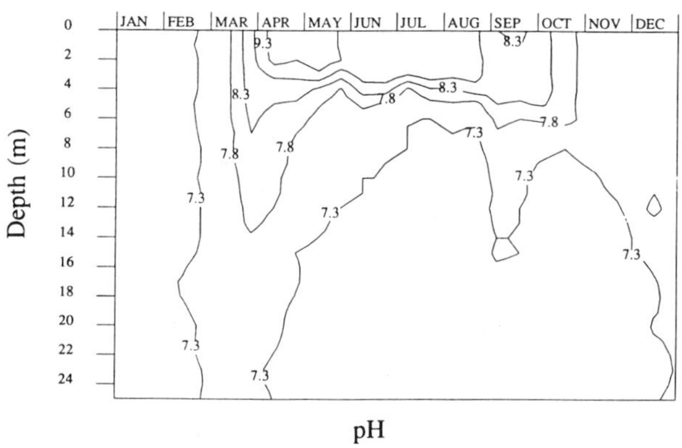

Figure 1.5. Annual variation of the pH values of Plußsee water in 1992.

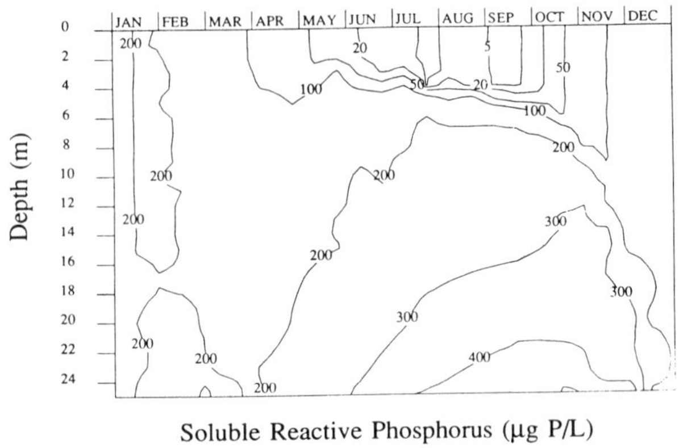

Figure 1.6. Distribution of soluble reactive phosphorus (SRP) in the entire water column of the Plußsee in 1992.

and bacteria were the most important planktonic organisms responsible for depletion of this nutrient in the lake epilimnion during summer stratification (Chróst 1988), when inorganic phosphorus content was below 5 μg P/L (Fig. 1.6). During the spring circulation of the lake water, the content of soluble reactive phosphorus (SRP) in the water column was about 200 μg P/L. The amount of SRP decreased rapidly in the euphotic zone as a result of phytoplankton blooms during spring. In the hypolimnion the SRP concentrations were high and reached values up to 400 μg P/L. This was caused by remineralization of organic phosphorus and exchange between sediment and hypolimnetic water as a result of solubilization of inorganic phosphorus complexes in sediments under anoxic conditions. After lake water mixing in autumn, the concentrations of SRP in the water column in the winter were relatively high, about 200 μg P/L (Fig. 1.6).

Inorganic Nitrogen

After phosphorus, inorganic nitrogen is the second most important nutrient for phytoplankton growth in the Plußsee. During spring plankton development the content of nitrate (similarly to SRP) in the euphotic zone of the Plußsee rapidly decreased from 100–250 μg N/L in March to below 25 μg N/L at the end of April and remained at similar concentrations until December (Fig. 1.7). The highest concentrations (400–600 μg N/L) of nitrate in the water column were often found from the end of March until August at depths below 14 m (March) to 10 m (July). In autumn the whole water column contained less than 20 μg N-NO_3^-/L.

Nitrite concentrations in the euphotic zone of the Plußsee were low and oscillated around 2 μg N/L (Fig. 1.8). From the autumn lake water mixing (November) until the end of the spring homothermy of the lake (May), the amounts of nitrite in the water column were low (below 2 μg N/L). During the summer stratification, from July to the end of September, nitrite content in the hypolimnion increased up

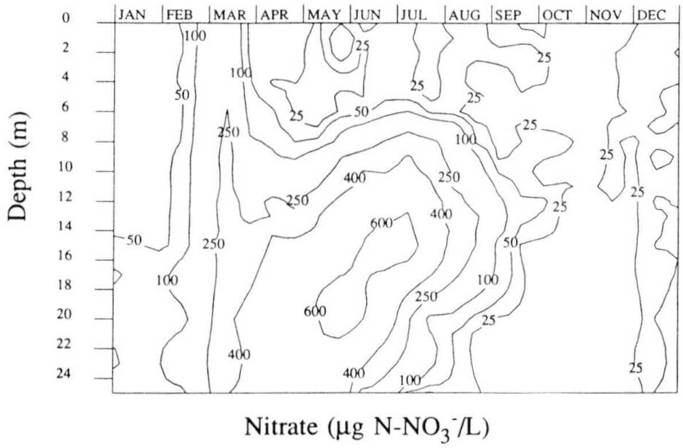

Figure 1.7. Concentration of nitrate in the water column of the Plußsee in 1992.

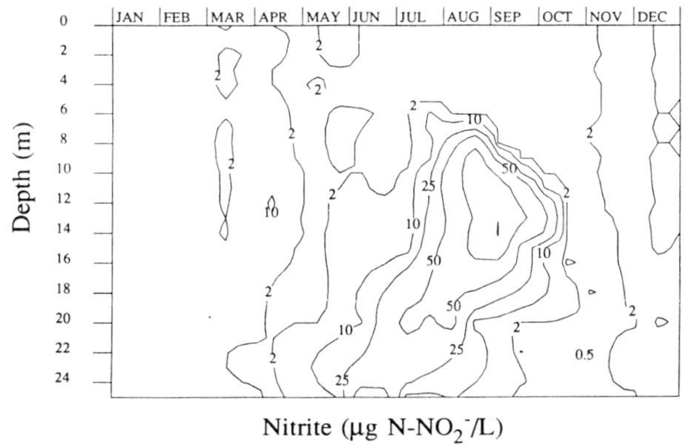

Figure 1.8. Concentration of nitrite in the water column of the Plußsee in 1992.

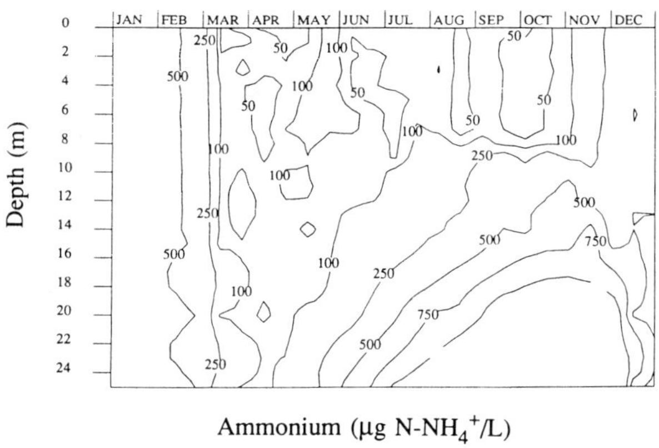

Figure 1.9. Concentration of ammonium ions in the water column of the Plußsee in 1992.

to 50 μg N/L and reached the maximum concentration of 150 μg N-NO_2^-/L at the beginning of September.

The highest concentrations of ammonia (500–700 μg N/L) in the Plußsee water column occurred after December, after the fall turnover. Because of nitrification, the content of ammonia decreases continuously during the winter, resulting in high amounts of nitrate. This microbial process proceeds slowly at the low temperatures of the water (Fig. 1.9). During the summer stratification of the Plußsee, ammonia content in the epilimnion varied between 25 and 50 μg N-NH_4^+/L. Hypolimnetic waters during the summer stagnation were rich with ammonia (500–1000 μg N/L), and the highest concentrations of ammonia 1500–2000 μg N-NH_4^+/L were usually found at the sediment–water interface (below 25 m).

Silicon

During the early part of spring, dissolved silicate in the water column varied between 350 and 400 μg Si/L (Fig. 1.10). The mass development of diatoms caused a rapid decrease of silicon dioxide to a concentration of about 20 μg Si/L at the end of March until the middle of April. During the summer period the epilimnion contained about 100 μg Si/L. Hypolimnetic water had very high concentrations of silicon (500–800 μg Si/L), which very often exceeded 1000 μg Si/L at depths below 20 m.

Conductivity

Conductivity of Plußsee water (because of the smaller amounts of magnesium and calcium) was smaller than that of other lakes of Schleswig-Holstein. In the Plußsee conductivity of the water column did not change significantly throughout the year and varied between 240 and 280 μS/cm (average 260 μS/cm).

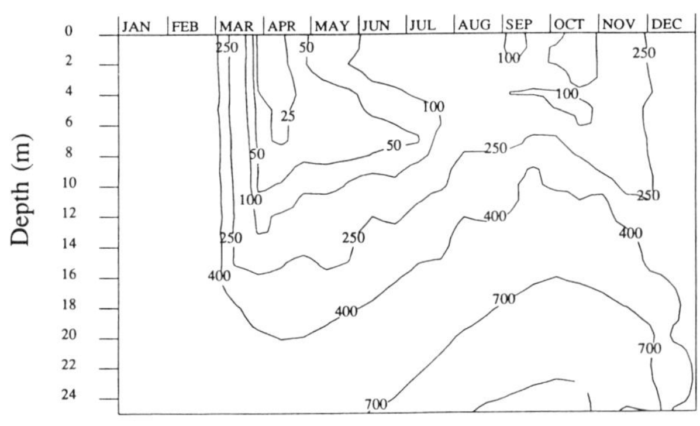

Figure 1.10. Dissolved silicon dioxide distribution in the water column of the Plußsee in 1992.

Phytoplankton

Phytoplankton Assemblage

Early studies of the phytoplankton species composition of the Plußsee were undertaken by Lemmermann (1893, 1895, 1903) and Utermöhl (1925). Later Overbeck (1968, 1972, 1973) studied the seasonal and vertical distribution of the phytoplankton. Long-term changes in the species composition since the earliest studies were analyzed by Hickel (1975).

The phytoplankton of the Plußsee is composed of at least 180 species common in eutrophic lakes of the temperate zone. Chlorophytes, chrysophytes, and cyanobacteria (cyanophytes) were each represented by more than 20 species. Long-term study of the phytoplankton (1972–1992) revealed marked changes in the species composition, periodicity, and relative abundance of the phytoplankton (Hickel 1988a). Years that were dominated by filamentous cyanophytes (*Limnothrix redekei, Aphanizomenon gracile, Planktothrix agardhii, P. rubescens*) were followed (after a year of transition) by blooms of dinoflagellates (*Ceratium furcoides, C. rhomvoides, C. hirundinella*) accompanied by *Anabaena flos-aquae* and *A. mendotae*. An increase in number of species, especially chroococcal cyanophytes and chlorophytes, was observed during the years after the disappearance of the filamentous cyanophytes. Some planktonic algae, common in other lakes of the area, such as species of *Microcystis* and *Aulacoseira,* are rare in the Plußsee.

An important component of the spring diatom bloom is *Stephanodiscus alpinus* along with other diatoms (*S. hantzschii, S. minutulus, Asterionella formosa, Cyclotella* spp.). This was frequent in 9 out of 19 years, and auxosporulation was observed several times (Hickel and Håkansson 1993). *S. alpinus* has been present throughout the postglacial period in Plußsee. By transmission electron microscopy, the largest number of species of chrysophytes in the lakes of East-Holstein was observed in the Plußsee (Hickel and Maaß 1989), and it had a remarkable abundance of species of *Paraphysomonas,* minute heterotrophic flagellates. Studies of the morphology, life cycle, sexual reproduction, and population structure of dinoflagellates confirmed *C. furcoides* as a distinct species forming blooms in the Plußsee (Hickel 1985, 1988b) and revealed *C. rhomvoides* as a new species (Hickel 1988c). Sexual reproduction by anisogamic copulation was reported for the first time in situ for *C. furcoides* and *C. rhomvoides* in the Plußsee. In contrast to *C. hirundinella,* sexual reproduction is a frequent process in the life history of *C. furcoides* and *C. rhomvoides,* with implications for the seasonal succession.

Seasonal Succession

Conspicuous year-to-year variations in the sequence and abundance of algal species are characteristic for the Plußsee. The development of the phytoplankton in spring is influenced by climatic conditions as years with and without ice cover alternate. The phytoplankton assemblages in spring are dominated by diatoms

(*Stephanodiscus, Cyclotella, Asterionella* species) and flagellates (cryptophytes, the haptophyte *Chrysochromulina parva*). In some years chlorophytes such as *Chlorella minutissima, Chlamydomonas, Carteria,* and *Pascherina* (Hickel 1980) are abundant during spring. During the years when cyanophages were dominant, a clear-water phase did not develop. After the shifts in species composition it appeared again. After a clear-water phase, an increase in cyanophyte abundance (*Anabaena, Aphanizomenon*) is followed by species of *Ceratium,* which form blooms. Large colonial chlorophytes have their maximum abundance in summer. In autumn, flagellates (*Ceratium,* cryptophytes) are dominant and accompanied by desmids (*Staurastrum, Cosmarium, Closterium*) and occasionally by diatoms (*Asterionella formosa, Stephanodiscus alpinus*). The filamentous cyanophytes *Planktothrix rubescens, P. agardhii,* and *Limnothrix redekei* were also dominant during winter and spring in those years when they dominated in summer. Cryptophytes, chrysophytes (*Mallomonas caudata*), desminds (*Closterium acutum* var. *variabile*), and diatoms (*Asterionella formosa, Stephanodiscus alpinus*) each prevailed in different years during winter after the shift in species composition. The largest number of algal units was produced by *Limnothrix redekei,* the nanoflagellates *Chrysochromulina parva* (Haptophyta), and *Rhodomonas minuta* (Cryptophyta). Among the picoplankton, the chlorophyte *Chlorella minutissima* and the cyanophyte *Synechococcus* sp. were most abundant.

Vertical Distribution

The vertical distribution of the phytoplankton in the water column is closely related to the development of thermal stratification and the oxygen distribution, which are stable from year to year (Overbeck 1968; see above). During homothermy the algae are distributed homogeneously throughout the water column. With the onset of the stable stratification from the end of March to November, resulting in a very shallow epilimnion, the phytoplankton is distributed in the upper 4-m layer (Hickel 1978), and phytoplankton maxima are generally found at the depths of 1 and 2 m. Occasionally a small maximum is developed in the metalimnion. With the destratification and mixing in the lake, the phytoplankton is gradually distributed from the surface layer to deeper depths.

Phytoplankton Biomass

The total biomass of the phytoplankton in the Plußsee exhibits marked annual (Fig. 1.11) and interannual variability (Fig. 1.12). The lowest values of biomass were observed during the winter period (November–February) and late fall, when primary production also reached its lowest values. Higher crops were produced by overwintering populations of cyanophytes, e.g., *Planktothrix rubescens,* in winter 1972–1973. Highest values of biomass are associated with highest values of primary production (Chapter 3, this volume). Maximal values of algal biomass were reached in summer and early autumn (June–September). The exceptions were the years 1978, 1979, and 1982, when the maximum was found during the spring bloom in March–April. The highest biovolume in the epilimnion was

Figure 1.11. Biomass of phytoplankton in the Plußsee in 1981 (A) and 1972 (B), when dinoflagellates and cyanobacteria formed summer blooms, respectively.

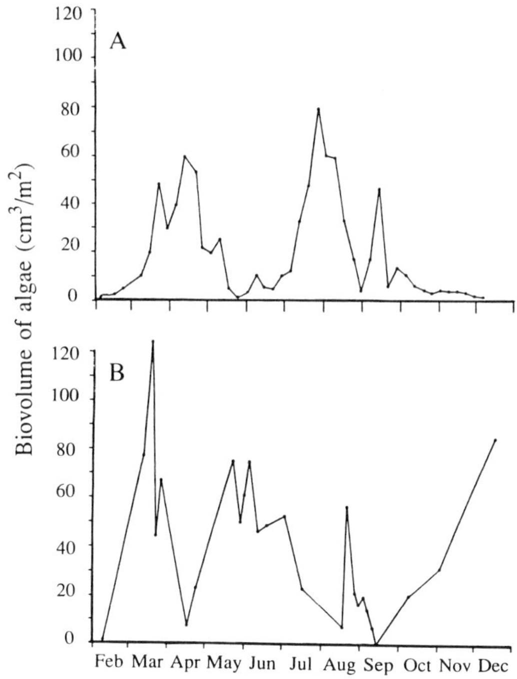

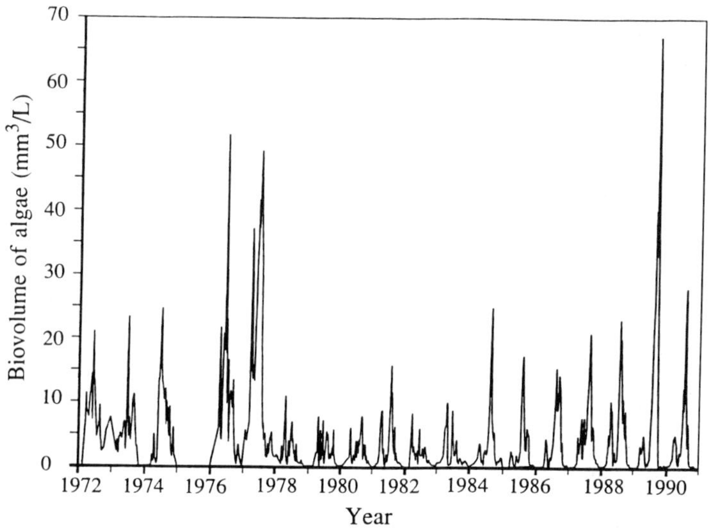

Figure 1.12. Long-term variations in phytoplankton biomass in the Plußsee from 1972 to 1990.

produced by the dinoflagellate *Ceratium hirundinella* in September 1989. The next highest values were from *Aphanizomenon gracile* in June 1976 and *Limnothrix redekei* in June 1977. The mean annual total biomass ranged between 1.8 mm^3/L in 1982 and 15 mm^3/L in 1977. The contribution of the cyanophytes to the total biomass ranged from 24% to 94% during the period 1972–1978 and decreased to 2–41% between 1979 and 1990 after the shift in species composition. In contrast, the dinoflagellates (*Ceratium, Peridinium*) represented 0.5–20% of the total biomass during those years, with dominance of Cyanophyta, and increased to 12–87% in the following years.

Macrozooplankton

The Zooplankton Assemblage

The list of the abundant plankton rotifers, cladocerans, and copepods of the Plußsee comprises the stock of zooplankton taxa typical for eutrophic north German lakes (Table 1.2). In the Plußsee and other lakes of the region the varieties

Table 1.2. Abundant and Representative Species[a] of Zooplankton in the Plußsee

Rotifera	Cladocera	Copepoda
Keratella quadrata	*Diaphanosoma brachyurum*	*Eudiaptomus gracilis*
Keratella hiemalis		*Diacyclops bicuspidatus*
Keratella cochlearis	*Daphnia cucullata*	*Mesocyclops leuckarti*
Kellicottia longispina	*Daphnia galeata*	*Thermocyclops oithonoides*
Notholca squamula	*Daphnia hyalina*	*Cyclops abyssorum*
Euchlanis dilatata	*Ceriodaphnia quadrangula*	*Cyclops vicinus*
Trichocerca capucina	*Bosmina longirostris*	
Trichocerca cylindrica	*Bosmina coregoni*	
Trichocerca similis	*Chydorus sphaericus*	
Ascomorpha ecaudis		
Asplanchna girodi		
Asplanchna priodonta		
Synchaeta oblonga		
Synchaeta pectinata		
Polyarthra dolichoptera		
Polyarthra vulgaris		
Polyarthra major		
Filinia hofmanni		
Filinia terminalis		
Conochilus natans		
Conochilus unicornis		

[a]Taxonomy refers to Koste (1978, 1980), Flößner (1972), and Kiefer (1978); identification of the *Cyclops* species is according to Einsle (1968).

tecta and *hispida* of *Keratella cochlearis* appear to represent separate units. They are not connected with *cochlearis* by morphological variation in the sense of "Lauterborn cycles" (Hofmann 1983). The abundance of *tecta* varies considerably from year to year independent of the dynamics of *K. cochlearis*. In the Plußsee, Wolf and Mort (1986) found hybrids among all three *Daphnia* species discernible by their allozyme patterns. *Bosmina coregoni* corresponds to Lieder's (1983) *B. c. coregoni*. The Plußsee is different from most of the other lakes of the region in that species of the genus *Brachionus* were found only on a few occasions and in very small numbers (*B. calyciflorus*). This particular characteristic of the Plußsee rotifer plankton had been mentioned by Voigt (1903), who did not find any *Brachionus* species.

Seasonal Succession

The seasonal dynamics of the zooplankton are characterized by two peaks following the spring and summer phytoplankton maxima (Krambeck 1981). The spring peak is caused by exponential growth phases occurring almost simultaneously in several rotifer populations. The most abundant species during this period are *Keratella cochlearis, Filinia terminalis, Polyarthra vulgaris (dolichoptera), Synchaeta pectinata,* and *Kellicottia longispina*. Some years *Conochilus natans* and *Asplanchna priodonta* were also found. The population of *Conochilus unicornis* typically increases toward the end of this period. During the summer peak cladocerans—*Diaphanosoma brachyurum, Daphnia cucullata, Ceriodaphnia quadrangula, Bosmina coregoni,* and *Chydorus sphaericus*—contribute to a major part of the zooplankton biomass. Among the rotifer plankton, the forms of *Keratella cochlearis* and *Polyarthra* spp. are predominant, whereas in strongly eutrophic lakes of this region taxa such as *Filinia longiseta* and *Brachionus angularis* play a major role (Hofmann 1981).

The seasonal succession of the Plußsee plankton rather closely follows the PEG model, as was shown on the basis of the 1969 data (Sommer et al. 1986). With respect to the zooplankton the following major events occurred. (1) There was an exponential increase of small planktonic herbivores, e.g., rotifers and *Bosmina longirostris,* in the spring. (2) Reduction of the phytoplankton biomass by zooplankton community feeding led to clear-water phase. (3) There was a subsequent decrease of the zooplankton biomass. (4) As a result of the size distribution of the following summer's phytoplankton bloom, e.g., the presence of large numbers of inedible algae, small cladoceran species predominated during the second zooplankton biomass peak: *Diaphanosoma brachyurum, Daphnia cucullata, Ceriodaphnia quadrangula, Bosmina coregoni,* and *Chydorus sphaericus*.

Hence, both zooplankton peaks are caused by small herbivores—rotifers in spring and cladocerans in summer. Large herbivores such as *Daphnia galeata* are most abundant toward the end of the spring peak and during the period between the maxima.

Apart from these repeated patterns of succession on the level of zooplankton compartments, considerable year-to-year variation may occur on the species level.

Maximum abundance of *Synchaeta pectinata* during the rotifer spring maximum varied in 4 years between 30 and 1300 individuals per liter. The same range of variation was observed in *Bosmina longirostris*. In *Asplanchna priodonta* the abundance varied from 0 to 2200 individuals per liter. In some years it was completely replaced by *Asplanchna girodi* (Hofmann 1983).

Vertical Distribution

The alternation of circulation and stagnation periods determines the vertical dispersion of the zooplankton. In spring and late autumn the whole water column is occupied; during summer the zooplankton habitat is confined to the upper 5 m of the water column (Hofmann and Krambeck 1991). The lower boundary of zooplankton distribution moves up and down with the oxygen boundary, which may reach the 5-m water layer during a long period of ice cover in winter (Hofmann 1982).

During the spring period, when the zooplankton are distributed in the whole water column, the assemblage exhibits a clear vertical structure in terms of species composition, as was shown for the rotifer plankton by cluster analysis of percentage similarities (Hofmann 1985). The rotifer community divided the water column into an epilimnetic and a hypolimnetic section with a clear break below 4 or 5 m water depth (Fig. 1.15, top). The former was characterized by a large number of species, such as *Keratella cochlearis*, *Kellicottia longispina*, *Synchaeta pectinata*, *Polyarthra vulgaris*, and *Asplanchna priodonta*, and by considerable vertical variation within short intervals; the latter was composed of only three taxa, *Keratella hiemalis*, *Filinia terminalis*, and *Filinia hofmanni*. An upper section (6–9 m) with predominance of *Filinia terminalis* was separated from the lower hypolimnion (10–25 m) where the other two species were most abundant and where the vertical structure of the assemblage was clearly more homogeneous than in the layers above.

The latter three taxa represent specific hypolimnetic populations that start their development in early spring in the lowest part of the hypolimnion just above the sediment with an increase in birth rates and abundances (Fig. 1.13). In the subsequent period the populations shift upwards because of the vertical movement of the oxycline. Only *Filinia terminalis* regularly establishes a separate epilimnetic subpopulation, which leaves this section of the water column when water temperature reaches about 12°C (Hofmann 1987, Loose 1987). There is only one species of crustacean, *Diacyclops bicuspidatus*, that prefers the hypolimnion.

The hypolimnetic assemblage also consists of populations that start their development in the epilimnion and then shift downward, maintaining high abundances for a short time period, e.g., the rotifers *Keratella cochlearis* and *Kellicottia longispina* and the cladocerans *Daphnia hyalina* and *Bosmina longirostris*. In the summer, when the hypolimnion is free of zooplankton, the true hypolimnetic populations totally disappear from the pelagic zone (*Keratella hiemalis*) or in some years may appear in a thin metalimnetic layer (*Filinia* spp.).

Because of the wind-sheltered site of the Plußsee, spring circulation is incomplete if the ice cover does not break up before early spring, as was the case in 1970.

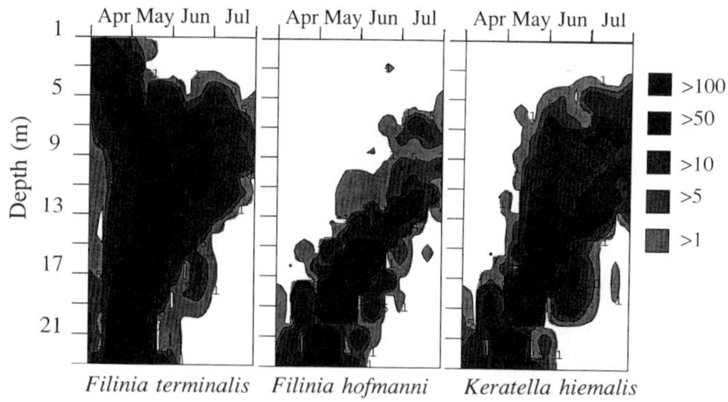

Figure 1.13. Standardized abundances (sum of abundances was fixed to 10,000) of hypolimnetic rotifer populations *Filinia terminalis* (left), *Filinia hofmanni* (middle), and *Keratella hiemalis* (right panel) in the Plußsee in 1972 (according to Loose 1987).

In that year only the upper 10 m of the water column was mixed. Thus, the zooplankton was confined to this zone. This situation particularly affected the dynamics of the true hypolimnetic elements, which were locked out of the lower hypolimnion and were crowded into a thin layer near the oxycline directly beneath the epilimnetic populations (Hofmann 1975).

Dynamics of Zooplankton Biomass

The reduction of the pelagic zone as a zooplankton habitat by oxygen stratification had a considerable effect on the zooplankton biomass. In 1981 the biomass peak occurred around the middle of May, when the major contribution came from the hypolimnetic zooplankton (Fig. 1.14). A comparison with the development in the upper 5 m shows that the epilimnetic peak occurred 1 week earlier (May), thus demonstrating the shift of maximum biomass from the epilimnion to the hypolimnion. Consequently, when spring circulation is incomplete, the hypo-

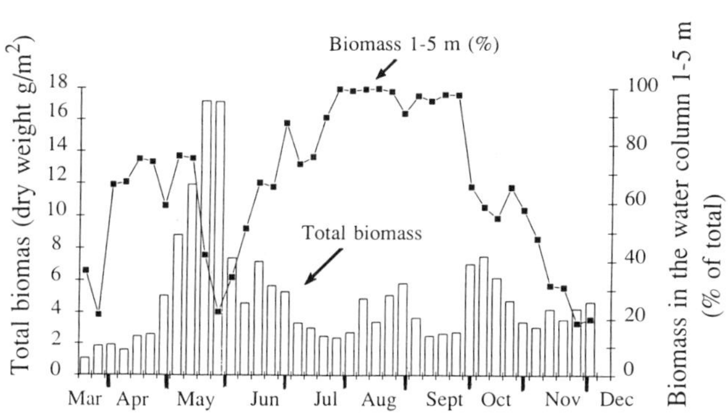

Figure 1.14. Zooplankton total biomass and the percentage contribution of biomass of the top 5 m of water to the total biomass in the Plußsee in 1981.

limnetic portion of zooplankton biomass is missing, and the biomass curve indicates only the epilimnetic response and therefore has a more slender shape (Hofmann 1975).

Unlike the 1969 succession, in 1981 a series of three biomass maxima were observed during the summer. The first one, distinct from preceding maxima, was produced not by rotifers but by a high abundance of adult *Eudiaptomus gracilis*. Hence, it was determined by the generation time of this species and the conditions the copepodides were exposed to in the preceding period rather than by the conditions in the middle of June when the adults appeared. Because of the high abundance of small cladocerans, *Chydorus sphaericus* (middle of August) and *Ceriodaphnia quadrangula* (beginning of October), the remaining two peaks again followed the PEG model.

There is some evidence that the relatively low summer biomass level as compared with the spring situation results not only from the specific food supply provided by the phytoplankton but also from the limited space available for the zooplankton. In 1986 epilimnetic water was pumped into the upper hypolimnion, and a new vertical structure of water column was established by the existence of a second thermocline at about 12 m and by the presence of oxygen between the thermoclines, e.g., in the upper hypolimnion (Hofmann and Krambeck 1991; see below).

Zooplankton populations responded to the new situation. The majority of epilimnetic rotifers and crustaceans extended their habitat vertically, occupying the upper hypolimnion, and the hypolimnetic elements maintained their position below the second thermocline. Hence, the zooplankton community divided the

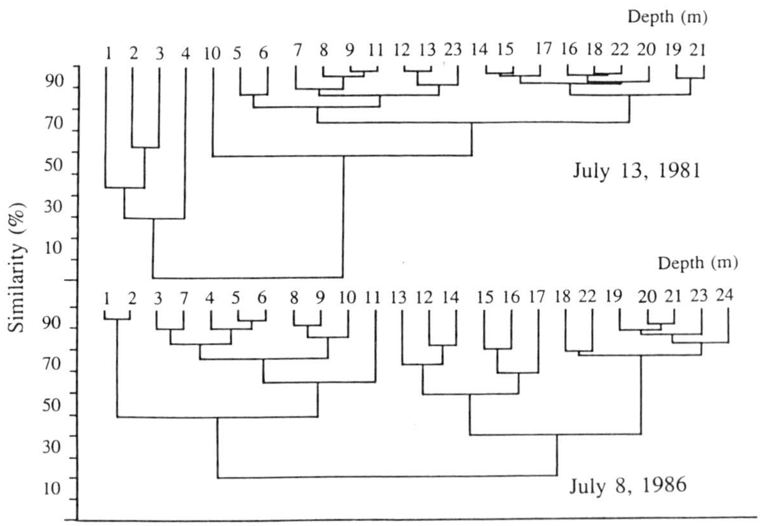

Figure 1.15. Percentage similarities between rotifer assemblages in the entire water column sampled on July 13, 1981 (1–23 m) and July 8, 1986 (1–24 m) in the Plußsee (according to Hofmann and Krambeck 1991).

water column in accordance with the new situation indicated by a shift of the boundary between the epilimnetic and the hypolimnetic assemblage from 4 m down to 12 m (Fig. 1.15). The vertical extension of the zooplankton habitat led to a considerable increase in zooplankton biomass from about 5 g dry weight/m^2 to 15 g dry weight/m^2, thus reaching the level of the spring biomass maximum in 1981.

Fish Fauna

Because of its characteristic narrow and deep basin and sheltered location, the Plußsee is strongly stratified from May to October with depletion of O_2 below a depth of 6 m (see above). Fish, therefore, tend to stay in the epilimnion throughout the summer. The epilimnion comprises approximately 52% (689,955 m^3) of the total water body. As a result of the anoxic hypolimnion, a large area of the lake sediment is also unreachable for both fish and many benthos organisms. Only 43% (61,753 m^2) of the entire benthal can be used by fish for habitation, feeding, and spawning during the summer.

Moreover, the steep littoral of the lake is not favorable to the development of emergent and floating macrophytes. Submerged macrophytes, specifically a 500-m^2 patch of *Elodea* sp., only occur in a small, shallow bight. The littoral sediment consists predominantly of sand and gravel down to a depth of 5 m. It is often covered with leaves and branches from the surrounding beech forest. Below a depth of 5 m, the bottom of the Plußsee consists of a dark brown humic sediment ("Dy-Gyttja," Koppe 1924). Lundbeck (1926) also noted the poor benthos fauna in the Plußsee: "Benthic fauna: largely nonexistent and only occasionally to be found near the littoral (less than 100, approx. 0.5–1.0 g to 1 m^2)."

The only connection between the Plußsee and neighboring lakes is a concrete pipe that is 50 cm in diameter and functions as an overflow. After a 200-m subterranean stretch it feeds into a small brook that runs into a lake below. This is the only "natural" entrance or exit for fish. The unfavorable location of this pipe means that it is often clogged with branches, leaves, and other plant matter. The attractant current that guides the fish is very slow. The majority of fish probably do not use this passage except for small eels (*Anguilla anguilla*), 12–15 cm in length, which can be found occasionally in the lake. It is presumed that they come from the plentiful stocks in the lower lakes and enter the lake through the brook and the pipe. In general, the fish fauna of Lake Plußsee can be regarded as isolated.

In the beginning of the 1960s, commercial fishing in the Plußsee came to an end. Until now, mainly eel (*Anguilla anguilla*) and pike (*Esox lucius*) were stocked every year. The marginal yield and difficulties involved in fishing in the lake were largely responsible for the decision to stop intensive cultivation. Subsequently the lake was managed by private leaseholders who continued to stock high-value commercial fish and game fish. Annually, between 200 and 300 pike and 25 kg of eel (approximately 1000 eel) were introduced into the lake. However, fish yield and productivity were very low for the type of fish being cultivated. In April 1983, the Plußsee was leased by the Max-Planck Institute for Limnology in Plön. Since that time there has been no fishing, and the fish fauna have not been interfered with in any way, apart from poaching.

The fish fauna of the Plußsee consists of 13 species, predominantly roach (*Rutilus rutilus*) and perch (*Perca fluviatilis*). Together, both species amount to approximately 70% of the standing stock. Roach is mainly detritovorous, whereas perch is benthivorous and later piscivorous. Young fish of both species feed on zooplankton for the first year of their life. In the Plußsee, the littoral zone is the home range of roach and perch in the summer. Although older roach and older perch are oriented to a pelagic environment and prefer a greater depth, the young fish of both species are more abundant nearer the shore. Cyprinid fry, especially roach, rudd (*Scardinius erythrophthalmus*), and bream (*Abramis brama*), live concealed between the rare emergent macrophytes (*Phragmites, Typha, Carex,* and *Glyceria* spp.) in the upper littoral zone, whereas 0-group perch prefer the water in front of the emergent vegetation. Larger perch remain in a greater depth of about 4–6 m. They are frequently observed devouring their own fry. The average growth of the Plußsee roach is retarded compared with the average size of other stocks; perch growth appears to be normal.

Even ruff (*Gymnocephalus cernua*) and bream (*Abramis brama*) appear regularly in the Plußsee. Numerically bream is not particularly abundant, but the body size and weight of the older fish contributes essentially to the total biomass. Bream, once they have reached a total length of about 30 cm, face no further risk from predators. The body length of these fish excludes them from predation by pike, the exclusive top predator in Lake Plußsee. Bream, as a benthivorous species, also grows slowly in this environment. In the summer they remain near the bottom of the lower littoral zone at a depth of 3–6 m. As the oxygen content of the hypolimnion increases in winter, the bream move to greater depths.

In the Plußsee, the benthivorous ruff seldom reaches a body length over 13 cm. They also exhibit the retardation of growth observed in fish from this lake. Until its second year the ruff lives in the shallow water near the shore. They shelter themselves within roots forming the natural bank of the lake and lie underneath the bedding of branches and leaves located on the bottom. Adult ruff prefer deeper water. Eel, the second most important predator of Lake Plußsee, inhabit the same environment as the ruff, their potential prey. Once eel have reached a length of 25–30 cm, their feeding habits become increasingly piscivorous. This is largely because of the poor development of the benthos fauna. More than 90% of the eel investigated were found to be "bighead," the ecological predator form of eel. In addition to ruff, the eel also prey on perch and roach.

When ground otoliths were used to determine age, the Plußsee eel were found to be retarded in growth compared to eel from other inshore waters. Despite this fact, in the Plußsee very old eel can reach a total body length of up to 1 m and weigh up to 2 kg. Two specimens of this size (which had obviously not found the pipe leading from the lake) have been documented. The present abundant eel population is the result of stocking activities in the past. Roughly estimated, the total eel population of Lake Plußsee at present is approximately 5000 animals. The majority of eel examined were female (74.5%); 24.6% were juvenile, and only 0.9% were male.

Pikes are the main predators in the Plußsee. They prey on roach, perch, and rudd and are very often cannibalistic. The number of pike is limited by the scarcity of hiding places in the littoral. Thus, it is possible to find individual pike in the pelagial, although this environment is not particularly favorable for pike. However, the pike will remain there as long as there is no interspecific competition with other "pelagic predators" such as the pike perch (*Stizostedion lucioperca*). The individual growth of the pike appears to be normal. Pike are the only species of Lake Plußsee that is selectively fished by poachers. It is estimated that at least 20 to 30 pike 60–80 cm in length and 1.5–3.5 kg in weight are removed annually. Because of the lack of suitable spawning grounds and the limited space available for young fish, cultivation of pike in the Plußsee is problematic. Nevertheless, the pike population appears to have adapted well to the general situation of the fish fauna in the lake.

Neither herbivorous rudd (*Scardinius erythrophthalmus*) nor benthivorous tench (*Tinca tinca*) are very abundant, but they continue to reproduce successfully each year. Minor species found in the lake are *Leucaspius delineatus*, common carp (*Carassius carassius*), bleak (*Alburnus alburnus*), carp (*Cyprinus carpio*), and spined loach (*Cobitis taenia*). They make up less than 0.1% of fish fauna of the Plußsee.

Impact of Physical Conditions on Limnological Processes

During the year the pooled amount of autotrophic (algae) and heterotrophic (zooplankton, bacteria) biomass shows a simple pattern. A succession of different species is more flexible than one particular genetically fixed organism, and thus the always varying composition of nutrients, light, and temperature conditions is utilized by a succession of different algae. The development of algal communities forms a single peak in spring, in summer, and in autumn (see above). The same is found for the pooled heterotrophic biomass, except that the peaks exhibit a phase shift to the autotrophic data (see above; Chapter 7, this volume). This pattern of organic carbon flow is repeated every year, and it was possible to simulate the processes properly by means of mathematical models (Krambeck and Krambeck 1980) that adopted environmental conditions (light, temperature, nutrients) and limnological processes such as uptake kinetics and predator–prey relationships. According to calculations from the model, exchange rate A, which controls the vertical transport processes of matter, plays a most important role with respect to the biomass production in the Plußsee. The exchange rate is so small that the limiting nutrient, phosphorus, which is removed from the epilimnion during the collapse of the spring bloom, cannot return to the epilimnion by means of internal loading.

In order to investigate the role of the exchange processes, a pump was installed in the Plußsee in 1986 to increase the exchange rate (Krambeck 1988). Figure 1.16

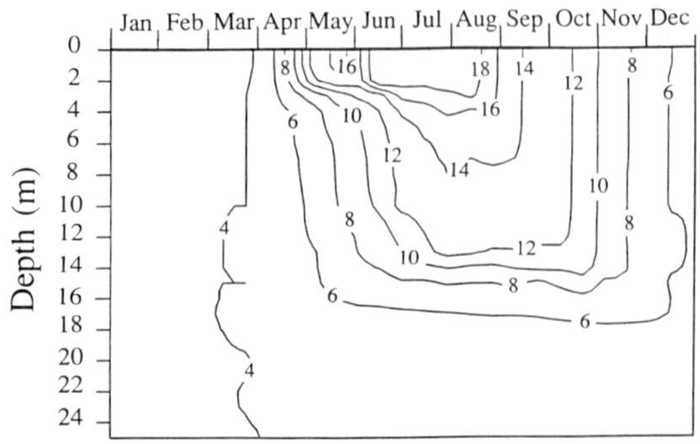

Figure 1.16. The thermal stratification of the Plußsee after the pumping experiment in 1986. The lake had two thermoclines at depths of 4 and 14 m.

demonstrates the impact of the pumping process on lake water stratification. The lake had two thermoclines in 1986, one at a depth of 4 m and the second at a depth of about 14 m. Changes in the biomass production were not obvious from looking at the lake, but in fact they were dramatic. When the annual biomass (Fig. 1.18B) was compared with phytoplankton biomass in 1986 (Fig. 1.17A), there was a 20%

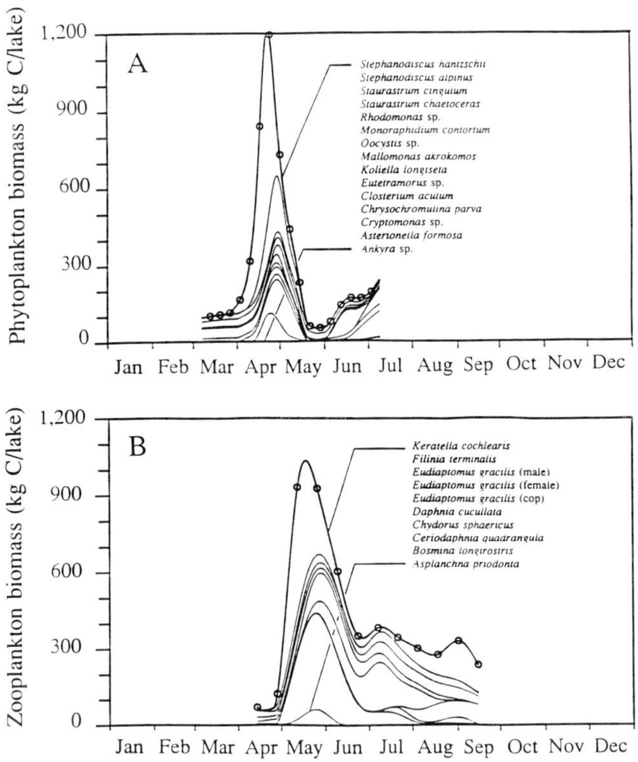

Figure 1.17. The total biomass of A phytoplankton and (B) zooplankton in the Plußsee in 1986 after the pumping experiment.

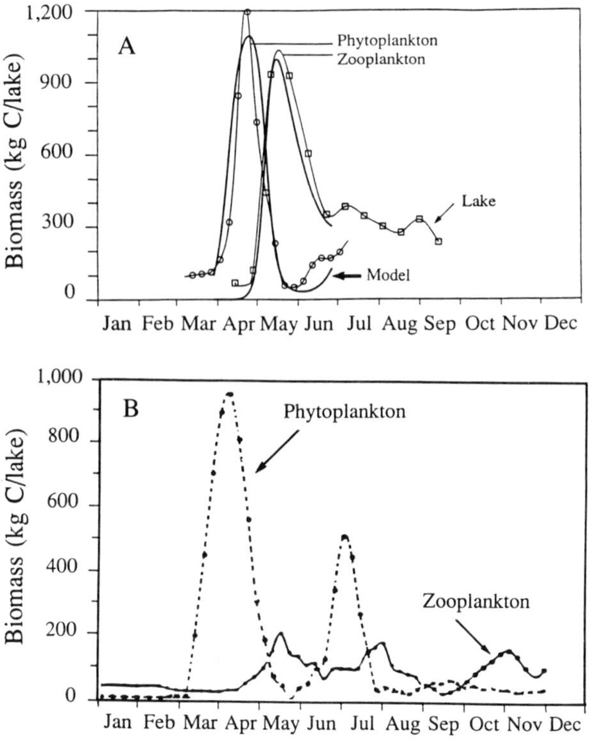

Figure 1.18. (A) Biomass of phytoplankton and zooplankton after the pumping experiment in 1986. The zooplankton biomass was about 500% higher than in previous years. (B) Annual pattern of biomass of phytoplankton and zooplankton in the Plußsee before the pumping experiment. The zooplankton biomass is usually equal to about 20% of the algal biomass.

increase, but if one looks at the zooplankton biomass (Fig. 1.18B, lower curve) there was an increase of about 500% (Fig. 1.17B). The explanation is that the pump recycled phosphate from the rich hypolimnetic water to the euphotic zone which increased phytoplankton primary production. At the same time the pumping increased the biotope for the zooplankton as well as their food availability and concentration. Thus all the extra primary production was transformed into heterotrophic biomass. The quantitative analysis by means of mathematical models was described by Krambeck (1988). The data of biomass of plankton and the fitted model data are shown in Figure 1.18.

References

Albrecht D (1973) Verbesserte Methoden zur automatischen Wasseranalyse in der Limnologie. Vom Wasser 41:129–135

Albrecht D, Overbeck J (1969) Wasseranalysen in der Limnologie mit Hilfe des Technicon AutoAnalyzers. Technicon Wassercolloquium, Bad Nauheim, Sonderheft 855:1–15

Chróst RJ (1988) Phosphorus and microplankton development in an eutrophic lake. Acta Microbiol Polon 37:205–225

Einsle U (1968) Cytologisch-taxonomische Studien an *Cyclops*-Populationen Schleswig-Holsteins. Gewässer Abwässer 47:31–40

Flößner D (1972) Kiemen- und Blattfüßer, Branchiopoda, Fischläuse, Branchiura. Tierwelt Dtschl 60:1–499

Grasshoff K, Ehrhardt M, Kremling K (eds) (1983) *Methods of Seawater Analysis.* Verlag Chemie, Weinheim

Hickel B (1975) Changes in phytoplankton species composition since 1894 in two lakes of East-Holstein, Germany. Verh Int Ver Limnol 19:1229–1240

Hickel B (1978) Phytoplankton population dynamics in Plußsee (East-Holstein, Germany). Verh Ges Ökol 1977:119–126

Hickel B (1980) Massenentwicklung von *Pascherina tetras* (Korsch.) Silva, einer seltenen Spondylomoracee (Chlorophyta) im Frühjahrsplankton des Plußsees (Ostholstein). Nova Hedw 33:913–918

Hickel B (1985) The population structure of *Ceratium* in a small eutrophic lake. Verh Int Ver Limnol 22:2845–2849

Hickel B (1988a) Unexpected disappearance of cyanophyte blooms in Plußsee (North Germany). Arch Hydrobiol Suppl 80:545–554

Hickel B (1988b) Sexual reproduction and life cycle of *Ceratium furcoides* (Dinophyceae) *in situ* in the lake Plußsee (F.R.G.). Hydrobiologia 161:41–48

Hickel B (1988c) Morphology and life cycle of *Ceratium rhomvoides* nov. sp. (Dinophyceae) from the lake Plußsee (F.R.G.). Hydrobiologia 161:49–54

Hickel B, Håkansson H (1993) *Stephanodiscus alpinus* in Plußsee, Germany, ecology, morphology and taxonomy in combination with initial cells. Diatom Res 8(1) (in press)

Hickel B, Maaß I (1989) Scaled chrysophytes, including heterotrophic nanoflagellates from the lake district in Holstein, northern Germany. Nova Hedw 95:233–257

Hofmann W (1975) The influence of spring circulation on zooplankton dynamics in the Plußsee. Verh Int Ver Limnol 19:1241–1250

Hofmann W (1981) Limnologische Untersuchungen an Seen des Kreises Plön. Jb Heimatkd Kreis Plön 11:159–176

Hofmann W (1982) On the coexistence of two pelagic *Filinia* species (Rotatoria) in Lake Plußsee. I. Dynamics of abundance and dispersion. Arch Hydrobiol 95:125–137

Hofmann W (1983) On temporal variation in the rotifer *Keratella cochlearis* (Gosse): The question of "Lauterborn cycles." Hydrobiologia 101:247–254

Hofmann W (1985) Dynamics of vertical zooplankton community structure in the Plußsee: Cluster analysis. Verh Int Ver Limnol 22:2983–2986

Hofmann W (1987) Population dynamics of hypolimnetic rotifers in the Pluss-see (North Germany). Hydrobiologia 147:197–201

Hofmann W, Krambeck H-J (1991) Manipulation of the thermocline in a eutrophic lake: The response of the zooplankton. Verh Int Ver Limnol 24:786–790

Hutchinson GE (1957) *A Treatise on Limnology,* vol 1. John Wiley & Sons, New York

Kiefer F (1978) Freilebende Copepoda. In Elster H-J, Ohle W (eds) *Die Binnengewässer,* vol 26(2). Schweizerbart, Stuttgart, pp 1–343

Koppe F (1924) Die Schlammflora ostholsteinischer Seen und des Bodensees. Arch Hydrobiol 14:619–627

Koste W (1978) Rotatoria. Die Rädertiere Mitteleuropas. Überordnung Monogononta. Borntraeger, Berlin

Koste W (1980) Über zwei Plankton-Rädertiertaxa *Filinia australiensis* n. sp. und *Filinia hofmanni* n. sp., mit Bemerkungen zur Taxonomie der *longiseta-terminalis*-Gruppe. Genus *Filinia* Bory De St. Vincent, 1824, Familie Filiniidae Bartos 1959 (Überordnung Monogononta). Arch Hydrobiol 90:230–256

Krambeck HJ (1974) Energiehaushalt und Stofftransport eines Sees. Beispiel einer mathematischen Analyse limnologischer Prozesse. Arch Hydrobiol 73:137–192

Krambeck H-J (1981) Longterm monitoring and modelling of the nutrient cycle in a baltic lake. Verh Int Ver Limnol 21:460–465

Krambeck HJ (1988) Partial destratification of eutrophic lakes: A tool for "ecosystem modelling." Verh Int Ver Limnol 23:801–806

Krambeck HJ, Krambeck C (1980) Modelling carbon and phosphorus in a small hypertrophic north german lake. In Barica J, Mur LR (eds) *Hypertrophic Ecosystems*. W. Junk bv Publishers, The Hague, pp 159–169

Lemmermann E (1893) Verzeichnis der in der Umgegend von Ploen gesammelten Algen. Forsch Ber Biol Stn Plön 1:18–72

Lemmermann E (1895) Zur Algenflora des Plöner Seengebietes. Zweiter Beitrag. Forsch Ber Biol Stn Plön 4:134–188

Lemmermann E (1903) Beiträge zur Kenntnis der Planktonalgen. XV. Das Phytoplankton einiger Plöner Seen. Forsch Ber Biol Stn Plön 10:116–171

Lieder U (1983) Die Arten der Untergattung *Eubosmina* SELIGO, 1900 (Crustacea: Cladocera, Bosminidae). Mitt Zool Mus Berl 59:195–292

Loose C (1987) Populationsökologische Untersuchungen an hypolimnischen Rädertieren (Rotatoria) des Plußsees. MSc Thesis, Kiel University, Germany

Lundbeck J (1926) Die Bodentierwelt norddeutscher Seen. Arch Hydrobiol Suppl 7:1–473

Overbeck HJ (1968) Bakterien im Gewässer—Ein Beispiel für die gegenwärtige Entwicklung in der Limnologie. Mitt Max-Planck-Ges 3:165–182

Overbeck J (1972) Zur Struktur und Funktion des aquatischen Ökosystems. Ber Dtsch Bot Ges 85:553–577

Overbeck J (1973) Über die Kompartimentierung der stehenden Gewässer—ein Beitrag zur Struktur und Funktion des limnischen Ökosystems. Verh Ges Ökol 1973:211–223

Sommer U, Gliwicz ZM, Lampert W, Duncan A (1986) The PEG-model of seasonal succession of planktonic events in fresh waters. Arch Hydrobiol 106:433–471

Utermöhl H (1925) Limnologische Phytoplanktonstudien. Die Besiedlung ostholsteinischer Seen mit Schwebpflanzen. Arch Hydrobiol Suppl 5:1–527

Voigt M (1903) Das Zooplankton des Kleinen Uklei- und Plus-Sees bei Plön. Forsch Ber Biol Stat Plön 10:105–115

Wolf HG, Mort MA (1986) Inter-specific hybridization underlies phenotypic variability in *Daphnia* populations. Oecologia 68:507–511

2. Dissolved Organic Matter: Analysis of Composition and Function by a Molecular-Biochemical Approach

Uwe Münster and Dieter Albrecht

Introduction

General consensus has emerged among aquatic scientists that dissolved organic matter (DOM) is a key component in the interplay of aquatic food web processes (Hobbie 1992). Bacteria play a central role in this interplay (Azam and Cho 1987): they contribute to nutrient dynamics, carbon flux, and food web interactions. Potentially, about 40–50% of photosynthetically fixed carbon may be rapidly channeled via dissolved organic carbon (DOC) and the microbial loop to higher trophic levels (Azam et al. 1983, Williams 1990). However, the efficiency of this direct metabolic coupling between primary production and heterotrophic utilization of organic matter depends mainly on the "organic matter reactivity" (OMR). This OMR is poorly understood, especially with respect to the chemical composition and structures of the DOM faction that is recalcitrant to microbial degradation. Ecological function of those recalcitrant organic substrates is understood even less (Hobbie 1992, Münster 1993).

Studies of the pelagial water and the sediments could show that there are at least two to three distinct functional fractions of organic matter with different importance for microheterotrophs. According to the biological lability of these different organic matter fractions, two first-order rate models, the "G model" (Westrich and Berner 1984) and the "mixed model" (Burdidge 1991), have been formulated and tested in marine sediments. Billén and Servais (1989) have developed and tested a similar approach for pelagic water systems, the "HSB model," for studies of DOM decomposition and utilization by bacterioplankton in freshwater and marine environments. However, the main drawbacks of these approaches are the lack of information about the chemical characterization and identification of the labile and the refractory DOC fractions and their structural-chemical properties, which may help to explain their differences in the OMR. Combined chemical and microbial physiological approaches in eutrophic and humic lakes have revealed that the labile dissolved organic carbon pool (LDOCP) comprises mostly low-molecular-weight dissolved organic carbon (LMW DOC) compounds and represents only 1–20% of the DOC (Chróst et al. 1989, Münster and Chróst 1990, Münster 1991, 1993). However, the majority of the DOC is of a polymeric nature, and 80–90% of DOC is comprised of high-molecular-weight dissolved organic carbon (HMW DOC). This fraction of HMW DOC is a mixture of different biopolymers (proteins, polynucleotides, polysaccharides, lipids, lignins) and geopolymers, e.g., dissolved humic matter (DHM). The HMW DOC fraction is not directly available to bacterial metabolism and comprises much of the recalcitrant dissolved organic carbon pool (RDOCP). The trophic role of this RDOCP in energy and food web processes is still poorly understood (Hobbie 1992). Physical and chemical methods characterized part of this RDOCP in marine waters with turnover times of about 2000–6000 years (Druffel and Williams 1990). However, microbiological approaches found much shorter turnover times (about 1–10 h) for such DOC bulk material (Kirchman et al. 1991). These discrepancies in the exact chemical and biological characterization of the DOM reflect the poor knowledge about the trophic role of refractory or recalcitrant organic carbon pools in aquatic environments.

In the following sections results of a chemical and biochemical approach are presented to characterize the composition, dynamics, processing, and utilization of DOM in the eutrophic Plußsee. Special attention is paid to enrichment, fractionation, and characterization techniques of DOM and the microbial utilization of the labile DOC pool, e.g., monomeric substrates such as dissolved free amino acids (DFAA), carbohydrates (DFCHO), and typical excretion products by photoautotrophs (e.g., glycolic acid), and the recalcitrant DOC pool, e.g., polyphenols, the main constituents of DHM. Further, interactions of microbial hydrolytic enzymes (MHE) with natural biopolymers and their role in aquatic foodweb processes are discussed.

Characterization and Composition of the DOM

The input of DOC and particulate organic carbon (POC) into aquatic environments follows many different routes. According to Odum (1963) and Wetzel (1984), DOC and POC have been defined as detritus with a definite function in aquatic ecosystems. This detritus can be either of autochthonous or allochthonous origin (Wetzel 1983). They differ in their chemical composition and structures, and this is of fundamental concern to understanding the function of detritus as carbon, nutrient, and energy sources for microorganisms (Allen 1977, Rich and

Wetzel 1978). The amount of detritus input into aquatic environments depends mainly on the productivity and nonpredatory losses of organic matter. The detritus pools are primarily exploited by heterotrophic microorganisms. Osmotrophic microorganisms (e.g., bacteria) are the most efficient utilizers of DOC (Münster and Chróst 1990). Detritus and bacteria bridge the carbon, nutrient, and energy flow between primary producers and the classical grazing food chain via the microbial loop (Azam et al. 1983).

Dissolved organic matter is a larger fraction of detritus in aquatic environments and comprises large fractions of biopolymers with the biologically important elements carbon, nitrogen, phosphorus, and sulfur. DOM is a highly complex and dynamic compartment and requires a multidisciplinary approach for characterization and identification. Nondestructive analytical methods include fluorescence and light absorbance; the molecular weight composition, ion exchange, and complexation capacity are important bulk-phase properties that can be measured relatively easily. Destructive chemical methods are needed to ascertain the elemental composition and some structural information. Enrichment techniques are valuable tools for tracing specific compounds at low concentration levels with specific key functions. We have applied some of those listed tools to characterize Plußsee DOM at the bulk-phase level and also to get some structural information on the polymeric and monomeric levels. All our approaches were primarily directed to understand the role of DOM as substrate for microheterotrophs. Pure chemical structural aspects were followed only if practical methods were available and when valuable information could be obtained for ecological questions.

Bulk-Phase Characterization

In most cases lakes are small with large catchment-to-surface ratios (Wetzel 1991). Land–water interfaces are the main routes of DOM supply to open waters (Rasmussen et al. 1989). During its transport to lakes DOM is microbially and chemically transformed and modified. The remaining fraction is mostly recalcitrant and mixed with autochthonously produced DOM in the lake water column. Autochthonous DOM is assumed to be less recalcitrant because of lower contents of lignins and celluloses (Hedges et al. 1988), but heterotrophic microorganisms can degrade and utilize those recalcitrant biopolymers as well (Alexander 1975, Benner et al. 1986). Attempts have been made to characterize the DOM fractions with respect to chemical composition and microbial degradation. The contributions of different organic solutes of DOM from autochthonous or allochthonous origin to microbial growth can be chemically differentiated by stable-isotope analyses (Peterson and Fry 1987). Stable-isotope data can be used to trace sources and sinks and their regulating processes in aquatic environments by measuring the δ-$^{12}C/^{13}C$ ratio in DOC and POC samples. Heterotrophs from land–water interfaces have higher δ ratios than those from open waters (Coffin et al. 1990). However, the values of δ-$^{12}C/^{13}C$ can vary over a large range and are sometimes difficult to understand. More such precise chemical studies are needed to understand biogeochemical processes in aquatic environments.

Dissolved Organic Carbon

According to filtration techniques and the measurements of the carbon content of organic matter by wet chemical oxidation (WCO), two fractions have been defined: dissolved organic carbon (DOC) and particulate organic carbon (POC). The ratio of DOC to POC can vary over a large range between 10:1 to 100:1 in different aquatic environments (Ohle 1933, Wetzel 1983). Over a long period of time the particulate and dissolved phases in aquatic systems have been traditionally defined by using 0.2 to 0.45-μm pore size filtration, but recent data show that small particles and colloids can pass through 0.2 to 0.45-μm pore size filters also (Moran and Buesseler 1992). Therefore, the definitions of DOC or POC are still more or less operational. The ecological function of these fractions is not described very well by such methods. However, both carbon sources are important substrates for osmotrophic microheterotrophs.

In the past, studies on DOC have been done in many different aquatic environments ranging from temperate lakes to oceans, tropical waters, estuaries, rivers, groundwater, and rain. For most of those measurements, wet chemical oxidation methods (WCO) were applied (Menzel and Vaccaro 1964). Recent results by Sugimura and Suzuki (1988) showed higher DOC values when they applied a high-temperature catalytic oxidation (HTCO) method for DOC measurements. With HTCO, DOC concentrations 2–2.5 times higher were measured from deep ocean water samples compared to the WCO method. These results have a serious aspect for global carbon balance calculations. However, such differences between the methods have not been found in freshwater DOC, where the HTCO method yielded values only 3–6% higher than the values for the WCO method (Kaplan 1992). The DOC data presented in this contribution were obtained with the WCO method for samples from 1972–1978. DOC data from 1981–1989 were obtained with the HTCO method at 960°C catalytic oxidation and IR detection.

DOC in the Plußsee Analyzed by the WCO Method (1974–1978)

Examples of DOC data from the Plußsee are shown in Figure 2.1. In 1974, the DOC within the water column varied between 4.1 and 29.7 mg C/L (mean 9.97 ± 1.04 mg C/L) with highest values (25–29.7 mg C/L) during early (June–July) and late summer (July–August) in the epilimnion (0–4 m). Below these depths the DOC values showed less variation (Fig. 2.1). In 1976, the DOC was measured from April to December with maxima of 18.2–19.5 mg C/L in early summer (June–July), again in the epilimnion, and mean values of 9.1 ± 0.8 mg C/L throughout the whole sampling period and water column. A continuation of the DOC measurements from 1977–1978 revealed that the same distribution pattern of DOC from the former time series in 1974–1976 in the epilimnion could not been repeated. During these last measurements the DOC varied between 5.1 and 20.6 mg C/L (mean 9.5 mg C/L), but there was no significant DOC maximum during the sampling period. The DOC was more equally distributed throughout the water column. The change of this DOC distribution pattern can be explained by the change of phytoplankton species composition from 1976 to 1977–1978 (Chap-

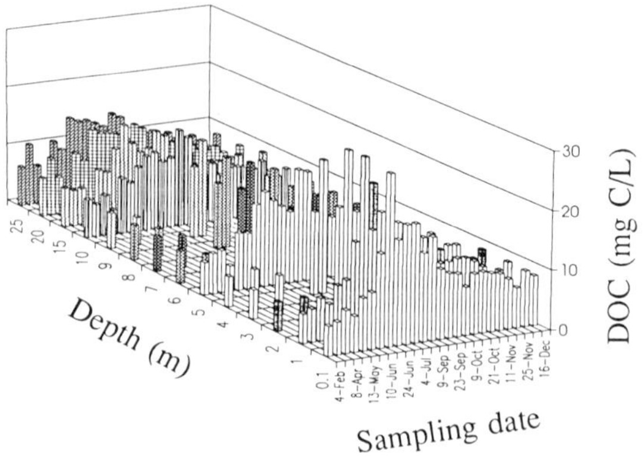

Figure 2.1. Concentrations and distribution of dissolved organic carbon (DOC) in the Plußsee from 1974 determined by the wet chemical oxidation method (WCO).

ter 1, this volume). According to Hickel (1988), cyanobacteria dominated in the Plußsee from 1972 to 1977 with blooms of *Limnothrix redekei* and *Aphanizomenon gracile*. After an intermediate period during 1978 the second period (1978–1985) was dominated by flagellate blooms. The phytoplankton biomass decreased in the first period from annual averages of 2.9–10.4 mm^3/L down to 0.2–1.0 mm^3/L in the epilimnion (0–4 m depth) during the second period. Maxima of DOC appeared about 1 week after the phytoplankton biomass maxima, which indicates the important role of autochthonous DOC input into the DOM pool. This effect can be seen in the DOC distribution patterns during 1974–1976 (Fig. 2.1) but decreased significantly during the change in the phytoplankton community.

DOC in the Plußsee Analyzed by the HTCO Method (1981–1989)

The DOC measurements for the Plußsee with the HTCO method did not show twice the carbon content as found by the WCO method, as had been reported for marine waters (reviewed by Hedges and Lee 1993). We started using HTCO methods for DOC measurements in 1981, and our data showed minor differences between WCO and HTCO methods for DOC measurements. The DOC data from 1981 and 1987 are summarized in Figure 2.2, and the DOC values with the HTCO method are within similar ranges to those obtained with the WCO method (Fig. 2.1). Although DOC was measured in more special phytoplankton productivity time series in the Plußsee (Fig. 2.2), the DOC values from 1974–1978 did not differ from those from 1981–1989. DOC data from 1981 and 1987 varied between 7.3 and 15.3 mg C/L (mean 10.3 mg C/L) during a diurnal study (Fig. 2.2A) and between 9.5 and 12.3 mg C/L (mean 10.9 mg C/L) during a spring bloom development (Fig. 2.2B). During an annual sampling period in 1989, the DOC varied between 6.5 and 18.1 mg C/L (mean 11.1 mg C/L). Although there are certainly short-term, seasonal, and annual variances in the DOC input, which may affect the level of DOC, we did not find DOC values from WCO measurements to be double those from the HTCO measurements. These results correspond to those from

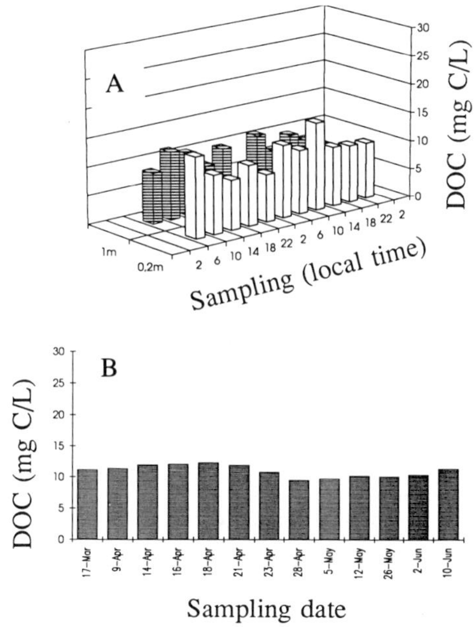

Figure 2.2. Concentrations and distribution of dissolved organic carbon (DOC) in the Plußsee determined by the high-temperature catalytic oxidation method (HTCO) in 1981 and 1987 during spring bloom development.

Kaplan (1992), who obtained values only 3–4% higher with HTCO compared to the WCO method for DOC measurements in different freshwater samples. Except for DOC data during 1974–1976, there was less variance in the DOC level over a 10 to 12-year period. One might assume that DOC measurements are unsuitable for DOM characterizations, but we consider DOC measurements to be important background data for carbon flux and carbon balance calculations.

Dissolved Organic Nitrogen

Nitrogen compounds in aquatic environments can be found in many different oxidation states. The most widely studied forms of nitrogen in aquatic environments are those found in inorganic compounds (e.g., NO_3^-, NO_2^-, N_2, NH_4^+). These inorganic nitrogen compounds have been studied intensively with respect to their nutrition, interactions, and utilization by microplankton. The dissolved organic nitrogen (DON) pool in aquatic environments is, however, less studied than DOC. Although the DON pool can constitute more than 50–60% of total dissolved nitrogen sources (Wetzel 1983), there is still less knowledge about the chemical composition and structure of the DON in marine and fresh water. Recent measurements by Suzuki et al. (1985) have further shown that, similar to DOC, the DON pool may be underestimated in marine waters by a factor of 2. Such methodological problems in the DON measurements in the past make it more difficult to compare former and recent DON values (Hedges and Lee 1993). The best-known component of the DON is the dissolved free amino acid (DFAA) pool and its contribution to the DON and DOC pools and microbial utilization. The determination and analysis of the dissolved combined amino acid (DCAA) fraction has already shown some methodological problems with respect to the hydrolysis step. There is a significant loss of DON and DOC during hydrolysis of proteins to amino acids with standard methods (Robertson et al. 1987). Some protein amino acids (e.g., glutamine, tryptophan, cysteine) are very sensitive to degradation reactions during standard hydrolysis conditions with 6 N HCl under N_2 atmosphere. Recently, methodological developments with gas-phase hydrolysis techniques under vacuum conditions combined with microwave heating systems could yield better and more reproducible analytical results for protein hydrolysis and the resulting amino acids (Meltzer et al. 1987, Knecht and Chang 1986). From such careful analytical studies on measurements of DOC and DON components in aquatic environments, important analytical recommendations for DOC and DON determinations in marine and freshwater have recently emerged (Hedges and Lee 1993).

DON in the Plußsee 1974–1977

DON measurements of Lake Plußsee were done by standard WCO methods. In Figure 2.3 an example of the distribution of DON during the sampling period 1974 is shown. The sampling scale is identical to that for DOC in Figure 2.1 but less frequent than the DOC measurements. The concentration of DON was generally lower than that of DOC by a factor of 10–50. The DON concentration varied between 0.1 and 2.63 mg N/L (mean 0.66 ± 0.13 mg N/L) throughout the whole sampling period and vertical profiles from spring 1974 to summer 1977, with no significant seasonal or vertical maxima throughout the entire 3 years. The highest values (1.24 to 2.63 mg N/L) were found in the hypolimnion (Fig. 2.3) between 20 and 25 m depths, where mostly anaerobic conditions exist. Reduced nitrogen sources may be more abundant in the DON pool at this depth. The DON seemed to be more evenly distributed throughout the water column compared to the DOC. Although reduced nitrogen is an important nutrient source (e.g., NH_4^+) for phytoplankton and bacteria (LaRoche 1983, Dugdale and Goering 1967, Zehr et al. 1985), there was less variance observed in the DON pool, which might indicate some degradation of DON and its utilization in the water column. This might be

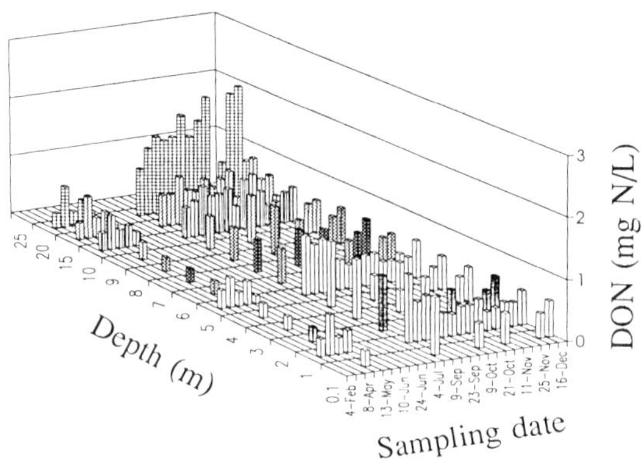

Figure 2.3. Vertical and annual distribution of dissolved organic nitrogen (DON) in the Plußsee from 1974.

explained by a balance between input and output (production and utilization) or by the unavailability of the majority of the DON pool, similar to DOC. Such long sampling intervals may not encounter a highly dynamic variation of a biologically labile small DON pool fraction as comprised by DFAAs (see below, this chapter). The majority of the DON pool may be degraded over much longer time intervals, which can not be measured by conventional methods.

Dissolved Organic Phosphorus

In many aquatic environments phosphorus is the most important element encountered that limits the growth and production of planktonic organisms (Dillon and Rigler 1974, Schindler 1977). Most research efforts were directed to the role and dynamics of inorganic phosphorus (Lean 1973), which is the major form in which phosphorus is translocated into microbial cells. The largest fraction of dissolved phosphorus in the trophogenic zone in many freshwater systems arises in the dissolved organic phosphorus (DOP) fraction during summer stagnation (Francko and Heath 1979, Chróst and Overbeck 1987, Wetzel 1983; Chapter 6, this volume). DOP can be regarded as an essential constituent of the DOM and may be used in this context as a valuable parameter to characterize the nutritional and labile value of the DOM.

DOP in the Plußsee 1975–1976

Figure 2.4 shows the distribution of DOP in the water column during 1976. The highest amounts of DOP were found in spring and summer, with DOP maxima of 0.119 mg P/L in the surface water and the lowest value of 0.002 mg P/L (mean 0.025 mg P/L) in the hypolimnion. Although our data set is small, it can be clearly seen that during winter less DOP is observed throughout the whole water column. From spring to summer DOP increases and becomes an important P pool in this lake (Chróst 1988). DOP can be regarded as an important nutritional substrate pool for P-limited organisms in the epilimnion of the Plußsee. The utilization of this

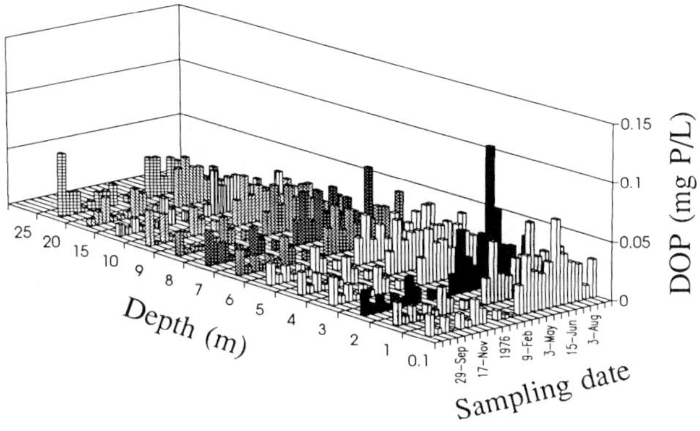

Figure 2.4. Vertical and annual distribution of dissolved organic phosphorus (DOP) in the Plußsee in 1975–1976.

DOP pool requires an extracellular hydrolysis step by alkaline or acid phosphatase (Chróst and Overbeck 1987, Francko 1984; Chapter 6, this volume).

Elemental Composition of DOM

The relative abundance of carbon, nitrogen, and phosphorus in suspended and dissolved forms in aquatic environments has been the subject of research and discussion for a quarter century (Vollenweider 1968, Levine and Schindler 1992). Most of the research has focused on the ratio of those elements in inorganic constituents and their relative contribution to eutrophication and lake water productivity. However, fewer studies have looked at the carbon, nitrogen, and phosphorus composition in the DOM. DOM can include large amounts of all three elements, and they are essential substrate sources for microheterotrophs. Knowledge of the elemental composition of the DOM and the pool sizes of C, N, and P in the organic carbon pool may help us to understand better the nutrient cycles and heterotrophic processes in aquatic environments. In the following paragraphs these C:N:P ratios are used for further characterization of the DOM and as indicators of potential substrate sources of the important elements C, N, and P for microheterotrophs.

C:N Ratio 1974–1977

C:N ratios are often used as indicators for the degradation or humification of organic matter in aquatic and terrestrial studies (Schnitzer and Khan 1972). Knowledge of the variation of those C:N values are of general concern for DOM characterization. The C:N ratio in the Plußsee water body varied between 8:1 and 58:1 (mean 16:1), with higher C:N ratios in the deep hypolimnion (15- to 20-m depth) and above the sediments (25-m depth). In the trophogenic zone (0- to 6-m depth) there are no differences in the C:N ratios over time and depth. C:N ratios in the spring–summer period are slightly higher than those in autumn and winter. The opposite can be seen in the deeper hypolimnion samples. The narrower C:N range in the upper 0- to 6-m depth with C:N ratios between 5:1 and 10:1 clearly shows that the bulk material of DOM is less degraded than the bulk material in the deep hypolimnion with C:N ratios between 30:1 and 58:1.

C:P Ratio in 1976–1977 and in 1981

A distribution pattern similar to the C:N ratios can also be seen in the C:P ratio. The C:P ratios varied in depth and time between 29:1 and 1100:1 (mean 381:1), with lowest values in the surface water (0–5 m) and highest values in the deep hypolimnion (10–25 m). Unfortunately, the data basis is much smaller than those for DOC and DON, but even this small data sample shows that there is an increase from surface to deep water samples in carbon contribution to the DOM pool.

N:P Ratio

The N:P ratio in living and nonliving sources on earth can vary over a large range from >200:1 to <0.1:1 (Downing and McCauley 1992). In comparison, N:P ratios in the DOM of Plußsee water varied between 75:1 to 2.5:1 (mean 30:1). Lowest

N:P ratios were found in surface water (0–5 m), and highest ratios were found in deeper hypolimnion water. This distribution pattern is similar to those found for the C:N and C:P ratios in the DOM. This elemental composition may reflect the general degradation status of the DOM in this lake. More labile DOM, narrower C:N:P ratios in surface water, a higher contribution of DOM production from autotrophic processes, and an increase in carbon, larger C:N:P ratios, and a higher contribution by heterotrophic processes in deeper water layers yield a permanent increase in carbon in the DOM skeleton. These results show how DOM changes with depth depending on microbial degradation processes in the water column. Such elemental composition analysis of DOM reflects the changes in DOM better than DOC measurements alone.

Chemical Characterization of the DOM

In lacustrine environments DOM is a mixture of thousands of different organic compounds with different structures, sizes, concentrations, and reactivities. Among the organic solutes in the DOM, dissolved amino acids (DAA), dissolved carbohydrates (DCHO), and dissolved organic acids (DOAc) are the most extensively studied. Knowledge of their quantities and contribution to the DOC pool is

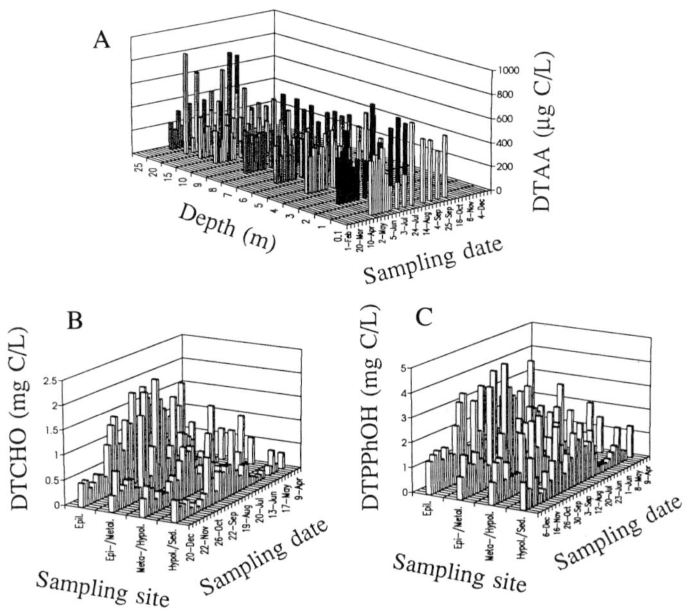

Figure 2.5. Annual and vertical distribution of (A) dissolved total amino acids (DTAA), (B) dissolved total carbohydrates (DTCHO), and (C) dissolved total polyphenols (DTPPhOH) in the Plußsee in 1989 and 1976.

necessary for carbon flux calculations, degradation studies, and a basic understanding of microbial heterotrophy in aquatic environments.

The chemical composition of the DOM was studied by gas-liquid chromatography (GLC) and by high-performance liquid chromatography (HPLC) to identify the main building blocks of DOM and their contribution to the variation and distribution of the DOC pool. The results of these studies are summarized in Figure 2.5.

Dissolved Amino Acids

Amino acids are important molecules in protein synthesis, and enzyme structures, activities, and regulation (Kirchman et al. 1985). They are important tracers in studies on the dynamics of substrate utilization and turnover in aquatic environments (Kirchman and Hodson 1986). Dissolved amino acids in the Plußsee have been studied by Brehm (1967). We continued DAA analysis in the Plußsee in 1981, 1987, and 1989. Dissolved total amino acids (DTAA) varied during the summer of 1989 between 2.5 to 16.5 μM (mean 7.15 ± 0.98 μM). Highest DTAA were found in the hypolimnion (20–25 m) during spring and in the epilimnion (0–6 m) during summer (Fig. 2.5A). There was no clear distribution gradient of DTAA in the water column. The carbon content in the DTAA varied between 151.3 and 861.1 μg C/L (mean 371.9 ± 0.40 μg C/L) and contributed 1.24–9.9% of organic carbon to the DOC.

Dissolved Carbohydrates

Carbohydrates are one of the first carbon and energy storage products after photosynthesis. They participate in many metabolic processes and serve cell recognition and communication functions. Because of their central role in cell growth and metabolism they have received the largest attention in aquatic microbial substrate studies. Dissolved carbohydrates in East Holstein lakes were analyzed by Weinmann (1970) and Stabel (1977). We have continued these studies in 1976–1977, 1981, 1984, and 1991 in the Plußsee.

Dissolved total carbohydrates (DTCHO) varied during an annual sampling period in 1976 between 0.09 and 2.28 mg C/L (mean 0.65 ± 0.28 mg C/L), with highest values in spring (May), late summer (August), and autumn (September–October) in the epilimnion (Fig. 2.5B). Below the thermocline (4–6 m) there was a sharp decrease in the DTCHO concentration in the DOM; the lowest values were from the deep hypolimnion. DTCHO contributed 3.6% to 34.6% of its organic carbon to the DOC (mean 16.3 ± 5.6% of DOC). The highest percentage contribution was found in the epilimnion, and lower values were found in the hypolimnion. The DTCHO distribution showed patterns that were similar to the phytoplankton biomass distribution. Autochthonous production seemed to be an important input source of DTCHO for Plußsee DOM.

Dissolved Polyphenols

Polyphenolic compounds are regarded as the main building blocks in aquatic dissolved humic matter and in the past were a main concern in soil and aquatic

Table 2.1. Mean Concentration of the Dissolved Organic Carbon Fractions in Plußsee Water Sampled Weekly between Spring and Winter[a]

	Organic Carbon (mg C/L)			
DOM Fraction	Epilimnion	Epilimnion/ Hypolimnion	Metalimnion/ Hypolimnion	Hypolimnion
Total DOC[b]	4.28	4.13	3.21	3.56
Dissolved polyphenolic compounds (DPhOH)				
Total (DTPhOH)	1.89 (44.2)	1.93	1.26	1.10
Free (DFPhOH)	0.44 (10.3)	0.41	0.34	0.22
Combined (DCPhOH)	1.45 (33.9)	1.70	0.92	0.75
Dissolved amino acids (DAA)				
Total (DTAA)	0.61 (14.3)	n.d.[c]	n.d.[c]	n.d.[c]
Free (DFAA)[d]	0.005 (0.1)	0.006	0.005	0.003
Free (DFAA)	0.004 (0.09)	n.d.[c]	n.d.[c]	n.d.[c]
Peptide DAA	0.23 (5.4)	n.d.[c]	n.d.[c]	n.d.[c]
Dissolved carbohydrates (DCHO)[e]				
Total (DTCHO)	0.94 (21.9)	0.83	0.44	0.36
Free (DFCHO)	0.12 (2.8)	0.10	0.05	0.05
Combined (DCCHO)	0.82 (19.2)	0.73	0.39	0.31
Fatty acids[f]	0.05 (1.2)	n.d.[c]	n.d.[c]	n.d.[c]
Organic acids	0.001–0.23 (0.02–5.4)	n.d.[c]	n.d.[c]	n.d.[c]

[a]Modified from Münster (1993). Percentage contribution of fraction to the total dissolved organic carbon is given in parentheses. Samples were taken from the epilimnion (0–1 m) and from the interfaces between epilimnion and metalimnion (3–5 m) and the metalimnion and hypolimnion (5–8 m) and from the deep hypolimnion (20–25 m).

[b]Determined by WCO method.

[c]Not determined.

[d]According to Brehm (1967).

[e]According to Münster (1985).

[f]According to Poltz (1972).

chemistry studies (Frimmel and Christman 1988). Ohle (1934) stated that the Plußsee can be classified as a typical "Waldhumussee," which carries higher amounts of DHM than other East Holstein lakes. We established a specially designed analytical procedure in our laboratory (Münster 1982, 1985) that was adopted to analyze those phenolic compounds that might contribute to the large bulk material, often called dissolved humic matter or "Gelbstoffe" (Kalle 1966). For this reason 54 different naturally occurring polyphenolic compounds were analyzed by capillary GLC in Plußsee water during annual sampling. The results are summarized in Figure 2.5C. The dissolved total polyphenols (DTPPhOH) varied between 0.2 and 4.7 mg C/L (mean 1.55 ± 0.43 mg C/L), with highest values during autumn (September–October) in the thermocline and hypolimnion (Fig. 2.5C) and lower values in surface water. DTPPhOH contributed 8–71% of organic carbon to the DOC (mean 39 ± 8% of DOC). The highest contribution was found in the thermocline and upper hypolimnion, and lower contributions in the surface water and above sediments. The mechanisms behind this distribution are not quite clear, but degradation and condensation processes might be higher in the epi-/metalimnion and meta-/hypolimnion interfaces as can be seen from bacteria distributions and their activities (see Chapter 7, this volume).

On the basis of these analyses the amounts of DOC contributed by DTPPhOH, DTAA, and DTCHO to the DOC are summarized in Table 2.1. All substrates together contributed 6.7% to 98.9% (mean 44.9 ± 5.3%) of organic carbon to the DOC pool. Highest values were found in the surface water and thermocline during spring (May), summer (August), and autumn (September–October). The amount of identified compounds in the DOC was highest in surface water and the thermocline (on average 60–90%) and decreased to 30–40% in the hypolimnion and sediments above. Similar DOC identification problems were observed by marine chemists (Hedges and Lee 1993).

Other organic compounds, including organic acids, vitamins, pigments, lipids, steroids, and gases, have been found in lower concentrations and seemed to be of minor importance for DOC balance calculations in the water column (Münster 1993). Their metabolic and ecological functions may be as important as, or even more important than, previously believed.

Enrichment Experiments

Studies on the chemistry of DOM and their structural building blocks often require sample isolation and enrichment steps (Thurman 1985) because many of the organic molecules that comprise DOM are too dilute to be easily analyzed and identified. Methods for DOM enrichment from aquatic environments have been evaluated (reviewed by Aiken 1988). For DOM studies in the Plußsee an enrichment and preseparation procedure was developed (Münster 1982, 1985) similar to that described by Leenheer (1981). Three different macroporous resins were useful for DOM studies: Amberlite XAD-2, polyvinylpyrrolidone (PVP, both from Serva, F.R.G.), and Polyamide-6 (from Macherey-Nagel, F.R.G.). The goal of these adsorption studies was to get more information about the functional and chemical character of the DOM (Münster 1982, 1985). The results of these studies are presented in Table 2.2 and summarize data from lake surface water samples (0–3 m depth) of Plußsee DOM during April, June, July, November, and December, 1979. A detailed description of these experiments is given by Münster (1982, 1985).

The XAD-2 adsorption experiments clearly showed the pH dependence of DOC adsorption from the Plußsee, as previously described by Mantoura and Riley (1975) for marine DOC. Our results clearly support the assumption that lake water

Table 2.2. Results of DOM Adsorption Experiments on Amberlite-XAD-2 and Polyvinylpyrrolidone (PVP)[a]

	Amberlite XAD-2						Polyvinylpyrrolidone					
	Adsorption		Recovery		Loss		Adsorption		Recovery		Loss	
Percentage	DOC	UV	DOC	UV	DOC	UV	DOC	UV	DOC	UV	DOC	UV
Maximum	77.8	86.5	96.6	98.4	3.3	3.3	70.1	76.6	97.6	97.6	4.1	2.7
Minimum	23.8	29.5	93.4	90.7	1.2	1.4	25.0	27.9	87.1	87.9	1.0	1.5
Average	55.5	63.2	95.0	95.7	2.2	2.2	51.7	55.9	93.9	94.8	2.5	2.2

[a]Water samples taken in April–July and November–December 1979 were adsorbed at different pH values on XAD-2. The percentage of DOM adsorption was determined by DOC measurements and UV absorbance. All data are the mean values of triplicates. Modified according to Münster (1985).

DOM has a highly polar character and good water solubility at natural pH conditions (e.g., pH 8.0). Acidification lowers the polarity or hydrophilicity and increases the lipophilicity after protonation of the polar functional groups (e.g., -C-O^{-1}, –CO_2^{-1}). This forces lipophilic interactions between the polystyrene XAD-2 matrix and the DOM compounds. On average 23–78% of Plußsee DOM from surface water can carry organic building blocks with such structural functional groups.

Studies by Olsson et al. (1976) on the adsorption mechanisms between polyphenolic compounds and polyvinylpyrrolidone (PVP) revealed that the pattern of substitution of aromatic ring systems with -OH or -CO_2H groups significantly affected the binding enthalpy between the polyphenolic compounds and the PVP matrix. The main binding adsorption mechanisms of polyphenols on PVP were identified as hydrogen bridge bonds with 12.6–16.8 kJ/mol.

A summary of the results of our studies with PVP adsorption experiments is given in Table 2.2. Similar to the XAD-2 studies, DOC adsorption on PVP showed significant pH dependence with adsorption maxima at pH 2–3. The DOC adsorption levels were 4–5% lower on PVP compared to XAD-2. This can be explained by the more specific adsorption mechanisms of hydrogen bridge bonding on PVP compared to the nonspecific hydrophobic interaction mechanisms on XAD-2. Both macroporous resins were suitable for DOC enrichment studies. However, PVP has some disadvantages compared to XAD-2 because of its much lower flow rate and intensive precleaning procedures. Still, on average 45% (XAD-2) to 49% (PVP) of the DOC was not isolated by these methods (Münster 1982, 1985). Thus, we could not identify nearly half of the DOC. We still have no definite chemical explanation for this observation, which is in agreement with earlier studies of DOM composition by GLC and HPLC (see Table 2.1). There is at present about 50% of DOC we cannot identify with conventional available methods. More advanced physical-chemical methods are needed to identify those unknown DOM fractions.

Polyphenols are known to have strong binding activities to proteins (McManus et al. 1985). Such interactions have a fundamental impact on enzymatic and biochemical processes at the cellular and organism levels (Spencer et al. 1988). As a continuation of our bulk-phase characterization of the Plußsee DOM, we used the synthetic protein Polyamide-6 (Macherey-Nagel, F.R.G.) to search for the presence of such fundamental interactions, which might affect enzyme–substrate interactions under in situ conditions. For these adsorption studies of the Plußsee DOM with Polyamide-6 we have used thin-layer chromatography (TLC) combined with specific detection reagents for direct polyphenol, carbohydrate, and amino acid detection on the plates (Münster 1982, 1985). The results revealed that Plußsee DOM contained strong interactions with the Polyamide-6, which indicated the presence of polyphenolic building blocks in the DOM similar to those found with PVP.

Based on the resolution power of modern GLC with two parallel WCOT fused silica columns, 25 and 50 m long, 0.32 mm I.D., and coated with CP^{tm}SIL-5 and CP^{tm}SIL-8 from Chrompack (Netherlands) and FID detection, we calibrated our gas chromatographic system with 54 different naturally occurring polyphenols (as TMS-derivatives) and analyzed the Plußsee DOM from XAD-2 and PVP adsorption experiments for the occurrence of polyphenolic compounds (Münster 1982). With this technique we could quantify the amount of dissolved free polyphenols (DFPPhOH) adsorbed by the XAD-2 and PVP resins. The DFPPhOH concentration varied between 12.5 and 500 µg C/L (mean 194 µg C/L) for the XAD-2 resin. The DFPPhOH concentrations on PVP varied between 32 and 362 µg C/L (mean 216 µg C/L) with the highest amounts at pH 2. These results supported our former adsorption results obtained with the carbon and UV absorbance measurements and supported our results from Polyamide–DOM interaction studies.

Our next interest was to identify the main functional groups in the DFPPhOH fraction that we could isolate at different pH with XAD-2 and PVP. From the amounts of identified DFPPhOH we calculated how many functional groups were present in the DOM isolates, which were expressed as relative numbers. Carboxyl and hydroxyl groups were the dominant functional groups in the DFPPhOH fraction isolated on XAD-2 and PVP. Ketone, methoxy, and methyl groups were found in relatively lower amounts (Münster 1982, 1985). It was further evident that the numbers of -OH and -CO_2H groups increased with the decrease of the pH in the applied water sample.

We then combined the adsorption technique with a size fractionation technique by gel permeation chromatography (GPC with Sephadex G15) to localize the functional groups of the polyphenolic fraction (DPPhOH) in the DOM matrix that are responsible for the adsorption mechanisms on XAD-2 and PVP. For this purpose we performed two adsorption experiments with hydrolyzed and non-hydrolyzed Plußsee DOM. We applied both DOM sources to PVP adsorption and fractionated the PVP eluates of the Plußsee DOM on Sephadex G15 (according to their residual molecular structures) into a macromolecular (F_1), an oligomeric (F_2), and a monomeric (F_3) fraction. Thereafter, from all three fractions the DFPPhOH were analyzed by GLC, and the individual DPPhOH was identified (Münster

1982, 1985). These results showed that with hydrolysis treatment (1 N HCl under nitrogen at 110°C for 10 h) there was an increase of the relative amounts of all functional groups in the macromolecular fraction F_1 (> 1500 daltons), in the oligomer fraction F_2 (600–800 daltons), and in the monomeric fraction F_3. The highest increases in -OH and the -CO_2H groups compared to the untreated water (no hydrolysis) sample were found only in the fraction F_3, which was characterized by large amounts of the -OH group, and in fractions F_1 and F_2, where we found equal levels of -OH, -CO_2H, and -$COCH_3$ groups. According to the distribution pattern of the GPC chromatogram we could obviously not completely cleave the DOM matrix into monomeric or at least oligomeric compounds. There was still a macromolecular fraction F_1 present, even after 10 h of hydrolysis. On the other hand, we could identify higher amounts of -OH and -CO_2H groups in fractions F_1 and F_2.

Dissolved Organic Carbon: Still a Black Box?

The characterization and availability of DOC in lacustrine environments is mainly based on its utilization by microheterotrophs, predominately bacteria. According to this microbial physiological approach, the DOC pool can be divided into a labile DOC pool (LDOCP) and a recalcitrant DOC pool (RDOCP). Parallel to the physiological approach, analytical chemical measurements are more exceptional then regular in freshwater ecological studies (Münster and Chróst 1990, Münster 1991, 1993). This is fundamentally different from marine studies, where geo- and marine chemists have already established a fundamental chemical inventory of marine DOM. There is still a large gap in our basic understanding of the chemical composition and structure of aquatic DOM. According to Williams (1986) only 10–15% of the marine DOC were easily chemically identified. The majority of the DOC require more sophisticated advanced analytical and chemical approaches for final identifications.

For Plußsee DOM, Münster (1982, 1985, 1993) could identify about 84% of the DOC in Plußsee surface water samples based on GLC and HPLC analysis. This is certainly higher than what is achieved by microbial physiological approaches alone (Overbeck 1979, Münster 1991, 1993). Surface-water DOM is more labile and easier to identify than deep-water samples. This can also be seen in further results from Plußsee DOM identification data as summarized in Table 2.1. Based on the identification of DTAA, DTPPhOH, DTCHO, and DOAc, 41–84% (mean 61 ± 18%) of the DOM could be chemically identified in the whole water column (Table 2.1). Calculated on the DOC basis, the highest level of identification was achieved in the epilimnion (84%), a lower level in the thermocline (53–67%), and the lowest at the sediment/hypolimnion interface (41%). This DOC identification level was higher than that from the microbial physiological approach when measuring substrate uptake by bacterioplankton (Overbeck 1979). About 16–49% of the DOC was not identified by these analytical and chemical procedures.

Labile DOC Pool: A Highly Dynamic Compartment

From 21% to 67% (average 43%) of daily phytoplankton production flows through bacteria in the euphotic zone of the Plußsee (Chapter 5, this volume). In the Plußsee, Chróst et al. (1989) and Sell and Overbeck (1992) found a close coupling among primary production, extracellular released organic carbon (EOC), and bacteria substrate uptake. Because of the short turnover times of this EOC only small molecules may comprise this labile EOC pool (Chróst and Faust 1983), which can be easily assimilated and transferred into bacteria cells. Quantification and chemical identification of this labile DOC pool (LDOCP) is of fundamental concern for basic understanding of microbial processes in aquatic environments. In the following paragraphs some examples are presented for the distribution and dynamics of LDOCP compounds in the Plußsee.

Dissolved Free Amino Acids and Carbohydrates

Dissolved free amino acids (DFAA) and dissolved free carbohydrates (DFCHO) are regarded as important tracer molecules for studies of the dynamics of microbial substrate uptake, utilization, substrate turnover and bacteria production (Overbeck 1979, Azam and Cho 1987, Jørgensen 1987, Simon and Azam 1988, Coffin 1989). Knowledge about their distribution, variation and dynamics are of fundamental concern in aquatic microbial studies. Results from our DFAA analysis in the Plußsee will be presented from three case studies: DFAA and DFCHO distribution during annual distribution, spring bloom developments and during a diurnal cycle.

Annual DFCHO concentrations varied in the Plußsee between 0.5 and 616.3 μM (mean 246.5 μM). Higher values were found in spring, late summer, and autumn. DFCHO concentrations were higher (maximum 0.103 mg C/L) in the epilimnion and decreased rapidly below the meta-/hypolimnion interface to 0.002 mg C/L above sediment. The average DFCHO concentration was 0.04 mg C/L in the whole water column. DFCHO carbon contributed 0.06–3.1% (mean 1.1%) of organic carbon to the DOC (Fig. 2.6A). During spring bloom development with 2-day sampling intervals from March to April 1984 (Fig. 2.6B), DFCHO varied between 0.11 and 1.17 μM (mean 0.31 μM) in lake surface water (0–3 m depth). The DFCHO distribution pattern corresponded with the algal carbon biomass contribution to total POC (Chróst et al. 1989). The highest DFCHO concentration (0.84 mg C/L) appeared 1–2 days before chlorophyll$_a$ maxima (Chróst et al. 1989). The lowest DFCHO concentration with 0.006 mg C/L was found in March (Fig. 2.6B), when chlorophyll$_a$ and primary production were low (Chróst et al. 1989).

A second important organic compound group in the LDOCP are the DFAA. Three years later, in 1987 (Fig. 2.6C), the DFAA pool size during a spring bloom development in the Plußsee was analyzed by the HPLC method according to Lindroth and Mopper (1979). DFAA varied between 0.15 and 1.85 μM with three maxima of 0.7, 1.8, and 0.8 μM (mean 0.6 μM). The first small peak appeared in early spring (April 21); a second, the largest one, in late spring (May 12); and a

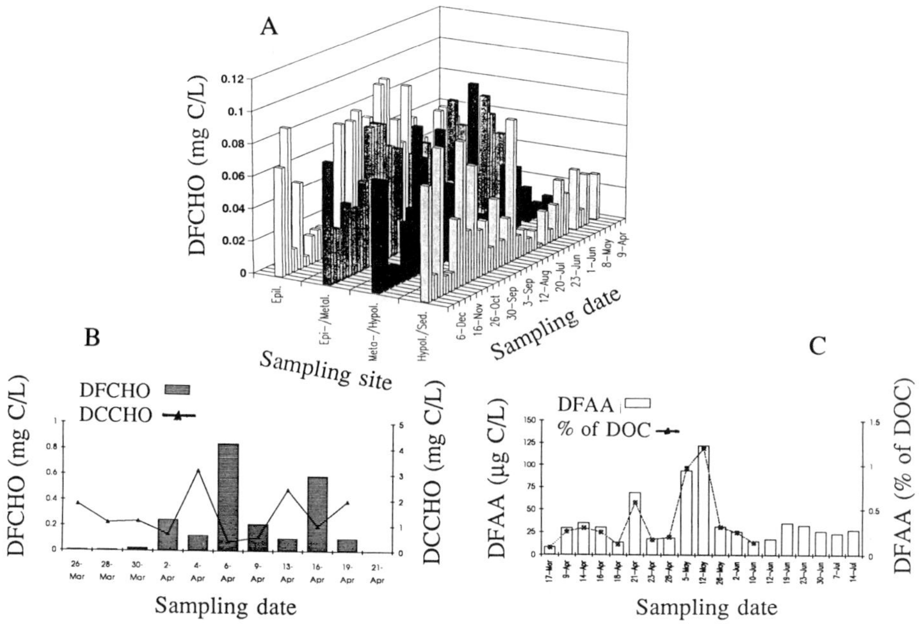

Figure 2.6. Annual and seasonal distribution of (A,B) dissolved free carbohydrates (DFCHO) and (C) dissolved free amino acids (DFAA) in the Plußsee from 1976, 1984, and 1987. Modified according to Münster (1985, 1993).

third small one in early summer (June 19). The contribution of the DFAA to the total DOC varied between 0.1 and 1.2% (mean 0.3%) of organic carbon to the DOC. This carbon content of the DFAA pool of Lake Plußsee is in a range similar to that found by Jørgensen et al. (1983) in eutrophic Danish lakes.

Low-molecular-weight organic compounds in the LDOCP pool of marine and freshwater systems varied slightly on a seasonal or annual scale but more intensively on a diurnal scale (Münster 1984, 1985, Münster and Chróst 1990, Jørgensen et al. 1983, Carlucci et al. 1984). Labile organic substrates such as DFAA or DFCHO are usually taken up rapidly and have short turnover times (reviewed by Münster 1993). Most researchers measured only the EOC and its utilization by bacterioplankton during diel studies. Fewer studies followed the variation of specific compounds such as DFAA or DFCHO and their release and uptake under natural conditions (Burney 1986, Münster 1984, 1985, Münster and Chróst 1990, Carlucci et al. 1991). However, such studies are very useful for a better understanding of the dynamics of LDOCP compounds and their close coupling to their producers (phytoplankton, macrophytes, grazers) and consumers (bacteria).

One example of such diurnal variation and the dynamics of LDOCP compounds (DFAA and DFCHO) is given in Figure 2.7. The DFAA varied between 0.20 and 116.5 µg C/L (mean 29.5 µg C/L) in the euphotic zone (0–5 m). Maxima

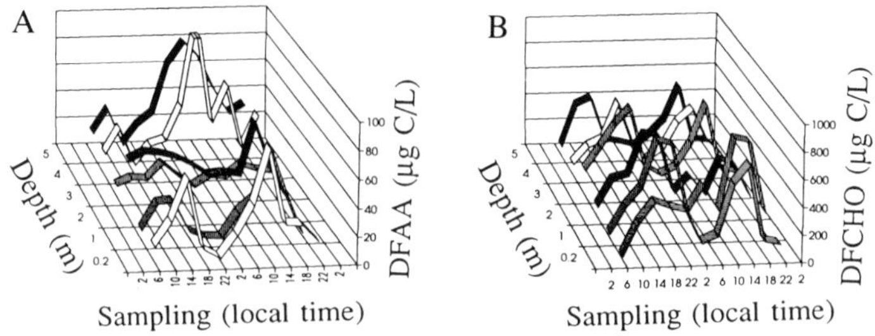

Figure 2.7. Diurnal and vertical distribution of (A) dissolved free amino acids (DFAA) and (B) dissolved free carbohydrates (DFCHO) in the period March 31 to April 1, 1981. Modified from Münster (1985, 1991, 1993).

in surface water were found from 1000 to 1400 h during the light period, and minima during the dark period from 2200 to 0200 h (Fig. 2.7A). This distribution pattern changed, however, in deeper water layers (4–5 m), where maxima appeared in the dark period (2200-0400 h), and smaller DFAA peaks during the light period (0600-1400 h). The reasons for these differences are not clear, but the light–dark synchronicity in the upper layers may change with depth along with a decrease of available photosynthetic light and a change in the microbial community structure. Similar diurnal variations of DFAA were found by Jørgensen et al. (1983) in two Danish lakes.

The DFCHO pool also varied (Fig. 7B) between 28.9 and 700.3 μg C/L (mean 290.3 μg C/L) with a maximum of 700.3 μg C/L in the upper water layer (0.2–1.0 m) in the afternoon. Below this depth, the concentration maxima decreased to 511.2–669.6 μg C/L. In the upper 2-m water layer a significant diurnal light–dark cycle could be measured. This light–dark cycle was also detectable in deeper water layers. Maxima of DFCHO followed 2–4 h after the photosynthesis maxima. The measured DFCHO concentrations were higher compared to former DFCHO results (Münster 1985) and are assumed to be caused by the predominance of a diatom bloom that had just reached its biomass maximum and decreased rapidly within the next 2–5 days (Hickel 1978). The diurnal DFCHO values varied more than when weekly or even longer sampling intervals were used (Münster 1985), but this data set clearly showed that the dynamic character of the LDOCP is an important aspect for water sampling and DOC measurements.

Excreted Glycolic Acid by Phytoplankton: A Tracer for Metabolic Coupling

Among biologically derived dissolved organic acids (DOAc), glycolic acid may have a biomarker function for the release of DOC or excreted organic carbon (EOC) by phytoplankton (Fogg 1983). The biochemical production pathway for glycolic acid follows the reaction of ribulose bisphosphate with oxygen or CO_2 as substrate in a competitive manner (Tolbert 1974). Following this reaction, glycolic acid can be excreted or metabolized via glyoxalate to serine or glycine (Tolbert and Zill 1956). In natural waters, glycolic acid concentrations were observed to be

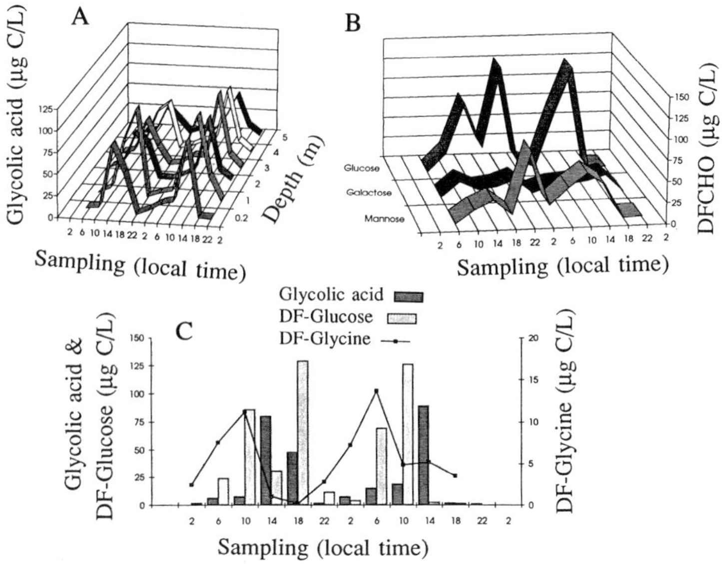

Figure 2.8. Vertical and diurnal distribution of (A) glycolic acid, (B) dissolved free hexoses, and (C) glycolic acid, dissolved free glucose (DF-glucose), and dissolved free glycine (DF-glycine) in the period from March 31 to April 1, 1981. Modified from (Münster 1985, 1991, 1993).

relatively high compared to other low-molecular-weight dissolved organic carbon compounds (Shah and Wright 1974, Al-Hasan et al. 1975). Bacteria from upper lake water could assimilate glycolic acid and may use this carbon source for their energy requirements (Wright 1970, 1975).

During the same diurnal studies as reported above for DFAA and DFCHO, glycolic acid was studied as a tracer compound for phytoplankton EOC in an experiment on the release and composition of photosynthetically fixed carbon (PhDOC) by phytoplankton (Münster and Chróst 1990, Münster 1991). Glycolic acid showed distinct light and dark variation in water samples in the upper 5 m (Fig. 2.8A). This biphasic distribution pattern was expressed more distinctly in the upper 2 m and decreased down to a depth of 5 m. Glycolic acid concentrations varied between 1.1 and 104 μg C/L (mean 15.5–36.2 μg C/L) throughout the whole water column. Maxima (52–104 μg C/L) appeared mostly between 1000 and 1400 h, and most of the glycolic acid produced disappeared during the dark period (2200–0400 h). Many bacteria can assimilate glycolic acid but do not metabolize it for growth or biosynthesis. They preferentially oxidize it to CO_2 and use it as an energy source in the respiration pathway (Wright 1975). According to Wright and Shah (1975), energy from glycolic acid oxidation may be coupled to active transport of multiple mixed substrates. The key enzyme glycolate oxidase is a membrane-bound enzyme and couples the energy to transport systems of other carbon substrates (Lord 1972). The existence and dynamics of such a mixed-substrate situation of cooccurring LDOCP solutes in Plußsee DOM from surface water

(0–1 m depth) are shown in Figure 2.8B and C. From the most abundant dissolved free hexoses (DF-glucose, DF-galactose, and DF-mannose), DF-galactose showed the lowest concentration and varied between 1.1 and 21.7 (mean 11.2) μg C/L. DF-mannose varied between 1.1 and 83.1 (mean 23.8) μg C/L, and DF-glucose showed higher concentrations and varied between 1.4 and 189.1 (mean 30.8) μg C/L. Only DF-mannose and DF-glucose showed distinct diurnal variations, whereas DF-galactose varied less. A further example of the rapid variation of LDOCP solutes is shown in Figure 2.8C, where we have plotted the coexistence and variation of DF-glucose, DF-glycine, and the EOC tracer glycolic acid. DF-glycine was present in a much lower concentration than the DF-hexoses and glycolic acid. DF-glycine showed a distinct diurnal variation but inverse to that of glycolic acid. DF-glycine contributed 0.2–13.7 (mean 5.4) μg C/L to the DOC. Although we have not measured all detailed metabolic biochemical steps involved in the glycolic acid processing, we believe that cooccurring substrates may affect substrate utilization under natural conditions. The existence of such mixed substrate conditions in aquatic environments and the effects on their utilization by microbes is still not well recognized in natural environments. We consider chemical and microbial physiological approaches to be important tools in improving our understanding of microbial processes and mixed substrate interactions.

Recalcitrant Dissolved Organic Carbon: A Heavy Fuel for Microheterotrophs

We could easily identify chemically about 10–20% of the Plußsee DOC (see Table 2.1). The largest part of this DOM fraction can be characterized as the recalcitrant dissolved organic carbon pool (RDOCP), which includes polyphenolic compounds. The trophic role of this RDOCP in energy and food web processes is still poorly understood (Wetzel 1992, Hobbie 1992).

Polyphenols, Lignins, Lignocellulose, and DHM: A Continual Carbon Pool for Microbial Heterotrophs

An attempt was made to characterize with a chemical method the part of the DOM fraction that is recalcitrant to rapid microbial degradation and utilization: dissolved polyphenolic compounds present in the DOC pool were analyzed and identified by capillary gas chromatography as described above for enrichment and functional group analysis (Münster 1985). The DTPPhOH pool was separated into dissolved free polyphenols (DFPPhOH) and dissolved combined polyphenols (DCPPhOH). The DFPPhOH fraction varied between 0.01 and 0.81 (mean 0.35) mg C/L (Fig. 2.9A). Higher DFPPhOH amounts were found in autumn throughout the whole water column. We could not see any typical vertical distribution pattern that might have been related to biological activities. The DFPPhOH pool contributed with 0.4–22.3% (mean 9.2%) of organic carbon to the DOC. The combined polyphenolic fraction (DCPPhOH) varied in Plußsee water (Fig. 2.9B) between 0.07 and 4.1 mg C/L (mean 1.2 mg C/L) and contributed 2.1–83.9% (mean 44.5%) of organic carbon to the DOC pool. Eighty to 90% of these polyphenols was bound

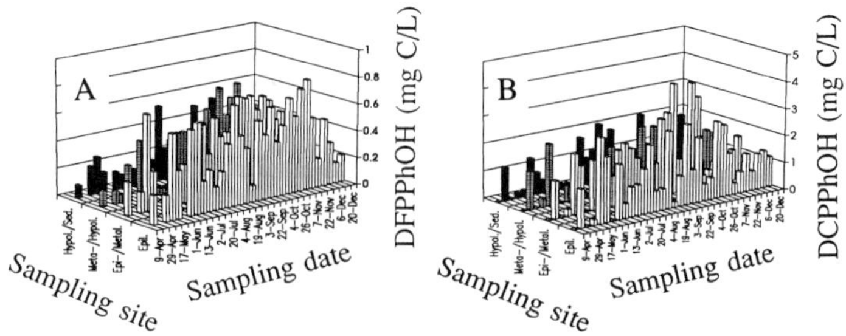

Figure 2.9. Annual and vertical distribution of (A) dissolved free polyphenols (DFPPhOH) and (B) dissolved combined polyphenols (DCPPhOH) in 1976. Modified from Münster (1985).

to high-molecular-weight fractions (> 1,500), and only 2.5–20.1% (mean 10.8%) was found in dissolved free monomeric polyphenolic compounds (DFPhOH).

We have not addressed the microbial uptake and utilization of polyphenolic compounds in these experiments. There is enough evidence in the literature about aquatic bacteria's ability to degrade polyphenolic humic matter and utilize it as a carbon source (Tranvik 1988, 1992, Hessen 1992). In East-Holstein lakes, Tranvik and Höfle (1987) could show that bacteria from humic waters were able to metabolize added [^{14}C]phenol and reach higher biomass compared to bacteria from clearwater lakes. RDOCP substrates such as DFPPhOH and DCPPhOH may contribute to the RDOCP, which is slowly utilized by bacterioplankton and may contribute an important continual carbon source for microheterotophs (Hessen 1985, Tranvik 1989, 1990).

Biopolymers: A Neglected DOM Component in the Past

In nature, the majority of organic matter is present as geopolymers and biopolymers. Cleavage and degradation of this material to smaller molecules is done mainly by microheterotrophs. There is increasing research effort and interest in the mechanisms of biodegradation under natural conditions (Chróst 1991b). Human activities release many synthetic polymers into the environment, but little is known about their biodegradation (Swift 1992, Schink et al. 1992).

Large molecules can not pass bacteria cell membranes. They have to be hydrolyzed to smaller molecules before they are assimilated. For a long time this rate-limiting step for DOC utilization was not considered sufficiently in studies of bacterioplankton utilization of DOC that modeled the carbon flow (Billén 1984). Recently these important processes have been recognized in aquatic microbial studies (reviewed by Chróst 1990b, 1991a, Billén 1991, Hoppe 1991, Wetzel 1991; Chapter 6, this volume). Correlations have been found between bacterial biomass and exoproteolytic activities (Billén 1991, Hollibaugh and Azam 1983, Hoppe et al. 1988, Meyer-Reil 1986). Protein (BSA) turnover times (1–1000 h) and monomeric substrates (e.g., leucine; Billén et al. 1990) were found to be similar. The majority of those enzymes are cell surface bound and followed Michaelis–Menten kinetics (Chróst 1989, 1990b, 1991a; Chapter 6, this volume). Significant correlations were found between GLCase activities and the concentration of DC-glucose ($r^2 = 0.85$, $P \leq 0.001$, $n = 16$) and between aminopeptidase activity and DC-leucine ($r^2 = 0.93$, $P \leq 0.01$, $n = 14$) in the polyhumic lakes (Münster 1991). Many aspects of the function of those enzyme systems in humic-rich or acid environments are not well understood (Wetzel 1991, Münster 1991). In many cases these enzymes may be inhibited or adsorbed onto particles and immobilized and translocated to other water bodies where reactivation may occur (Wetzel 1991). For a better understanding of the utilization and processing of biopolymers in aquatic environments, it is necessary to characterize and identify distinct molecular weight fractions in the DOM responsible for microbial enzyme–substrate interactions and microbial nutrition. We have followed a combined analytical chemical/microbial physiological approach to search for some relationships between the molecular structures of the DOM and microbial activities.

Biopolymer Fractionation Experiments

Because of increasing evidence of the importance of the high-molecular-weight organic carbon fraction as a carbon, nitrogen, and phosphorus source for microbial heterotrophs, we fractionated Plußsee DOM by two different techniques and compared the DOM fractions with microbial parameters. The results for this approach are summarized in the following paragraphs.

Fractionation Techniques

Biopolymer fractionation techniques have received increasing attention in molecular and biochemical studies for isolation of proteins, polysaccharides, lipids, and polynucleotides. During the last 10–15 years these techniques have been introduced to aquatic studies. Ultrafiltration techniques are applied most commonly for DOM fractionation, followed by GPC. Polyacrylamide gel electrophoresis and isoelectric focusing (PAGE/PAGIF) are very seldom used in aquatic DOM characterizations. This method has very high resolution, especially when combined with detergents like SDS; it is a state-of-the-art method in biochemistry and molecular biology. We have used both methods, GPC and PAGE, for DOM characterization in the Plußsee, to gather more information about the nature of the HMWDOC fraction. A detailed method and application description is given by Münster (1982, 1985).

Sephadex Gel Permeation Chromatography

Sephadex-GPC has occupied a key position among the techniques for purification and isolation of biologically important macromolecules such as enzymes, polysaccharides, proteins, nucleic acids, and others (Determann 1967). We have used Sephadex G-75 for the fractionation of the original sample of 10–50 ml Plußsee

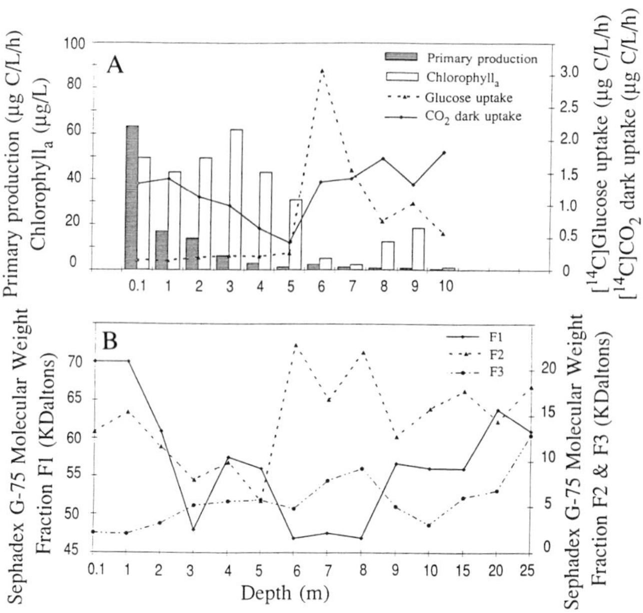

Figure 2.10. (A) Vertical distribution of rates of primary production, chlorophyll$_a$ concentration, and the rates of [^{14}C]glucose uptake and [^{14}C]CO_2 dark fixation. (B) Vertical distribution of Sephadex G-75 molecular weight fractions (F_1, F_2, F_3) of DOM in the Plußsee. Modified from Münster (1985).

DOM without any further preconcentration steps and monitored the eluate at 254 nm with a UV detector. Calibration settings and chromatographic conditions and calculations are described by Determann (1967) and Münster (1982, 1985). Typical results from such GPC experiments are summarized in Figure 2.10.

Sephadex G-75 is specified for its exclusion limits between 3,000 and 80,000 daltons for globular proteins and peptides and for 500 to 10,000 daltons for dextrans (Pharmacia, Sweden). With Sephadex G-75 we received three fractions for the Plußsee DOM. The molecular weight ranges were between 46,854 and >70,000 (mean 56,962) for F_1; for F_2 the molecular weight sizes varied between 5,600 and 22,675 (mean 14,555), and for F_3 2,012 to 12,853 (mean 5,651) throughout the whole water column (Fig. 2.10B). Fraction F_1 showed highest values in the epilimnion from 0 to 1 m and declined rapidly within a depth of 3 m. In the thermocline (5–7 m) F_1 increased slightly and decreased again in the hypolimnion. Above the sediment–water interface F_1 increased again. The second fraction, F_2, showed similar distributions, but the highest values were below the thermocline at 6–9 m in depth and decreased down to the sediment. The lowest-molecular-weight fraction, F_3, showed little variability and reached its highest values above the sediments (Fig. 2.10B.). Parallel measured primary production and heterotrophic activities are summarized in Figure 2.10A. Primary production decreased rapidly from surface water with 63 µg C L^{-1} h^{-1} down to 0.4 µg C L^{-1} h^{-1}

at 5-m depth and followed the light gradient in the water. Chlorophyll$_a$ started at 49.4 μg/L for the surface water and reached maximum values of 61.8 μg/L at 3 m, then decreased in the thermocline down to 1.1 μg/L. The heterotrophic activities measured as [^{14}C]glucose uptake and [^{14}C]CO_2 dark fixation (Overbeck 1979) showed only a high activity peak with 3.1 μg C L^{-1} h^{-1} in glucose uptake in 6- to 7-m depths, whereas the CO_2 dark fixation showed its highest values in the thermocline (1.22–1.41 μg C L^{-1} h^{-1}) and its lowest value (0.42 μg C L^{-1} h^{-1}) at 5 m depth in the thermocline. Below 5 m there was again an increase in CO_2 dark fixation with a maximum at 8 m (1.72 μg C L^{-1} h^{-1}). Both data sets showed the dynamic character of DOM macromolecules in the water column of the Plußsee and their close relation to autotrophic and heterotrophic processes. The F_1 fraction corresponded to primary production in surface water but not in the thermocline, whereas the F_2 fraction corresponded to the heterotrophic activities in the thermocline. It is not possible from these data to specify which fraction belongs to which process. It only shows the variation of autotrophic and heterotrophic activities and their coupling with the distribution and variation of macromolecular fractions in the DOM.

Polyacrylamide Gel Electrophoresis

In addition to the Sephadex technique we applied another, less used fractionation technique in aquatic DOM characterization, polyacrylamide gel electrophoresis (PAGE). PAGE is a powerful technique for protein fractionation and characterization (Münster 1982).

With PAGE we identified two DOM fractions (F_1 and F_2) with different electrophoretic mobility, molecular weight, radius, and net charge (Münster 1982, 1985). The molecular weight varied between 7,660 and 45,370 (mean 21,770) with the highest values in the surface water (0–1 m) for F_1 and between 7,120 and 90,190 (mean 37,270) for F_2 with the highest values in the surface water at 1-m depth. The mean radius displayed similar distribution patterns. The net charge for F_1 varied between 7.1 and 20.1 × 10^{-12} (mean 12.5 × 10^{-12}) Coulomb and between 10.2 and 23.3 × 10^{-12} (mean 15.6 × 10^{-12}) Coulomb for F_2. The highest variation of net charge was observed in the thermocline (3–7 m depth) and above the sediments. Parallel measurements of autotrophic and heterotrophic activities (Münster 1985) showed the highest primary production in surface water (24.90 μg C L^{-1} h^{-1}) and a rapid decrease within the epilimnion down to 0.4 μg C L^{-1} h^{-1}. Glucose uptake was more evenly distributed throughout the water column and varied between 0.14 and 0.32 (mean 0.24) μg C L^{-1} h^{-1} with no significant maxima. CO_2 dark fixation showed a maximum (3.80 μg C L^{-1} h^{-1}) in surface water and varied between 0.65 and 3.80 (mean 1.49) μg C L^{-1} h^{-1}. Molecular weight distribution of F_1 and F_2 and the microbial activities like primary production and [^{14}C]CO_2 dark fixation corresponded with each other. [^{14}C]Glucose uptake did not match the molecular weight structures in the DOM, whereas the CO_2 dark fixation seemed to follow patterns that were similar to the molecular weight structures in the DOM. The net charge distribution was similar to the patterns we observed for [^{14}C]glucose uptake activities (Münster 1985). According to Overbeck (Chapter 9, this volume), the anaplerotic CO_2 fixation may be a more useful parameter for displaying general degradation and organic substrate utilization processes than a highly specific substrate uptake such as DF-glucose. This is partially true in relation to the high-molecular-weight DOC distributions we received from PAGE biopolymer fractionation.

Specific staining of both fractions for the presence of the major building blocks revealed that in both fractions (F_1 and F_2) proteins, carbohydrates, and DHM were positively detected in the same band of the gel with a predominance of DHM (Münster 1982, 1985). A more advanced application of this PAGE technique (Ferguson plot) revealed that both fractions (F_1 and F_2) displayed an unusual electrophoretic mobility in gradient gels, with less retardation in increasing gel densities. That means that decreasing pore size in the gel did not reduce the mobility of Plußsee DOM fractions in the gel. From these observations we assumed that both DOM fractions had small globular structures but behaved more like stretched flexible rods when placed in an electrophoresis gel system. Such weakly structured organic aggregates with less rigid bonding within the macromolecules may explain their relatively higher electrophoretic mobility in gradient gels. These highly flexible structures may give them flexible ecophysiological functions in aquatic environments (Münster 1982, 1985).

Biopolymer Distribution and Dynamics

Five different major groups of biopolymers are present in freshwater DOM: proteins, polysaccharides, polynucleotides, lipids, and lignins and lignocelluloses. All these polymers have definite structures with definite repeating building blocks or units. Polyphenolic compounds are mostly involved as building blocks in dissolved humic matter (DHM) formations. During our studies on Plußsee DOM we have analyzed the distribution of polysaccharides, proteins, and DHM compounds. Lipids are less soluble in water and are often adsorbed onto surfaces of particles and concentrated in microlayers at water surfaces. Dissolved polynucleotides (DNA, RNA) have not yet been studied by us but may receive increasing attention in the future, especially with respect to the viral loop and the cycling of biopolymers in aquatic food chains (Proctor and Fuhrman 1990, 1992, Heldal and Bratbak 1991, Reisser et al. 1993; Chapter 13, this volume).

Seasonal Distribution and Variation of High-Molecular-Weight DOC

During seasonal and annual sampling in Lake Plußsee, concentrations of dissolved combined carbohydrates (DCCHO) and dissolved proteins (combined amino acids, DCAA) were 10–50 times higher than the DFAA and DFCHO and showed much higher variations (Brehm 1967, Münster 1984, 1985, Münster and Chróst 1990). Maxima of DCCHO were found in spring (May 17), in mid- and late summer (July 20, August 8), and in early and late autumn (September 9, Octo-

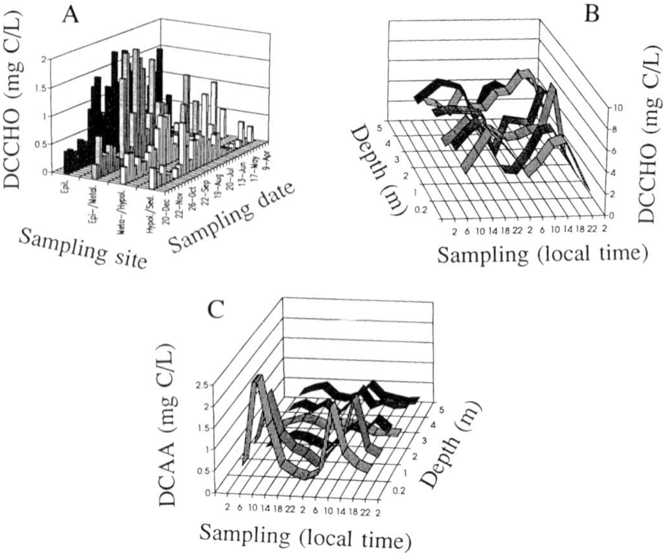

Figure 2.11. Annual and diurnal distribution of (A,B) polysaccharides and (C) proteins in the DOM pool of the Plußsee. Modified from Münster (1985).

ber 19). DCCHO varied between 0.01 and 2.22 mg C/L (mean 0.6 mg C/L) during an annual sampling period in 1976 (Fig. 2.11A) and contributed 0.3–45.9% (mean 15.4%) of organic carbon to the DOC. There was a sharp decline in the DCCHO concentration below the thermocline (Fig. 2.11A). Such a distribution of polysaccharides may be closely related to autotrophic and heterotrophic processes. Data from phytoplankton biomass (Hickel 1978) correlated with DCCHO ($r^2 = 0.67$, $P \leq 0.001$, $n = 35$) and suggested that autochthonously produced DOC was a major contributor to the DCCHO pool (Münster 1985). Bacterial activity measurements revealed that within the upper 4–5 m the majority of photosynthetically fixed carbon is already recycled (Chapter 5, this volume). Recent results from DFCHO, DCCHO, [^{14}C]glucose uptake, and GLCase activity measurements support the earlier data from the Plußsee that DCCHO are important substrates for bacterioplankton in this lake (Chróst et al. 1989, Münster 1991). DCCHO can be therefore be regarded as an important carbon source in the high-molecular-weight DOC pool for microbial nutrition.

Diurnal Variation of High-Molecular-Weight DOC

Studies on short-term variation of protein (DCAA) and polysaccharide (DCCHO) pool sizes in aquatic environments are rather scarce but are important for LDOCP compounds (DFAA and DFCHO) when degraded by microbial hydrolytic enzymes (MHE). Results of a diurnal study of the variation of DCAA and DCCHO in the Plußsee are shown in Figure 2.11B and C. In the euphotic zone DCAA was significantly stratified on a vertical and temporal scale (Fig. 2.11C). The highest concentrations were found during the light period (0600–1000 h), and the lowest during the dark phase (1800–0200 h). DCAA concentrations decreased rapidly in the upper 2 m and varied between 3.2 and 54.6 μmol/L (mean 8.9–16.5 μmol/L); in the lower 3–5 m it remained between 3.3 and 16.3 μmol/L (mean 6.04–8.86 μmol/L). The carbon concentration of DCAA varied between 0.17 and 2.84 mg C/L (mean 0.55 mg C/L). The highest concentrations were found in surface water (0.2 m) after sunrise (0600–1000 h) on both days. There was a significant reduction in the DCAA concentration as depth increased (Fig. 2.11C). This was different from what we found for the DFAA pool (Fig. 2.7A). In the polysaccharide distribution, measured as the DCCHO pool (Fig. 2.11B), there was no clear vertical profile with a sharp decrease in the DCCHO concentration between 1 and 2 m (Fig. 2.11B). DCCHO concentration varied between 7.3 and 115.8 μmol/L (mean 59.8 μmol/L). Carbon concentration varied between 0.53 and 8.36 mg C/L (mean 4.3 mg C/L). The highest values (74–115 μmol/L) were found during late afternoon (1400–1800 h), and lowest values (42.2–73 μmol/L) during the dark period (2200–0200 h). The DCCHO values were two to three times higher than the DCAA values (Fig. 2.11B and C). Both polymeric substrate pools displayed a diurnal variation that was similar to that of the monomeric (DFAA and DFCHO) compounds (Fig. 2.7A and B) and may be utilized efficiently by bacterioplankton after hydrolysis by MHE (Chróst 1990b, 1991b, 1992; Chapter 6, this volume).

Microbial Response to DOM Solutes: DOM as a Target of Microbial Heterotrophic Activities

On daily and seasonal scales, pulses of specific compounds in the DOC pool were observed with different intensities. This DOC input is mainly from the release of organic carbon compounds by photosynthesizing phytoplankton (Chapter 4, this volume), from lysis and "sloppy feeding" by macrozooplankton (Lampert 1978), and from phagotrophic micrograzers (Nagata and Kirchman 1991, 1992). Bacteria are regarded as sinks for organic carbon and do not contribute directly to DOC input (Ducklow et al. 1986, Ducklow 1991), although this last view of bacterial function in aquatic environments may have to be modified in the future following recent observations on the role of viruses and bacteriophages (Chapter 13, this volume). More DOM, especially of a polymeric nature, may cycle in aquatic food webs, as has been assumed and calculated in the past (Münster 1993).

Monomeric Substrates

In the previous sections we showed that LDOCP compounds such as DFCCHO and DFAA can vary over long periods (weeks), during annual sampling (Fig. 2.6A–C), and over short time scales, e.g., within a few days or hours (Fig. 2.7A and B). The question arises, how do natural bacterial populations respond to such substrate pulses? Another, less recognized, problem in DOC degradation studies arises from the dilution effect of added radiolabeled substrates by unlabeled natural substrates during uptake measurements (Chróst and Overbeck 1989, Chróst 1990a). We have looked for such "isotope dilution effects" and possibilities of correcting them.

Labile dissolved organic carbon pools (LDOCP) are found mostly in low concentration levels compared to the recalcitrant DOC pool fractions (Münster 1993). Natural bacterioplankton populations are therefore concerned with highly diluted substrate media and are assumed to live partially under substrate limitations (Overbeck 1979). It is further argued that during low substrate availability bacteria become dormant or are starving (Morita 1982, Kjelleberg et al. 1987). Other studies, especially those conducted in highly oligotrophic marine waters, showed that marine bacteria can very efficiently scavenge labile organic solutes from a highly diluted LDOCP (Azam and Cho 1987). We have concentrated our studies on the composition and dynamics of DOM solutes. These data were related to available substrate uptake data from the Plußsee.

Dissolved Free Amino Acids

During a spring bloom development in 1987 DFAA uptake measurements were done in integrated water samples (0–3 m depth) with a [^{14}C]amino acid mixture (Simon 1985). In parallel samples, the DFAA concentration was measured by HPLC. After corrections for isotope dilution both data sets are plotted in Figure 2.12. A correlation plot between uncorrected [^{14}C]amino acid uptake and DFAA concentrations for both uptake data sets (dark and light) revealed correlations between them [$r^2 = 0.31$, $n = 10$, $P \leq 0.001$ (dark uptake); $r^2 = 0.50$, $n = 10$, $P \leq 0.001$ (light uptake)]. After correcting for isotope dilution by natural DFAA concentrations, we obtained correlations of $r^2 = 0.91$, $n = 10$, $P \leq 0.001$ ($y = 0.0140x - 0.0860$) for light uptake and $r^2 = 0.90$, $n = 10$, $P \leq 0.001$ ($y = 0.0118x - 0.0465$) for dark uptake. Corrected uptake rates varied between 0.14 and 2.36 (mean 0.54) μg C L^{-1} h^{-1} for light uptake and between 0.13 and 1.89 (mean 0.49) μg C L^{-1} h^{-1} for dark uptake. Both corrected [^{14}C]amino acid uptake rates followed the distribution and concentration patterns of the chemically analyzed DFAA

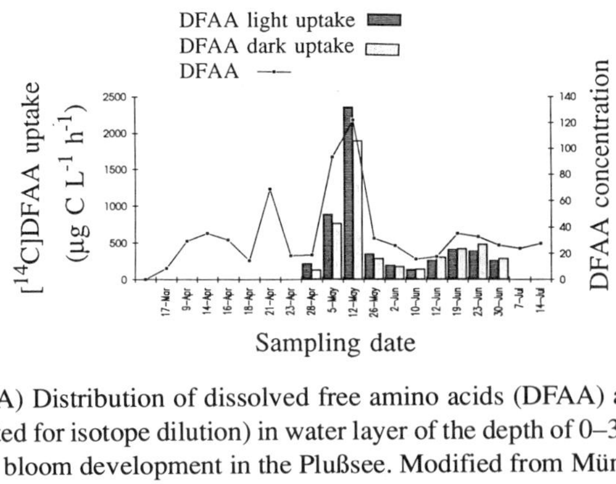

Figure 2.12. (A) Distribution of dissolved free amino acids (DFAA) and [^{14}C]amino acid uptake (corrected for isotope dilution) in water layer of the depth of 0–3 m during the spring phytoplankton bloom development in the Plußsee. Modified from Münster (1993).

concentrations (Fig. 2.12), which varied between 8.6 and 122.3 (mean 35.6) μg C/L. Specific corrected uptake rates for [^{14}C]amino acids varied between 0.07 and 1.07 (mean 0.37) fg C cell^{-1} h^{-1} for light uptake and 0.05 to 1.05 (mean 0.37) fg C cell^{-1} h^{-1} for dark uptake. These data show that there is a linear response of DFAA uptake systems in the microbial community to the DFAA concentration range between 0.07 and 1.85 μmol/L. The highest uptake rates corresponded to highest DFAA availability. There is a significant change in uptake rate values after corrections for isotope dilutions, and there are severe effects on carbon flux calculations if it is not considered in radioisotope assays.

Dissolved Free Carbohydrates

A similar approach was followed for [^{14}C]glucose uptake measurements during a spring bloom development in the Plußsee surface water (0–3 m depth) during 1984. According to Chróst et al. (1989), tight metabolic coupling was observed among photosynthetic production of labile DOM, exoenzymatic activities, and heterotrophic substrate assimilation. DF-glucose concentrations varied between 0.004 and 0.12 (mean 0.06) mg C/L. Corrected [^{14}C]glucose uptake varied in parallel measurements between 0.001 and 0.67 (mean 0.27) μg C L^{-1} h^{-1}. Simple linear regressions between DF-glucose concentrations and [^{14}C]glucose uptake gave a positive correlation ($r^2 = 0.96$, $P < 0.001$, $n = 6$; $y = 4.7064x - 0.0007$) for this spring bloom aspect (Münster 1993).

During diurnal studies [^{14}C]glucose uptake measurements were used to follow the heterotrophic activities of bacterioplankton (Fig. 2.13A and B). DF-glucose was measured in parallel by GLC (Münster 1985). [^{14}C]Glucose uptake measurements were done according to Overbeck (1979) and varied between 0.06 and 0.95 μg C L^{-1} h^{-1} (mean 0.25 μg C L^{-1} h^{-1}). DF-glucose concentrations varied between 1.4 and 189.1 μg C/L (mean 30.8 μg C/L). Statistical analysis between DF-glucose and [^{14}C]glucose data sets gave no correlations. When we corrected the [^{14}C]glucose uptake for dilution by available cold DF-glucose, as measured by GLC, [^{14}C]glucose uptake was larger and varied between 0.2 and 102.7 (mean 11.3) μg C L^{-1} h^{-1}. Linear regression analysis gave a correlation of $r^2 = 0.87$, $n =$

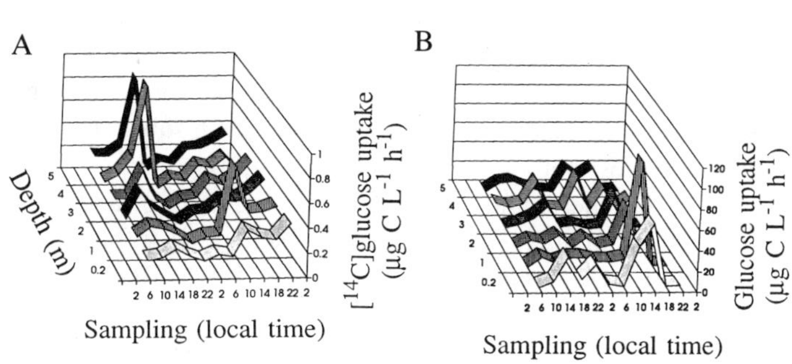

Figure 2.13. Diurnal distribution of [^{14}C]glucose uptake uncorrected (A) and corrected for isotope dilution (B). Modified from Münster and Chróst (1990) and Münster (1993).

66, $P \leq 0.001$ ($y = 0.0758x + 0.2919$). The distribution of [^{14}C]glucose uptake changed significantly after correction for isotope dilution (Fig. 2.13B). The highest activities were now observed in surface water in the afternoon (1400–1800 h) on the second day compared to highest activities at 4 and 5 m depth on the first day for uncorrected uptake. Again we observed a linear response of substrate uptake to a short DF-glucose pulse for a range of 1.4 to 189.1 μg C/L. That means that for each bacteria cell the uptake rates varied between <0.1 and 7.5 fg C cell^{-1} h^{-1} during the diurnal experiment. DF-glucose values were from 10 to 100 times higher during diurnal studies than in early spring studies. In both substrate uptake experiments lower and higher DF-glucose availability was encountered by the microbial uptake/transport systems. The substrate–uptake relationship covers a large range of naturally observed DF-glucose concentrations and may show how aquatic bacteria can respond metabolically to low and high pulses of different amounts of LDOCP substrates. However, without correction for "isotope dilution" no correlation was obtained in all experiments. We believe that an analytical chemical approach is a valuable tool to prove the presence of isotope dilution and correct for it if necessary. Otherwise data from radio assays may not represent real microbial activities or rates.

Polymeric Substrates

We have a rather fragmentary knowledge of the role of dissolved biopolymers in aquatic environments. This holds true for their chemical compositions and structures and their ecological functions. During our studies of DOM we were able to gain some knowledge of this complex world of organic substrates, but our chemical data base showed that we have characterized no more than 50% of the DOM. Even less well understood are their interactions with the microbial population. A combined analytical chemical and biochemical approach may improve our understanding. Microbial hydrolytic enzymes play an important role in such an approach.

Microbial Hydrolytic Enzyme Interactions with Biopolymers

As delineated above, aquatic environments can be regarded as highly dilute growth media for bacteria, and they may not favor rapid growth because they lack sufficient availability of LDOCP solutes (Figs. 2.6–2.8). Biopolymers such as proteins, polysaccharides, and humic matter are present in concentrations 10 to 100 times higher (Fig. 2.11), and bacteria can utilize such substrates by synthesizing hydrolytic enzymes that cleave the polymers in a sequence of reactions by different MHE to oligomeric and monomeric substrates. Exo- and endo-cleaving mechanisms are the dominant degradation steps to release biologically labile substrates into the LDOCP (Chróst 1990b, 1991a, 1992, Münster 1991, 1993). Such microbial hydrolytic enzymes are important tools for microbes to get access to high-molecular-weight DOC pools for their subsequent utilization (Chapter 6, this volume). According to Chróst (1989) and Chróst et al. (1989), the synthesis of β-glucosidase (GLCase) and β-galactosidase (GALase) in Plußsee was under

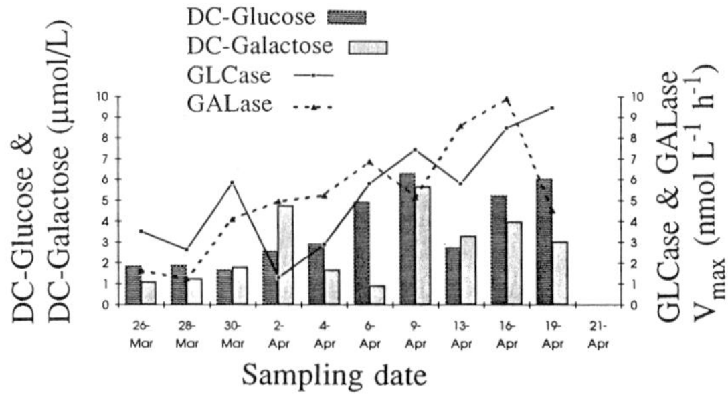

Figure 2.14. Distribution of dissolved combined glucose (DC-glucose), dissolved combined galactose (DC-galactose), β-glucosidase activity (GLCase), and β-galactosidase activity (GALase) during the spring phytoplankton bloom development in the Plußsee surface water.

control of induction and derepression mechanisms and was dependent on the presence of LDOCP and high-molecular-weight DOC substrates.

With respect to the presence of such MHE–biopolymer interactions in Plußsee, the dissolved polysaccharide fraction in the HMWDOC pool was analyzed and related to MHE during a spring bloom development in 1984. DC-glucose concentrations varied between 1.7 and 6.3 µmol/L (mean 3.6 µmol/L), and DC-galactose between 0.9 and 5.6 µmol/L (mean 2.7 µmol/L). GLCase and GALase partially displayed an inverse distribution pattern (Fig. 2.14): GLCase varied between 1.3 and 9.5 nmol L^{-1} h^{-1} (mean 5.3 nmol L^{-1} h^{-1}), and GALase varied between 1.3 and 9.9 nmol L^{-1} h^{-1} (mean 5.2 nmol L^{-1} h^{-1}). Statistical analysis of both data sets gave significant linear correlations between DC-glucose and GLCase ($r^2 = 0.71$, $n = 10$, $P \leq 0.005$, $y = 1.2434x + 0.0774$) and between DC-galactose and GALase ($r^2 = 0.59$, $n = 10$, $P \leq 0.005$, $y = 0.9803x + 1.4414$), but with lower significance. The ability of aquatic bacteria to synthesize and control such MHE–substrate interactions may become an essential competition and fitness criteria for growth and survival during low LDOCP availabilities (Münster 1991, 1993; Chapter 6, this volume). We believe that a combined approach of analytical chemical and biochemical methods may provide a deeper insight into and understanding of biopolymer–microbe interactions in the future. It may also advance knowledge of LDOCP dynamics and understanding of the succession of microbial populations, nutrient cycling, and carbon fluxes.

Present and Future Aspects of DOM Research

According to our present data from DOM studies in the Plußsee, we can characterize about 5–10% of the DOC as LDOCP with highly dynamic variance and fluxes between producers (phytoplankton) and consumers (bacteria). This labile carbon pool is mostly comprised of low-molecular-weight DOC compounds such as DFAA, DFCHO, and some other organic trace compounds. Under certain environmental and physiological conditions EOC compounds such as glycolic acids may enhance the carbon flux between autotrophs and heterotrophs. Such short substrate pulses are an occasional bonus for aquatic bacteria but not a steady diet. Aquatic bacteria can respond quickly to such short substrate pulses. However, such short LDOCP pulses can also enhance and stimulate the degradation of the RDOCP fraction, which represents a continual carbon source for slowly growing microorganisms. Bottom-up processes are of much greater importance for bacterial nutrition with energy and carbon substrates than top-down processes, which rather control bacteria mortality and their link to higher trophic levels. Biopolymers are the major carbon pools in the Plußsee (and in most other aquatic systems), and their utilization requires an extracellular enzymatic cleavage process. At present only 40–50% of the DOC can be chemically identified and characterized from Plußsee. We have still less information about the chemical structures of the unidentified fractions and their impact on microbial degradation and utilization. In the future research will likely be focused on the high-molecular-weight DOC pool. There is increasing evidence that viruses and bacteriophages can significantly modify and shorten the energy and carbon flux in aquatic environments, bypassing the classical carbon flow of the grazing food chain and the microbial loop. Within this "viral loop" more polymeric substrates will participate in the cycling of nutrients and carbon than have been considered in the past.

In most cases during our excursion into the word of dissolved organic matter in the Plußsee we have looked at DOM–microbe interactions. However, there is increasing evidence that light may play an important role in the degradation of DOM in aquatic environments (Mopper et al. 1991). Light–DOM interactions become more relevant in respect to global climate changes and the increase in UV radiation. More advanced analytical chemical methods are required to characterize the structures of important substrates and to quantify their contribution to the carbon flux and their interactions with abiotic and biotic components. Molecular biochemical approaches may provide more specific and sensitive methods to trace certain enzymatic and metabolic activities at the cellular and species level. Integrated analytical and physiological biochemical approaches are required in future DOM research to advance our understanding of DOM composition and processing in aquatic environments.

Acknowledgments. This work has benefited substantially from the support of and discussions with our colleagues at the Department of Microbial Ecology at the Max-Planck-Institute for Limnology in Plön. We acknowledge the contribution by Dr. B. Meyer of amino acid uptake data, Dr. H. Rai for DOC measurements and glucose uptake data, Dr. B. Hickel for phytoplankton data, Prof. Dr. R. J. Chróst for enzyme activity data, Prof. Dr. J. Overbeck for glucose uptake data and dark CO_2 fixation data, and many others who helped us to take water samples and provided technical assistance and technical support. Our special thanks are directed to Prof. Dr. J. Overbeck for his encouragement and patience during critical periods in the preparation of this manuscript. The critical comments, recommendations, and improvements of the manuscript by Prof. Dr. R. J. Chróst are also acknowledged.

References

Aiken GR (1988) A critical evaluation of the use of macroporous resins. In Frimmel FH, Christman RF (eds) *Humic Substances and Their Role in the Environment.* John Wiley & Sons, New York, pp 15–28

Alexander M (1975) Environmental and microbiological problems arising from recalcitrant molecules. Microb Ecol 2:17–27

Al-Hasan RH, Coughlan SJ, Aditipant A, Fogg GE (1975) Seasonal variations in phytoplankton and glycolate concentrations in the Menai Straits, Anglesey. J Mar Biol Assoc UK 55:557–565

Allen HL (1977) Experimental studies of dissolved organic matter in a soft-water lake. In Cairns J (ed) *Aquatic Microbial Communities.* Garland, New York, pp 440–476

Azam F, Cho BC (1987) Bacterial utilization of organic matter in the sea. In Fletcher M, Gray TRG, Jones JG (eds) *Ecology of Microbial Communities.* Cambridge University Press, Cambridge, pp 261–281

Azam F, Fenchel T, Field JG, Gray JS, Meyer-Reil LA, Thingstad F (1983) The ecological role of water column microbes in the sea. Mar Ecol Prog Ser 10:257–263

Benner R, Moran MA, Hodson RE (1986) Biogeochemical cycling of lignocellulosic carbon in marine and freshwater ecosystems: Relative contributions of procaryote and eucaryotes. Limnol Oceanogr 31:89–100

Billén G (1984) Heterotrophic utilization and regeneration of nitrogen. In Hobbie JE, Williams LeBPJ (eds) *Hetrotrophic Activity in the Sea.* Plenum Press, New York, pp 313–355

Billén G (1991) Protein degradation in aquatic environments. In Chróst RJ (ed) *Microbial Enzymes in Aquatic Environments.* Springer-Verlag, New York, pp 123–143

Billén G, Servais P (1989) Modelisation des procès de dégradation bactérienne de la matière organique en milieu aquatique. In Bianchi M, Coll E (eds) *Micro-Organismes dans les Ecosystèmes Oceaniques.* Masson, Paris, pp 219–245

Billén G, Servais P, Becquevort S (1990) Dynamics of bacterioplankton in oligotrophic and eutrophic aquatic environments: bottom-up or top-down control. Hydrobiologia 207:37–42

Brehm J (1967) Untersuchungen über den Aminosäure-Haushalt holsteinischer Gewässer, insbesondere des Plußsees. Arch Hydrobiol Suppl 32:313–435

Burdidge DJ (1991) The kinetics of organic matter mineralization in anoxic marine sediments. J Mar Res 49:727–761

Burney CM (1986) Diel dissolved carbohydrate accumulation in coastal water of South Florida, Bermuda and Oahu. Estuar Coast Shelf Sci 23:197–203

Carlucci AF, Craven DB, Henrichs SM (1984) Diel production and microheterotrophic utilization of dissolved free amino acids in waters off Southern California. Appl Environ Microbiol 48:165–170

Carlucci AF, Craven DB, Wolgast DM (1991) Microbial populations in surface films and subsurface waters: Amino acid metabolism and growth. Mar Biol 108:329–339

Chróst RJ (1988) Phosphorus and microplankton development in an eutrophic lake. Acta Microbiol Pol 37:205–225

Chróst RJ (1989) Characterization and significance of β-glucosidase activity in lake water. Limnol Oceanogr 34:660–672

Chróst RJ (1990a) Application of the isotope dilution principle to the determination of substrate incorporation by aquatic bacteria. Arch Hydrobiol Ergeb Limnol 34:111–117

Chróst RJ (1990b) Ectoenzymes in aquatic environments: Origin, activity and ecological significance. In Overbeck J, Chróst RJ (eds) *Aquatic Microbial Ecology: Biochemical and Molecular Approaches.* Springer-Verlag, New York, pp 47–78

Chróst RJ (1991a) Environmental control of the synthesis and activity of aquatic microbial ectoenzymes. In Chróst RJ (ed) *Microbial Enzymes in Aquatic Environments.* Springer-Verlag, New York, pp 29–59

Chróst RJ (ed) (1991b) *Microbial Enzymes in Aquatic Environments*. Springer-Verlag, New York

Chróst RJ (1992) Significance of bacterial ectoenzymes in aquatic environments. Hydrobiologia 243:61–70

Chróst RJ, Faust, MA (1983) Organic carbon release by phytoplankton: Its composition and utilization by bacterioplankton. J Plankton Res 5:477–493

Chróst RJ, Overbeck J (1987) Kinetics of alkaline phosphatase activity and phosphorus availability for phytoplankton and bacterioplankton in lake Plußsee (north German eutrophic lake). Microb Ecol 13:229–248

Chróst RJ, Overbeck J (1989) Application of the isotope dilution principle to the determination of [^{14}C]glucose incorporation by aquatic bacteria. Acta Microbiol Pol 38:75–89

Chróst RJ, Münster U, Rai H, Albrecht D, Witzel KP, Overbeck J (1989) Photosynthetic production and exoenzymatic degradation of organic matter in euphotic zone of an eutrophic lake. J Plankton Res 11:223–242

Coffin RB (1989) Bacterial uptake of dissolved free and combined amino acids in estuarine waters. Limnol Oceanogr 34:531–542

Coffin RB, Velinsky DJ, Devereux R, Price WA, Cifuentes LA (1990) Stable isotope analysis of nucleic acids to trace sources of dissolved substrates used by estuarine bacteria. Appl Environ Microbiol 56:2012–2020

Determann H (ed) (1967) *Gelchromatographie, Gelfiltration, Gelpermeation, Molekülsiebe: Ein Laboratoriumshandbuch*. Springer-Verlag, Heidelberg

Dillon PJ, Rigler FH (1974) The phosphorus–chlorophyll relationship in lakes. Limnol Oceanogr 19:767–773

Downing JA, McCauley E (1992) The nitrogen:phosphorus relationship in lakes. Limnol Oceanogr 37:936–945

Druffel ERM, Williams PM (1990) Identification of a depp marine source of particulate organic carbon using bomb ^{14}C. Nature 347:172–174

Ducklow HW (1991) The passage of carbon through microbial foodwebs: Results from flow network models. Mar Microb Food Webs 5:129–144

Ducklow HW, Purdie DA, Williams LeBPJ, Davis JM (1986) Bacterioplankton: A sink for carbon in a coastal marine plankton community. Science 332:865–867

Dugdale RC, Goering JJ (1967) Uptake of new and regenerated forms of nitrogen in primary production. Limnol Oceanogr 12:196–206

Fogg GE (1983) The ecological significance of extracellular products of phytoplankton photosynthesis. Botan Mar 26:3–14

Francko DA (1984) Relationship between phosphorus functional classes and alkaline phosphatase activity in reservoir lakes. J Freshwat Ecol 2:541–547

Francko DA, Heath RT (1979) Functionally distinct classes of complex phosphorus compounds in lake water. Limnol Oceanogr 24:463–473

Frimmel FH, Christman RF (eds) (1988) *Humic Substances and Their Role in the Environment*. John Wiley & Sons, New York

Hedges JI, Lee CL (1993) Measurement of dissolved organic carbon and nitrogen in natural waters. Mar Chem 41:1–290

Hedges JI, Clark WA, Conoric GL (1988) Organic matter sources to the water column and surfical sediments of marine bay. Limnol Oceanogr 33:1116–1136

Heldal M, Bratbak G (1991) Production and decay of viruses in aquatic environments. Mar Ecol Prog Ser 72:205–212

Hessen DO (1985) The relation between bacterial carbon and dissolved humic compounds in oligotrophic lakes. FEMS Microb Ecol 31:215–223

Hessen DO (1992) Dissolved organic carbon in a small humic lake: Effects on bacterial production and respiration. Hydrobiologia 229:115–123

Hickel B (1978) Phytoplankton population dynamics in Plußsee (East-Holstein, Germany). Verh Ges Ökol Kiel 1977:119–126

Hickel B (1988) Unexpected disappearance of cyanophyte blooms in Plußsee (North Germany). Arch Hydrobiol Suppl 80:545–554

Hobbie JE (1992) Microbial control of dissolved organic carbon in lakes: Research for the future. Hydrobiologia 229:115–123

Hollibaugh JT, Azam F (1983) Microbial degradation of dissolved proteins in seawater. Limnol Oceanogr 28:1104–1116

Hoppe HG (1991) Microbial extracellular enzyme activity: A new key parameter in aquatic ecology. In Chróst RJ (ed) *Microbial Enzymes in Aquatic Environments.* Springer-Verlag, New York, pp 60–83

Hoppe HG, Kim SJ, Gocke K (1988) Microbial decomposition in aquatic environments: Combined processes of extracellular enzyme activity and substrate uptake. Appl Environ Microbiol 54:784–790

Jørgensen NOG (1987) Free amino acids in lakes: Concentrations and assimilation rates in relation to phytoplankton and bacterial production. Limnol Oceanogr 32:97–111

Jørgensen NOG, Søndergaard M, Hansen HJ, Bosselman S, Riemann B (1983) Diel variation in concentration, assimilation and respiration of dissolved free amino acids in relation to planktonic primary and secondary production in two eutrophic lakes. Hydrobiologia 107:107–122

Kalle K (1966) The problem of the Gelbstoffe in the Sea. Oceanogr Mar Biol Annu Rev 4:91–104

Kaplan LA (1992) Comparison of high-temperature and persulfate oxidation methods for determination of dissolved organic carbon in freshwaters. Limnol Oceanogr 37:1119–1125

Kirchman DL, Hodson R (1986) Metabolic regulation of amino acid uptake in marine waters. Limnol Oceanogr 31:339–350

Kirchman DL, K'Nees E, Hodson R (1985) Leucine incorporation and its potential as a measure of protein synthesis by bacteria in natural aquatic systems. Appl Environ Microbiol 49:599–607

Kirchman DL, Suzuki Y, Garsdide C, Ducklow HW (1991) High turnover rates of dissolved organic carbon during a spring phytoplankton bloom. Nature 352:612–614

Kjelleberg S, Hermansson M, Mården P, Jones GW (1987) The transient phase between growth and nongrowth of heterotrophic bacteria, with special emphasis on the marine environment. Annu Rev Microbiol 41:25–49

Knecht R, Chang JY (1986) Liquid chromatographic determination of amino acids after gas-phase hydrolysis and derivatization with (dimehtylamino)azobenzenesuflfonyl chloride. Anal Chem 58:2375–2379

Lampert W (1978) Release of dissolved organic carbon by grazing zooplankton. Limnol Oceanogr 23:831–834

LaRoche J (1983) Ammonium regeneration: Its contribution to phytoplankton nitrogen requirements in a eutrophic environment. Mar Biol 75:231–240

Lean DRS (1973) Phosphorus dynamics in lakewater. Science 179:678–680

Leenheer JA (1981) Comprehensive approach to preparative isolation and fractionation of dissolved organic substances from large volumes of waters and wastewaters. Environ Sci Technol 15:578–587

Levine SN, Schindler DW (1992) Modification of the N:P ratio in lakes by in situ processes. Limnol Oceanogr 37:917–935

Lindroth P, Mopper K (1979) High performance liquid chromatographic determination of subpicomole amounts of amino acids by precolumn fluorescence derivatization with *o*-phthaldehyde. Anal Chem 51:1667–1674

Lord JM (1972) Glycolate oxidoreductase in *Escherichia coli.* Biochim Biophys Acta 267:227–237

Mantoura RFC, Riley JP (1975) The analytical concentration of humic substances from natural waters. Anal Chim Acta 76:97–106

McManus JP, Davis KG, Beart JE, Gaffey SH, Lilley TH, Haslam E (1985) Polyphenol interactions. Part 1. Introduction; some observations on the reversible complexation of polyphenols with proteins and polysaccharides. J Chem Perkin Trans 2:1429–1438

Meltzer MN, Tous GI, Gruber S, Stein S (1987) Gas-phase hydrolysis of proteins and peptides. Anal Biochem 160:356–361

Menzel DW, Vaccaro RF (1964) The measurement of dissolved organic and particulate carbon in seawater. Limnol Oceanogr 9:138

Meyer-Reil LA (1986) Measurements of hydrolytic activity and incorporation of dissolved organic substrates by microorganisms in marine sediments. Mar Ecol Prog Ser 31:143–149

Mopper K, Xianliang Z, Kieber RJ, Kieber DJ, Sikorski RJ, Jones RG (1991) Photochemical degradation of dissolved organic carbon and its impact on the oceanic carbon cycle. Nature 353:60–62.

Moran BS, Buesseler KO (1992) Short residence time of colloids in the upper ocean estimated from ^{238}U–^{234}Th disequilibria. Nature 359:221–223

Morita RY (1982) Starvation survival of heterotrophs in the marine environment. Adv Microb Ecol 6:171–198

Münster U (1982) Physikalisch-chemische Untersuchungen zur Anreicherung, Fraktionierung und Strukturaufklärung monomerer und polymere organischer Substrate im DOC vom Plußsee. PhD Dissertation, University of Kiel, Germany

Münster U (1984) Distribution, dynamic and structure of free dissolved carbohydrates in the Plußsee, a North German eutrophic lake. Verh Int Ver Limnol 22:929–935

Münster U (1985) Investigations about structure, distribution and dynamics of different organic substrates in the DOM of lake Plußsee. Arch Hydrobiol Suppl 70:429–480

Münster U (1991) Extracellular enzyme activity in eutrophic and polyhumic lakes. In Chróst RJ (ed) *Microbial Enzymes in Aquatic Environments.* Springer-Verlag, New York, pp 96–122

Münster U (1993) Concentrations and fluxes of organic substrates in the aquatic environment. Antonie van Leeuwenhoek (in press)

Münster U, Chróst RJ (1990) Dissolved organic matter (DOM) in aquatic environments: Origin, distribution, composition and microbial utilization. In Overbeck J, Chróst RJ (eds) *Aquatic Microbial Ecology: Biochemical and Molecular Approaches.* Springer-Verlag, New York, pp 8–46

Nagata T, Kirchman DL (1991) Release of dissolved free and combined amino acids by bacteriovorous marine flagellates. Limnol Oceanogr 36:433–443

Nagata T, Kirchman DL (1992) Release of macromolecular organic complexes by heterotrophic marine flagellates. Mar Ecol Prog Ser 83:233–240

Odum EP (1963) Primary and secondary energy flow in relation to ecosystem structure. Proc Int Cong Zool 4:336–338

Ohle W (1933) Über organische Stoffe in Binnenseen. Verh Int Ver Limnol 6:249–262

Ohle W (1934) Chemische und physikalische Untersuchungen norddeutscher Seen. Arch Hydrobiol 26:386–464,585–685

Olsson L, Renne N, Samuelson O (1976) Chromatography of aromatic compounds on cross-linked polyvinylpyrrolidone and on anion-exchange resins. J Chromatogr 123:355–365

Overbeck J (1979) Studies on the heterotrophic function and glucose metabolism of microplankton in Plußsee. Arch Hydrobiol Ergeb Limnol 13:56–76

Peterson BJ, Fry B (1987) Stable isotopes in ecosystem studies. Annu Rev Ecol Syst 18:293–320

Poltz J (1972) Untersuchungen über das Vorkommen und den Abbau von Fettsäueren in Seen. Arch Hydrobiol Suppl 40:315–399

Proctor LM, Fuhrman JA (1990) Viral mortality of marine bacteria and cyanobacteria. Nature 343:60–62

Proctor LM, Fuhrman JA (1992) Mortality of marine bacteria in response to enrichments of the virus size fraction from seawater. Mar Ecol Prog Ser 87:283–293

Rasmussen JB, Godbout L, Schallenberg M (1989) The humic content of lake water and its relationship to watershed and lake morphometry. Limnol Oceanogr 34:1336–1343

Reisser W, Grein S, Krambeck C (1993) Extracellular DNA in aquatic ecosystems may be in part due to phycovirus activity. Hydrobiologia 252:199–201

Rich PH, Wetzel RG (1978) Detritus in the lake ecosystem. Am Nat 112:57–71

Robertson KJ, Williams PM, Bada JL (1987) Acid hydrolysis of dissolved combined amino acids in sea water: A precautionary note. Limnol Oceanogr 32:996–997

Schindler DW (1977) Evolution of phosphorus limitation in lakes. Science 195:260–262

Schink B, Janssen PH, Frings J (1992) Microbial degradation of natural and of new synthetic polymers. FEMS Microb Rev 103:311–316

Schnitzer M, Khan SU (1972) *Humic Substances in the Environment.* Marcel Dekker, New York

Sell A, Overbeck J (1992) Exudates: Phytoplankton–bacterioplankton interactions in Plußsee. J Plankton Res 14:1199–1215

Shah NM, Wright RT (1974) The occurrence of glycolic acid in coastal sea water. Mar Biol 24:121–124

Simon M (1985) Specific uptake rates of amino acids by attached and free-living bacteria in a mesotrophic lake. Appl Environ Microbiol 49:1254–1259

Simon M, Azam F (1988) Protein content and protein synthesis rates of planktonic marine bacteria. Mar Ecol Prog Ser 51:201–213

Spencer CM, Cai Y, Martin R, Gaffney SH, Goulding PN, Magnolato D, Lilley TH, Haslam E (1988) Polyphenol complexation—some thoughts and observations. Phytochemistry 27:2397–2409

Stabel HH (1977) Gebundene Kohlenhydrate als stabile Komponenten im Schöhsee und in *Scenedesmus*-Kulturen. Arch Hydrobiol Supl 53:159–254

Sugimura Y, Suzuki Y (1988) A high temperature catalytic oxidation method for the determination of non-volatile dissolved organic carbon in seawater by direct injection of a liquid sample. Mar Chem 24:105–131

Suzuki Y, Sugimura Y, Itoh T (1985) A catalytic oxidation method for the determination of total nitrogen dissolved in seawater. Mar Chem 16:83–97

Swift G (1992) Biodegradability of polymers in the environment: Complexities and significance of definitions and measurements. FEMS Microbiol Rev 103:339–346

Thurman M (ed) (1985) *Organic Geochemistry of Natural Waters.* Martinus Nijhoff/Junk, Boston

Tolbert NE (1974) Photorespiration by algae. In Stewart WDP (ed) *Algal Physiology and Biochemistry.* Blackwell, Oxford, pp 474–504

Tolbert NE, Zill LP (1956) Excretion of glycolic acid by algae during photosynthesis. J Biol Chem 22:895–906

Tranvik L (1988) Availability of dissolved organic carbon for planktonic bacteria in oligotrophic lakes of differing humic content. Microb Ecol 16:311–322

Tranvik L (1989) Bacterioplankton growth, grazing mortality and quantitative relationship to primary production in a humic and clearwater lake. J Plankton Res 11:985–1000

Tranvik L (1990) Bacterial growth on fractions of dissolved organic carbon of different molecular weights from humic and clear waters. Appl Environ Microbiol 56:1672–1677

Tranvik L (1992) Allochthonous dissolved organic matter as energy source for pelagic bacteria and the concept of the microbial loop. Hydrobiologia 229:107–114

Tranvik L, Höfle M (1987) Bacterial growth in mixed cultures on dissolved organic carbon from humic and clear waters. Appl Environ Microbiol 53:482–488

Vollenweider RA (1968) The scientific basis of lake and stream eutrophication, with particular reference to phosphorus and nitrogen as eutrophication factors. Tech Rep DAS/CSI 6827 Organization of Economic Cooperative Development (OECD), Paris

Weinmann G (1970) Gelöste Kohlenhydrate und andere organische Stoffe in natürlichen Gewässern und in Kulturen von *Scenedesmus quadricauda*. Arch Hydrobiol Suppl 37:164–242

Westrich JT, Berner RA (1984) The role of sedimentary organic matter in bacterial sulfate reduction: The G model tested. Limnol Oceanogr 29:236–249

Wetzel RG (1983) *Limnology*, 2nd ed. WB Saunders, Philadelphia

Wetzel RG (1984) Detrital dissolved and particulate organic carbon functions in aquatic ecosystems. Bull Mar Sci 35:503–509

Wetzel RG (1991) Extracellular enzymatic interactions in aquatic ecosystems: Storage, redistribution, and interspecific communication. In Chróst RJ (ed) *Microbial Enzymes in Aquatic Environments*. Springer-Verlag, New York, pp 6–28

Wetzel RG (1992) Gradient-dominated ecosystems: Sources and regulatory functions of dissolved organic matter in freshwater ecosystems. Hydrobiologia 229:181–198

Williams PJLeB (1990) The importance of losses during microbial growth: Commentary on the physiology, measurement and ecology of the release of dissolved organic material. Mar Microb Food Webs 4:175–206

Williams PM (1986) Chemistry of the dissolved and particulate phases in the water column. In Eppley RW (ed) *Plankton Dynamics of the Southern California Bight*. Lect Notes Coastal Estuarine Stud 15:53–83

Wright RT (1970) Glycolic acid uptake by planktonic bacteria. In Hood DW (ed) *Symposium on Organic Matter in Natural Waters*. University of Alaska, College of Marine Science, Occasional Publication No. 1. Institute of Marine Science, Alaska, pp 521–536

Wright RT (1975) Studies on glycolic acid metabolism by freshwater bacteria. Limnol Oceanogr 20:626–633

Wright RT, Shah N (1975) The trophic role of glycolic acid in coastal seawater. I. Heterotrophic metabolism in seawater and bacterial cultures. Mar Biol 33:175–183

Zehr, JP, Axler, RP, Goldman CR (1985) Heterotrophic mineralization of amino acid nitrogen in subalpine Castle Lake, California. Mar Chem 16:343–350

3. Primary Production, Photosynthesis, and Chlorophyll$_a$ in the Plußsee

Maria-Elisabeth Meffert and Jürgen Overbeck

Introduction

The numerous naturally eutrophic baltic lakes of the East-Holstein district have similar origins but differ in morphometry and trophic levels. Plußsee, a small wind-sheltered forest lake (Utermöhl 1925), is notable because of its surface area/depth ratio and resulting stable summer stratification (Krambeck 1974). These features in particular have led to numerous and comprehensive studies, among which chlorophyll and primary production have been examined (Ohle 1962, 1964, Overbeck 1979) along with the vertical distribution of chlorophyll (Rai 1975, Krambeck 1981).

In this chapter the results of weekly investigations of algal growth and decline stages, characterized by the dynamics of chlorophyll levels and photosynthesis, are presented. Special emphasis is placed on 1981; 1981 data from the Plußsee are compared with data of other years and of several other lakes of the East-Holstein area (Table 3.1).

Methods

In situ measurements were described in detail by Meffert and Overbeck (1985a,b). Temperature, Secchi disk visibility, chlorophyll and phaeopigments, and the rates of photosynthesis were assessed for the phytoplankton in the Plußsee, the Edebergsee, and the northern basin of the Große Plöner See. Water samples were collected weekly at 1–m intervals from 0 to 10 m or from 0 to 7 m (Edebergsee).

Table 3.1. Characteristics of Studied Lakes

Lake	Location		Altitude (m)	Area (km^2)	Depth (m)	
	East	North			Max	Mean
Schluensee	10°28′	54°11′	23	1.28	50	18
Stocksee	10°21′	54°05′	28	2.07	30	12
Selenter See	10°26′	54°19′	37	22.40	34	15
Tresdorfer See	10°28′	54°14′	25	1.15	15	6
Belauer See	10°15′	54°06′	29	1.26	29	9
Dobersdorfer See	10°18′	54°19′	19	3.54	20	5
Passader See	10°19′	54°21′	19	2.91	12	4
Schierensee	10°13′	54°07′	28	0.32	7	4
Stolper See	10°14′	54°08′	28	1.49	15	6.5
Postsee	10°15′	54°14′	21	3.35	9	3.5
Gr. Binnensee	10°38′	54°20′	−0.1	5.28	3	2
Rottensee	10°29′	54°13′	25	0.46	3.8	2.3
Bothkamper See	10°08′	54°13′	25	1.58	4.5	
Plußsee	10°27′	54°11′	24	0.14	29	9.3
Edebergsee	10°27′	54°09′	22	0.09	11	7
Gr. Plöner See	10°23′	54°10′	21	29.11	60	16

[a]Data from Meffert and Wulff (1987).

Chlorophyll$_a$ and phaeopigments of the 10 deepest lakes listed in Table 3.1 were determined for the main bloom periods in April, August, and October. Water samples for all ten lakes were collected from two layers of the epilimnion, at depths of 1 m and 5 or 8 m. Only one sample per lake was taken, at the depth of 1 m, from each of three shallow lakes: Großer Binnensee, Rottensee, and Bothkamper See.

Global radiation was continuously measured at 30-s intervals using an L-200 S pyranometer sensor positioned at the center of the Plußsee (Krambeck 1979). The data were given in joules per square centimeter per hour and transformed to quantum flux in terms of unit energy (E) per square meter per unit of time following Morel and Smith (1974). The photosynthetically active radiation (PAR) was computed using the factors given by Talling (1957a), which take reflection into account. An underwater quantum sensor (Li-92 S) connected to a Q/R/P meter (Li-185 A) was used to measure underwater light. In the Edebergsee underwater scalar quantum irradiance was determined using a spherical sensor (Li-193 SB) in conjunction with a Q/R/P meter, Li-185 B (all instruments from Lambda Instruments Corp., U.S.A.).

Quantum yields were computed according to the equations and methods of Bannister and Weidemann (1984). The specific light attenuation of chlorophyll was estimated from the linear regression of the vertical extinction coefficient against phytoplankton chlorophyll concentration. Data were corrected for phaeopigments according to Tyler (1975) and Tilzer (1983). Dark respiration and release of photosynthetic metabolites were not considered.

The following abbreviations and units are used in this chapter: chla, chlorophyll$_a$; I, photon flux densities; P, specific photosynthetic rate [mg C (mg chla)$^{-1}$ h^{-1}]; P v I, photosynthesis-versus-light intensity curve; I_k, the saturation onset parameter—the light intensity at which the linear part of the P versus I plot crosses P_m values (in μE m^{-2} s^{-1}); α, the slope of the linear part of the P v I plot [mg C (mg chla)$^{-1}$ mE^{-1} m^2]; the linear part of the P v I plot = $(P_2 - P_1)$ $(I_2 - I_1)$ = maximal and constant; LCP, light compensation point of photosynthesis, the light intensity at which photosynthesis balances respiration = the positive abscissa intercept of the P v I plot at zero P (μE m^{-2} s^{-1}) calculated from the negative ordinate intercept; k_c, m^2 mg^{-1} chla; $\Phi_m k_c$, moles C m^2 E^{-1} absorbed mg chla; P_m, maximal P measured.

Results and Discussion

Light Penetration and Temperature

Seasonal variation in Secchi disk visibility and transparency of euphotic zone depth (1% level subsurface light intensity at 400–700 nm), and the average extinction coefficient of the euphotic zone (Åberg and Rodhe 1942) were closely related to chlorophyll concentrations (Fig. 3.1). Thus, the average euphotic zone depth of 5.5 m in 1981 concurred with the production zone of 5 m that was measured in 1960 by Ohle (1962), which was within the range of variation observed over the previous 20 years. The range of average extinction coefficients of the euphotic zone corresponded with that of mesotrophic lakes (Wetzel 1983). On average, about 90% of the incident radiation (PAR) had been absorbed by the

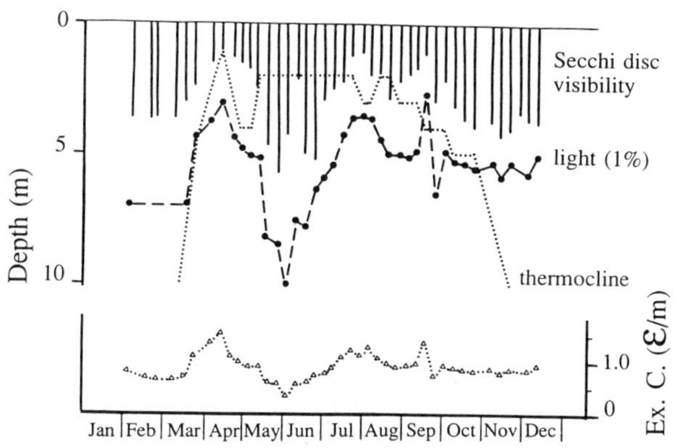

Figure 3.1. Variation of Secchi disk visibility, depth of euphotic zone (1% of light), and mixed-layer and vertical extinction coefficient (Ex.C.).

2.4-m-deep water layer. Mean Secchi disk visibility depths represented about 13% of light transmission.

The surface water temperature was 2.4°C at the end of February and reached a maximum temperature of 21°C in July. The temperature of the euphotic zone was above 15°C from July to mid-September and 17–19°C for 4 weeks in July–August. Thermal stratification began in April; the autumn overturn occurred in September. Isothermal conditions prevailed from November to March. During winter mixing, the ratio of mixed zone to euphotic zone, z_m/z_{eu} (Talling 1971), was relatively large, and algal cells were suspended for longer periods of time in water layers where phytoplankton growth was limited by associated effects of the seasonal radiation minimum. This may have selected for a particular algal species and resulted in the relatively late onset of the spring bloom. Between May and the autumn overturn, the thermocline was usually above the lower level of the euphotic zone. This means that the upper part of the metalimnion (rich with nutrients) was illuminated.

Phytoplankton Primary Production

Vertical Distribution Pattern

About 97% of phytoplankton primary production occurred in the euphotic zone of the Plußsee, i.e., in the upper 5-m water layer of the investigated 10-m water column. The data presented refer to this upper layer. The vertical distribution pattern of primary production showed that maxima developed mostly in the upper 2-m layer throughout the entire year, often deviating from the chlorophyll maxima. On several occasions (between March and August) subsurface primary production was light inhibited. During mass development of *Ceratium* in August the corresponding chlorophyll$_a$ and phytoplankton primary production maxima indicated that motile cells avoided the subsurface water layers on sunny days. From April to September mean primary production was 17.5 mg C m^{-3} h^{-1} and increased >50 mg C m^{-3} h^{-1} in the upper water layer during April and July. The characteristic vertical stratification pattern of the Plußsee (Anagnostidis and Overbeck 1966) developed between August and October. The maximum of primary production decreased from the subsurface layer to a depth of 10 m. Pronounced chlorophyll$_a$ and phaeopigment maxima occurred in the metalimnion, at a depth of about 9 m, when pigment epilimnetic concentrations were low. CO_2 dark fixation and glucose uptake indicated increased heterotrophic activity.

Seasonal Cycles

In 1981 three bloom periods occurred, each consisting of several algal maxima, and each was characterized by similar increase in chlorophyll$_a$ concentration and primary production (Fig. 3.2). In the nonstratified lake, primary production and chlorophyll maxima occurred simultaneously. During thermal stratification chlorophyll maxima usually followed the primary production maxima with a lag of about 1 week, which indicated changing rates of photosynthesis. The spring phytoplankton bloom (which started in February–March) passed its maximum by

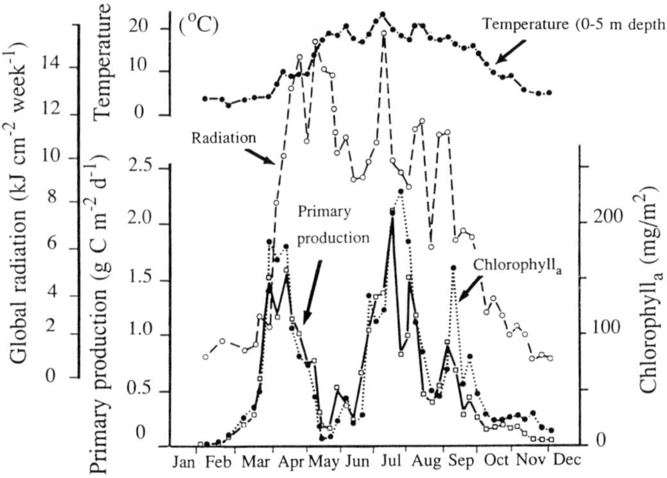

Figure 3.2. Seasonal variation of primary production, chlorophyll$_a$ concentration, global radiation, and temperature.

April. In May a pronounced minimum of primary production and chlorophyll$_a$ coincided with a period of high radiation that lasted for 6 weeks. During this period incident light energy exceeded 2 kJ cm^{-2} d^{-1}. The summer algal bloom followed and lasted about 6 weeks. The last phytoplankton bloom occurred in late summer and was distinctly smaller than previous blooms. Highest production rates corresponded to those of hypereutrophic lakes (Wetzel 1983), but growth periods and number of algal maxima were limited. Production rates were low from October to February. The monthly means for daily photosynthetic CO_2 assimilation increased from about 9 mg C (mg chla)$^{-1}$ d^{-1} in April to about 15 mg C (mg chla)$^{-1}$ d^{-1} in May and June, and decreased again to 9 mg C (mg chla)$^{-1}$ d^{-1} in July. Highest rates of phytoplankton primary production [about 26 mg C (mg chla)$^{-1}$ d^{-1}] were recorded in June.

In summer the monthly mean of daily phytoplankton primary production varied between 500 and 1500 mg C m^{-2}; the highest and lowest production values were measured in July and May, respectively (Table 3.2). In winter values ranged between 17 and 400 mg C m^{-2}. The half-year means for daily phytoplankton primary production were about 900 and 140 mg C m^{-2} for summer and winter, respectively. This corresponded to 85% and 15% of the annual phytoplankton production. Primary production rates of the Plußsee were determined for the first time in 1958–1959 by Ohle (1959) with a mean value of 284 mg C m^{-2} d^{-1} during the six summer months. But even in summer 1960 (Ohle 1962, 1975) production rates were comparable to our data, although monthly means differed markedly (Table 3.2). In 1977–1978 production rates also corresponded to those we determined in 1981 (Krambeck and Krambeck 1980, Krambeck 1981). The lower chlorophyll$_a$ for 1979 suggested that primary production scarcely exceeded that of 1981 (Krambeck 1981). Thus, over the last 20 years mean daily phytoplankton

Table 3.2. Phytoplankton Production Rates in the Water Column (0–5 m Depth) of Plußsee[a]

Time Period (1981)	Production ($mg\ C\ m^{-2}\ d^{-1}$)	Time Period (1960)[b]	Production ($mg\ C\ m^{-2}\ d^{-1}$)
January	17		
February	96		
March	403		
April	1241		
May	507	11 May–30 May	1207
June	602	31 May–23 June	2180
July	1493	24 June–8 August	1220
August	713	9 August–16 September	793
September	722	17 September–4 October	437
October	204		
November	93		
December	18		
Daily mean for 6 months:			
April–September	883 (1–2077)	11 May–4 November	1033
January–March + October–December	139 (0–1496)		
Daily mean for the entire year	512		
Annual productivity ($g\ C\ m^{-2}\ year^{-1}$)	187		

[a]Data from Meffert and Overbeck (1985b).
[b]Data from Ohle (1962).

production for the entire year was about 500 mg C m^{-2}, which corresponded to about 25 tons of organic carbon per year for the whole lake. Annual and mean daily primary production (Table 3.2) correspond to published values for mesotrophic lakes (Wetzel 1983) and also agree with extinction coefficients and chlorophyll$_a$ concentrations. Utermöhl (1925) characterized the Plußsee as a moderately eutrophic lake. The data published by Ohle (1959, 1962) showed that compared to other lakes in the Schleswig-Holstein region, phytoplankton primary production of the Plußsee varied between the lower and mean values.

Phytoplankton production rates in the Edebergsee were higher than those measured in the Plußsee, especially during winter. In summer, when phytoplankton growth was mainly nutrient limited, the difference between these two lakes could be a result of the lower nutrient content in the Plußsee. However, during winter mixing the total phosphorus concentration (90–95% is mineral) was from 100 to 200 μg P L^{-1} (Albrecht et al. 1978, Golachowska 1979), which could stimulate algal growth when light was not the limiting factor. The lakes differ in their morphometry (Table 3.3). During periods of near-isothermal mixing in winter, the ratio of mixed to euphotic depth (Talling 1971) was higher in funnel-shaped Plußsee than in the shallow Edebergsee. This morphometric factor

Table 3.3. Comparison of Morphometric Lake Characteristics, Rates of Primary Production, and Chlorophyll$_a$ Concentration in Two East Holstein Lakes[a]

	Plußsee		Edebergsee	
	Winter[b]	Summer[c]	Winter[b]	Summer[c]
Z_m/Z_{ln}[d]	5		3	
Z_{Hyp}/Z_{Epi}[e]		7–10		3
Primary production (mg C m^{-2} d^{-1})	139	883	618	1336
Primary production (% of the mean)	27	172	63	137
Chlorophyll$_a$ (µg L^{-1})	5	20	49	57

[a]Data from Meffert and Overbeck (1985b).
[b]January–March + October–December.
[c]April–September.
[d]Ratio mixed to euphotic depth.
[e]Ratio (hypolimnion + metalimnion):(epilimnion).

accentuated the limitation of phytoplankton growth associated with the seasonal radiation minimum. The delayed onset of the spring phytoplankton population increase could also reflect this morphometry. In summer the high depth ratio of (hypolimnion + metalimnion)/epilimnion could influence nutrient cycling in the Plußsee. This idea was further supported by the stable stratification pattern and the low exchange coefficients (Krambeck 1974). In addition to differing nutrient levels, differences between the two lake basins were largely a result of contrasting responses, induced by morphometry and light penetration, to a common surface light climate. The occasional occurrences of winter blooms in the Plußsee correspond with these conclusions. Optimal conditions can allow the growth of gas-vacuolated species of *Limnothrix* and *Planktothrix* (Meffert 1987, Anagnostidis and Komarek 1988) during winter months as described by Ohle (1962) and found later in 1974, 1977, and 1978. During the last 10 years winter phytoplankton blooms sometimes appeared in the Plußsee but regularly in the Edebergsee.

In Situ Photosynthetic Characteristics

Quantum Yields

Quantum yields estimation for water bodies requires in situ measurement of the chlorophyll$_a$ specific extinction coefficient. This essential parameter varies considerably and depends on algal species composition and their physiology (Welschmeyer and Lorenzen 1981, Dubinski and Berman 1979) and algal cell geometry (Kirk 1983). It is also influenced by the spectral composition of light (Atlas and Bannister 1980) and determination methods (Bannister and Weidemann 1984).

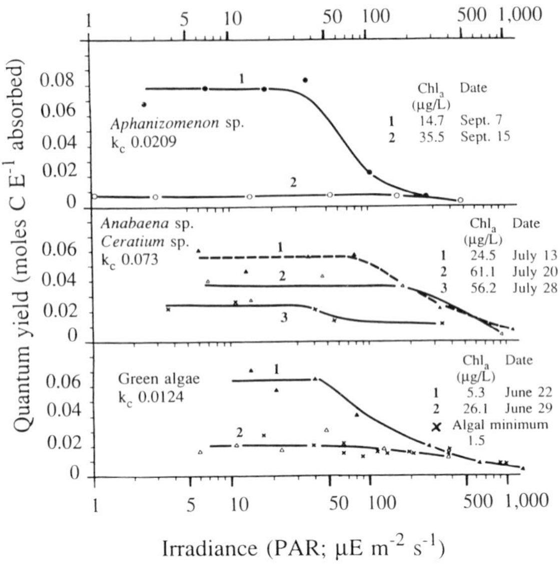

Figure 3.3. Quantum yields of three algal assemblages growing in the stratified Plußsee. Note the time-dependent changes in maximum quantum yield and chlorophyll$_a$.

In this study the determination of photosynthesis throughout the entire growth period of each algal assemblage allowed for differentiation between growing and declining stages (Fig. 3.3). Data for the calculations were selected from five growing algal assemblages excluding data obtained during decline of the algal populations. Chlorophyll$_a$ concentrations ranged between 5 and 50 mg m^{-3}. Observations of spring and winter populations of *Asterionella* in the Plußsee and *Limnothrix redekei* (Meffert 1989b) in the Edebergsee, respectively, are the most complete and unequivocal as far as growth stages, chlorophyll concentration, interference from declining algal populations, and nutrient supply are concerned. The lakes were homothermal and well mixed. During thermal stratification algal assemblages dominated by chlorococcalean algae of the *Sphaerocystis* type in June and *Aphanizomenon* in September followed chlorophyll minima. The July *Anabaena*/*Ceratium* population followed the chlorococcalean algae, so interference from declining cells cannot be excluded.

Photosynthetic rates were maximal in the upper 1 m (P_m), but subsurface photosynthesis could be light inhibited. Light-limited photosynthesis could occur between 2 and 5 m of depth. Compared with larger water bodies this is a narrow range within which photosynthetic rates are proportional to light intensity according to Tallings (1957b) model (Rodhe 1965). Limiting light intensity corresponding to about 50% of saturation intensity was less than 10% of surface illumination (Figs. 3.5 and 3.6, Table 3.4) as described by Rodhe (1965) for several European lakes. The values of $\Phi_m k_c$ (moles C m^2 E^{-1} absorbed/mg chla) were between 4.2 and 8.1 $\times$ 10^{-4}. For the *Asterionella* and *Anabaena*/*Ceratium* populations it is

Table 3.4. In Situ Photosynthetic Characteristics of Algal Assemblages[a]

	Winter	Spring	Summer	Irondequoit Bay[b]
Algal assemblages dominated by	*Limnothrix redekei*	*Asterionella* sp.	*Anabaena/Ceratium*	Mixed[c]
Irradiance, measured	Scalar	Cosine	Cosine	Scalar
Photosynthesis profiles: number	4	2	1	1
Data recorded	January–February, November 1983, February 1984	March–April 1981	July 1981	July 1981
$\mu E\ m^{-2}\ s^{-1}$, subsurface	201	611	1160	2000
Temperature euphotic zone (°C)	4.4	5.2	18.6	25
Light-proportional photosynthesis, ranges:				
$\mu E\ m^{-2}\ s^{-1}$	<1–16	4–40	6–80	40–350
mg C (mg chla)$^{-1}$ h^{-1}	0.03–0.5	0.14–0.5	0.2–1.5	0.4–5.1
k_c; m^2 (mg chla)$^{-1}$	0.0135	0.0124	0.0073	~0.01
P_m; mg C (mg chla)$^{-1}$ h^{-1}	1.4	2.7	2.4	5.1
Φm; mol C E^{-1} abs.[d]	0.049 ± 0.004	0.074 ± 0.007	0.056 ± 0.006	0.04

[a]Data from Meffert and Overbeck (1985b).

[b]From Bannister and Weidemann (1984).

[c]*Microcystis, Coelosphaerium, Phacotus, Pandorina.*

[d]Respiration not considered.

assumed that scalar irradiance was about 20% higher than the measured cosine irradiance (Morel 1978). Corrected values were 3.4 to 6.5 × 10^{-4} and corresponded to values estimated for several algal cultures (Welschmeyer and Lorenzen 1981).

For *Asterionella*- or *Sphaerocystis*-type algae in the Plußsee, as well as for *Limnothrix redekei* in the Edebergsee, the chlorophyll specific light attenuation (k_c) values corresponded within 10% (Table 3.4, Fig. 3.4). A distinctly lower k_c value of 7.3 × 10^{-3}, corresponding to 6.7 × 10^{-3} for *Peridinium* (Dubinsky and Berman 1979), was found in algal assemblages dominated by *Ceratium*. According to Talling (1971), *Ceratium* populations may be characterized by considerably lower extinction coefficients than other algae. The k_c values of 0.0135 and 0.02 that were obtained for populations dominated by *Limnothrix redekei* and *Aphanizomenon flos aquae* corresponded with k_c values estimated from published data (Kirk 1983). The k_c values for *Asterionella* populations in Lake Windermere (Talling 1971) and the Plußsee were different.

Light-proportional photosynthesis, which depends on the season and degree of latitude (Table 3.4), was distinctly lower at 54° North in Germany than at 43°

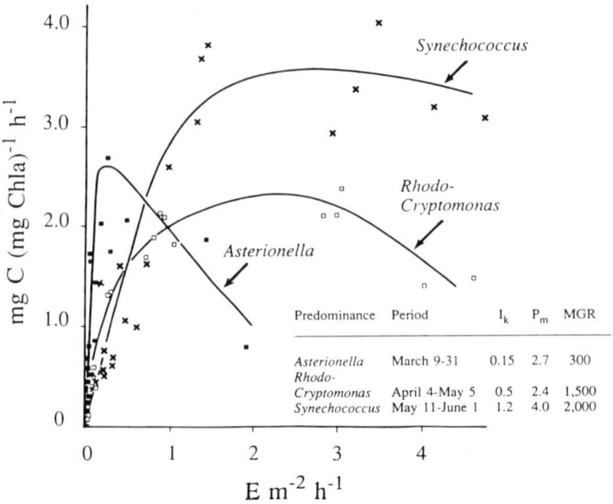

Figure 3.4. Light-saturation curves of three succeeding algal populations during spring 1981 in the Plußsee. I_k (E m^{-2} h^{-1}); $P_m = P_{max}$; MGR, mean global radiation (J cm^{-2} d^{-1}).

North in the United States. It is known that the compensation point and the onset of light saturation vary within species, and they are influenced by lower temperatures in the winter. In addition, there is reason to assume interspecific differences exist. Maximum quantum yields were constant for light-proportional photosynthesis, at about 0.05, 0.06, and 0.07 mol C E^{-1} abs. for *Limnothrix redekei-*, *Anabaena/Ceratium-*, and *Asterionella*-dominated assemblages, respectively. Corrected for both cosine irradiance and dark respiration, maximum quantum yields of about 0.06 mol C E^{-1} abs. were at the upper range of in situ yields (Bannister and Weidemann 1984, Tilzer 1984). Maximum quantum yields of *Limnothrix redekei* were relatively constant, as shown by repeated measurements over several winter months.

Quantum yields are closely related to the initial slope of the light-saturating curve (Platt and Jassby 1976, Tilzer 1984), which varied with photosynthetic capacity depending on the growth stage (Fig. 3.5). In the Plußsee from June to September, when thermal stratification was stable, the changing photosynthetic rates are shown by the diverging primary productivity and chlorophyll$_a$ maxima (Fig. 3.2). Maximum quantum yields changed correspondingly, as shown for several algal populations (Fig. 3.3). Following primary production maxima, quantum yields decreased from 0.08–0.07 to 0.02–0.01 mol C E^{-1} abs. with periods of chlorophyll$_a$ maxima between 25 and 60 mg m^{-3}. Low quantum yields of about 0.02 also occurred during the extended algal minimum. This differing response to light resembles that of "sun and shade" phytoplankton (Talling 1966, Yentsch and Lee 1966, Falkowsky 1981), whereas quantum yields produced by shade-adapted populations of *Limnothrix redekei* during winter were distinctly higher. Thus, during the chlorophyll maximum the physiological state of these algal populations

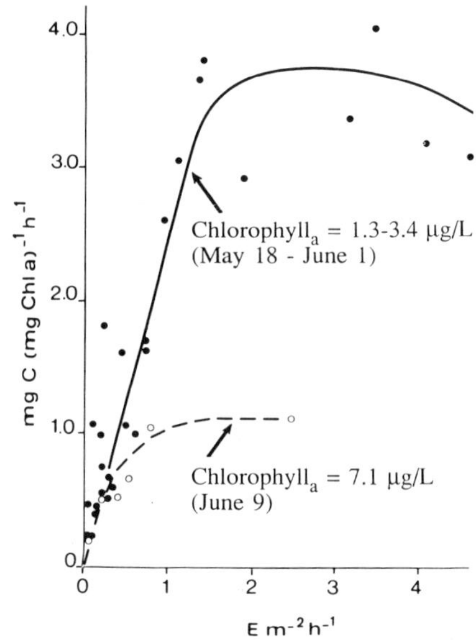

Figure 3.5. Light-saturation curves of a *Synechococcus* population at different growth stages.

was characterized by an increasing rate of phaeopigment production (Meffert and Overbeck 1985a) as well as by reduction of both the specific photosynthetic rates and the maximum quantum yields. These features point to the beginning of the population's decline. Maximum quantum yields determined in the Plußsee and the Edebergsee ranged between 0.01 and 0.08 mol C E^{-1} abs. and were dependent chiefly on algal growth stages. They exceeded the variation range given for algal species.

Photosynthetic Rates of Algal Assemblages

The curves of specific photosynthetic rates against light intensity show the successive types of photosynthesis: light-proportional, light-saturated, and light-inhibited photosynthesis. Data for the photosynthetic capacity, P_m, and the onset of light saturation, I_k, were taken from those curves where light was saturating. The initial slopes of the light-saturating curve, α (Platt and Jassby 1976), or the initial gradient at low light intensities (Talling 1969) was computed by regression analysis from the specific photosynthetic rate and the limiting light intensity or by P_m/I_k (Table 3.5; Talling 1957a, 1966).

Growth and decline of the first spring population, dominated by *Asterionella,* showed successive response to increasing global irradiance (Fig. 3.4). Exponential growth started in March from temperatures of about 3°C to 5°C. The chlorophyll$_a$ was almost uniformly distributed throughout the euphotic zone (0–5 m) and declined slightly between 3 and 5 m depth on 31 March, although there was no thermal stratification of the lake. The slope α increased from 7.4 to 15 within 7 days. Chlorophyll$_a$ doubling time decreased from 23–12 to 3–4 days. Photosynthetic capacity increased proportionally to the slope α, indicating a relative con-

Table 3.5. In Situ Photosynthetic Characteristics of Growth Stages of Successive Algal Populations in the Plußsee[a]

Period, Dominating Algae	Global Radiation (J cm $^{-2}$ d^{-1})	Temperature (°C)[b]	P-PO_4^{3-} (μg/L)[b]	Chlorophyll (μg/L)[b]	α (r)[c]	P_m	I_k (μE m^{-2} s^{-1})
March 9–31 *Asterionella* sp.	300	3.6–6.0	181.0	17.8	12 (0.99)	2.0	49
April 6–May 4 *Rhodo-Cryptomonas* sp.	1500	6.4–8.5	83.8	26.5	5 (0.98)	2.4	140
May 11–June 1 *Synechococcus* sp.	2000	11.0–15.6	73.3	3.7	2.5 (0.98)	4.0	320
June 15–July 6 *Spaerocystis* sp.	1400	15.8–18.3	28.1	14.4	4.3 (0.98)	2.7	170
July 13–20 *Anabaena*/*Ceratium* sp.	1800	18.6–17.5	4.9	42.8	4.8 (0.98)	2.4	130
July 29–August 25 *Ceratium*/*Phacotus* sp.	1400	17.0–16.0	10.4	28.3	4.9 (0.87)	1.2/3.2	70
August 31–September 28 *Aphanizomenon* sp.	1100	16.0–14.6	2.9	17.3	5.4 (0.76)	2.2	110

[a]Data from Meffert and Overbeck (1985b).
[b]Mean values for the euphotic zone.
[c]μg C (μg chla)$^{-1}$ μE^{-1} m^2.

stant I_k (Harris 1978) of about 49 μE m^{-2} s^{-1}. The general level of activity was about 2.0 mg C (mg chla)$^{-1}$ h^{-1}, confirming the P_m and I_k values obtained for *Asterionella* populations in Lake Windermere (Talling 1966), i.e., 6 mg O_2 or 2.2 mg C (mg chla)$^{-1}$ h^{-1} and 2.8 klux or 49 μE m^{-2} s^{-1}, respectively. Photosynthesis was further characterized by the narrow light saturation range and the clear susceptibility to photoinhibition that commenced at 60–120 μE m^{-2} s^{-1}.

In April mean global radiation was about 1500 J cm^{-2} d^{-1}, but water temperature was low (6–9°C). Apart from temporary subsurface warming, the water was vertically mixed. Diatom dominance was superseded by Cryptophyceae (Fig. 3.4, Table 3.5) with differing photosynthetic characteristics. The slope α decreased from about 12 to 5, P_m was about 2.4, and I_k correspondingly increased to 140 μE m^{-2} s^{-1}. Photoinhibition commenced at about 600 μE m^{-2} s^{-1}, indicating a relatively broad light saturation range. Similar photosynthetic patterns were measured on several occasions in the summer.

Collapse of the spring phytoplankton bloom resulted in a pronounced chlorophyll minimum during an 8-week period of fine weather when the relatively high global radiation (10–16 kJ cm^{-2} $week^{-1}$) was measured but water temperature was about 10°C. Concentrations of P-PO_4^{3-} allowed algal growth (Table 3.5). The low chlorophyll$_a$ content was largely a result of the presence of small blue-green algae of the *Synechococcus* sp. The trend toward lower photosynthetic efficiency as radiation increased continued. The slope α fell to the lowest level for the Plußsee in 1981 (Fig. 3.2, Table 3.5). Maximal values for photosynthetic capacity and onset of light saturation were 4 mg C (mg chla)$^{-1}$ h^{-1} and 320 μE m^{-2} s^{-1}, respectively. Photosynthetic capacity decreased distinctly when the chlorophyll$_a$ concentration reached maximum (Fig. 3.3). Surface phytoplankton was photoinhibited at about 800 μE m^{-2} s^{-1}.

Global radiation decreased to 1000–1500 J cm^{-2} d^{-1} over the following weeks with one exception. Water temperature increased to about 17°C. The summer bloom started with green algae of *Sphaerocystis* sp. Algal populations with special qualities coincided with nutrient depletion as P-PO_4^{3-} concentration decreased to the annual minimum (3–10 μg L^{-1}) between July and September (Table 3.5). Nitrogen-fixing *Anabaena* species formed a short-term maximum that was superseded by the motile algae *Ceratium* and *Phacotus* sp. Four chlorophyll$_a$ measurements taken at 1-h intervals showed that, as a result of diurnal migration, the motile cells formed nighttime maxima below 3 m depth, thus encountering the nutrient-rich upper metalimnion. Corresponding vertical diurnal migrations are well known for several motile algal species (Ruttner 1962, Nauwerk 1963, Soeder 1967). During the day chlorophyll$_a$ maxima appeared at 1–3 m depth. Chlorophyll$_a$ concentrations were lowest at the surface of Plußsee, where a predominance of *Phacotus* sp. was found. Maximal photosynthetic capacities were 3.2 and 1.2 mg C (mg chla)$^{-1}$ h^{-1} for the *Phacotus* and the *Ceratium* species, respectively (Fig. 3.6). Large-celled *Ceratium* sp. that accumulated in the reduced-light-intensity zone showed the lowest photosynthetic capacity. Photosynthesis was photoinhibited at about 600 μE m^{-2} s^{-1}.

The slopes (α), onset of light saturation (I_k), and photosynthetic capacity (P_m) showed corresponding behavior in April and July to September that was

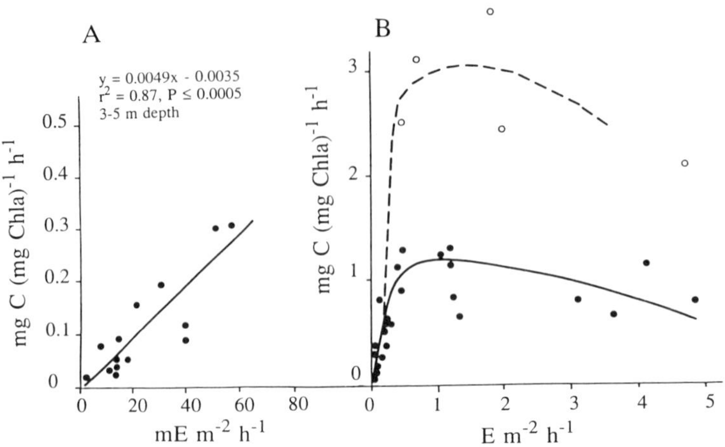

Figure 3.6. Light-saturation curves of the *Ceratium/Phacotus* populations. (A) Linear light-limited part and (B) complete curves.

independent of the algal species, temperature, and nutrient supply. Over both periods the mean global radiation was about 1500 J cm^{-2} d^{-1}, indicating its controlling role.

Chlorophyll$_a$ Primary Production of the Plußsee and Other East-Holstein Lakes

Lakes can be classified on the basis of their chlorophyll$_a$ concentration, which can be used as a production indicator (Fig. 3.7) as represented by the first 13 lakes of Table 3.1 (Meffert and Wulff 1987). The first group of these lakes is

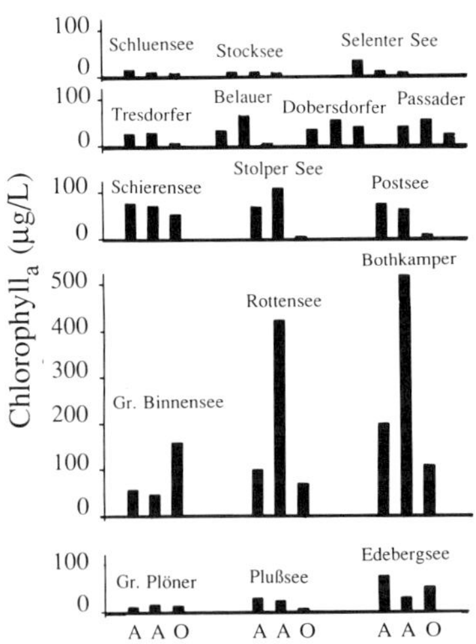

Figure 3.7. Mean values of epilimnetic (1–5 or 1–8 m depth) chlorophyll$_a$ concentration of 16 East-Holstein lakes in April (A), August (A), and October (O).

characterized by relatively low chlorophyll$_a$ maxima of about 30 μg L^{-1}, which are produced only once a year. In the second and third groups, chlorophyll$_a$ maxima of 30–60 and 65–100 μg L^{-1} repeatedly occurred. The three shallow lakes (Gr. Binnensee, Rottensee, Bothkamper See) had high chlorophyll$_a$ concentrations of 150–500 μg L^{-1}. The chlorophyll$_a$ concentrations of the Plußsee and of the Grosser Plöner See, the largest East-Holstein lake, corresponded with the first and second groups, respectively. Data from the small Edebergsee fit with the data of the third group.

The correlation between Secchi disk visibility and chlorophyll$_a$ concentration ($y = 32.24x - 0.83$, $r^2 = -0.93$, $P \leq 0.0005$) showed that water transparency decreases as chlorophyll$_a$ increases. Decreasing chlorophyll$_a$ concentrations correlated with increasing mean depths of lakes, as shown by their linear regression analysis ($y = 296.5x - 1.17$, $r^2 = -0.89$, $P \leq 0.0005$). According to Thienemann (1921), the relationship between the mean lake depth and the chlorophyll$_a$ concentration is based on the dependence of production on the area of the lake surface and the volume-dependent amount of nutrient. The relationship presupposes light- and nutrient-limited algal growth. The uniform lake type and the corresponding geographic and climatic area explain the clear relationship between biomass production and lake morphometry. Under these conditions the mean depth appeared to be a factor controlling the production.

The comparison of chlorophyll maxima and minima of the Plußsee, the Grosser Plöner See, and the Edebergsee and their relation to global radiation is shown in Figure 3.8 (Meffert 1989a). The chlorophyll$_a$ maxima of the three lakes were all about 200 mg m^{-2} (euphotic depth) during three different years, but the summer minima differed distinctly. Plußsee had a pronounced chlorophyll$_a$ minimum in contrast to the Edebergsee. Increase and decrease of chlorophyll$_a$ concentrations were mainly controlled by the seasonal radiation. Chlorophyll$_a$ minima and radiation maxima of 2.0–2.3 kJ cm^{-2} d^{-1} occurred simultaneously in May 1981 (Plußsee), June 1984 (Grosser Plöner See), and in August 1983 (Edebergsee). Chlorophyll maxima coincided with a mean global radiation of about 1.5 kJ cm^{-2} d^{-1}.

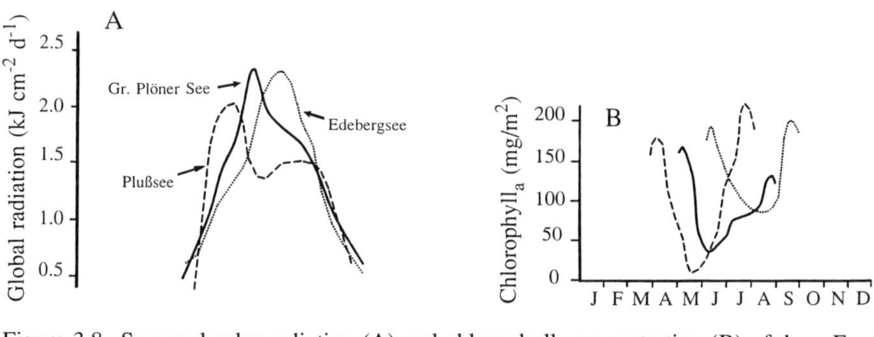

Figure 3.8. Seasonal solar radiation (A) and chlorophyll$_a$ concentration (B) of three East Holstein lakes.

Conclusions

The production data classify the Plußsee as a moderately eutrophic lake. The controlling role of global radiation was illustrated by many features of algal maxima and successions. In the Plußsee highest production rates occurred in April and July 1981 at different temperatures and nutrient concentrations, but levels of global radiation were about 1.5 kJ cm^{-2} d^{-1}. An increase of global radiation to about 2 kJ cm^{-2} d^{-1} coincided with minor production rates when temperatures were relatively low. Similar findings with other East-Holstein lakes confirmed the relationship between radiation and phytoplankton primary production. Motile algae that were able to migrate to areas of lower light intensities and higher nutrient levels represented an exception, as has been shown by the high production rates in July in the Plußsee.

Photosynthetic characteristic data were measured in situ for successive algal populations over several growth stages. Temporal and vertical shifts were rare and were usually identifiable. Most of the Plußsee in situ $chlorophyll_a$ specific extinction coefficients were in the narrow range of 0.012–0.014, but the occurrence of extreme values (0.007 and 0.02) revealed that $chlorophyll_a$ specific extinction coefficients should be determined for each algal population, preferably during lag and exponential stages. Both *Ceratium* and *Peridinium* spp. had low k_c values as measured in different water bodies. These low $chlorophyll_a$ specific extinction coefficients may be caused by the low light interception efficiency of large pigmented cells.

Estimated maximum quantum yields for natural algal populations were between 0.04 and 0.09 mol C E^{-1} abs., with a mean about 0.06 mol C E^{-1} abs. Quantum yields measured in situ in the Plußsee were in range of 0.01–0.08 moles C E^{-1} abs. The photosynthetic characteristics of successive algal growth stages in the Plußsee varied considerably during summer stratification, whereas three different algal assemblages had quantum yields within a narrower range.

Seasonal changes in the photosynthesis parameters were shown by the global radiation-controlled light adaptation (Table 3.5). During winter shade-adapted algal cells had high photosynthetic efficiencies at low but increasing global radiation (Harris 1978), as shown by the Plußsee data for March. During periods of permanent high radiation the slope α was markedly depressed. Constant low photosynthetic efficiency was measured in summer, independent of algal species, and this increased slightly when global radiation decreased.

Photosynthetic capacity data in the Plußsee varied over the relatively narrow range of 1–4 mg C (mg chla)$^{-1}$ h^{-1}, and frequently the maximum was about 2.5 mg C (mg chla)$^{-1}$ h^{-1}. Corresponding determinations from East-Holstein lakes of similar trophic type showed that the upper range of 4 mg C (mg chla)$^{-1}$ h^{-1} was seldom exceeded. The relationship between photosynthetic rate and $chlorophyll_a$ concentrations basically corresponded to that reported by Javornicky (1980). Photosynthetic rates decreased when chlorophyll concentration increased.

References

Albrecht D, Münster U, Stabel HH (1978) Untersuchungen zum Stoffhaushalt des Plußsees. Verh Ges Ökol 7:109–118

Anagnostidis K, Komarek J (1988) Modern approach to the classification system of cyanophytes. 3 Oscillatoriales. Arch Hydrobiol Algol Stud 50–53:327–472

Anagnostidis K, Overbeck J (1966) Methanoxydierer und hypolimnische Schwefelbakterien. Studien zur ökologischen Biocönotik der Gewässermikroorganismen. Ber Deutsch Bot Ges 79:163–174

Atlas D, Bannister TT (1980) Dependence of mean spectral extinction coefficient of phytoplankton on depth, water color and species. Limnol Oceanogr 25:157–159

Bannister TT, Weidemann AD (1984) The maximum quantum yield of phytoplankton photosynthesis in situ. J Plankton Res 6:275–294

Dubinsky Z, Berman T (1979) Seasonal changes in the spectral composition of downwelling irradiance in Lake Kinneret (Israel). Limnol Oceanogr 24:652–653

Falkowsky PG (1981) Light–shade adaptation and assimilation numbers. J Plankton Res 3:203–216

Golachowska JB (1979) Phosphorus forms and their seasonal changes in water and sediments of lake Plußsee. Arch Hydrobiol 86:217–241

Harris GP (1978) Photosynthesis, productivity and growth: The physiological ecology of phytoplankton. Arch Hydrobiol Beih 10:1–171

Javornicky P (1980) Density dependent effects. In Le Cren ED, Lowe-McConnell RH (eds) *The Functioning of Freshwater Ecosystems.* Cambridge University Press, Cambridge, pp 170–176

Kirk, JTO (ed) (1983) *Light and Photosynthesis in Aquatic Ecosystems.* Cambridge University Press, Cambridge

Krambeck HJ (1974) Energiehaushalt und Stofftransport eines Sees. Beispiel einer mathematischen Analyse limnologischer Prozesse. Arch Hydrobiol 73:137–192

Krambeck HJ (1979) Programmgesteuerte in-situ Meßwerterfassung limnologischer Umweltparameter. GIT Fachz Lab 23:842–844

Krambeck HJ (1981) Longterm monitoring and modelling of the nutrient cycle in a Baltic lake. Verh Int Ver Limnol 21:460–465

Krambeck HJ, Krambeck C (1980) Modelling carbon and phosphorus in a small hypertrophic lake. In Barica J, Mur LR (eds) *Hypertrophic Ecosystems.* W. Junk, The Hague, pp 159–169

Meffert ME (1987) Planktonic unsheathed filaments (Cyanophyceae) with polar and central gas vacuoles. I. Their morphology and taxonomy. Arch Hydrobiol Suppl 76:315–346

Meffert ME (1989a) Planktic unsheathed filaments (Cyanophyceae) with polar and central gas vacuoles. II. Biology, population dynamics and biotopes of *Limnothrix redekei* (Van Goor) Meffert. Arch Hydrobiol 116:257–282

Meffert ME (1989b) In situ and in vitro photosynthesis of *Limnothrix redekei* (Cyanophyta). Arch Hydrobiol 116:401–414

Meffert ME, Overbeck J (1985a) Dynamics of chlorophyll and photosynthesis in natural phytoplankton associations. I. Seasonal and annual cycles of chlorophyll and phaeopigments, their relation to euphotic depth and their vertical distribution in small North German lakes. Arch Hydrobiol 104:219–234

Meffert ME, Overbeck J (1985b) Dynamics of chlorophyll and photosynthesis in natural phytoplankton and associations. II. Primary productivity, quantum yields and photosynthesis rates in small North German lakes. Arch Hydrobiol 104:363–385

Meffert ME, Wulff WR (1987) Morphometrie und Chlorophyllproduktion von ostholsteinischen Seen. Z Wasser Abw Forsch 20:13–15

Morel A (1978) Available, usable and stored radiant energy in relation to marine photosynthesis. Deep Sea Res 25:673–688

Morel A, Smith RC (1974) Relation between total quanta and total energy for aquatic photosynthesis. Limnol Oceanogr 19:561–600

Nauwerk A (1963) Die Beziehungen zwischen Zooplankton und Phytoplankton im Erkensee. Symb Bot Upsalienses 17:1–163

Ohle W (1959) Die Seen Schleswig-Holsteins, ein Überblick nach regionalen, zivilisatorischen und produktionsbiologischen Gesichtspunkten. Wasser 26:16–41

Ohle W (1962) Der Stoffhaushalt der Seen als Grundlage einer allgemeinen Stoffwechseldynamik der Gewässer. Kieler Meeresforsch 68:107–120

Ohle W (1964) Interstitiallösungen der Sedimente, Nährstoffgehalt des Wassers und Primärproduktion des Phytoplanktons in Seen. Helgol Wiss Meer 10:411–429

Ohle W (1975) General consideration on environmental problems of lakes. Proc Int Cong Hum Environ pp 383–390

Overbeck J (1979) Studies on heterotrophic functions and glucose metabolism of microplankton in Plußsee. Arch Hydrobiol Beih 13:56–76

Platt T, Jassby AD (1976) The relationship between photosynthesis and light for natural assemblages of coastal marine phytoplankton. J Phycol 12:421–430

Rai H (1975) Pigment cycle in a north German lake. Verh Int Ver Limnol 19:1220–1228

Rodhe W (1965) Standard correlations between pelagic photosynthesis and light. Mem Ist Ital Idrobiol 18:367–381

Ruttner F (1962) *Grundriß der Limnologie*. Walter de Gruyter, Berlin

Soeder CJ (1967) Tagesperiodische Vertikalwanderung bei begeisselten Planktonalgen. Umsch Wiss Tech 12:388

Talling JF (1957a) Photosynthetic characteristics of some freshwater plankton diatoms in relation to underwater radiation. New Phytol 56:29–50

Talling JF (1957b) The phytoplankton population as a compound photosynthetic system. New Phytol 56:133–149

Talling JF (1966) Photosynthetic behavior in stratified and unstratified lake populations of a planktonic diatom. J Ecol 54:99–127

Talling JF (1969) The incidence of vertical mixing and some biological and chemical consequences, in tropical African lakes. Verh Int Ver Limnol 17:998–1012

Talling JF (1971) The underwater light climate as controlling factor in the production ecology of freshwater phytoplankton. Mitt Int Ver Limnol 19:214–243

Thienemann A (1921) Seetypen. Naturwissenschaften 18:87–91

Tilzer M (1983) The importance of fractional light absorption by photosynthetic pigments for phytoplankton productivity in Lake Constance. Limnol Oceanogr 28:833–846

Tilzer M (1984) The quantum yield as a fundamental parameter controlling vertical photosynthetic profiles of phytoplankton in Lake Constance. Arch Hydrobiol Suppl 69:169–198

Tyler JE (1975) The in situ quantum yield efficiency of natural phytoplankton populations. Limnol Oceanogr 20:976–980

Utermöhl H (1925) Limnologische Phytoplanktonstudien. Die Besiedlung ostholsteinischer Seen mit Schwebpflanzen. Arch Hydrobiol Suppl 5:1–527

Welschmeyer NA, Lorenzen CJ (1981) Chlorophyll-specific photosynthesis and quantum efficiency at subsaturating light intensities. J Phycol 17:283–293

Wetzel RG (1983) *Limnology*. WB Saunders, Philadelphia

Yentsch CS, Lee RW (1966) A study of photosynthetic light reactions, and a new interpretation of sun and shade phytoplankton. J Mar Res 24:319–337

4. Phytoplankton-Excreted Organic Carbon

Anne F. Sell

Introduction

The classical food chain model views phytoplankton as the connecting link between abiotic factors such as nutrients and light and further biotic beneficiaries that live on its cellular substances. Carbon is transferred via herbivorous and carnivorous zooplankton into biomass of planktivorous and piscivorous fish. Not only is pelagic primary production an advantage to grazing processes, but photosynthetically fixed carbon is also released as soluble organic compounds (Fogg 1958, 1966, Anderson and Zeutschel 1970, Hellebust 1974) that bypass grazing activity and directly support heterotrophic microbial metabolism (Overbeck 1972, Overbeck 1974 Fig. 11, Nalewajko 1977, Chróst 1978, 1981, 1986, Larsson and Hagström 1979, Meffert and Overbeck 1979, Bell 1980, Joint and Morris 1982, Chróst and Faust 1983). This alternative pathway of carbon was first characterized in marine systems as a "microbial loop" (Azam et al. 1983), and it described bacterial activity as consuming as well as decomposing (cf. also Chróst 1983, Chróst et al. 1989; Chapter 12, this volume).

Because of its effective uptake and short generation times, bacterial growth may influence the rapid turnover of carbon into a particulate fraction that can be consumed by protozoa (Jørgensen et al. 1983, Goldman and Caron 1985, Porter et al. 1985, Güde 1985, 1989, Caron et al. 1988, Vadstein et al. 1989, Fukami et al. 1991, Meier 1991), phagotrophic phytoflagellates (Bird and Kalff 1986, Goldman et al. 1987, 1989, Porter 1988, Tranvik et al. 1989, Sanders et al. 1990, Schnepf and Elbrächter 1992), and macrozooplankton (Bird and Kalff 1986, Sanders et al. 1989; Chapters 11 and 12, this volume).

Sharp (1977) was the first to pose the question, "Excretion of organic matter by . . . phytoplankton: Do healthy cells do it?" which has been changed to "Why do healthy cells do it?" (Bjørnsen 1988). Release of organic compounds is a process accompanying normal photosynthetic activity of algae (only those compounds are considered to be extracellular organic carbon, EOC, sensu stricto). There are a variety of possibilities regarding the mechanisms leading to release of cellular contents: "release of simple compounds, e.g., sugars, amino acids, and organic acids, by actively growing cells probably occurs mainly by diffusion through the plasmalemma" (Hellebust 1974). This passive loss of substances from the cell would need a concentration gradient directed toward the medium. That means that loss rates would be enhanced as the intracellular pool of these compounds increases. It is plausible that algal cells use passive diffusion for low-molecular-weight substances. For example, glycolate release is used as a type of overflow outlet for products that accumulate when high light intensities lead to fixation rates of carbon that are higher than the simultaneous anabolic turnover via the Calvin cycle. In this case, the oxygenase function of ribulose-bisphosphate carboxylase (RUBISCO) would dominate the carboxylase reaction. Instead of running the complete pathway of photorespiration, as do higher terrestrial plants, algae should be able to directly eliminate small molecules into their liquid medium (Wood and Van Valen 1990, J. Benndorf, personal communication).

Nevertheless, further biochemical and mechanical processes contribute to the release of organic substances. Active mechanisms can produce extracellular organic compounds, e.g., the excretion of toxins (Jüttner 1981, Gross et al. 1991). Nutrient depletion or other kinds of stress may promote exudation (Zlotnik and Dubinsky 1989). Chelators released by cyanobacteria enhance their access to iron, which favors iron-limited cyanobacteria over eukaryotic algae (Murphy et al. 1976, Wetzel 1983).

Passive loss occurs through membranes that changed their permeability during aging or as a result of parasite infection of the cell (Daft and Stewart 1973, Van Donk 1989, Cottrell and Suttle 1991) or mechanical injury by "sloppy feeding" of herbivorous zooplankton (Lampert 1978).

Quantification of EOC sensu stricto, as stated above, is only possible if any kind of stress is excluded. Since this cannot be guaranteed during incubation of the complete phytoplankton community in situ, EOC as measured here can be defined as "extracellular organic carbon that is exuded from healthy phytoplankton cells as well as released by other mechanisms except for 'sloppy feeding'" (zooplankton were removed from the samples).

As a consequence, quality as well as quantity of EOC differ with seasonal succession and physiological status of the phytoplankton. This leads to changes in the dependence on EOC for pelagic bacterial growth.

In the Plußsee the maximum EOC release was up to 250%, compared to the carbon found in the algal cells after incubation. Also, it is clear that quantity of EOC dissolved is not a suitable measure for EOC produced, because high-quality exudates have a high turnover in the microbial metabolism (Sell and Overbeck 1992).

Experimental Design

Organic Carbon Release and Its Utilization

In the Steemann-Nielsen technique (1952, 1977) to measure photosynthesis with ^{14}C, we use radioactivity to trace the pathways of carbon within planktonic food webs (Harris et al. 1989).

From May to September 1989, primary production and EOC exudation were recorded at weekly intervals. Bottles were filled with water from each of the depths investigated in the epilimnion, metalimnion, and hypolimnion after water had been filtered (56 µm) to exclude zooplankton grazing. For each depth, two light and two dark bottles were filled. Samples were labeled with sodium [^{14}C]-bicarbonate (ca. 0.09 µCi/ml sample represented total activity in each bottle) and incubated 4 h (1000–1400 h) at in situ depths before incubation was stopped with formalin. Separation of the different compartments of the food chain was possible with filtration through 0.2- and 3.0-µm pore-size Nuclepore filters. The fraction >3.0 µm reflected algal incorporation of carbon, whereas the size class between 0.2 and 3.0 µm almost exclusively consisted of bacteria. EOC was measured in the filtrate < 0.2 µm after acidification and bubbling with air to expel the surplus inorganic tracer. Respiration cannot be followed since the respired $^{14}CO_2$ is identical to the tracer and has to be removed as well in order to quantify the dissolved organic products (Fig. 4.1). Uptake rates for each of these three fractions were calculated from liquid scintillation counting and took into account that ^{14}C was discriminated from ^{12}C by 6% (for further details see Sell and Overbeck 1992).

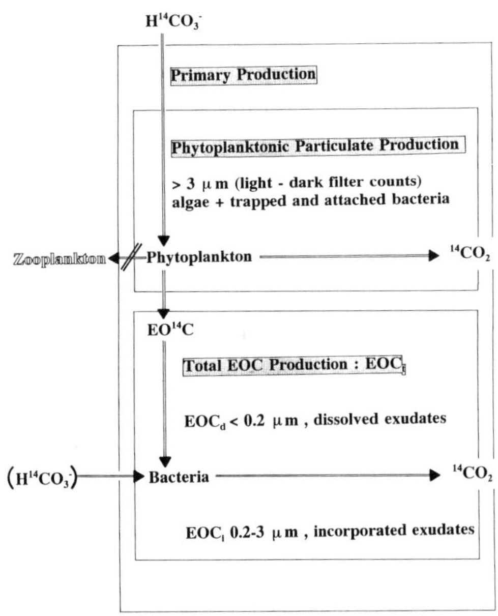

Figure 4.1. Scheme of separation of the different ^{14}C-labeled fractions of phytoplankton primary production: filtration through 0.2-µm and 3-µm pore-size filters. Surplus inorganic $^{14}C\text{-}HCO_3^-$ was removed from the 0.2-µm filtrate before determination of EOC.

The natural concentration of bicarbonate at each depth was obtained from the alkalinity and pH.

Although it is never possible to separate all trophic levels completely by dividing them into different size classes, the filtration through 0.2 µm for bacteria and 3.0 µm for phytoplankton was appropriate for the Plußsee. As shown by Chróst (1989), 98% of total chlorophyll$_a$ could be collected on 3.0-µm filters, whereas 95.2% of the pelagic bacteria passed through 3.0-µm and were found on 0.2-µm filters (Chróst and Overbeck 1987). This separation does not work well in samples that have a high percentage of autotrophic picoplankton or have many microorganisms attached to algal cells (for "trapped and attached bacteria" see Riemann and Søndergaard 1986, Pedrós-Alió 1989). For the neighboring mesotrophic Schöhsee, Rai (1984) reported that on average 45–65% of microbial activity was found in the 3-µm "algal size" fraction. The conclusion that picoplankton were not of significant influence during the studies of the Plußsee corresponds with the more general observation that eutrophic lakes tend to have less autotrophic picoplankton than oligotrophic lakes (Gliwicz 1969, Søndergaard et al. 1985, Riemann and Søndergaard 1986) and marine systems (Johnson and Sieburth 1979, 1982, Waterbury et al. 1979, Joint and Pomroy 1983, Joint et al. 1986, Lewis et al. 1986, Shapiro and Guillard 1986).

In addition to the heterotrophic uptake of carbon compounds, bacteria have the ability to assimilate inorganic HCO_3^- (and $H^{14}CO_3^-$) by anaplerotic reactions if there is a demand for substitution of depleted substrates in the citric acid cycle, e.g., the Wood–Werkman reaction in which pyruvate is carboxylated (Kornberg 1966, Lignell 1990, Overbeck and Sako 1990). As this fixation is not light dependent, it should reach the same rates under light and dark conditions.

Definitions

A number of terms are used to describe autotrophic and heterotrophic processes. "Phytoplankton particulate production" is defined by the difference between the 3.0-µm filters from the light and dark bottles.

With respect to the basic function of photosynthesis to support the aquatic community with organic carbon, "primary production" is considered to be the sum of the particulate fraction (algae and bacteria) found on 0.2-µm pore-size filters plus the dissolved EOC. Total EOC production (EOC_t) is divided into dissolved free EOC (EOC_d) and preexisting EOC that is already incorporated into the microorganisms (EOC_i). EOC production is described separately for light and dark metabolism.

Excreted Organic Carbon: Functional Coupling between Phytoplankton and Bacteria

In the Plußsee, three phases could be distinguished within the study period, which started after the breakdown of the phytoplankton spring bloom and ended after the

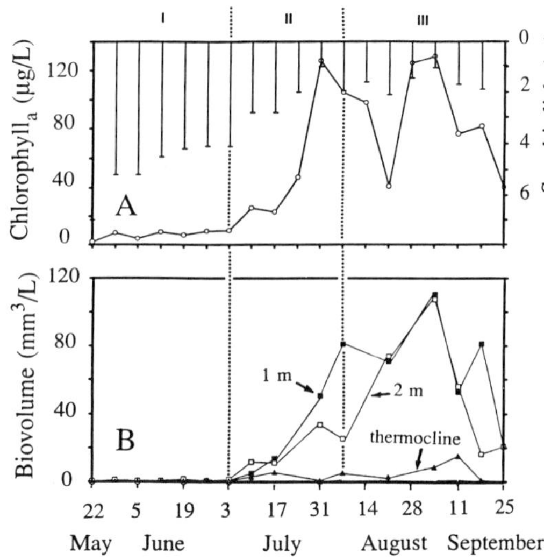

Figure 4.2. (A) Maximal concentration of chlorophyll$_a$ (curve with open circles) and Secchi disk visibility (vertical bars) in the water column, and (B) phytoplankton biovolume at depths of 1 m, 2 m, and thermocline (5–6 m) during the study periods: May–June (I), July (II), and August–September (III) in the Plußsee.

decline in autumn (Fig. 4.2). The importance of exudates as loss factors for the algal cells and their availability as carbon source for microbial growth varied for each period.

The *clear-water phase* (Lampert et al. 1986) was expressed by a drop in chlorophyll$_a$ concentration from over 20 µg/L to 2 µg/L within 1 week (May 22–29, 1-m depth) and a corresponding rise in water transparency to more than 7 m as phytoplankton biomass was decimated by herbivorous zooplankton. The end of this period was distinguished by the increase in grazing-resistant phytoplankton such as dinoflagellates (*Ceratium hirundinella,* nonedible) and cyanobacteria (*Aphanothece clathrata,* barely edible), which profited from the decrease in competitors (PEG model; Sommer et al. 1986). Whereas phytoplankton biomass was low during the spring clear-water phase (Fig. 4.2), photosynthetic activity was high (Fig. 4.3), resulting in high growth rates. Except for one day (June 12, 1989) exudation rates were low throughout the entire period.

The fact that there were few released exudates (EOC sensu stricto) does not necessarily mean that dissolved organic compounds were not available and not important to bacteria. On the contrary, since zooplankton controlled phytoplankton biomass and reached high ingestion rates, a high percentage of algal cells were injured and lost intracellular compounds as a result of "sloppy feeding" of the herbivores (Lampert 1978). It is obvious that these substances are the dominant source of dissolved carbon for pelagic bacteria during this season.

In *early summer,* the biomass of inedible phytoplankton increased and algae (dinoflagellates: *Ceratium, Peridinium, Peridinopsis;* cyanobacteria: *Aphanothece, Anabaena, Limnothrix* spp.) that interfered with the filter systems of zooplankton were present. Weeks of low primary production after the breakdown of the spring phytoplankton bloom were followed by a period of high carbon fixation rates (Fig. 4.3). (Note, however, that dinophytes or filamentous algae were abun-

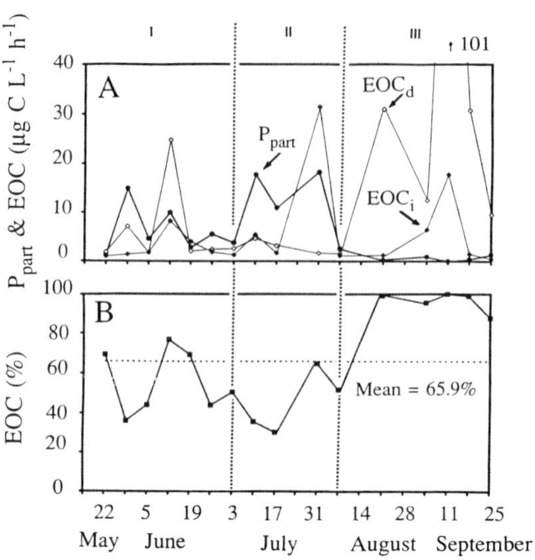

Figure 4.3. (A) Phytoplankton particulate production and excreted organic carbon (EOC) at the depth of 1 m in the Plußsee. Total EOC is separated into EOC that remained dissolved (EOC_d) and EOC incorporated by bacteria (EOC_i). (B) Percentage of the total extracellular release of EOC_t related to phytoplankton primary production.

dant when absolute values of primary production might be underestimated because of prefiltration. This was accepted, as zooplankton should be excluded.) On July 31, chlorophyll$_a$ concentration was 127 µg/L (1-m depth), and phytoplanktonic particulate production 18 µg C L^{-1} h^{-1}. Still, almost no dissolved exudates were found in the water: 2 µg C L^{-1} h^{-1} (light bottles) and 1 µg C L^{-1} h^{-1} (dark bottles). In contrast to the spring phytoplankton, early summer plankton exuded EOC at high rates. EOC_d was negligible in the samples, as it was immediately taken up by bacteria, showing that EOC consisted of easily degradable, probably low-molecular-weight compounds that represented a high-quality carbon source (Nalewajko et al. 1976, Itturiaga 1981, Chróst and Faust 1983, Jørgensen et al. 1983, Kato and Stabel 1984, Chróst 1986). Although 33 µg C L^{-1} h^{-1} EOC_t was produced (light bottles), 31 µg C L^{-1} h^{-1} was incorporated during incubation (i.e., EOC_i). Apparently, a specialized population of free-living bacteria developed parallel to the cyanobacteria bloom (Meffert and Overbeck 1979, Witzel et al. 1982, Van Donk 1989, pp. 309–318). Microbial growth, the demand for carbon, and consequently the turnover rate of EOC were enhanced by the high water temperatures in July (up to 23.6°C), which allowed fast metabolism.

In *late summer,* the phytoplankton decline was paralleled by minimal productivity as measured by carbon incorporated into the cells. Release processes of "real EOC" were superimposed by the loss of intracellular carbon compounds through senescence of phytoplankton (*Ceratium hirundinella*). Cell membranes changed to higher permeability during the aging process, leading to leakage of large-molecular-weight substances (Nalewajko and Lean 1972, Sundh 1992) that could not pass the plasmalemma of healthy, actively growing algal cells. Therefore, release rates were beyond the scope of real exudation and amounted to 63–95% of total production that remained dissolved in the water for the most part. Incorporation of

EOC by microorganisms did not reach the level of July, although the supply was higher. The reason was that it was a mixture of few real exudates (mainly low-molecular-weight compounds that were easily incorporated by bacteria) and a high percentage of polymeric, refractory products of leakage and lysis—not belonging to EOC in the strict sense—that were resistant to rapid bacterial utilization.

Conclusions

The following questions arise: (1) How important is EOC as a means of carbon loss in regard to phytoplankton metabolism? and (2) How much do they contribute to the carbon demand of pelagic bacteria?

There was no doubt that release of EOC occurs as a byproduct of healthy photosynthesis. It was obvious that the significance of the release of EOC as a mechanism of carbon loss, as compared to the carbon incorporated into algal cells during the same period of time, was considerable in early summer. I postulate that this phenomenon is caused by warm water temperatures and high light intensities, leading to carbon assimilation rates that exceed the rates of further intracellular biosynthesis. The cellular pathway of carbon is dependent on enzymatic reactions; the first reaction is the carboxylation of ribulose-1,5-bisphosphate. This reaction is catalyzed by ribulose-bisphosphate carboxylase, but this enzyme can change to the reverse oxygenase function, which dominates when there are high concentrations of molecular oxygen in the medium and hence at high rates of photosynthetic activity. Whereas terrestrial plants run a complicated cycle of photorespiration, aquatic plants should not need to drive this costly overflow mechanism but should be able to release small molecules (e.g., glycolate) directly to the surrounding water by diffusion (Wood and Van Valen 1990).

When bacterial production was compared to phytoplanktonic exudation (cf. also Chapter 5, this volume), it was clear that bacteria profited from EOC—mostly in spring and early summer during the time of blue-green blooms and the onset of dinoflagellate blooms. Bacteria can also contribute to their advantage of phytoplankton exudation. Ammerman and Azam (1985) showed that bacterial 5′-nucleotidase regenerated phosphorus from organic compounds dissolved in the lake water and made phosphorus available to algal cells. Also Chróst (1986) proposed that the microorganisms, which are limited by the supply of easily assimilable organic carbon, stimulate the EOC release of phytoplankton by themselves. In addition to other ectoenzymes, they can produce an alkaline phosphatase (APase) that has higher specific activity and lower K_M values than that of the phytoplankton. Thus, the bacteria would be able to hydrolyze organic phosphorus compounds and increase the PO_4^{3-} pool, which in turn promotes primary production of phosphorus-limited phytoplankton and exudation of low-molecular-weight carbon compounds. This means, when phytoplankton growth rates are phosphorus limited, a functional coupling with pelagic bacteria could be given by microbial phosphohydrolases and by phytoplankton exudates as well (Chróst and Overbeck 1987, Sherr et al. 1988, Lignell 1990; Chapter 6, this volume).

References

Ammerman JW, Azam F (1985) Bacterial 5′-nucleotidase in aquatic ecosystems: A novel mechanism of phosphorus regeneration. Science 277:1338–1340

Anderson GC, Zeutschel RP (1970) Release of dissolved organic matter by marine phytoplankton in coastal and offshore areas of the North-East Pacific Ocean. Limnol Oceanogr 15:402–407

Azam F, Fenchel T, Field JG, Gray JS, Meyer-Reil LA, Thingstad F (1983) The ecological role of water-column microbes in the sea. Mar Ecol Prog Ser 10:257–263

Bell WH (1980) Microbial ecology of algal extracellular products. The specificity of algal–bacterial interactions. Biol Bull 159:456

Bird DF, Kalff J (1986) Bacterial grazing by planktonic lake algae. Science 231:493–495

Bjørnsen PK (1988) Phytoplankton exudation of organic matter: Why do healthy cells do it? Limnol Oceanogr 33:151–154

Caron DA, Goldman JC, Dennett MR (1988) Experimental demonstration of the roles of bacteria and bacterivorous protozoa in plankton nutrient cycles. Hydrobiologia 159:27–40

Chróst RJ (1978) The estimation of extracellular release by phytoplankton and heterotrophic activity of aquatic bacteria. Acta Microbiol Pol 27:139–146

Chróst RJ (1981) The composition and bacterial utilization of DOC released by phytoplankton. Kiel Meeresforsch Sonderh 5:325–332

Chróst RJ (1983) Plankton photosynthesis, extracellular release and bacterial utilization of released organic carbon (RDOC) in lakes of different trophy. Acta Microbiol Pol 32:275–287

Chróst RJ (1986) Algal–bacterial metabolic coupling in the carbon and phosphorus cycle in lakes. In Megusar F, Gantar M (eds) *Current Perspectives in Microbial Ecology.* Slovene Society for Microbiology, Ljubljana, pp 360–366

Chróst RJ (1989) Characterization and significance of β-glucosidase activity in lake water. Limnol Oceanogr 34:660–672

Chróst RJ, Faust MA (1983) Organic carbon release by phytoplankton: Its composition and utilization by aquatic bacterioplankton. J Plankton Res 5:477–493

Chróst RJ, Overbeck J (1987) Kinetics of alkaline phosphatase activity and phosphorus availability for phytoplankton and bacterioplankton in lake Plußsee (north German eutrophic lake). Microb Ecol 13:229–248

Chróst RJ, Münster U, Rai H, Albrecht D, Witzel KP, Overbeck J (1989) Photosynthetic production and exoenzymatic degradation of organic matter in the euphotic zone of a eutrophic lake. J Plankton Res 11:223–242

Cottrell M, Suttle C (1991) Wide-spread occurrence and clonal variation in viruses which cause lysis of a cosmopolitan, eucaryotic marine phytoplankter *Micromonas pusilla.* Mar Ecol Prog Ser 78:1–9

Daft MJ, Stewart WDP (1973) Light and electron microscope observations on algal lysis by bacterium CP-1. New Phytol 72:799–802

Fogg GE (1958) Extracellular products of phytoplankton and the estimation of primary production. Rapp P-V Reun Cons Int Explor Mer 144:56–60

Fogg GE (1966) The extracellular products of algae. Oceanogr Mar Biol Annu Rev 4:195–212

Fukami K, Meier B, Overbeck J (1991) Vertical and temporal changes in bacterial production and its consumption by heterotrophic nanoflagellates in a north German eutrophic lake. Arch Hydrobiol 122:129–145

Gliwicz ZM (1969) Studies on the feeding of pelagic zooplankton in lakes with varying trophy. Ekol Pol 36:663–708

Goldman JC, Caron DA (1985) Experimental studies on an omnivorous microflagellate: Implications for grazing and nutrient regeneration in the marine microbial food chain. Deep-Sea Res 32:899–915

Goldman JC, Caron DA, Dennett MR (1987) Nutrient cycling in a microflagellate food chain: IV. Phytoplankton–microflagellate interactions. Mar Ecol Prog Ser 38:75–87

Goldman JC, Dennett MR, Gordin H (1989) Dynamics of herbivorous grazing by the heterotrophic dinoflagellate *Oxyrrhis marina*. J Plankton Res 11:391–407

Gross EM, Wolk CP, Juttner F (1991) Fischerellin, a new allelochemical from the freshwater cyanobacterium *Fischerella muscicola*. J Phycol 27:686–692

Güde H (1985) Influence of phagotrophic processes on the regeneration of nutrients in two-stage continuous culture systems. Microb Ecol 11:193–204

Güde H (1989) The role of grazing on bacteria in plankton succession. In Sommer U (ed) *Plankton Ecology: Succession in Plankton Communities*. Springer-Verlag, Berlin, pp 337–364

Harris GP, Griffiths FB, Thomas DP (1989) Light and dark uptake and loss of ^{14}C: Methodological problems with productivity measurements in oceanic waters. Hydrobiologia 173:95–105

Hellebust JA (1974) Extracellular products. In Stewart WDP (ed) *Algal Physiology and Biochemistry*. Blackwell, Oxford, pp 838–863

Iturriaga R (1981) Phytoplankton photoassimilated extracellular products: Heterotrophic utilization in marine environment. Kiel Meeresforsch Sonderh 5:318–324

Johnson PW, Sieburth JM (1979) Chroococcoid cyanobacteria in the sea: A ubiquitous and diversive phototrophic biomass. Limnol Oceanogr 24:328–335

Johnson PW, Sieburth JM (1982) In situ morphology and occurrence of eucaryotic phototrophs of bacterial size in the picoplankton of estuarine and oceanic waters. J Phycol 18:318–327

Joint IR, Morris, JR (1982) The role of bacteria in the turnover of organic matter in the sea. Oceanogr Mar Biol Annu Rev 20:65–118

Joint IR, Pomroy AJ (1983) Production of picoplankton and small nanoplankton in the Celtic Sea. Mar Biol 77:19–27

Joint IR, Owens NJ, Pomroy AJ (1986) Seasonal production of photosynthetic picoplankton and nanoplankton in the Celtic Sea. Mar Ecol Prog Ser 28:251–258

Jørgensen NOG, Søndergaard M, Hansen HJ, Bosselman S, Riemann B (1983) Diel variation in concentration, assimilation and respiration of dissolved free amino acids in relation to planktonic primary and secondary production in two eutrophic lakes. Hydrobiology 107:107–122

Jüttner F (1981) Biological active compounds released during algal blooms. Verh Int Ver Limnol 21:227–230

Kato K, Stabel H-H (1984) Studies on the carbon flux from phyto- to bacterioplankton communities in Lake Constance. Arch Hydrobiol 102:177–192

Kornberg HL (1966) Anaplerotic sequences and their role in metabolism. In Campbell PN, Greville GD (eds) *Essays in Biochemistry*. Academic Press, New York, pp 1–31

Lampert W (1978) Release of dissolved organic carbon by grazing zooplankton. Limnol Oceanogr 23:831–834

Lampert W, Fleckner W, Rai H, Taylor BE (1986) Phytoplankton control by grazing zooplankton: A study on the spring clear-water phase. Limnol Oceanogr 31:478–490

Larsson U, Hagström A (1979) Phytoplankton exudate release as an energy source for growth of pelagic bacteria. Mar Biol 52:199–206

Lewis MR, Warnock RE, Platt T (1986) Photosynthetic response of marine picoplankton at low photon flux. In Platt T, Li WKW (eds) *Photosynthetic Picoplankton*. Can Bull Fish Aquat Sci 214:235–250

Lignell R (1990) Excretion of organic carbon by phytoplankton: Its relation to algal biomass, primary productivity and bacterial secondary productivity in the Baltic Sea. Mar Ecol Prog Ser 68:85–99

Meffert ME, Overbeck J (1979) Regulation of bacterial growth by algal release products. Arch Hydrobiol 87:118–121

Meier BG (1991) Untersuchungen zur "Microbial loop" im Plußsee. PhD Thesis, University of Kiel, Germany

Murphy TD, Lean DRS, Nalewajko C (1976) Blue-green algae: Their excretion of iron-selective chelators enables them to dominate other algae. Science 192:900–902

Nalewajko C (1977) Extracellular release in freshwater algae and bacteria: Extracellular products of algae as a source for heterotrophs. In Cairns J (ed) *Aquatic Microbial Communities*. Garland, New York, pp 589–629

Nalewajko C, Dunstall TG, Shear H (1976): Kinetics of extracellular release in axenic cultures and in mixed algal-bacterial cultures: Significance in estimation of total (gross) phytoplankton excretion rates. J Phycol 12:1–5

Nalewajko C, Lean DRS (1972) Growth and excretion in planktonic algae and bacteria. J Phycol 8:361–366

Overbeck J (1972) Bakterien in der Nahrungskette im See. Umschau 72:358

Overbeck J (1974) Microbiology and biochemistry. Mitt Int Ver Limnol 20:198–228

Overbeck J, Sako Y (1990) Ecological aspects of enzyme regulation in aquatic bacteria. Arch Hydrobiol Ergeb Limnol 34:81–92

Pedrós-Alió C (1989) Toward an autecology of bacterioplankton. In Sommer U (ed) *Plankton Ecology: Succession in Plankton Communities*. Springer-Verlag, New York, pp 297–336

Porter KG, Sherr EB, Sherr BF, Pace M, Sanders RW (1985) Protozoa in planktonic food webs. J Protozool 32:409–415

Porter KG (1988) Phagotrophic phytoflagellates in microbial food webs. Hydrobiologia 159:89–97

Rai H (1984) Magnitude of heterotrophic metabolism of photosynthetically fixed dissolved organic carbon (PDOC) in Schöhsee, West Germany. Arch Hydobiol 102:91–103

Riemann B, Søndergaard M (eds) (1986) *Carbon Dynamics in Eutrophic Temperate Lakes*. Elsevier Science Publishers BV, Amsterdam

Sanders RW, Porter KG, Bennet SJ, DeBiase AE (1989) Seasonal patterns of bacterivory by flagellates, ciliates, rotifers, and cladocerans in a freshwater planktonic community. Limnol Oceanogr 34:673–687.

Sanders RW, Porter KG, Caron DA (1990) Relationship between phototrophy and phagotrophy in the mixotrophic chrysophyte *Poterioochromonas malhamensis*. Microb Ecol 19:97–109

Schnepf E, Elbrächter M (1992) Nuritional strategies in dinoflagellates. A review with emphasis on cell biological aspects. Eur J Protistol 28:3–24

Sell AF, Overbeck J (1992) Exudates: Phytoplankton–bacterioplankton interactions in Plußsee. J Plankton Res 14:1199–1215

Shapiro LP, Guillard RRL (1986) Physiology and ecology of the marine eucaryotic ultraplankton. In Platt T, Li WKW (eds) *Photosynthetic Picoplankton*. Can Bull Fish Aquat Sci 214:371–389

Sharp JH (1977) Excretion of organic matter by marine phytoplankton: Do healthy cells do it? Limnol Oceanogr 22:381–399

Sherr BF, Sherr EB, Hopkinson CS (1988) Trophic interactions within pelagic microbial communities: Indications of feedback regulation of carbon flow. Hydrobiologia 159:19–26

Sommer U, Gliwicz ZM, Lampert W, Duncan A (1986) The PEG-model of seasonal succession of planktonic events in fresh waters. Arch Hydrobiol 106:433–471

Søndergaard M, Riemann B, Jørgensen NOG (1985) Extracellular organic carbon (EOC) released by phytoplankton and bacterial production. Oikos 45:323–332

Steemann Nielsen E (1952) The use of radioactive carbon (^{14}C) for measuring organic production in the sea. J Cons Int Explor Mer 18:117–140

Steemann Nielsen E (1977) The carbon-14 technique for measuring organic production by phytoplankton algae. A report on the present knowledge. Folia Limnol Scand 17:45–48

Sundh I (1992) Biochemical composition of dissolved organic carbon derived from phytoplankton and used by aquatic bacteria. Appl Environ Microbiol 58:2938–2947

Tranvik LJ, Porter KG, Sieburth JMcN (1989) Occurence of bacterivory in *Cryptomonas,* a common freshwater phytoplankter. Oecologia 78:473–476

Vadstein O, Harkjerr BO, Jensen A, Olsen Y, Reinertsen H (1989): Cycling of organic carbon in the photic zone of a eutrophic lake with special reference to the heterotrophic bacteria. Limnol Oceanogr 34:840–855

Van Donk E (1989) The role of fungal parasites in phytoplankton succession. In Sommer U (ed) *Plankton Ecology: Succession in Plankton Communities.* Springer-Verlag, New York, pp 171–194

Waterbury JB, Watson SW, Guillard RR, Brand LE (1979) Widespread occurrence of a unicellular, marine, planktonic, cyanobacterium. Nature 277:293–294

Wetzel RG (1983) *Limnology.* Saunders College Publishing, Philadelphia

Witzel K-P, Overbeck J, Moaledj K (1982) Microbial communities in Lake Plußsee—An analysis with numerical taxonomy of isolates. Arch Hydrobiol 94:38–52

Wood AM, Van Valen LM (1990) Paradox lost? On the release of energy-rich compounds of phytoplankton. Mar Microb Food Webs 4:103–116

Zlotnik I, Dubinsky Z (1989) The effect of light and temperature on DOC excretion by phytoplankton. Limnol Oceanogr 34:831–839

5. Bacterial Secondary Production

Ryszard J. Chróst and Hakumat Rai

Introduction

During the past decade ecologists have become more aware of the significant role that heterotrophic bacteria play in aquatic ecosystems. In both the water column and sediments of freshwater and marine systems, bacteria are thought to be the major decomposers of organic matter (Point and Morris 1982, Azam and Cho 1987, Münster and Chróst 1990, Chróst 1990; Chapter 6, this volume). Nowadays we accept that the production of bacterial biomass represents an important link among detritus, dissolved organic matter (DOM), and higher trophic levels in aquatic ecosystems. Important recent findings demonstrate that DOM → heterotrophic bacteria → protozoa → metazoa process as much energy as the classical grazing food chains. By virtue of their abundance, low substrate affinities, and potentially rapid growth rates, bacteria are capable of rapidly converting energetically low labile DOM into their biomass, i.e., high-quality bacterial particulate organic matter that can be utilized by bacterivorous protozoans and zooplankton (Pedrós-Alió and Brock 1983, Hessen and Andersen 1990, Pace et al. 1990). Moreover, bacterivores feeding on bacteria liberate a substantial amount of macromolecular DOM that is enzymatically degradable and utilizable by bacteria (Jumars et al. 1989, Nagata and Kirchman 1991, 1992; Chapter 6, this volume). Consequently, heterotrophic bacteria occupy a pivotal position in the aquatic trophic structure and act as the primary agents by which elements (e.g., carbon, nitrogen, phosphorus, sulfur) bound to organic matter are recycled or, in the absence of predation, act as the primary sink of organic matter within an ecosystem (Azam et al. 1983, Ducklow et al. 1986, Cho and Azam 1988, Simon et al. 1992). The recognition that heterotrophic bacteria mediate a major pathway for the flux of organic matter in natural waters has fundamentally altered the current concept of the structure and roles of aquatic food webs (Fenchel 1988).

Organic matter in general, and DOM in particular, constitute a basis for metabolism of heterotrophic bacteria. Bacteria are incapable of directly ingesting particulate organic matter (POM); they have no mechanism of phagocytosis or pinocytosis, and all organic matter must be soluble before it can be transported across the cell membrane. Not all DOM can, however, be readily utilized by bacteria. A variety of studies have demonstrated that the greatest part of the dissolved organic matter in aquatic environments consists of polymers, high-molecular-weight compounds (Ogura 1974, Allen 1976) that are too large to be transported into the cell. Only a very small fraction of DOM (<5%) is composed of low-molecular-weight substrates, which are molecules that are readily transportable by the bacterial active transport systems; therefore, they constitute readily utilizable DOM (UDOM) for bacteria (Allen 1978, Münster and Chróst 1990, Münster 1991; Chapter 2, this volume). The concentrations of UDOM in aquatic environments, however, are very low, often in the pico- to nanomolar range (Jørgensen 1987, Chróst et al. 1989, Münster 1991, Chróst and Rai 1993; Chapter 2, this volume), thereby limiting the growth and production of heterotrophic bacteria (Chróst et al. 1988, Gardner et al. 1986, 1989). Of course, bacteria do utilize polymers and POM, but these complex organic molecules must be hydrolyzed first by ectoenzymes to produce readily utilizable organic solutes (Chróst 1989, 1990, 1991a,b, 1992; Chapter 6, this volume). The UDOM can, therefore, be considered to be the preferred pool of substrates for aquatic bacteria since they can be utilized without any investment of energy for the production of ectoenzymes.

The transformation of organic substrates from the UDOM pool into bacterial biomass is the crucial step in organic matter processing by heterotrophic bacteria and its flux to higher trophic levels in aquatic ecosystem (Fig. 6.21, Chapter 6, this volume). Therefore, the interactions between bacteria and organic matter are of utmost importance for the functioning of lakes. These interactions are also important from another point of view, because the composition of organic matter together with other physicochemical conditions of the environment directly control the activity and structure of microbial communities (Witzel 1990). It has long been noted that nonliving organic matter plays an important role in the structure and function of aquatic ecosystems (Krogh and Lange 1932, Saunders 1977). The nonpredatory loss of organic matter was found to be essential for the function of the whole aquatic ecosystem. Over the past 25 years there was considerable interest, therefore, in estimating bacterial heterotrophic activity and their utilization of organic matter in natural waters (Wright and Hobbie 1965, Romanenko et al. 1972, Wetzel et al. 1972, Overbeck 1975, 1979a,b, 1981, Nalewajko 1977, Chróst 1978, 1981, 1983, Azam et al. 1983, Bell and Kuparinen 1984, Riemann and Søndergaard 1986). This remarkable collection of experimental data about microbial processing of organic matter in aquatic ecosystems was, to a large degree, the result of improvements in methods used for assessing bacterial activity and production. To quantify the flow of organic matter through an aquatic ecosystem, however, the studies require precise measurements of the biomass content of heterotrophic bacteria, their growth efficiency, and the rates of bacterial biomass production. In the last decade, several methods have been developed to measure bacterial biomass (Bratbak and Dundas 1984, Bratbak 1985, Lee and Fuhrman 1987), bacterial growth rates (Hagström et al. 1979, Karl 1981, Kirchman et al. 1982, Scavia et al. 1986), and bacterial secondary production (Fuhrman and Azam 1980, 1982, Moriarty and Pollard 1981, Kirchman et al. 1986, Chin-Leo and Kirchman 1988, Simon and Azam 1989, Kirchman 1992).

Of the various methods available to measure bacterial secondary production (BP), [^{3}H-methyl]thymidine ([^{3}H]TdR) incorporation into DNA [precipitated with cold trichloroacetic acid (TCA) as an insoluble material collected on membrane filters] has been the most widely used in aquatic studies (Fuhrman and Azam 1982). Although the [^{3}H]TdR approach was used for the first time in freshwaters in the 1970s (Chróst and Siuda 1978), and during the last decade has been tested extensively and improved, it still poses problems that must be resolved. One of the most serious problems is the need to calculate a conversion factor to translate [^{3}H]TdR incorporation into bacterial cell production (Riemann et al. 1987, Bell 1990). In spite of several limitations in the method (Chin-Leo and Kirchman 1988, Hollibaugh 1988, Jeffrey and Paul 1990), [^{3}H]TdR estimates of the rates of bacterial production have been instrumental in determinations of the importance of heterotrophic bacteria as biomass producers and as mineralizers of organic matter in a variety of aquatic ecosystems. The amount of bacterial biomass potentially available to grazers and thus to higher trophic levels can now be determined from rates of bacterial production (Barcina et al. 1992). Coupled with information about assimilation efficiency, production rates can also be used to estimate the total flux of organic matter through bacteria (Cole et al. 1988, Baines and Pace 1991). In addition, rates of bacterial production can be used to determine the average growth rate of bacterial assemblages (Chróst et al. 1988) and as an indicator of the response of bacterial communities to fluctuations in environmental conditions (Letarte and Pinel-Alloul 1991).

This chapter reports estimates of bacterial secondary production throughout the entire water column of the Plußsee determined by the incorporation of thymidine into bacterial DNA. We present the spatial and temporal variations of BP in lake water and discuss some environmental variables that influence the rates of production of bacterial biomass. Two independent methods for determining the rates of bacterial production (i.e., thymidine and leucine incorporation) were applied to estimate the bacterial organic carbon demand (BOCD) in the Plußsee. Finally, we compare the standing stock of algal and bacterial biomass and the rates of phytoplankton primary production and bacterial organic carbon demand to estimate the flux of photosynthetically produced organic matter from algae to bacteria in lake water.

Experimental Design

Water samples were collected with a 2.5-L Ruttner sampler from the entire water column (sampling site maximum depth 29 m) in the pelagic zone of the Plußsee. Spatial and temporal variations in bacterial production rates, bacterial abundance, chlorophyll$_a$, and chemistry were examined in water samples taken from depths of 0, 2, 4, 6, 8, 10, 15, 20, and 25 m and transported in plastic (polyethylene) containers to the laboratory within 1 h. Immediately after collection of the lake water, 50-ml subsamples were fixed with pH-neutral glutaraldehyde (3% final concentration) for counting of bacteria. Phytoplankton primary production and algal excreted organic carbon (EOC) were determined using samples taken at 1-, 2-, 5-, and 9-m depths.

Bacterial Secondary Production

Bacterial secondary production throughout the entire water column was determined by the [^{3}H]thymidine ([^{3}H]TdR) incorporation method (Fuhrman and Azam 1982), which was modified according to Chróst et al. (1988). Triplicate (10-ml) samples were dispensed to a series of glass tubes and supplemented with [^{3}H-methyl]thymidine (final concentration in assay 15 nmol/L; specific activity 84.9 Ci/mmol; New England Nuclear). Samples were incubated at the in situ temperature for 30 min and fixed with pH-neutral formalin (4% final concentration). The samples were then placed in an ice-water bath, and cold (0°C) 50% trichloroacetic acid (TCA) solution (Sigma) was added to a final concentration of 5% (wt/vol). After 20–30 min the TCA precipitate was collected on 0.2-μm pore-size membrane filters (cellulose nitrate; Sartorius) rinsed twice with 5 ml cold 5% TCA and once with 80% ice-cold ethanol and placed in scintillation vials. Filters were dissolved with 0.75 ml of Cellosolve (Sigma) and after 24 h 4.5 ml of scintillation cocktail (Filter-Count, Packard) was added to the vials.

In parallel incubations, we examined the incorporation rates of [^{3}H]TdR and [^{3}H]leucine (Chin-Leo and Kirchman 1988) to determine bacterial production in water samples taken from the euphotic zone (0–2 m depth). [^{3}H]Leucine ([^{3}H]Leu; specific activity 120 Ci/mmol; Amersham) was added to triplicate 10-ml samples ([^{3}H]Leu final concentration 20 nmol/L) and incubated at the in situ temperature for 60 min and then fixed with formalin. Incubated samples were supplemented with 50% TCA (final concentration 10%) and held in a water bath for 30 min at 80°C to hydrolyze all macromolecules except proteins. Following extraction, samples were filtered through 0.2-μm pore-size filters (Sartorius) and then processed as described above for [^{3}H]TdR incorporation. Triplicate 10-ml samples fixed with formalin (10 min before adding [^{3}H]TdR or [^{3}H]Leu) were incubated and treated as blanks of nonbiological adsorption of radioisotopes.

Rates of [^{3}H]TdR and [^{3}H]Leu incorporation were converted to bacterial cell production using the conversion factors $0.65–2.85 \times 10^{18}$ cells/mol [^{3}H]TdR and $0.021–0.185 \times 10^{18}$ cells/mol [^{3}H]Leu, respectively, and they were determined separately on each sampling day. Bacterial cell production was transformed to production of bacterial organic carbon using the conversion factor 19.8 fg C/cell (Lee and Fuhrman 1987).

Concentration of [^{3}H]TdR and [^{3}H]Leu for Measurements of Bacterial Production

To determine appropriate concentrations of the radiolabel that saturates thymidine or leucine uptake incorporation systems of bacterial assemblages in lake water, incorporation was measured for [^{3}H]TdR and [^{3}H]Leu additions in the range from 1 to 30 and 1 to 50 nmol/L (final concentration), respectively. Incorporation was measured as described above.

Conversion Factors for [^{3}H]TdR and [^{3}H]Leu Incorporation

To determine the factors necessary to convert the rates of [^{3}H]TdR and [^{3}H]Leu incorporation to rates of bacterial cell production we used the dilution-growth scheme (Kirchman et al. 1982, Scavia et al. 1986, Chróst et al. 1988). Conversion factor experiments for [^{3}H]TdR incorporation were performed using integrated water samples from the euphotic zone (0–2 m depth) and the profundal zone (10–25 m depth) of the lake. [^{3}H]Leu conversion factors were determined in integrated water samples (0–2 m depth) collected from the euphotic zone only. Samples were gravity-filtered through 1.2-µm pore-size polycarbonate filters (90 mm diameter, Nuclepore) and diluted 1:2 with water filtered through 0.2-µm pore-size filters (50 mm diameter, Nuclepore). Triplicate (1.25-L) diluted samples were incubated in sterile 2-L Erlenmeyer flasks on a rotary shaker in the dark at in situ temperature for 24–30 h. The flasks were subsampled every 5–6 h to measure changes in bacterial abundance and the incorporation of [^{3}H]TdR and [^{3}H]Leu. Changes in bacterial abundance and incorporation of [^{3}H]TdR or [^{3}H]Leu were calculated for each sampling interval. The conversion factors were determined by comparing incorporation of radiolabels with simultaneous increases in bacterial cell numbers according to Kirchman et al. (1982) assuming exponential growth of bacteria in diluted water samples.

Bacterial Abundance, Biovolume, and Biomass

Epifluorescence microscopy was used to determine bacterial cell numbers and biovolume. Glutaraldehyde-preserved and 4,6-diamidino-2-phenylindole (DAPI)-stained samples were counted for bacterial abundance according to Porter and Feig (1980). Epifluorescence photomicrography was used to examine cell size (Lee and Fuhrman 1987). Photographs (Kodak Ektachrome, 400 ASA) were taken from five randomly selected fields of each microscope slide, photographic images were projected onto a screen (final magnification 10,000), and 250–300 cells per photographic slide were measured to determine bacterial cell size and to calculate biovolume. Biomass of bacteria in water samples was calculated using a biovolume-to-biomass conversion factor, 350 fg C/µm^3, assuming mean cell volume of the Plußsee bacteria to be 0.06 µm^3 (Lee and Fuhrman 1987).

Other Analyses

Phytoplankton primary production, excretion of organic carbon (EOC), and bacterial utilization of EOC were determined by the [^{14}C]bicarbonate method (Sell and Overbeck 1992; Chapter 4, this volume). Inorganic phosphorus and nitrogen in samples filtered through precombusted glass fiber filters (GF/F Whatman) were determined according to Strickland and Parsons (1972). Chlorophyll$_a$ was extracted with 90% acetone and measured by spectrophotometry (Golterman and Clymo 1969). Phytoplankton biomass was calculated from chlorophyll$_a$ content using an organic carbon-to-chlorophyll$_a$ ratio of 47 (Cho and Azam 1990). Dissolved organic carbon (DOC) in GF/F-filtered (Whatman) samples was analyzed with the use of an organic carbon computational system (Tocomaster model 915B, Beckman). Radioactivity was assayed by liquid scintillation counting in a Beckman Computational System LS-1801. The statistical relationships (multiple regression analysis and ANOVA) between studied variables were analyzed using the IBM PC software program STATISTIX (Analytical Software, USA).

Incorporation of [^{3}H]Thymidine and [^{3}H]Leucine as a Measure of Bacterial Secondary Production

Recent reviews have emphasized that bacterial secondary production in aquatic ecosystems as determined by several different methods is roughly comparable (Cole et al. 1988, White at al. 1991, Ducklow and Carlson 1992). Most of the studies reported estimates of bacterial production based on the [^{3}H]TdR incorporation method. More recently, several reports presented BP data determined by means of [^{3}H]Leu incorporation (Kirchman et al. 1985, Chin-Leo and Kirchman 1988, Kirchman 1992). The [^{3}H]Leu incorporation method, which is generally similar to the [^{3}H]TdR approach, requires addition of an appropriate exogenous concentration of the radiolabel to inhibit de novo synthesis of leucine in bacteria and to minimize [^{3}H]Leu dilution by external and internal pools of leucine in the water sample (Simon and Azam 1989).

Concentration of Added Radiolabel

To test whether additions of [^{3}H]TdR (15 nM) and [^{3}H]Leu (20 nM) were appropriate, we measured incorporation rates of [^{3}H]TdR and [^{3}H]Leu at different concentrations ranging from 1 to 30 nM and from 1 to 50 nM for thymidine and leucine, respectively. The rates of incorporation of [^{3}H]TdR in the TCA-insoluble macromolecular fraction were constant when the concentration of thymidine was 15 nM or higher (Fig. 5.1A). [^{3}H]Leu incorporation rates for concentrations ranging between 1 and 15 nM were significantly lower than that measured for 20 nM (Fig. 5.1B), and higher concentrations of leucine did not affect [^{3}H]Leu incorporation rates. These results indicate that isotope dilution could lead to underestimating BP if the concentrations of radiolabels are lower than 15 and 20 nM for [^{3}H]TdR and [^{3}H]Leu, respectively.

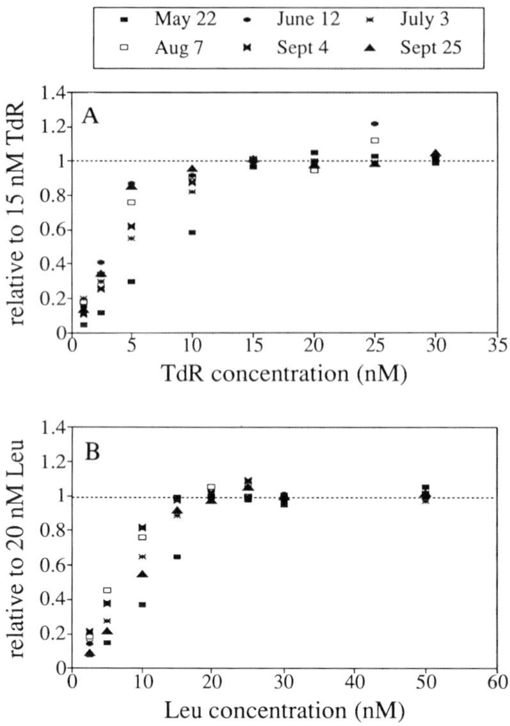

Figure 5.1. Incorporation rates at different added concentrations of radiolabeled [^{3}H]TdR (A) and [^{3}H]Leu (B), normalized for the rates at 15 nM TdR or 20 nM Leu. For example, the rate at 5 nM was divided by the rate at 15 nM TdR or 20 nM Leu.

[^{3}H]TdR and [^{3}H]Leu Conversion Factors

Conversion factors for [^{3}H]TdR and [^{3}H]Leu were estimated by comparing incorporation of radiolabels with simultaneous increases in bacterial cell numbers after minimizing grazing in 1.2-μm prefiltered and 1:2 diluted lake water. Filtration and dilution procedures decreased bacterial abundance by about 25–35% and eliminated 80–90% of bacterivores present in untreated lake water. We applied the approach of Kirchman et al. (1982) to determine conversion factors for [^{3}H]TdR and [^{3}H]Leu incorporation. They derived a relationship between the changes in [^{3}H]TdR incorporation (V_{TdR}) and bacterial cell number (*BN*) that takes advantage of the exponential rate of change of both variables during growth of the diluted bacterial populations. If plots of natural logarithms of *BN* and V_{TdR} ln(*BN*) and ln(V_{TdR}) versus incubation time are linear, then the slopes determined from regression of those variables on time yield two estimates of growth (μ_{BN} and μ_{TdR}). The intercept of the TdR-incorporation regression (V_0) can be combined with an estimate of the growth rate (μ_{BN} or μ_{TdR}) and the intercept of the bacterial abundance regression (BN_0) to yield the conversion factor (*CF;* cells produced per picomole of TdR incorporated):

$$CF = \mu \times BN_0 \times V_0^{-1} \tag{1}$$

Actually, any estimates of growth rate and simultaneous measure of [^{3}H]TdR or [^{3}H]Leu incorporation and *BN* can be used in Equation 1 if the above regressions are significant and $\mu_{BN} = \mu_{\text{TdR}} = \mu_{\text{Leu}}$. Natural population growth rates (μ) can then be calculated from:

$$\mu = CF \times V \tag{2}$$

where V is the rate of [^{3}H]TdR or [^{3}H]Leu incorporation in freshly collected, unaltered lake water.

Tests for the presence of nondividing cells were accomplished through nonlinear regression of BN on time (t) using the model:

$$BN = P_{\text{nd}} + BN_0 \exp(\mu t) \tag{3}$$

where P_{nd} is the population of nondividing cells and BN_0 is now a population of dividing cells. Results for the nonlinear regression were compared with those of the linear regression of ln (BN) versus time, and the existence of a sizable population of nondividing cells was discounted when the nonlinear model did not improve the explained variance.

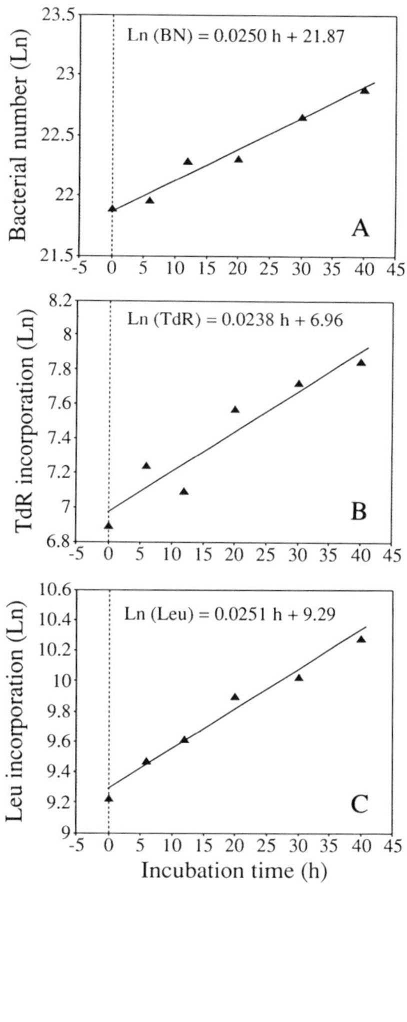

Figure 5.2. Example of linear regression plots of natural logarithms (Ln) of changes in (A) bacterial number, (B) [^{3}H]thymidine (TdR), and (C) [^{3}H]leucine (Leu) incorporation during incubation of diluted water samples. Integrated samples (0–2 m depth) from the pelagial zone of the Plußsee were taken on September 4, 1989. Growth rates estimated from changes in bacterial cell number and incorporation of thymidine and leucine were: $\mu_{\text{bn}} = 0.0250\ \text{h}^{-1}$, $\mu_{\text{TdR}} = 0.0238\ \text{h}^{-1}$, and $\mu_{\text{Leu}} = 0.0251\ \text{h}^{-1}$, respectively. Conversion factors estimated from Equation 2 were: 2.1×10^6 cells/pmol for [^{3}H]TdR and 0.185×10^6 cells/pmol for [^{3}H]Leu.

An example of the determination of the conversion factor for [^{3}H]TdR and [^{3}H]Leu described above is shown in Figure 5.2. The slopes of linear regressions of natural logarithms of three independent estimates, ln(*BN*), ln(TdR), and ln(Leu), versus incubation time were almost identical. Growth rates of diluted bacterial populations determined from incorporation experiments varied between 0.0238 h^{-1} and 0.0251 h^{-1} for [^{3}H]TdR and [^{3}H]Leu, respectively (Fig. 5.2B and C). Direct observation of the change in the bacterial cell number showed their exponential growth in diluted water samples and revealed a growth rate of 0.0250 h^{-1} (Fig. 5.2A). Conversion factors estimated from this experiment were 2.1×10^6 and 0.185×10^6 cells/pmol of incorporated [^{3}H]TdR and [^{3}H]Leu, respectively.

Conversion factors for [^{3}H]TdR and [^{3}H]Leu incorporation in the euphotic zone of the Plußsee were determined simultaneously in integrated (0–2 m depth) water samples, whereas conversion factor experiments for [^{3}H]TdR incorporation in the profundal zone of the lake were performed in integrated samples taken from depths of 10, 15, 20, and 25 m. The [^{3}H]TdR incorporation method was routinely

Table 5.1. Conversion Factors for [^{3}H]TdR and [^{3}H]Leu Incorporation and Growth Rates and Doubling Time of Natural Bacterial Assemblages in the Euphotic Zone in the Plußsee

Sampling Date, 1989	Conversion Factor (10^6 cells/pmol)		Growth Rate[a] (h^{-1})	Doubling Time[a,b] (h)
	[^{3}H]TdR	[^{3}H]Leu		
May 22	1.28	0.0729	0.014	49.5
May 29	1.43	0.0915	0.012	57.8
June 5	0.95	0.0567	0.016	43.3
June 12	1.42	0.1152	0.018	38.5
June 19	1.62	0.1345	0.095	7.3
June 26	1.09	0.0690	0.054	12.8
July 3	1.05	0.0375	0.105	6.6
July 10	1.95	0.1350	0.108	6.4
July 17	0.65	0.0210	0.092	7.5
July 31	1.92	0.1053	0.088	7.9
August 7	1.35	0.0523	0.065	10.7
August 21	1.62	0.1048	0.052	13.3
September 4	2.10	0.1850	0.024	28.9
September 11	1.76	0.0963	0.026	26.6
September 18	1.96	0.1500	0.030	23.1
September 26	1.85	0.1186	0.018	38.5
Mean	1.50	0.0966	0.051	23.7
SD	±0.41	±0.0423	±0.034	±16.7
n	16	16	16	16

[a]Average growth rate and doubling time, estimated from [^{3}H]TdR and [^{3}H]Leu incorporation rates multiplied by conversion factors (Equation 2).

[b]Doubling time was calculated by dividing ln(2) by average growth rate.

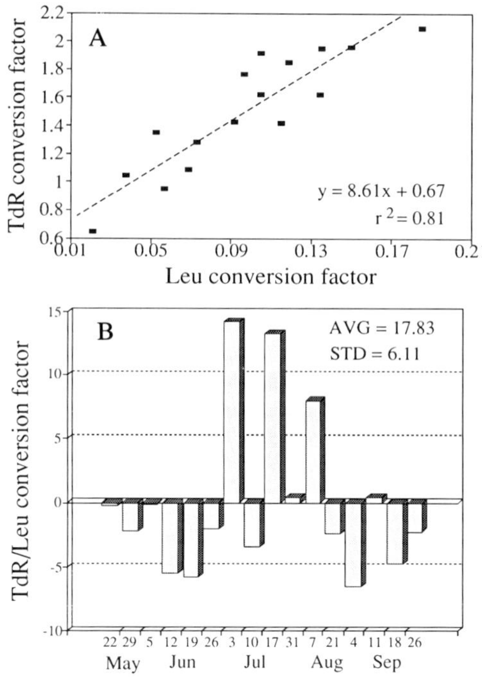

Figure 5.3. (A) Correlation between thymidine (TdR) and leucine (Leu) conversion factors and (B) deviation of ratios of TdR to Leu conversion factors from average ratio (AVG) during the study period in the Plußsee. Conversion factors unit is 10^6 cells/pmol.

used for bacterial production measurements throughout the entire water column of the Plußsee during the study period. Incorporation of [^{3}H]Leu was measured additionally with [^{3}H]TdR only in the upper layer of the photic zone during primary production experiments because we wanted to have precise BP measurements based on two independent estimates. Table 5.1 summarizes [^{3}H]TdR and [^{3}H]Leu conversion factors for BP estimates and growth rates and doubling time of bacterial assemblages in the water from the euphotic zone of the Plußsee during the study period. The [^{3}H]TdR conversion factors ranged from 0.65 to 2.1 × 10^6 cells/pmol (average 1.5 ± 0.4 × 10^6) and correlated positively with those determined for [^{3}H]Leu incorporation, which varied between 0.021 and 0.185 × 10^6 (average 0.0966 ± 0.042 × 10^6) cells/pmol (Fig. 5.3A). The ratio of TdR to Leu conversion factors varied from 11.35 to 32 (average 17.83 ± 6.11; Fig. 5.3B). Kirchman (1992) found a slightly lower average (16) *CF* ratio. Growth rate and doubling time of bacteria in the Plußsee euphotic zone varied between 0.012 and 0.108 (average 0.051 ± 0.034) h^{-1} and from 6.4 to 57.8 (average 23.7 ± 16.7) h, respectively. High growth rates and short doubling times of bacterial assemblages were found from the middle of June until the end of August (Table 5.1). Bacteria in the euphotic zone of the Plußsee during summer (water temperature ranged between 18.5 and 21°C) grew and divided faster than those in Lake Arlington (Chrzanowski and Hubbard 1988). Analogous growth rates and doubling times to water from the Plußsee were found in Lake Michigan (Scavia et al. 1986) and in the Gulf of Calvi (Velimirov and Walenta-Simon 1992) at similar or higher water temperatures.

Table 5.2. Conversion Factors for [^{3}H]TdR Incorporation, Growth Rates, and Doubling Time of Natural Bacterial Assemblages in the Profundal Zone in the Plußsee

Sampling Date, 1989	Conversion Factor (10^6 cells/pmol)	Growth Rate[a] (h^{-1})	Doubling Time[a,b] (h)
April 22	1.58	0.004	173.3
May 8	1.65	0.011	63.0
May 22	1.89	0.009	77.0
May 29	1.38	0.008	86.6
June 5	1.25	0.013	53.3
June 19	0.96	0.006	115.5
July 3	2.12	0.015	46.2
July 17	1.85	0.017	40.8
July 31	2.85	0.016	43.3
August 21	1.95	0.022	31.5
September 4	1.65	0.018	38.5
September 11	1.48	0.008	86.6
October 2	2.2	0.006	115.5
Mean	1.75	0.011	74.7
SD	±0.46	±0.005	±39.3
n	13	13	13

[a]Growth rate and doubling time, estimated from [^{3}H]TdR incorporation rates, multiplied by conversion factor (Equation 2).
[b]Doubling time was calculated by dividing ln(2) by average growth rate.

Bacterial production in the profundal zone of the Plußsee was calculated using [^{3}H]TdR conversion factors derived from integrated water samples (10–25 m depths) and growth-dilution experiments (Table 5.2). Average *CF* ($1.75 \pm 0.46 \times 10^6$ cells/pmol) for [^{3}H]TdR incorporation in profundal lake water was not significantly higher than that estimated in the euphotic zone of the lake and varied between 0.96 and 2.85×10^6 cells/pmol. The growth rates and doubling times, however, were notably different for profundal bacteria in comparison to those living in the euphotic zone. Bacterial growth rates varied from 0.004 to 0.022 h^{-1} (average 0.011 ± 0.005 h^{-1}), which corresponds to doubling times that range between 31 and 173 (average 75 ± 39) h (Table 5.2). Much longer doubling times resulted at low water temperature (4–5°C) and perhaps were caused by distinctly lower concentration and availability of DOC for bacteria in the profundal lake water throughout the study period (Lovell and Konopka 1985a, Chróst et al. 1988, Robarts et al. 1991). Our observation that bacterial growth rates are very dependent on habitat temperature agrees well with the model for marine and freshwater systems tested by White et al. (1991). They found a strong relationship between growth rate and temperature ($r^2 = 0.71$, $P \leq 0.001$) that was also expressed as a Q_{10} value of 3.3 for all tested aquatic environments.

Figure 5.4. (A) Comparison of bacterial production estimated from [^{3}H]thymidine (TdR) and [^{3}H]leucine (Leu) incorporation rates, and (B) correlation between these two estimates of bacterial production in the euphotic zone of the Plußsee (0–2 m depth). Average bacterial production is arithmetic mean between values obtained from TdR and Leu incorporation measured simultaneously.

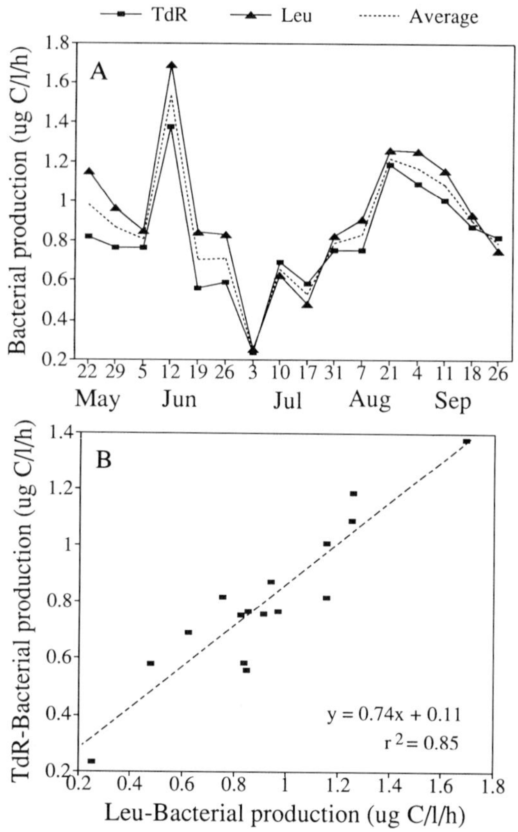

Rates of bacterial secondary production estimated by [^{3}H]Leu incorporation were on average 15% higher than BP simultaneously determined by the [^{3}H]TdR method. The BP measured by both methods displayed parallel patterns of change in the Plußsee water throughout the study period (Fig. 5.4A): BP determined by [^{3}H]TdR correlated significantly (r^2 = 0.85, $P \leq 0.005$) with estimates based on [^{3}H]Leu incorporation (Fig. 5.4B).

Our results indicate that [^{3}H]TdR and [^{3}H]Leu incorporation give equivalent estimates of bacterial secondary production in lake water. Small discrepancies between rates of BP simultaneously measured by these two methods (Fig. 5.4A) can indicate that in some samples (e.g., 22 May) growth of bacterial populations was unbalanced, and bacterial protein synthesis did follow DNA synthesis. Another possible explanation of observed differences in [^{3}H]TdR and [^{3}H]Leu incorporation rates may be an effect of isotope dilution by external and internal pools of TdR or Leu that strongly influences values of conversion factors (Chróst et al. 1988, Bell 1990, Simon 1991). Kirchman (1992) suggested that the average bacterial production calculated from these two methods can be more accurate than examining [^{3}H]TdR-based or [^{3}H]Leu-based BP estimates separately.

Bacterial Secondary Production in the Plußsee

There has been an explosion of research on aquatic microbial food web processes in the past decade. The introduction of convenient methods for the estimation of microbial biomass and activities in natural waters strongly stimulated this research. Moreover, many of these methods (taken from molecular biology and biochemistry and adapted to natural waters) made these studies possible (Overbeck and Chróst 1990). Discovery that bacteria produce biomass that can be equal to or exceed algal biomass, and that they constitute the most active fraction of planktonic organisms, brought a new attraction to the studies of microbial processing of organic matter in aquatic ecosystems.

One of the most fundamental problems facing microbial ecologists studying the Plußsee ecosystem has been the difficulty in specifying the identity and fluxes of organic matter to bacterial communities (Chapters 2 and 8, this volume). Several reports have been published on microbial uptake and metabolism of individual simple compounds such as amino acids and monosaccharides determined by heterotrophic uptake assays of ^{14}C- and ^{3}H-labeled substrates by microorganisms in the Plußsee (Overbeck 1965, 1974, 1975, 1979a, Stabel et al. 1979, Moaledj and Overbeck 1980, Chróst and Overbeck 1989, 1990, Chróst 1989, Chróst et al. 1989, Münster and Chróst 1990; Chapter 8, this volume). The main drawback of these studies is that they reveal the dynamics of only the target compounds, not the complete array of substrates actually being used at any given moment. One attraction of using various methods for estimating bacterial production is that the resulting data provide an integration of the total utilization of various substrates, even as they change in time and space. Heterotrophic uptake of organic solutes provides selective exact information on bacterial metabolism of individual, defined compounds. Therefore, it is possible to make only a very rough estimate of flux of organic matter to bacteria using substrate uptake data alone. Bacterial production assays give a better estimate of the output resulting from metabolism of undefined substrates that have been used by heterotrophic bacteria in an aquatic ecosystem (Ducklow and Carlson 1992). Heterotrophic uptake of substrates and BP are complementary approaches, and both are required to allow budgeting of the carbon flow, but better knowledge of the sources supporting BP is still needed.

Bacterial Production in the Entire Water Column

First attempts in assessing bacterial production in the Plußsee water column (0–10 m in depth) were made in 1968–1971 using the $^{14}CO_2$ dark uptake method (Overbeck 1972). This study also reported a significant correlation (r^2 = 0.75) between phytoplankton primary production and $^{14}CO_2$ dark uptake by bacteria. Here we present results of BP estimates for the entire Plußsee water column (sampled biweekly) throughout the study period, 22 April to 2 October 1989. [^{3}H]TdR incorporation assay was routinely applied for this study.

Bacterial secondary production in the water column varied from 0.037 to 1.693 (average 0.401 ± 0.431) μg C L^{-1} h^{-1} throughout the study period. There were two

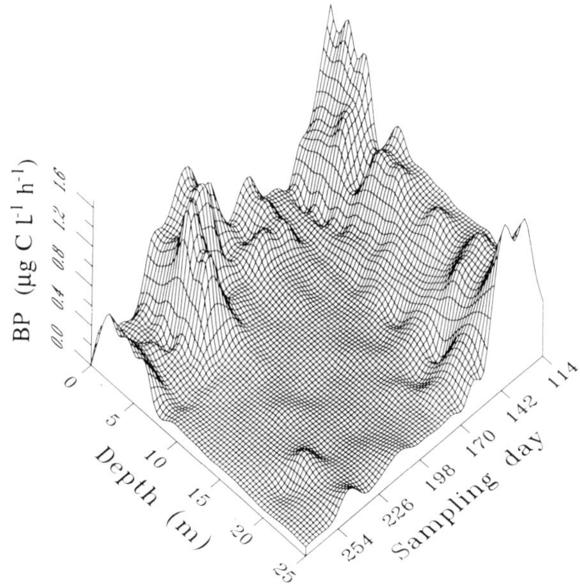

Figure 5.5. Bacterial production (BP) throughout the entire water column of the Plußsee. Number of sampling day is the number of the day on an annual scale.

different time periods and regions of high BP in the lake water (Fig. 5.5). Bacteria were most productive in the upper water layer between 0 and 4 m depth (0.087–1.674 µg C L^{-1} h^{-1}). The BP displayed the highest rates in early spring (1.483–1.674 µg C L^{-1} h^{-1}, April) and during the summer period (0.633–1.617 µg C L^{-1} h^{-1}, mid-July until mid-September). Generally, high values of BP rates corresponded well with enhanced rates of algal photosynthesis in the Plußsee (Chapters 3 and 4, this volume) during these periods. Average bacterial production (0.787 ± 0.467 µg C L^{-1} h^{-1}) in the euphotic zone was 2.5 times higher than BP determined in the profundal zone (0.037–1.693, average 0.309 ± 0.374 µg C L^{-1} h^{-1}) of the Plußsee. The BP was high and almost uniformly distributed (0.696–1.055 µg C L^{-1} h^{-1}) in the profundal zone during spring homothermy (April) of the lake (Fig. 5.5).

There was also a distinct seasonal distribution in bacterial cell abundance (Fig. 5.6). Bacterial cell numbers throughout the entire water column varied from 3.7 to 32.6 (average 12.4 ± 7.9) × 10^6 cells/ml during the study period. Bacterial biomass in lake water followed patterns of change in cell abundance and fluctuated between 85.7 and 745.2 (average 278.8 ± 157.7) µg C/L. Similar to BP, high bacterial cell numbers and their biomass were observed in spring. In April and May bacteria were the most abundant (21.1–33.2 × 10^6 cells/ml) in the lake; there was no marked variation in bacterial cell numbers and biomass (410.8–725.3 µg C/L) within the water column. Usually bacterial abundance in the epilimnion and the metalimnion was higher than in hypolimnetic water and varied between 4.2 and 19.3 × 10^6 cells/ml during stratification of lake water. At the end of August, however, bacterial abundance in the hypolimnion (6.6–10.6 × 10^6 cells/ml) was

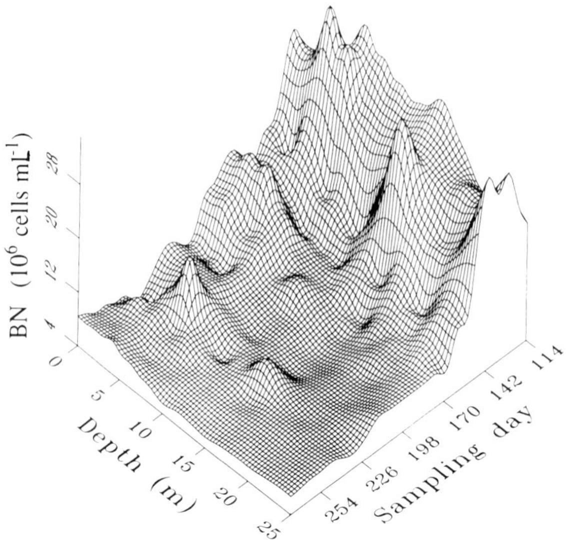

Figure 5.6. Number of bacterial cells (BN) in the entire water column of the Plußsee. Number of sampling day is the number of the day on an annual scale.

higher than in the upper layer (3.7–4.2 × 10^6 cells/ml) of the Plußsee. Bacterial abundance often reached its maxiumum in the metalimnion of the lake.

Bacterial biomass (BB) in the photic zone (0–4 m depth) varied between 131 and 593 (average 267.8 ± 135.8) μg C/L (Fig. 5.7) and from end of May until June 19, during the clear-water phase of the lake, bacterial biomass exceeded phytoplankton biomass (PB). Ratio of BB to PB varied from 0.043 to 1.580 (average 0.572 ± 0.539) during the study period (Fig. 5.7). This means that bacterial biomass constituted on average 29.5% (range 4.2–61.2%) of the total (bacteria plus algae) microplankton biomass in the euphotic zone of the Plußsee. Our results on bacterial and algal biomass in the Plußsee are very consistent with the previous (1983–1984) estimates of Plußsee microplankton biomass based on ATP measurements (Chróst 1988).

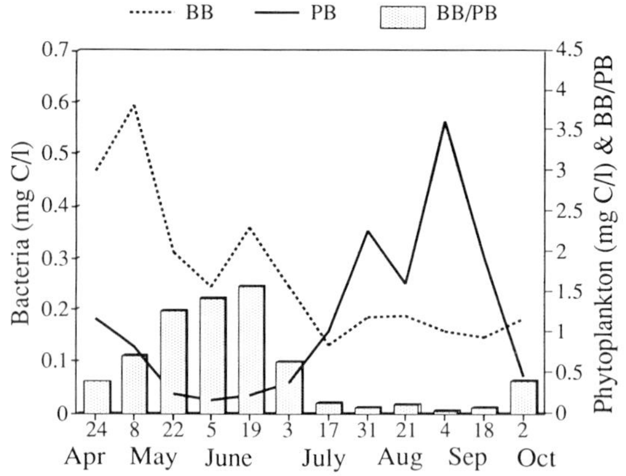

Figure 5.7. Biomass of bacteria (BB) and phytoplankton (PB) and their ratio in the euphotic zone (0–4 m) of the Plußsee during the study period.

The seasonal changes of bacterial production, bacterial abundance and their biomass found in the Plußsee are within the range of values published for freshwater lakes (Bell et al. 1983, Riemann 1983, Bell and Kuparinen 1984, Lovell and Konopka 1985b, McDonough et al. 1986, Edwards and Meyer 1986, Riemann and Søndergaard 1986, Scavia et al. 1986, Simon 1987, Simon and Tilzer 1987, Chrzanowski and Hubbard 1988, Cole et al. 1988). Our results showed that bacteria produce weighty biomass in the Plußsee water. This suggests that bacterial biomass production may represent an important link between detrital dissolved organic matter and higher trophic levels (Chapters 11 and 12, this volume).

Carbon Flux from Phytoplankton Production to Bacteria

Bacterial production and phytoplankton primary production follow similar patterns in many aquatic environments (Pedrós-Alió and Brock 1982, Scavia et al. 1986, Cole et al. 1988, Ducklow and Carlson 1992). This relationship between algal and bacterial production indicates that a significant fraction of photosynthetic products is fixed to bacterial biomass production and metabolism (Fig. 5.8).

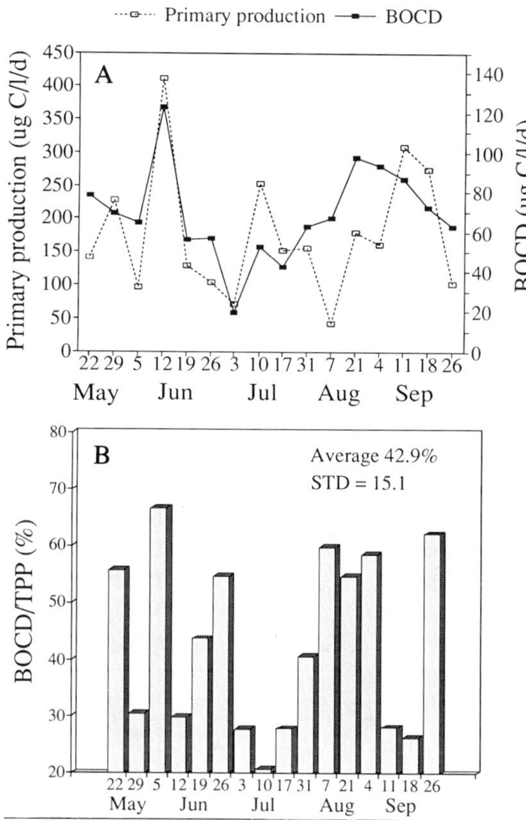

Figure 5.8. (A) Phytoplankton primary production and bacterial organic carbon demand (BOCD), and (B) percentage of carbon flux from total primary production (TPP) to bacteria in water at depth of 0–2 m in the Plußsee.

It is tempting to calculate the carbon flux from phytoplankton primary production to bacterial production and organic carbon demand.

In the upper layer (0–2 m) of the Plußsee water column total primary production (sum of particulate fraction and dissolved EOC) of phytoplankton averaged 175.3 ± 94.1 (range 41.8–419.6) µg C L^{-1} d^{-1}. Mean bacterial organic carbon demand estimated from simultaneous measurements of [^{3}H]TdR and [^{3}H]Leu incorporation (assuming 30% growth efficiency plus respiration) varied between 19.5 and 122.5 (average 69.1 ± 23.1) µg C L^{-1} d^{-1} (Fig. 5.8A). This means that from 20.8% to 66.7% (average 42.9 ± 15.1%) of phytoplankton daily production flows through bacteria in the euphotic zone of the Plußsee (Fig. 5.8B). Recent reviews have emphasized that estimates of bacterial production and conservative growth efficiencies imply that bacteria can consume a large, albeit variable (<10% to >100%), fraction of phytoplankton primary production (Cole et al. 1988, Ducklow and Carlson 1992). When we consider the entire water column, the carbon flux to bacterial production is even more significant, averaging 30% of primary production on an areal basis. Again, considering both the growth efficiency and respiration terms, for an average water column more than 60% of primary production would be expected to flow through bacteria (Cole et al. 1988).

What Regulates Bacterial Biomass and Production?

With many studies of microbial food web processes in the 1980s, microbial ecologists have begun to try to answer one of the fundamental question of ecology, "What controls bacterial populations in aquatic ecosystems?" Recently, several new concepts of population regulation have been introduced into plankton research. There are two opposing concepts regarding control mechanisms of bacterial populations: (1) "bottom-up" or "resource-limiting," i.e., controlled by the availability of resources required for maintenance and growth, and (2) "top-down" or "predator-limiting," i.e., quantitative changes in bacterial populations regulated by predation. Recent theoretical work has refined the debate regarding the relative importance of these opposing regulatory mechanisms in lacustrine plankton systems (Carpenter et al. 1985). Current theory states that the maximum or potential biomass of a given population is determined by "bottom-up" control (resource availability) but that the actual or realized biomass at any given time is set by predation ("top-down" control). The strengths of these controls are dependent on trophic level (McQueen et al. 1986). That is, bottom-up controls are strongest and most predictable near the base of food webs, whereas near the top of the food webs, the strength and predictability of bottom-up control by prey availability is replaced by top-down controls arising from predator pressure (Ducklow and Carlson 1992).

Generally, bacterial production in the Plußsee water column was positively correlated to the standing-stock biomass of bacteria. The correlation coefficient for the entire data set of linear regression between BP and bacterial biomass (BB) was statistically significant (Table 5.3). Contrary to a very strong relationship between

Table 5.3. Summary of Linear Regression Correlation Coefficients for Bacterial Production (BP) and Bacterial Biomass (BB) in Plußsee

Data Set, Pair Numbers (n)	Correlated Variables	Correlation Coefficient (r^2)	$P \leq$ Value
Water column			
$n = 108$	BP ↔ BB	0.40	0.0005
$n = 108$	BP ↔ DOC[a]	0.11	NS[b]
$n = 108$	BP ↔ AMPase[c]	0.41	0.0005
Euphotic zone			
$n = 36$	BP ↔ BB	0.22	NS
$n = 36$	BP ↔ DOC	0.42	0.005
$n = 14$	BP ↔ PP[d]	0.36	0.05
$n = 14$	BP ↔ EOC[e]	0.48	0.05
$n = 14$	BP ↔ EOC uptake	0.57	0.025
$n = 36$	BP ↔ AMPase	0.30	0.05
Profundal zone			
$n = 72$	BP ↔ BB	0.85	0.0001
$n = 72$	BP ↔ DOC	0.15	NS
$n = 72$	BP ↔ AMPase	0.76	0.0005

[a]Dissolved organic carbon.
[b]Nonsignificant.
[c]Aminopeptidase activity.
[d]Phytoplankton primary production.
[e]Excreted organic carbon by algae.

BP and BB found in the profundal zone (Table 5.3), there was statistically nonsignificant correlation between bacterial production and biomass in the euphotic zone of the Plußsee. We suggest that this lack of correlation between BP and BB in the euphotic zone of the Plußsee indicates very efficient removal of bacterial biomass by bacterivorous heterotrophic nanoflagellates (HNF), protozoans, and/or zooplankton in the course of bacterial production (Peterson et al. 1978, Sherr and Sherr 1987, McManus and Fuhrman 1988; Chapters 11 and 12, this volume). It is now commonly accepted that HNF are among the most important factors controlling bacterial biomass in natural waters (Azam et al. 1983, Kuuppo-Leinikki 1990, Wikner et al. 1990, Barcina et al. 1992). Moreover, our assumption is reasonable because direct proof of very efficient and rapid removal of bacteria by HNF in the Plußsee water column has been demonstrated (Fukami et al. 1991; Chapter 12, this volume). Estimates of HNF grazing rates on bacteria (using fluorescent-labeled cells) varied from 2.3 to 15.1 (average 8.4) bacteria HNF^{-1} h^{-1}, and total bacterial consumption rates by HNF fluctuated between 1.1 and 9.5 (average 5.0) $\times 10^7$ cells L^{-1} h^{-1}, which accounted for elimination of, on average, 81% of BP from the water in the upper layer (2-m depth) of the lake.

What regulates bacterial biomass production is still a matter of considerable speculation. The microbial literature contains the notion that it is controlled by, or directly related to, the supply of decomposable organic matter, but little quantitative data exist to support this opinion (Cole et al. 1988, Chróst 1989, 1990, 1991b, 1992). In the euphotic zone of the Plußsee, bacterial production correlated significantly with dissolved organic carbon (DOC). Phytoplankton are the main producers of DOC in the Plußsee, and they supply a majority of the substrates to bacteria (Chróst et al. 1989, Münster and Chróst 1990). There was a statistically significant, but not very strong, correlation between BP and phytoplankton primary production and total excreted organic carbon (EOC; Table 5.3). A close relationship was found between BP and bacterial uptake of EOC in the surface water of the Plußsee.

If we assume that our estimates of BP are unbiased and apply average bacterial growth efficiency to EOC, 30% (Cole et al. 1982, Chróst and Faust 1983), then the range of bacterial carbon demand (BOCD), i.e., range of carbon supplies required to balance bacterial secondary production, calculated for the euphotic zone of the Plußsee fluctuates between 0.81 and 5.10 (average 2.87 ± 0.96) μg C L^{-1} h^{-1} during May–September. Comparing EOC uptake and BOCD rates (Fig. 5.9A), we find that bacterial uptake of excreted organic carbon by algae fulfilled on average 71.5

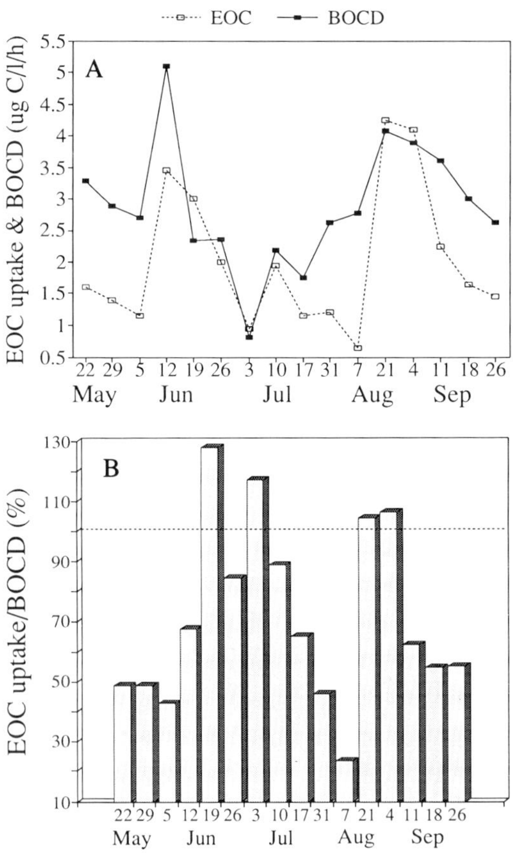

Figure 5.9. (A) Uptake of excreted organic carbon (EOC) by phytoplankton and bacterial organic carbon demand (BOCD), and (B) percentage of EOC uptake in total bacterial organic carbon demand for water at depths of 0–2 m in the Plußsee.

± 29.1% (range 23.4–128.1%) of BOCD (Fig. 5.9B). This means that only in some circumstances did EOC uptake not fully supply the bacteria with organic carbon. The excretion of dissolved organic substances by phytoplankton has long been recognized as an important source of high-quality substrates to bacteria (Nalewajko 1977, Chróst 1978, Nalewajko et al. 1980, Brock and Clyne 1984). It is well known that bacteria can very rapidly utilize low-molecular-weight EOC (Chróst 1981, Chróst and Faust 1983, Siuda et al. 1991, Sundh 1992). Recently Baines and Pace (1991) reviewed the patterns of production of dissolved organic matter by phytoplankton and its importance to bacterial production across marine and freshwater ecosystems. Testing the model they found that in most pelagic ecosystems EOC amounted to less than half of the organic carbon required for bacterial production.

Bacterial production throughout the entire water column was dependent on aminopeptidase activity, the model ectoenzyme studied (AMPase; Table 5.3). A very strong correlation between BP and AMPase was found in the profundal zone. In the euphotic zone of the Plußsee the correlation was not as strong which indicates that polymeric proteinaceous compounds were enzymatically depolymerized and then utilized in the course of bacterial production (Chapter 6, this volume). On the other hand, a statistically nonsignificant correlation between BP and DOC in the profundal zone suggests that DOC was mostly composed of refractory substances or high-molecular-weight enzymatically degradable compounds.

The correlations described above indicate that the potential control mechanisms of bacterial production and biomass in the Plußsee involve both types of regulation, i.e., the top-down predatory effect (HNF, protozoans, and zooplankton) and bottom-up regulation by availability of organic substrates to bacteria. Our studies of bacterial production did not include lysogenic activity of phages against bacteria in water from the Plußsee, possibly another mechanism of very efficient removal of biomass of bacterial host populations (Bergh et al. 1989, Bratbak et al. 1990, 1992, Heldal and Bratbak 1991, Proctor and Fuhrman 1992; Chapter 13, this volume).

Concluding Remarks

Our studies of bacterial secondary production clearly demonstrate that bacteria are very important components of the Plußsee pelagic food web and that bacteria produce much of the biomass in lake water. Bacterial biomass in the euphotic zone on average accounted for 268 μg C/L and, during the clear-water phase of lake, exceeded phytoplankton biomass. Biomass of bacteria constituted on average 29.5% of the total microplankton biomass in the euphotic zone of the Plußsee. Bacterial secondary production throughout the entire water column (0–25 m depth) varied from 63.8 to 590.4 (average 228) mg C m^{-2} d^{-1}. From 20.8% to 66.7% (average 42.9%) of phytoplankton daily production fluxes through bacteria in the euphotic zone of the Plußsee. Bacterial biomass production represents an important link between detrital dissolved organic matter and higher trophic levels

(Chapters 11 and 12, this volume). Bacterial biomass production in the Plußsee seems to be under top-down/bottom-up control mechanisms. Heterotrophic nanoflagellates, ciliates, and zooplankton remove a significant part of the bacterial biomass. On the other hand, the rates of bacterial production are regulated by the amount of excreted organic carbon from phytoplankton (in the euphotic zone) or by the flux of directly utilizable substrates from the pool of polymeric organic matter (the profundal zone) after its enzymatic degradation.

Some reflections on aquatic food-web structure suggest that bacterial biomass will be dependent not only on the levels immediately above (HNF, protozoans) and below (availability of organic substrates) but on more distant trophic levels as well (e.g., primary producers, zooplankton). This gives rise to the concept of the "trophic cascade" (Carpenter et al. 1985, see review by Ducklow and Carlson 1992), which suggests that we need to view the bacteria in the context of the plankton community and consider the entire food-web structure, not just the immediate substrates or bacterivores acting directly on the bacterial populations.

Studies of bacterial secondary production address fundamental questions in ecology regarding transfer of energy and nutrients within a microbial loop and throughout communities of the whole aquatic ecosystem. In recent years, research focused on the organisms of the microbial loop has resulted in a significant collection of data on microbial trophodynamics within pelagic food webs. From this mass of information, a revised concept of pelagic food-web structure is beginning to emerge. In this new concept, the microbial loop should now be considered as an integral component part of a larger microbial food web, which includes all pro- and eukaryotic unicellular organisms, both autotrophic and heterotrophic (Sherr and Sherr 1988). From this standpoint, the entire microbial food web, not only phytoplankton, supports the metazoan food web. Thus, the "link or sink" controversial debate initiated by Ducklow et al. (1986) about the relative importance of the microbial loop as a source of food for metazoan food webs or as a sink for fixed carbon via high bacterial respiratory losses is no longer an issue. Even though most of the primary production may be respired within the microbial food web, the microbial food web is at the same time the ultimate food resource for zooplankton (Sher and Sherr 1988; Chapters 11 and 12, this volume).

Acknowledgments. We thank A. Sell, M.Sc., for providing data of studies of primary production and excretion of organic carbon in the Plußsee.

References

Allen HL (1976) Dissolved organic matter in lake water: Characteristics of molecular weight size-fractions and ecological implications. Oikos 27:64–70

Allen HL (1978) Low molecular weight dissolved organic matter in five softwater ecosystems: A preliminary study and ecological implications. Verh Int Ver Limnol 20:514–524

Azam F, Cho BC (1987) Bacterial utilization of organic matter in the sea. In Fletcher M, Gray TRG, Jones JG (eds) *Ecology of Microbial Communities.* Cambridge University Press, Cambridge, pp 261–281

Azam F, Fenchel T, Field JG, Gray JS, Meyer-Reil LA, Thingstand F (1983) The ecological role of water column microbes in the sea. Mar Ecol Prog Ser 10:257–263

Baines SB, Pace ML (1991) The production of dissolved organic matter by phytoplankton and its importance to bacteria: Patterns across marine and freshwater systems. Limnol Oceanogr 36:1078–1090

Barcina I, Ayo B, Unanue M, Egea L, Iriberri J (1992) Comparison of rates of flagellate bacterivory and bacterial production in a marine coastal system. Appl Environ Microbiol 58:3850–3856

Bell RT (1990) An explanation for the variability in the conversion factor deriving bacterial cell production from incorporation of [^{3}H]thymidine. Limnol Oceanogr 35:910–915

Bell RT, Kuparinen J (1984) Assessing phytoplankton and bacterioplankton production during early spring in Lake Erken, Sweden. Appl Environ Microbiol 48:1221–1230

Bell RT, Ahlgren GM, Ahlgren I (1983) Estimating bacterioplankton production by measuring [^{3}H]thymidine incorporation in a eutrophic Swedish lake. Appl Environ Microbiol 45:1709–1721

Bergh Ø, Børsheim KY, Bratbak G, Heldal M (1989) High abundance of viruses found in aquatic environments. Nature 340:467–468

Bratbak G (1985) Bacterial biovolume and biomass estimations. Appl Environ Microbiol 49:1488–1493

Bratbak G, Dundas I (1984) Bacterial dry matter content and biomass estimations. Appl Environ Microbiol 48:755–757

Bratbak G, Heldal M, Norland S, Thingstad TF (1990) Viruses as partners in spring bloom microbial trophodynamics. Appl Environ Microbiol 56:1400–1405

Bratbak G, Heldal M, Thingstad TF, Riemann B, Haslund OH (1992) Incorporation of viruses into the budget of microbial C-transfer. A first approach. Mar Ecol Prog Ser 83:273–280

Brock TD, Clyne J (1984) Significance of algal excretory products for growth of epilimnetic bacteria. Appl Environ Microbiol 47:731–734

Carpenter SR, Kitchell JF, Hodgson JR (1985) Cascading trophic interactions and lake productivity. BioScience 35:634–639

Chin-Leo G, Kirchman DL (1988) Estimating bacterial production in marine waters from simultaneous incorporation of thymidine and leucine. Appl Environ Microbiol 54:1934–1939

Cho BC, Azam F (1988) Major role of bacteria in biogeochemical fluxes in the ocean's interior. Nature 332:441–443

Cho BC, Azam F (1990) Biogeochemical significance of bacterial biomass in the ocean's euphotic zone. Mar Ecol Prog Ser 63:253–259

Chróst RJ (1978) The estimation of extracellular release and heterotrophic activity of aquatic bacteria. Acta Microbiol Pol 27:139–146

Chróst RJ (1981) The composition and bacterial utilization of DOC released by phytoplankton. Kiel Meeresforsch Sonderh 5:325–332

Chróst RJ (1983) Plankton photosynthesis, extracellular release and bacterial utilization of released dissolved organic carbon (RDOC) in lakes of different trophy. Acta Microbiol Pol 32:275–287

Chróst RJ (1988) Phosphorus and microplankton development in a eutrophic lake. Acta Microbiol Pol 37:205–225

Chróst RJ (1989) Characterization and significance of β-glucosidase in lake water. Limnol Oceanogr 34:660–672

Chróst RJ (1990) Microbial ectoenzymes in aquatic environments. In Overbeck J, Chróst RJ (eds) *Aquatic Microbial Ecology: Biochemical and Molecular Approaches.* Springer-Verlag, New York, pp 47–78

Chróst RJ (ed) (1991a) *Microbial Enzymes in Aquatic Environments.* Springer-Verlag, New York

Chróst RJ (1991b) Ectoenzymes in aquatic environments: Microbial strategy for substrate supply. Verh Int Ver Limnol 24:2597–2600

Chróst RJ (1992) Significance of bacterial ectoenzymes in aquatic environments. Hydrobiologia 243/244:61–70

Chróst RJ, Faust MA (1983) Organic carbon release by phytoplankton: Its composition and utilization by bacterioplankton. J Plankton Res 5:477–493

Chróst RJ, Overbeck J (1989) Application of the isotope dilution principle to the determination of [^{14}C]glucose incorporation by aquatic bacteria. Acta Microbiol Pol 38:75–89

Chróst RJ, Overbeck J (1990) Substrate–ectoenzyme interaction: Significance of β-glucosidase activity for glucose metabolism by aquatic bacteria. Arch Hydrobiol Ergeb Limnol 34:93–98

Chróst RJ, Rai H (1993) Ectoenzyme activity and bacterial secondary production in nutrient-impoverished and -enriched freshwater mesocosms. Microb Ecol 25:131–150

Chróst RJ, Siuda W (1978) Some factors affecting the heterotrophic activity of bacteria in a lake. Acta Microbiol Pol 27:129–138.

Chróst RJ, Overbeck J, Wcisło R (1988) Evaluation of the [^{3}H]thymidine method for estimating bacterial growth rates and production in lake water: Re-examination and methodological comments. Acta Microbiol Pol 37:95–112

Chróst RJ, Münster U, Rai H, Albrecht D, Witzel KP, Overbeck J (1989) Photosynthetic production and exoenzymatic degradation of organic matter in euphotic zone of an eutrophic lake. J Plankton Res 11:223–242

Chrzanowski TH, Hubbard JG (1988) Primary and bacterial secondary production in a Southwestern reservoir. Appl Environ Microbiol 54:661–669

Cole JJ, Likens GE, Strayer DL (1982) Photosynthetically produced dissolved organic carbon: An important carbon source for planktonic bacteria. Limnol Oceanogr 27:1080–1090

Cole JJ, Findlay S, Pace ML (1988) Bacterial production in fresh and saltwater ecosystems: A cross-system overview. Mar Ecol Prog Ser 43:1–10

Ducklow HW, Carlson CA (1992) Oceanic bacterial production. Adv Microb Ecol 12:113–181

Ducklow HW, Purdie DA, Williams PJL, Davies JM (1986) Bacterioplankton: A sink for carbon in a coastal marine plankton community. Science 232:865–867

Edwards RT, Meyer JL (1986) Production and turnover of planktonic bacteria in two Southwestern blackwater rivers. Appl Environ Microbiol 52:1317–1323

Fenchel T (1988) Marine plankton food chains. Annu Rev Ecol Syst 19:19–38

Fuhrman JA, Azam F (1980) Bacterioplankton secondary production estimates for coastal waters of British Columbia, Antarctica, and California. Appl Environ Microbiol 39:1085–1095

Fuhrman JA, Azam F (1982) Thymidine incorporation as a measure of heterotrophic bacterioplankton production in marine surface waters: Evaluation and field results. Mar Biol 66:109–120

Fukami K, Meier B, Overbeck J (1991) Vertical and temporal changes in bacterial production and its consumption by heterotrophic nanoflagellates in a north German eutrophic lake. Arch Hydrobiol 122:129–145

Gardner W, Chandler JF, Laird GA, Scavia D (1986) Microbial response to amino acids additions in Lake Michigan: Grazer control and substrate limitation of bacterial populations. J Great Lakes Res 12:161–174

Gardner W, Chandler JF, Laird GA (1989) Organic nitrogen mineralization and substrate limitation of bacteria in Lake Michigan. Limnol Oceanogr 34:478–485

Golterman HL, Clymo RS (1969) *Methods for Chemical Analysis of Freshwaters. IBP Handbook No 8*. Blackwell Science Publishers, Oxford

Hagström A, Larsson U, Horstedt P, Normark S (1979) Frequency of dividing cells, a new approach to the determination of bacterial growth rates in aquatic environments. Appl Environ Microbiol 37:805–812

Heldal M, Bratbak G (1991) Production and decay of viruses in aquatic environments. Mar Ecol Prog Ser 72:205–212

Hessen DO, Andersen T (1990) Bacteria as a source of phosphorus for zooplankton. Hydrobiologia 206:217–223

Hollibaugh JT (1988) Limitations of the [^{3}H]thymidine method for estimating bacterial productivity due to thymidine metabolism. Mar Ecol Prog Ser 43:19–30

Jeffrey WH, Paul JH (1990) Thymidine uptake, thymidine incorporation, and thymidine kinase activity in marine bacterium isolates. Appl Environ Microbiol 56:1367–1372

Jørgensen NOG (1987) Free amino acids in lakes: Concentrations and assimilation rates in relation to phytoplankton and bacterial production. Limnol Oceanogr 32:97–111

Jumars PA, Penry DL, Baross JA, Perry MJ, Frost BW (1989) Closing the microbial loop: Dissolved carbon pathway to heterotrophic bacteria from incomplete ingestion, digestion and absorption in animals. Deep Sea Res 36:483–495

Karl DM (1981) Simultaneous rates of ribonucleic acid and deoxyribonucleic acid syntheses for estimating growth and cell division of aquatic microbial communities. Appl Environ Microbiol 42:802–810

Kirchman DL (1992) Incorporation of thymidine and leucine in the subarctic Pacific: Application to estimating bacterial production. Mar Ecol Prog Ser 82:301–309

Kirchman DL, Ducklow H, Mitchell R (1982) Estimates of bacterial growth from changes in uptake rates and biomass. Appl Environ Microbiol 44:1296–1307

Kirchman DL, K'Nees E, Hodson RE (1985) Leucine incorporation and its potential as a measure of protein synthesis by bacteria in natural aquatic systems. Appl Environ Microbiol 49:599–607

Kirchman DL, Newell SY, Hodson RE (1986) Incorporation versus biosynthesis of leucine: Implications for measuring rates of protein synthesis and biomass production by bacteria in marine systems. Mar Ecol Prog Ser 32:47–59

Krogh AA, Lange E (1932) Quantitative Untersuchungen über Plankton, Kolloide und gelöste organische und anorganische Substanzen in dem Furesee. Int Rev Ges Hydrobiol 26:20–53

Kuuppo-Leinikki P (1990) Protozoan grazing on planktonic bacteria and its impact on bacterial population. Mar Ecol Prog Ser 63:227–238

Lee S, Fuhrman JA (1987) Relationships between biovolume and biomass of naturally derived marine bacterioplankton. Appl Environ Microbiol 53:1298–1303

Letarte Y, Pinel-Alloul B (1991) Relationships between bacterioplankton production and limnological variables: Necessity of bacterial size considerations. Limnol Oceanogr 36:1208–1216

Lovell CR, Konopka A (1985a) The effects of temperature on bacterial production in a dimictic eutrophic lake. FEMS Microbiol Lett 31:135–140

Lovell CR, Konopka A (1985b) Primary and bacterial production in two dimictic Indiana lakes. Appl Environ Microbiol 49:485–491

McDonough RJ, Sanders RW, Porter KG, Kirchman DL (1986) Depth distribution of bacterial production in a stratified lake with an anoxic hypolimnion. Appl Environ Microbiol 52:992–1000

McManus GB, Fuhrman J (1988) Control of marine bacterioplankton populations: Measurement and significance of grazing. Hydrobiologia 159:51–62

McQueen DJ, Post JP, Mills EL (1986) Trophic relationships in freshwater pelagic ecosystems. Can J Fish Aquat Sci 43:1571–1581

Moaledj K, Overbeck J (1980) Studies on uptake kinetics of oligocarbophilic bacteria. Arch Hydrobiol 89:303–312

Moriarty DJW, Pollard PC (1981) DNA synthesis as a measure of bacterial productivity in seagrass sediments. Mar Ecol Prog Ser 5:151–156

Münster U (1991) Extracellular enzyme activity in eutrophic and polyhumic lakes. In Chróst RJ (ed) *Microbial Enzymes in Aquatic Environments*. Springer-Verlag, New York, pp 96–122

Münster U, Chróst RJ (1990) Origin, composition, and microbial utilization of dissolved organic matter. In Overbeck J, Chróst RJ (eds) *Aquatic Microbial Ecology: Biochemical and Molecular Approaches*. Springer-Verlag, New York, pp 8–46

Nagata T, Kirchman DL (1991) Release of dissolved free and combined amino acids by bacterivorous marine flagellates. Limnol Oceanogr 36:433–443

Nagata T, Kirchman DL (1992) Release of macromolecular organic complexes by heterotrophic marine flagellates. Mar Ecol Prog Ser 83:233–240

Nalewajko C (1977) Extracellular release in freshwater algae and bacteria: Extracellular products of algae as a source of carbon for heterotrophs. In Cairns J (ed) *Aquatic Microbial Communities*. Garland, New York, pp 589–626

Nalewajko C, Lee K, Fay P (1980) Significance of algal extracellular products to bacteria in lakes and in cultures. Microb Ecol 6:199–207

Ogura N (1974) Molecular weight fractionation of dissolved organic matter in coastal seawater by ultrafiltration. Mar Biol 24:305–312

Overbeck J (1965) Primärproduktion und Gewässerbakterien. Naturwissenschaften 6:145–146

Overbeck J (1972) Experimentelle Untersuchungen zur Bestimmung der bakteriellen Produktion in Seen. Verh Int Verein Limnol 18:176–187

Overbeck J (1974) Microbiology and biochemistry. Mitt Int Ver Limnol 20:198–228

Overbeck J (1975) Distribution pattern of uptake kinetic responses in a stratified eutrophic lake (Plußsee ecosystem study). Verh Int Ver Limnol 19:2600–2615

Overbeck J (1979a) Studies on heterotrophic functions and glucose metabolism of microplankton in Plußsee. Arch Hydrobiol Beih Ergeb Limnol 13:56–76

Overbeck J (1979b) Dark CO_2 uptake: Biochemical background and its relevance to in situ bacterial production. Arch Hydrobiol Beih Ergeb Limnol 12:38–47

Overbeck J (1981) A new approach for estimating the overall heterotrophic activity in aquatic ecosystems. Verh Int Ver Limnol 21:1355–1358

Overbeck J, Chróst RJ (eds) (1990) *Aquatic Microbial Ecology: Biochemical and Molecular Approaches*. Springer-Verlag, New York

Pace ML, McManaus GB, Findlay SEG (1990) Planktonic community structure determines the fate of bacterial production in a temperate lake. Limnol Oceanogr 35:795–808

Pedrós-Alió C, Brock TD (1982) Assessing biomass and production of bacteria in eutrophic Lake Mendota, Wisconsin. Appl Environ Microbiol 44:203–218

Pedrós-Alió C, Brock TD (1983) The impact of zooplankton feeding on the epilimnetic bacteria of a eutrophic lake. Freshwat Biol 13:227–239

Peterson BJ, Hobbie JE, Hancy JF (1978) *Daphnia* grazing on natural bacteria. Limnol Oceanogr 23:1039–1044

Point IR, Morris RJ (1982) The role of bacteria in the turnover of organic matter in the sea. Oceanogr Mar Biol Annu Rev 20:65–118

Porter KG, Feig YS (1980) The use of DAPI for identifying and counting aquatic microflora. Limnol Oceanogr 25:943–948

Proctor LM, Fuhrman JA (1992) Mortality of marine bacteria in response to enrichments of the virus size fraction from seawater. Mar Ecol Prog Ser 87:283–293

Riemann B (1983) Biomass and production of phyto- and bacterioplankton in eutrophic Lake Tystrup, Denmark. Freshwat Biol 13:389–398

Riemann B, Søndergaard M (eds) (1986) *Carbon Dynamics in Eutrophic, Temperate Lakes*. Elsevier, Amsterdam

Riemann B, Bjørnsen PK, Newell S, Fallon R (1987) Calculation of cell production of coastal marine bacteria based on measured incorporation of [^{3}H]thymidine. Limnol Oceanogr 32:471–476

Robarts RD, Sephton LM, Wicks RJ (1991) Labile dissolved organic carbon and water temperature as regulators of heterotrophic bacterial activity and production in the lakes of Sub-Antarctic Marion Island. Polar Biol 11:403–413

Romanenko VI, Overbeck J, Sorokin YI (1972) Estimation of production of heterotrophic bacteria using ^{14}C. In Sorokin YI, Kadota H (eds) *Techniques for the Assessment of Microbial Production and Decomposition in Freshwater. IBP Handbook 23*. Blackwell Science Publishers, Oxford, pp 82–85

Saunders GW (1977) Carbon flow in the aquatic systems. In Cairn J (ed) *Aquatic Microbial Communities*. Garland, New York, pp 417–440

Scavia D, Laird GA, Fahnenstiel GL (1986) Production of planktonic bacteria in Lake Michigan. Limnol Oceanogr 31:612–626

Sell AF, Overbeck J (1992) Exudates: Phytoplankton–bacterioplankton interactions in Plußsee. J Plankton Res 14:1199–1215

Sherr EB, Sherr BF (1987) High rates of consumption of bacteria by pelagic ciliates. Nature 325:710–711

Sherr E, Sherr B (1988) Role of microbes in pelagic food webs: A revised concept. Limnol Oceanogr 33:1225–1227

Simon M (1987) Biomass and production of small and large free-living and attached bacteria in Lake Constance. Limnol Oceanogr 32:591–607

Simon M (1991) Isotope dilution of intracellular amino acids as a tracer of carbon and nitrogen sources of marine planktonic bacteria. Mar Ecol Prog Ser 74:295–301

Simon M, Azam F (1989) Protein content and protein synthesis rates of planktonic marine bacteria. Mar Ecol Prog Ser 51:201–213

Simon M, Tilzer MM (1987) Bacterial response to seasonal changes in primary production and phytoplankton biomass in Lake Constance. J Plankton Res 9:535–552

Simon M, Cho BC, Azam F (1992) Significance of bacterial biomass in lakes and the ocean: Comparison to phytoplankton biomass and biogeochemical implications. Mar Ecol Prog Ser 86:103–110

Siuda W, Wcisło R, Chróst RJ (1991) Composition and bacterial utilization of photosynthetically produced organic matter in an eutrophic lake. Arch Hydrobiol 121:473–484

Stabel HH, Moaledj K, Overbeck J (1979) On the degradation of dissolved organic molecules from Plußsee by oligocarbophilic bacteria. Arch Hydrobiol Beih Ergeb Limnol 12:95–104

Strickland JDH, Parsons TR (1972) *A Practical Handbook of Seawater Analysis*. Fish Research Board of Canada, Ottawa

Sundh I (1992) Biochemical composition of dissolved organic carbon derived from phytoplankton and used by heterotrophic bacteria. Appl Environ Microbiol 58:2938–2947

Velimirov B, Walenta-Simon M (1992) Seasonal changes in specific growth rates, production and biomass of a bacterial community in the water column above a Mediterranean seagrass system. Mar Ecol Prog Ser 80:237–248

Wetzel RG, Rich PH, Miller MC, Allen HL (1972) Metabolism of dissolved and particulate detrital carbon in temperate, hard-water lake. Mem Ist Ital Idrobiol Suppl 29:185–243

White PA, Kalff J, Rasmussen JB, Gasol JM (1991) The effect of temperature and algal biomass on bacterial production and specific growth rate in freshwater and marine habitats. Microb Ecol 21:99–118

Wikner J, Rassoulzadegan F, Hagström Å (1990) Periodic bacterivore activity balances growth in the marine environment. Limnol Oceanogr 35:313–324

Witzel KP (1990) Approaches to bacterial population dynamics. In Overbeck J, Chróst RJ (eds) *Aquatic Microbial Ecology: Biochemical and Molecular Approaches*. Springer-Verlag, New York, pp 96–128

Wright RT, Hobbie JE (1965) The uptake of organic solutes in lake water. Limnol Oceanogr 10:22–28

6. Microbial Enzymatic Degradation and Utilization of Organic Matter

Ryszard J. Chróst

Introduction

The most significant contribution of microbial ecology, which altered the principles of a current concept in aquatic science, was the recognition that heterotrophic microorganisms play a dominant role in the cycling of organic and inorganic matter and that a large fraction of primary production is not consumed directly by herbivorous consumers but is transformed into bacterial biomass (Lovell and Konopka 1985, Chróst and Rai 1993; Chapter 5, this volume) and then channeled to phagotrophic microorganisms (Sanders and Porter 1988). The above studies led to the idea of the "microbial loop" (Azam et al. 1983; Chapter 12, this volume) and to the discovery that aquatic food chains include a higher number of trophic levels than hitherto believed (Fenchel 1987, Pomeroy and Wiebe 1988). Therefore, the view of the role of microorganisms in aquatic ecosystems has changed significantly in the last decade. In some respects, the change is sufficiently dramatic to make some authors speak about a "change in paradigm" (Williams 1981). Nowadays we recognize that heterotrophic bacteria not only are the major users of organic carbon but also form microbial food webs that transfer energy more efficiently through the aquatic ecosystems than do classical food chains.

Organic matter in general, and dissolved organic matter (DOM) in particular, constitutes a basis for metabolism of microheterotrophs whose various representatives are present at every level of aquatic food webs. Therefore, the interactions between bacteria and organic matter are of utmost importance for the functioning of lakes. These interactions are also important from another point of view, because the composition of organic matter together with other physicochemical conditions of the environment directly controls the activity and structure of microbial communities. DOM is the main fraction of the total pool of organic material existing in aquatic environments, and a minor part consists of particulate organic matter (Münster and Chróst 1990). The predominance of DOM in aquatic ecosystems is of great importance for heterotrophic bacteria because they are capable of assimilating solutes only. Therefore, the DOM of natural waters can be regarded as the principal trophic basis of microbial aquatic food webs. The next higher trophic level is occupied by the heterotrophic microorganisms dependent on DOM. There is now good evidence that in aquatic ecosystems the bacteria primarily occupy this trophic level. The flow of organic matter in lakes is dominantd by its photosynthetic production and subsequent transformation and decomposition by heterotrophic organisms (Chróst et al. 1989). Heterotrophic microorganisms (bacteria and fungi) control the movement of major part of organic matter because they are the only biological populations capable of significantly altering both DOM and POM (Chróst 1990, Hoppe et al. 1993). Moreover, bacterial catabolism of organic matter, which is also tightly coupled to its oxidation, liberates inorganic nutrients fixed in organic compounds and CO_2, thus making them available again for bacteria and photosynthetic micro- and macroorganisms in lacustrine ecosystems (Chróst 1990, 1992).

Organic matter in lakes obviously consists of innumerable organic compounds that may serve as microbial substrates (Chapters 2 and 4, this volume). Many studies have demonstrated that bacterial growth and production positively correlate with the concentration of DOM in lakes. Some recent studies suggest that bacteria are more dependent on the composition of substrates in the DOM pool than on the ambient concentration (Chróst 1986, Chróst and Rai 1993). DOM in lakes is a mixture of organic compounds and comprises both low- and high-molecular-weight solutes. Among them, organic acids, carbohydrates, amino acids, proteins, nucleic acids, lipids, phosphoric esters, and humic substances composed of polyphenolic subunits predominant in lake waters (Steinberg and Münster 1985, Münster and Chróst, 1990; Chapter 2, this volume). Only a small fraction (0.5% to 5%) of the total DOM in lake waters constitutes readily utilizable DOM (UDOM) for aquatic bacteria. The concentrations of UDOM in lacustrine environments are extremely low, often in the pico- to nanomolar range (Jørgensen and Søndergaard 1984, Chróst et al. 1989, Münster and Chróst 1990; Chapter 2, this volume), thereby limiting the growth and production of heterotrophic bacteria (Chróst et al. 1988, Gardner et al. 1986). A major fraction of DOM consists of polymeric compounds that cannot be directly transported to the cell because of their high molecular weight and the large size of the molecules (Rogers 1961).

The apparent "bacterial paradox" (Chróst 1991a) between the low concentration of the UDOM and high biomass and production of aquatic bacteria indicates that these microheterotrophs efficiently utilize polymeric organic compounds. However, in order to make polymeric substrates transferable across the microbial cell membranes and then available for metabolism, they must be transformed to smaller, low-molecular-weight subunits. The stepwise depolymerization of polymeric organic matter in lakes is mediated mainly by heterotrophic bacteria and fungi, and their hydrolytic enzyme activities are of the utmost importance in this process (Chróst 1990, 1991b, Bergbauer and Newell 1992, Gajewski et al. 1993). Biopolymers such as proteins, polysaccharides (cellulose, chitin, starch), and organophosphoric esters (nucleic acids, phospholipids), which dominant in DOM pool of nonpolluted aquatic environments, are very important nutritional sources for heterotrophic bacteria (Chróst 1991b).

A variety of aquatic microorganisms synthesize hydrolytic enzymes and among them heterotrophic bacteria are excellent producers of these catalysts (Chróst 1990). Hydrolysis of polymers is an acknowledged rate-limiting step in the microbial utilization of DOM and POM and inorganic nutrient regeneration in aquatic environments (Hoppe 1986, Hałemejko and Chróst 1986, Chróst et al. 1986a, 1989, Chróst, 1989, 1992, Meyer-Reil and Köster 1992). Most of the microbial enzymes responsible for degradation of polymeric organic matter are inducible catabolic enzymes that are associated with the microbial cell surface or located in the periplasmic space of gram-negative bacteria (Martinez and Azam 1993). They are called "ectoenzymes" and are involved in the degradation of nonpenetrating cell membrane substrates and polymers, which are not available continuously in the water (Chróst, 1990, 1991b). In addition to ectoenzymes, which are tightly associated with their producers, a fraction of hydrolytic enzymes in natural waters are found dissolved (Reichardt et al. 1967, Rego et al. 1985, Chróst and Overbeck 1987, Paul et al. 1987, Chróst 1989). They are termed as "extracellular enzymes" (Chróst 1990, 1991b). In contrast to ectoenzymes, extracellular enzymes occur in free form, dissolved in the water, and/or adsorbed to surfaces other than those of their producers (e.g., detrital particles, clay material). They may originate from the ectoenzyme pool as a result of washout from the cell surface or release from the periplasmic space after change of microbial cell permeability, and/or from the pool of intracellular enzymes after phage lysis or cell damage by grazers (Chróst 1991b, Nagata and Kirchman 1992, Proctor and Fuhrman 1992).

The nutritional conditions of the environment and physiological status of the microbial cells regulate the rates of microbial ectoenzyme synthesis and activity (Chróst 1991b, Meyer-Reil and Köster 1992). Ectoenzyme synthesis is under a catabolite repression/derepression control mechanism. The production of virtually all ectoenzymes in most aquatic bacteria is repressed when concentrations of UDOM fulfill bacterial requirements for metabolism and growth. This mode of regulation is called catabolite repression. The synthesis of ectoenzymes becomes derepressed only when the concentration of readily utilizable substrates in the water falls below a critically low level. Using the repression strategy for ectoenzyme synthesis, microorganisms can avoid the wasteful production of inducible enzymes that are not useful when their growth is not limited by carbon, nitrogen, or phosphorus sources (Bengtsson 1988, Chróst 1991b). It is also well documented that synthesis of many ectoenzymes in aquatic microorganisms is repressed by the end product that is derived from the substrate and accumulates in the cell or in the surrounding environment. Thus, ectoenzyme repression/derepression in microbial cells is very dependent on availability in natural waters of readily utilizable organic or inorganic substrates for metabolism (Chróst and Rai 1993).

The aim of this chapter is to present the spatial and temporal variations of hydrolytic activity and mechanisms of enzymatic decomposition of the predominant high-molecular-weight constituents in the pool of dissolved organic matter in lake waters. Significance of the model ectoenzyme systems—alkaline phosphatase, 5′-nucleotidase, β-glucosidase, and aminopeptidase for hydrolysis of organophosphoric compounds, polysaccharides, and proteins, respectively—is discussed. Special attention is given to the role of enzymatic transformations of organic matter by heterotrophic bacteria and the significance of microbial food webs in the functioning of the lake ecosystem.

Experimental Design

The sampling site (maximum depth 29 m) was in the pelagic zone of the Plußsee. Water samples were collected with a 2.5-L Ruttner sampler. Temporal variations in ectoenzyme activity, bacterial production and uptake of organic solutes, and chemistry of samples were analyzed in integrated water samples obtained from the euphotic zone of the lake, which was established by Secchi disk visibility. Samples, taken at 0.5-m intervals, were mixed together in equal volumes, and this integrated sample was used to represent the entire euphotic zone. Spatial fluctuations of the parameters studied in the lake were examined in the water column at depths of 0, 2, 4, 6, 8, 10, 15, 20, and 25 m. Immediately after collection, water samples were prefiltered through 100-μm pore-size nylon plankton net (to remove large organisms) and transported in plastic (polyethylene) containers to the laboratory within 1 h. Samples filtered through 0.2-μm pore-size polycarbonate Nuclepore filters (vacuum filtration <15 kPa) were used to determine water chemistry and enzyme extracellular activity.

Enzyme Assays

Alkaline phosphatase (APase) and β-glucosidase (GLCase) activities were measured by the increase in fluorescence as the nonfluorescent methylumbelliferyl substrates (4-methylumbelliferyl phosphate and 4-methylumbelliferyl-β-D-glucopyranoside, Sigma) were enzymatically hydrolyzed, leading to the production of the fluorescent molecule 4-methylumbelliferone (MUF). The amount of substrate cleaved was equivalent to the amount of highly fluorescent MUF anion generated by quenching the reaction with alkaline (pH 10.5, Chróst and Krambeck 1986) glycine-ammonium hydroxide buffer. Quantification was achieved by calibrating the spectrofluorometer (Kontron SFM-25, Switzerland) with a standard solution of MUF (Sigma) prepared with filtered (Whatman GF/F glass fiber filter) lake water buffered with glycine-ammonium hydroxide solution (pH 10.5). MUF fluorescence was measured at 460 nm under 365-nm excitation.

L-Leucine-4-methyl-coumarinylamide hydrochloride (Leu-MCA; Fluka) was used for fluorometric determination of leucine-aminopeptidase (AMPase). The

AMPase activity was equal to the amount of highly fluorescent 7-amino-4-methylcoumarin (AMC) produced after Leu-MCA hydrolysis. The spectrofluorometer was calibrated with an AMC standard (Fluka) prepared in filtered (GF/F Whatman) lake water, and AMC fluorescence was determined at 380 nm (excitation) and 440 nm (emission).

Stock solutions of the fluorogenic substrates were prepared to a concentration of 5 mM in deionized autoclaved water and stored at a temperature of –25°C. The stock substrate solutions were thawed at room temperature and diluted with deionized water to 0.10, 0.25, 0.50, 1.0, and 2.5 mM immediately before assay. For all enzyme assays, 0.5 ml of substrate solutions were added to 4.5-ml triplicate water samples, yielding final substrate concentrations in assays of 10, 25, 50, 100, 250, and 500 μM. Autoclaved lake water was used as a blank. Samples were incubated in the dark at in situ temperature for 1–6 h (depending on enzyme activity).

Adenosine 5′-[^{32}P]monophosphate (specific activity 3 Ci/mmol, Amersham) was applied as substrate in 5′-nucleotidase (5PNase) assays. 5PNase activity was determined in water preincubated for 30 min with β-glycerophosphate (final concentration 10 mmol/L) to inhibit APase activity, then 20-ml samples were supplemented with 5 nmol/L (final concentration) of [^{32}P]AMP and incubated for 20–30 min in light (ca. 150 μE m^{-2} s^{-1}) at in situ temperature. After incubation, samples were placed in an ice-cold water bath, and triplicate 5-ml subsamples were immediately filtered through 0.2- and 2.0-μm pore-size Nuclepore filters (that had been presoaked in 100 mM KH_2PO_4), and their radioactivity was determined. Four-milliliter portions of the 0.2-μm filtrate were immediately frozen and kept at –25°C until further analysis for the concentration of remaining [^{32}P]PO_4^{3-}, which was enzymatically released from [^{32}P]AMP but was not taken up by microorganisms during incubation. The filtrates were acidified with 10 N H_2SO_4 (final concentration ca. 0.5 N), passed through columns (150 × 4 mm) of activated charcoal (to adsorb AMP) into scintillation vials and vacuum freeze-dried. Once the radioactive samples passed the columns, they were washed with three 1.5-ml portions of 10 μM orthophosphoric acid to elute [^{32}P]orthophosphate totally, which could be partially adsorbed by charcoal. Dehydrated samples were redissolved in 0.5 ml of distilled water, mixed with 4.5-ml Insta-Gel (Packard) scintillation cocktail, and assayed for radioactivity. The 5PNase activity was calculated as the sum of [^{32}P]PO_4^{3-} taken up by microorganisms (retained on 0.2-μm pore-size filters) plus [^{32}P]PO_4^{3-} found in 0.2-μm filtrates. Efficiency of AMP adsorption and recovery of orthophosphate elution was tested in 0.2-μm-filtered and autoclaved lake water with [^{32}P]AMP and [^{32}P]PO_4^{3-} standards, respectively. On average 74 ± 13% of [^{32}P]AMP was removed by activated charcoal, and from 76% to 89% (average 85 ± 6%) of [^{32}P]orthophosphate was recovered after elution of the column.

Rates of [^{32}P]PO_4^{3-} release after hydrolysis of adenosine 5′-[γ-^{32}P]triphosphate (final concentration in assay 5 nmol/L; specific activity 5000 Ci/mmol, Amersham) by phosphohydrolytic enzymes and simultaneous uptake of [^{32}P]orthophosphate in water samples were analyzed as described above.

Microbial uptake of the adenyl moiety (i.e., second final product of 5PNase hydrolysis of AMP) was analyzed by incubating water samples with 5 nmol/L (final concentration) of [2-^{3}H]adenosine 5′-monophosphate (specific activity 20 Ci/mmol, Amersham) for 30 min in light at in situ temperature. After incubation, samples were placed in an ice-cold water bath, and 5-ml triplicates were immediately filtered through 0.2- and 2.0-μm pore size Nuclepore filters.

Uptake of enzymatically liberated [^{32}P]orthophosphate and [^{3}H]adenyl moiety by the size fraction larger than 2 μm and by the fraction 0.2–2.0 μm is considered to be algal and bacterial, respectively.

Enzyme Kinetic Data Analysis

Varying amounts of substrates were added to the samples to establish enzyme–substrate saturation and to enable the calculation of the kinetic parameters of APase, GLCase, and AMPase. The enzymatic reactions followed Michaelis–Menten kinetics, and the plot of the initial velocity of reaction (v) against increased concentrations of substrate ([S]) gave a rectangular hyperbola:

$$v = (V_{max}\,[S])/(^{a}K_m + [S]) \qquad (1)$$

The parameters characterizing this equation were calculated from the experimental data. They were V_{max}, the maximum velocity of enzyme reaction, which is theoretically attained when the enzyme has been saturated by an infinite concentration of substrate ([S]), and $^{a}K_m$, the apparent Michaelis constant, which is numerically equal to the concentration of substrate for the half-maximal velocity ($\frac{1}{2}V_{max}$). A nonlinear regression analysis on the original experimental data was applied to determine their best fit to the hyperbolic function of reaction velocity (v) versus substrate ([S]) concentration. The IBM PC computer software program "Enzfitter" (Elsevier-Biosoft, U.K.) was applied for this purpose and for calculation of V_{max} and $^{a}K_m$ values. Turnover time (T_t) of substrate hydrolysis was computed from the kinetic parameters according to the following equation:

$$T_t = {}^{a}K_m / V_{max} \qquad (^{a}K_m \text{ is an apparent Michaelis constant}) \qquad (2)$$

Other Analyses

Bacterial biomass was determined by the ATP method (Chróst and Overbeck 1987) or by the biovolume–biomass approach (Lee and Fuhrman 1987, Chróst and Rai 1993) and converted into the amount of organic carbon (μg org C/L) according to Holm-Hansen (1984) and Lee and Fuhrman (1987), respectively. Algal biomass was determined by the ATP method (Holm-Hansen 1984). Bacterial cell size and numbers were determined by the epifluorescence microscopic method after acridine orange or DAPI staining. Bacterial production was estimated by the [^{3}H-methyl]thymidine incorporation method according to Chróst et al. (1988) (see also Chapter 5, this volume). Phytoplankton primary production and excretion of organic carbon were determined as described by Chróst and Faust (1983) and Cole et al. (1982), respectively. Chlorophyll$_a$, extracted with 90% methanol, was measured fluorometrically (Wetzel and Likens 1979) with a Kontron spectrofluorometer SFM-25. Surplus phosphorus (i.e., the amount of phosphorus stored intracellularly in microplankton cells) was determined by the extraction method (Chróst and Overbeck 1987). Inorganic phosphorus, in samples filtered through precombusted glass fiber filters (GF/F Whatman), was determined according to Strickland and Parsons (1972). Dissolved organic carbon (DOC) in 0.2-μm-filtered samples (polycarbonate membrane filters, Nuclepore) was analyzed with the use of the organic carbon computational system Tocomaster (model 915B, Beckman). Dissolved combined carbohydrates (DCHO) and dissolved combined amino acids (DCAA) were analyzed in samples filtered through 0.2-μm Nuclepore filters according to Burney and Sieburth (1977) and Udenfriens et al. (1972), respectively. Radioactivity was assayed by liquid scintillation counting in a Beckman Computational System LS-1801. The statistical relationships between studied variables were analyzed according to Dixon and Massey (1969).

Decomposition of the Organophosphoric Compounds: Significance to Inorganic Phosphorus Regeneration

Organic Phosphorus

Although inorganic orthophosphate (PO_4^{3-}, HPO_4^{2-}, and $H_2PO_4^{3-}$) is the form of phosphorus that is biologically mobile and the principal state in which this element is interchanged among aquatic microorganisms, a major portion of the phosphorus pool in aquatic environments exists in organic forms. The majority of these compounds occur in the form of organic orthophosphates in which phosphorus is linked to an organic moiety via the ester bond, and all organophosphoric compounds in unpolluted waters originate from biological activity (Chapter 2, this volume).

In most unpolluted lakes, the sestonic component of the organic phosphorus (P_s) appears to be equal to or larger than the dissolved organic fraction (DOP). In the pelagic euphotic zone of the Plußsee the concentration of sestonic phosphorus correlated positively to the total biomass of microplankton ($r^2 = 0.82$, $P \leq 0.001$) and varied from 5 μg P/L in January to 49 μg P/L in March during the spring phytoplankton bloom. P_s constituted from approximately 5% (in winter) to 55% (July–August) and from 35% to 64% of the total phosphorus and organic phosphorus, respectively. The sestonic phosphorus of a series of Ontario lakes averaged 65.4% of total phosphorus. This was compared to 28.7% for DOP and 5.9% for inorganic phosphate (Hooper 1973). In a variety of Connecticut lakes sestonic phosphorus, DOP, and inorganic dissolved phosphorus (DIP) constituted on average 62%, 28.5%, and 9.5% of the total phosphorus, respectively (Hutchinson 1957). Higher P_s values were reported for the euphotic zone of the highly eutrophic Lake Głębokie (Poland) during summer stratification, where sestonic phosphorus, DOP, and DIP contributed 77–85%, 10–22%, and 0.5–1% to the total pool

Table 6.1. Phosphorus Content of Various Aquatic Organisms[a]

Component	P per Unit of Dry Weight (%)
Bacteria	3.0
Cyanobacteria	
Microcystis sp.	0.52
Anabaena sp.	0.53
Algae	
Cladophora sp.	1.4
Zooplankton	
Cyclops sp.	1.02
Daphnia pulex	1.60
Angiosperm plants	2.3
Macrophyte	
Myriophyllum sp.	0.52
Aquatic insects	0.64
Mixed plankton	
Clear Lake	0.18
Lower Greshan Lake	0.39
Tow-net zooplankton	
Turtle Lake	0.173
Crystal Lake	0.604

[a]Data from Hooper (1973).

of phosphorus, respectively (Chróst et al. 1984). The amount of organic phosphorus in the seston fraction is strongly dependent on the composition of planktonic organisms. Considerable variation of phosphorus content per unit of dry weight of various components of aquatic biota was reported by Hooper (1973) in his review on organic phosphorus in aquatic environments (Table 6.1). The sestonic phosphorus consists of cellular organophosphoric compounds, which are structural constituents, or they are involved in the metabolism of living organisms (e.g., sugar and polysaccharide phosphates, inositol hexaphosphates, phytin, phospholipids, phosphoproteins, nucleic acids, and organic condensed phosphates such as ATP or coenzyme A), and nonliving detritus particles undergoing decomposition. Moreover, a variety of planktonic microorganisms, including algae and bacteria, accumulate inorganic polyphosphates intracellularly. After the death of organisms and during the degradation process, many of these cellular organophosphoric compounds may be liberated into the water. Thus, the concentration and composition of the DOP fraction in natural waters can be strongly dependent on the phosphorus content of sestonic organisms.

In lakes it is difficult or impossible to determine the origin of DOP compounds that appear in the euphotic zone, since DOP constituents may arise by (1) decomposition of planktonic organisms themselves (Hałemejko and Chróst 1984), (2) excretion by zooplankton (Ejsmont-Karabin 1984), (3) internal loading from sediments and transport to the upper waters (Golterman 1977), (4) extracellular release by algae and bacteria (Lean and Nalewajko 1976), (5) grazing effect of herbivorous zooplankton (Sterner 1989) and bacterivorous microflagellates (Jürgens and Güde 1990, Turk et al. 1992), and (6) influx of DOP from littoral zone and/or watershed (Pieczyńska 1976). All these processes usually occur simultaneously in the lake ecosystem and are the cause of the high DOP concentration in the euphotic zone during summer stratification (DOP constitutes 80–99% of the total soluble phosphorus in water, Chróst 1988). However, despite the wealth of data on the quantity of DOP in aquatic environments, only fragmentary and isolated efforts have been made towards the identification of DOP compounds (Francko and Heath 1979).

Dissolved organic phosphorus was the major soluble P form in the euphotic zone during summer stratification of the Plußsee (Fig. 6.1A). The concentrations of dissolved inorganic phosphate (DIP), the only form of phosphorus that is readily utilizable by algae and bacteria, were very low and fluctuated between 0.5 and 2.0 μg P/L. In the period July–September, DIP constituted only 1.5–9.5% of the total soluble phosphorus in the euphotic zone, and 91.5–98.5% of dissolved phosphorus was found in the DOP fraction. However, at the period of strong DIP depletion in the water, a high total biomass of microplankton (230–400 μg C/L)

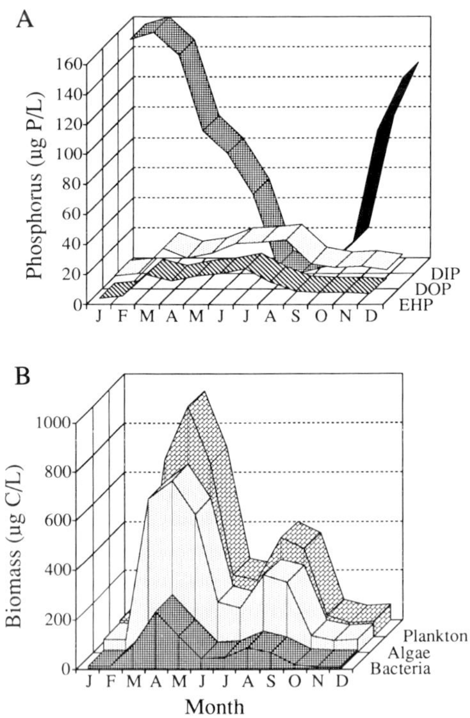

Figure 6.1. (A) Annual variation of concentration of dissolved inorganic phosphorus (DIP), dissolved organic phosphorus (DOP), and enzymatically hydrolyzable phosphate (EHP), and (B) biomass of total plankton and algal and bacterial size fraction of seston in the euphotic zone (0–2 m) of the Plußsee (January–December 1984).

was found in the euphotic zone (Fig. 6.1B). The efficiency of microplankton biomass production per unit of inorganic phosphate during its summer depletion varied from 46 to 205 (average 89 ± 42) μg C/μg P and was on average 12 times higher than during the spring phytoplankton bloom, when DIP concentrations were very high and did not limit the growth of microplankton (Fig. 6.1A). One explanation is that inorganic phosphate was turned over rapidly by microplankton in the summer. Alternatively, microplankton may have utilized phosphorus very efficiently from the DOP pool. It is generally believed that the ability to take up phosphorus from organophosphoric compounds is restricted to those microorganisms producing enzymes that hydrolyze the phosphoesters releasing orthophosphate ions (Siuda 1984, Chróst and Overbeck 1987, Ammerman 1991). DOP was the predominant fraction of soluble phosphorus in the euphotic zone of the Plußsee in summer, but all forms of organic phosphorus were not equally available for enzymatic hydrolysis. Data from freshwater show that two general types of DOP exist (Francko and Heath 1979): the high-molecular-weight compounds, which are slowly hydrolyzable by phosphohydrolases and are supposed to be mainly nucleic acids, and the second type of DOP, which is easily hydrolyzed, including all the phosphomonoesters. The latter are often found at low concentrations relative to total DOP (Solorzano 1978, Chróst et al. 1984, 1986b). Enzymatically hydrolyzable phosphate (EHP) in the Plußsee euphotic zone constituted on average 63% of the DOP and varied from 4 to 24 (average 13 ± 6.5) μg P/L. The highest EHP concentrations were found during summer (Fig. 6.1A). Obviously, the EHP concentrations effectively liberated to the water were much higher than that measured in the lake, because EHP release from the organic phosphorus pool and its enzymatic hydrolysis occurred simultaneously.

EHP constitutes a biologically important pool of labile organophosphoric compounds in natural waters that are utilized after enzymatic degradation and are regarded as an alternative phosphorus source for aquatic microorganisms when their growth and activity are limited by inorganic orthophosphate. A variety of enzyme systems capable of degradation of EHP compounds and release of inorganic phosphate were found in aquatic microorganisms, including algae and bacteria. One of the earliest recognized and most intensively studied enzymes in aquatic ecosystems was alkaline phosphatase (Overbeck 1961, 1962, Overbeck and Babenzien 1963); 5′-nucleotidase is the second ectoenzyme found in aquatic bacteria that is involved in the hydrolysis of specific (5′-mononucleotides) compounds in the EHP pool in marine and freshwater environments (Ammerman and Azam 1985).

Alkaline Phosphatase

Alkaline phosphatase (orthophosphoric monoester phosphohydrolase; E.C. 3.1.3.1) catalyzes the hydrolysis of a variety of phosphate esters (including esters of primary and secondary alcohols, sugar alcohols, cyclic alcohols, phenols, and amines), liberating inorganic phosphate. The enzyme also hydrolyzes polyphosphates, but phosphodiesters are not hydrolyzed. Alkaline phosphatase

(APase) is also known to catalyze a variety of transphosphorylation reactions; for instance, the enzyme from *E. coli* catalyzes both the hydrolysis of polyphosphates and the transfer of PO_4^{3-} group from polyphosphate, a number of nucleoside di- and triphosphates, and from mannose-6-phosphate to glucose, forming glucose-6-phosphate (Barman 1969).

The presence of APase has been demonstrated in filtrates of algal cultures and in fresh and seawater, as a dissolved (extracellular) enzyme, and in phytoplankton, periplasmic space of bacteria, zooplankton, and protozoans (Aaronson 1981, Wynne and Gophen 1981, Stewart and Wetzel 1982, Hałemejko and Chróst 1984, Chróst et al. 1984, 1986b, Chróst and Overbeck 1987, Nagata and Kirchman 1992, Martinez and Azam 1993). Most of the data in the literature indicate that APase is produced in copious amounts when inorganic phosphate becomes limiting in natural waters or in the culture medium, suggesting derepression of this enzyme (Siuda 1984, Chróst and Overbeck 1987, Siuda and Chróst 1987). Moreover, measurement of APase activity in algae has been suggested as a procedure for detecting phosphorus deficiency or restricted DIP availability in aquatic ecosystems (Healey and Hendzel 1980, Pettersson 1980, Vincent 1981, Gage and Gorham 1985).

Temporal Variation of Alkaline Phosphatase Activity in the Euphotic Zone

Activity of APase in the euphotic zone of the Plußsee displayed distinct variations throughout the annual study. High V_{max} activity (1721 nmol L^{-1} h^{-1}) of total APase in water samples was found at the beginning of May after the breakdown of a spring phytoplankton bloom in a lake (Fig. 6.2A). The highest V_{max} of total APase was found during orthophosphate depletion in August (5942 nmol L^{-1} h^{-1}) and September (3387 nmol L^{-1} h^{-1}). Algae were the major producers of APase (45–65% of total activity) in the euphotic zone of the lake in spring and summer (Fig. 6.2B). APase associated with bacterial size fraction of seston constituted from 17% (at the beginning of spring phytoplankton bloom) to 62% (December) of the total enzyme activity in water samples. Generally, the highest contribution of bacterial APase (45–62%) to the total activity of enzyme in the euphotic zone of the Plußsee was found during autumn and winter. Extracellular activity of free APase (fraction less than 0.2 μm) that was dissolved in the water varied between 15% and 34% (average 24 ± 6.4%).

Because the V_{max} of total APase is a function of both the specific activity of the enzyme and its total amount in the biomass of enzyme producers (i.e., algae and bacteria), it was necessary to estimate the specific APase activity and apparent K_m values of enzymes produced by the algal and bacterial size fractions of microplankton in order to evaluate the role of these microorganisms in DOP hydrolysis in lake water. By differential filtration, the microplankton assemblages were roughly separated into the algal (particles larger than 3 μm) and bacterial (0.2–3.0 μm seston) size fractions, and their APase kinetics were measured. The specific APase activity of the algal fraction displayed a pattern similar to the V_{max} of total APase and had minimum values (0.09–0.44 nmol μg C biomass^{-1} h^{-1})

Figure 6.2. (A) V_{max} of alkaline phosphatase activity (APase) and (B) contribution of algal and bacterial size fraction of seston and extracellular free enzymes in water to the total activity of alkaline phosphatase in the euphotic zone of the Plußsee (January–December 1984).

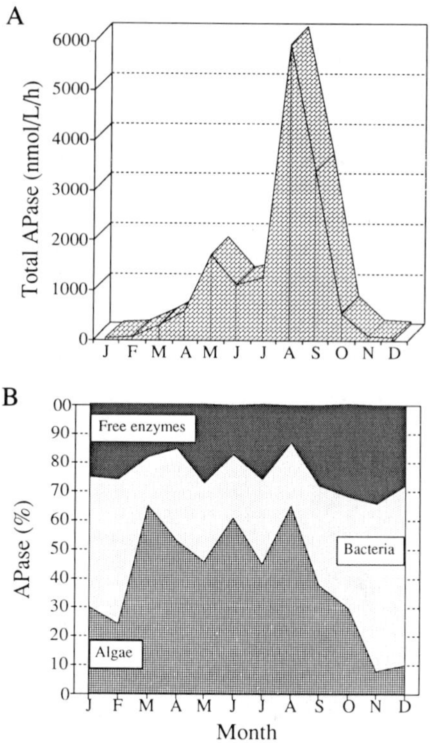

during the period from late autumn until early spring, i.e., when the ambient DIP concentrations in the lake were high (Fig. 6.3A). The highest APase specific activities produced by algae (3.1–13.9 nmol μg C biomass^{-1} h^{-1}) were observed during summer stratification of the lake, when DOP was the major fraction of soluble phosphorus in the water and DIP was nearly depleted and approached minimal concentrations (Fig. 6.1A). The specific activity of bacterial APase in all samples was higher than that found in the algal size fraction and varied from 0.53 to 17.42 (average 6.80 ± 5.41) nmol μg C biomass^{-1} h^{-1} (Fig. 6.3A). The lowest specific activity of the bacterial size fraction was observed during the spring phytoplankton bloom, when algae were actively growing and photosynthesizing. The apparent Michaelis constants (^aK_m) estimated for algal size fraction were in all cases higher (31.1–47.7, average 38.8 ± 4.7 μmol/L) than the bacterial ^aK_m (19.5–35.7, average 25.4 ± 4.9 μmol/L; Fig. 6.3B).

The activity of APase in algae has been reported to be directly related to inorganic phosphorus limitation (Smith and Kalff 1981, Vincent 1981). Several authors have indicated that algal APase synthesis and activity are repressed and inhibited by high DIP concentrations, respectively (Siuda 1984, Chróst and Overbeck 1987, Chróst 1990, 1991a). It is not quite correct, however, to conclude that APase synthesis is derepressed directly by low concentrations of inorganic phosphate in lake water. The mechanism of APase derepression is regulated by the size

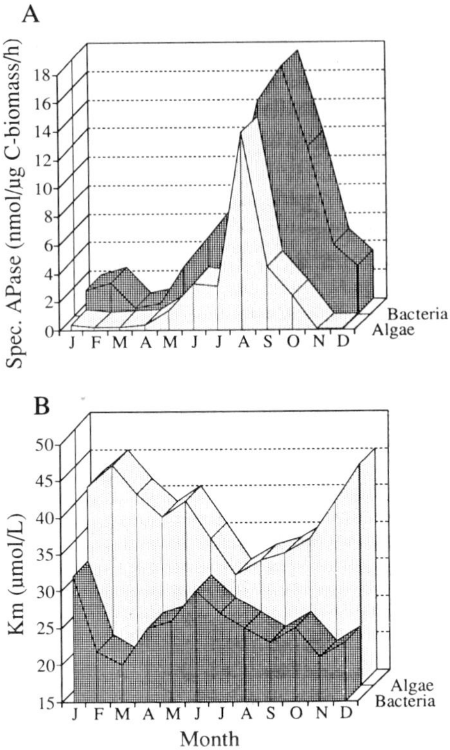

Figure 6.3. (A) Specific activity of alkaline phosphatase (APase) and (B) apparent Michaelis constants (K_m) of algal and bacterial size fraction of seston in the euphotic zone of the Plußsee (January–December 1984).

of the intracellular polyphosphate pool (P_{surpl}) in algal cells, which obviously depends on ambient DIP concentrations. In the euphotic zone of the Plußsee, the specific activity of algal APase significantly decreased when the ambient DIP concentrations were higher than 15–20 µg P/L (Chróst and Overbeck 1987). Derepression of APase synthesis in algal cells, however, did not occur immediately at the beginning of DIP depletion during summer because of the intracellular pool of P_{surpl} stored in the microbial cells. When ambient DIP concentrations in the Plußsee lake water were high (after spring and autumn turnover of the lake), algae accumulated phosphorus in the intracellular pool, and P_{surpl} constituted 23–34% (average 30%) of the total sestonic phosphorus. As DIP concentrations decreased in the water, a decrease of P_{surpl} content in algae was observed; i.e., part of the phosphorus stored intracellularly was utilized to support the cellular phosphate metabolism of microorganisms. Specific activity of APase was inversely related to the amount of P_{surpl} stored in algal cells (Fig. 6.4). When P_{surpl} constituted less than 15% of the total sestonic phosphorus (DIP concentrations were at the same time below 1 µg P/L), the algae produced APase with a high specific activity. Thus, when all external and intracellular phosphorus reservoirs were nearly depleted, algae began to synthesize APase with high specific activity to hydrolyze phosphate esters as an alternative P source to compensate for the lack of inorganic phosphate.

The apparent Michaelis constants, which indicate the affinity of an enzyme for its substrates (a lower $^{a}K_m$ indicates a higher affinity) of algal APase showed the direct positive relationship to DIP concentration in the water of the euphotic zone of the Plußsee. This relationship suggested that in many situations DIP was a strong competitive inhibitor of algal APase activity in the lake. Chróst and Overbeck (1987) observed a rapid and marked increase of the apparent Michaelis constants of APase of the algal size fraction when water samples were experimentally supplemented with inorganic phosphate. The hallmark of competitive inhibition of algal APase by inorganic phosphate is that the orthophosphate combines with the enzyme in such a way that it competes with a substrate for binding at the active site, thus decreasing the enzyme affinity for the substrate, and therefore inhibits the initial velocity of the reaction. Competitive inhibition of algal APase was reversible and was overcome by increased substrate (EHP) concentrations; therefore, the V_{max} of APase during spring was relatively high (Fig. 6.2A) in spite of high concentrations of DIP in the water (Fig. 6.1A). Contrary to algal APase, bacterial $^{a}K_m$ did not correlate to DIP concentrations in the euphotic zone of the lake, and both low and high values of bacterial $^{a}K_m$ were found in the water in low and high DIP concentrations.

Kinetic parameters (V_{max} and $^{a}K_m$) of the total APase in water and APase of bacterial and algal size fraction of microplankton were used to calculate the potential turnover time (T_t) of the DOP hydrolysis by APase in the euphotic zone of the Plußsee. The shortest T_t (2.1–4.3 h) was found during the summer (i.e., during DIP depletion in water), indicating that phosphorus from its dissolved organic fraction was rapidly regenerated by APase in lake water and utilized by microplankton to produce a high biomass (Fig. 6.5A). During the autumn and spring homothermy and mixing of lake waters, when DIP content in the euphotic zone was high and did not limit the microplankton growth, the turnover time of DOP hydrolysis by APase was the longest and varied from 7.7 to 45 days. Depending on the season, bacterial and algal size fractions of microplankton displayed different T_t of DOP hydrolysis by APase. Throughout spring and summer, algal APase had a shorter T_t than bacterial (Fig. 6.5B and C). During autumn and winter, however, the hydrolysis of DOP by bacterial APase was about five

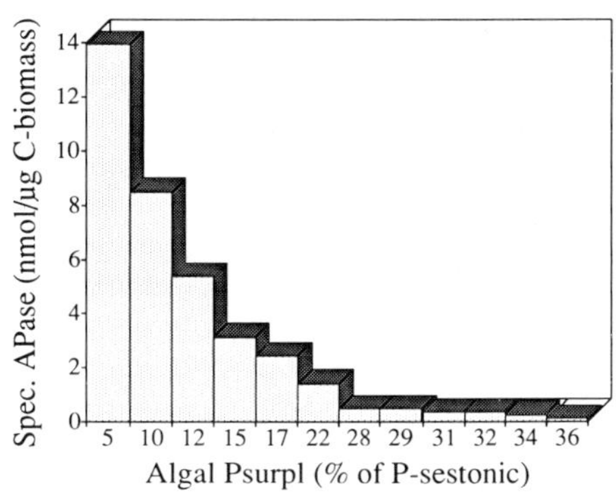

Figure 6.4. Relationship between specific activity of alkaline phosphatase and amount of surplus phosphorus (P_{surpl}) in algal size fraction of seston in the euphotic zone of the Plußsee (data collected in March–November 1984).

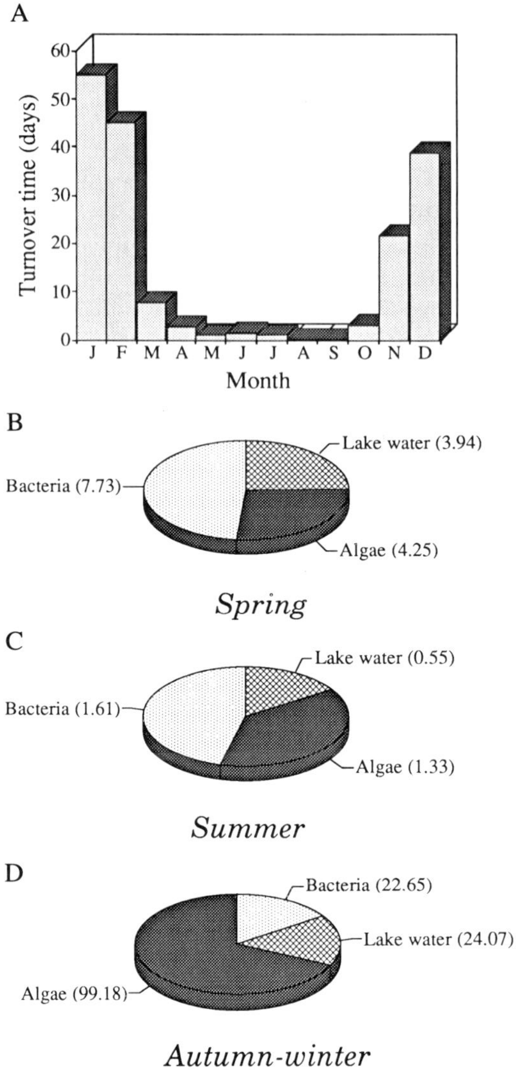

Figure 6.5. Turnover time of dissolved organic phosphorus hydrolysis by alkaline phosphatase in water (A), and an average turnover time of algal and bacterial size fraction and unfiltered lake water in spring (B), summer (C), and during the autumn–winter period (D) in the euphotic zone of the Plußsee in 1984. Turnover time in days is given in parentheses.

times shorter than algal (Fig. 6.5D), indicating that bacteria were the major component of plankton responsible for decomposition of organophosphoric compounds and inorganic phosphorus regeneration in the euphotic zone of the Plußsee.

Spatial Distribution of Alkaline Phosphatase Activity in the Water Column

The kinetic parameters of APase of algal and bacterial size fractions of microplankton displayed significant differences within the entire water column during summer stratification of the Plußsee. The highest V_{max} (3205–3801 nmol L^{-1} h^{-1}) of algal APase was found in the upper part of the epilimnion (0–4 m), i.e., in

Figure 6.6. (A) V_{max} of alkaline phosphatase (APase) activity and (B) apparent Michaelis constants (K_m) of algal and bacterial size fraction of seston in the summer (August 1986) stratified water column of the Plußsee.

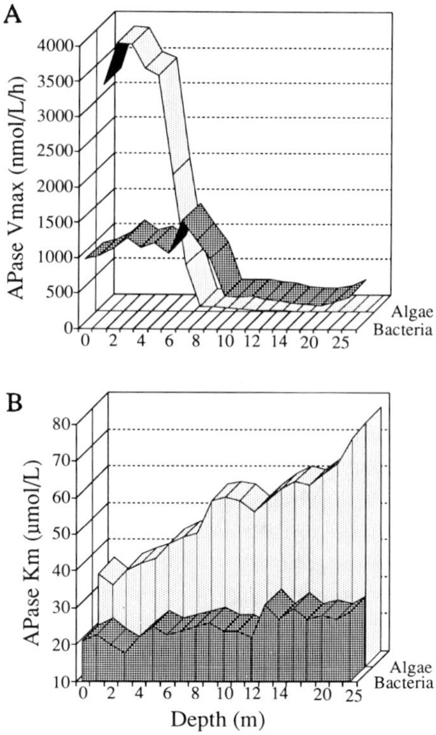

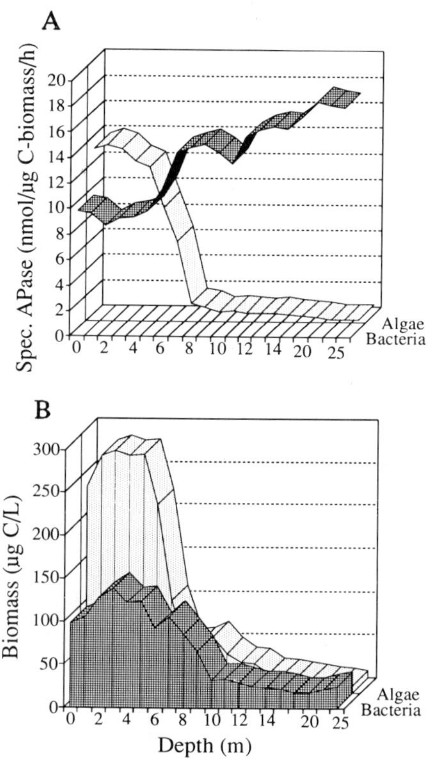

Figure 6.7. (A) Specific activity of alkaline phosphatase (APase) and (B) biomass of algal and bacterial size fraction of seston in the summer (August 1986) stratified water column of the Plußsee.

the euphotic zone (Fig. 6.6A), where the concentrations of DIP varied between 0.5 and 2 μg P/L. The V_{max} of algal APase dramatically decreased (from 3349 to 38 nmol L^{-1} h^{-1}) within a thermocline in the metalimnion and was very low (2–21 nmol L^{-1} h^{-1}) and almost uniformly distributed in the hypolimnion of the lake. High V_{max} values (970–1150 nmol L^{-1} h^{-1}) of bacterial APase were also measured in the epilimnion. However, the highest V_{max} (1255–1531 nmol L^{-1} h^{-1}) of APase in the bacterial size fraction was found at the boundary between metalimnion and hypolimnion of the Plußsee. Bacterial APase displayed relatively high and uniformly distributed V_{max} (341–463 nmol L^{-1} h^{-1}) values within the hypolimnion. APase of the algal size fraction of microplankton had the highest affinity for substrates (i.e., the lowest ${}^{a}K_{m}$) in the surface waters (0–4 m depth), and then the enzyme affinity gradually decreased within the water column, approaching the highest ${}^{a}K_{m}$ values in the deepest part of the lake's hypolimnion (Fig. 6.6B). Apparent Michaelis constants of APase of the bacterial size fraction were distinctly lower than those estimated for algae and did not fluctuate markedly within the water column.

The specific activity of APase of the algal size fraction was similar to its V_{max} distribution pattern in the vertical profile of the lake water and was highest (12.5–13.8 nmol μg C $biomass^{-1}$ h^{-1}) in the euphotic zone (Fig. 6.7A), where the highest algal biomass (240–276 μg C/L) was found (Fig. 6.7B). The specific activity of APase of the bacterial size fraction gradually increased within the water column and displayed lower activity (8.7–9.8 nmol μg C $biomass^{-1}$ h^{-1}) in epilimnetic waters than in the deep hypolimnion of the lake (16.3–18.4 nmol μg C $biomass^{-1}$ h^{-1}). A significant increase (from 9.8 to 15.2 nmol μg C $biomass^{-1}$ h^{-1}) of bacterial specific APase (Fig. 6.7A) was observed within the metalimnion. Specific activity of bacterial APase did not correlate to bacterial biomass (Fig. 6.7B), which was much higher (98–139 μg C/L) in the epilimnion than in hypolimnion (18–26 μg C/L) of the Plußsee.

Role of Alkaline Phosphatase in Microplankton Phosphorus Metabolism

Most of the published reports on APase activity in natural waters described the different relationships among enzyme activity, microplankton biomass (usually $chlorophyll_a$ concentration), and phosphorus content in the ambient environment (Heath and Cooke 1975, Pettersson 1980, Stewart and Wetzel 1982, Gage and Gorham 1985, Chróst et al. 1984, 1989, Siuda and Chróst 1987). Moreover, the majority of these studies dealt only with the surface waters during the spring and summer periods, where the phytoplankton are the predominant producers of APase (as shown in this study also). These data indicate that algal APase is produced in copious amounts when DIP becomes limited in the lakes. Therefore, measurement of APase activity in the lake and marine water has been suggested as a procedure for detecting phosphorus deficiency or restricted P availability in aquatic ecosystems (Perry 1972, Healey and Hendzel 1980, Vincent 1981). One difficulty in interpreting the significance of APase for phytoplankton phosphorus metabolism is that bacteria are also able to produce this enzyme and use organophosphoric compounds in natural waters (Kobori and Taga 1979, Halemejko and Chróst 1984). The potential role of bacteria in measurements of total APase activity makes using APase activity as an index of phosphorus limitation of the phytoplankton more difficult. Few attempts have been made to study the activity and significance of bacterial APase in aquatic ecosystems (Chróst and Overbeck 1987, Siuda and Chróst 1987, Cotner and Wetzel 1991a, Coveney and Wetzel 1992).

Bacterial APase in the Plußsee contributed significantly to the total activity of APase in the euphotic zone during spring and summer phytoplankton blooms. Moreover, from autumn until early spring, bacteria were the major APase producers in surface waters (Fig. 6.2B) and in the metalimnion and hypolimnion during summer stratification of lake (Fig. 6.7A). It must be emphasized that bacteria produced higher specific APase activity and had lower ${}^{a}K_{m}$ values than algae. Excluding only a short period during summer stratification, bacteria probably were not phosphorus-limited in the euphotic zone of the lake because of their high-affinity P-uptake system, which was efficient at very low ambient DIP concentrations (Chróst and Overbeck 1987, Chróst 1988). Therefore, the question arises, "why did bacteria produce highly efficient APase when DIP concentrations (during homothermy of the Plußsee or in anoxic waters during summer stratification) were high and exceeded their phosphorus requirement for uptake and metabolism?" Previous studies suggested that there are two possible explanations of this phenomenon (Chróst 1986, Chróst and Overbeck 1987). First, bacteria produce high APase activity in lake water to hydrolyze organophosphoric compounds and utilize the organic moiety as a carbon source, because their metabolism is limited by low concentrations of readily utilizable organic carbon substrates. Second, bacteria hydrolyze EHP but do not take up orthophosphate derived from hydrolyzed organic compounds; thus, their hydrolysis increased the DIP pool in the water available for phytoplankton metabolism. Consequently, more organic carbon compounds (EOC) were excreted during algal photosynthesis and served as readily utilizable substrates for bacteria (Cole et al. 1982, Chróst and Faust 1983, Münster and Chróst 1990).

The first hypothetical situation could occur in the euphotic zone in the Plußsee from autumn until spring, and it could also occur below the euphotic zone during summer stratification (when bacteria produced high specific APase activity) because of very low phytoplankton biomass and photosynthetic production of organic carbon substrates. The limitation of bacterial production and growth rates by availability of organic substrates in the hypolimnion of the Plußsee was reported by Chróst et al. (1988). The second explanation may be valid for bacterial APase production in the euphotic zone during summer phosphorus limitation of phytoplankton (Currie and Kalff 1984a,b). Moreover, recent studies on the regulation of synthesis and activity of bacterial APase in lake water from the Plußsee revealed that the enzyme was induced or repressed, depending on the organic compounds added to the samples (Chróst 1991a). Supplementation of lake water with additional phosphoester substrates (cAMP, ATP, glucose-6-phosphate) induced APase synthesis, and bacteria produced the enzyme with much higher specific activity and affinity for the substrate in comparison to the control samples. The highest specific APase activity was observed in samples when glucose-6-phosphate and ATP were added; cAMP only slightly increased the enzyme specific activity but markedly decreased the ${}^{a}K_m$ value. The repression of bacterial APase synthesis was observed, however, when samples were supplemented with easily utilizable natural substrates for bacteria in the Plußsee, such as glucose, cellobiose, and amino acids. The APase specific activity decreased significantly, and also the affinity of the enzyme for substrate decreased in comparison to the samples enriched with phosphoester compounds. Comparison of the results of experimental variants in which glucose-6-phosphate and glucose were added indicated that aquatic bacteria produced APase to supply their metabolism with organic carbon substrates (glucose, in this case), which (apart from orthophosphate ions) are also final products of phosphoester hydrolysis (Chróst 1991a).

Enzymatic Decomposition of Nucleotides

5′-Nucleotidase

5′-Nucleotidase (5′-ribonucleotide phosphohydrolase, E.C. 3.1.3.5) catalyzes the dephosphorylation of most nucleoside 5′-monophosphates to the corresponding nucleosides and orthophosphate. 5′-Nucleotidase (5PNase) activity has been detected both in a great number of animal tissues (Kreutzberg et al. 1986) and recently in natural assemblages of aquatic bacteria in freshwater and marine ecosystems (Ammerman and Azam 1985). Bacterial 5PNase is localized predominantly in the periplasmic space or is cytomembrane bound (Bengis-Garber 1985). 5PNase is inhibited by ADP, NAD, NADH, and concanavalin A (Heinz and Haeckel 1984). Manganese and magnesium ions activate 5PNase, but other metal cations (e.g., zinc, cobalt, mercury, nickel) and anions (e.g., fluoride, borate, arsenate) inhibit enzyme activity (van Husen and Gerlach 1984). The function of 5PNase in the cellular metabolism is not yet fully understood; speculations include a role in nucleic acid biosynthesis, in RNA-degrading systems, or in nucleotide metabolism. Recently, in freshwater and marine ecosystems, 5PNase activity has been found to be of great importance for organic phosphorus regeneration processes (Ammerman 1991)

Interference of Alkaline Phosphatase in Determination of 5′-Nucleotidase Activity

A special problem in 5PNase activity determination is derived from interference by nonspecific alkaline phosphatases (and other phosphohydrolytic enzymes), which simultaneously hydrolyze substrates for the 5PNase assay. Several methods have been proposed to distinguish between 5PNase and APase activity in samples. Correction for APase activity can be achieved by inhibition of 5PNase by nickel ions (Campbell 1962) or by suppression of APase activity using high concentrations of any phosphomonoester substrate that is not hydrolyzed by 5PNase and does not inhibit the nucleotidase activity. One of the most important issues of 5PNase assay in natural waters is a choice of substrate for 5PNase. Since Ammerman and Azam (1985) in their first report on the occurrence of 5PNase in natural bacterioplankton assemblages sampled from marine and freshwater environments have applied [γ-^{32}P]ATP for 5′-nucleotidase assay, in all the following published studies, 5PNase activity in aquatic ecosystems was determined with the use of this substrate (Tamminen 1989, Ammerman 1991, Ammerman and Azam 1991a,b, Cotner and Wetzel 1991b). [γ-^{32}P]ATP is not, however, the best substrate for 5PNase. The relative rates of 5PNase activity in animal tissues or venoms determined with ATP as substrate are 40–70% lower in comparison to 5′-AMP, 5′-CMP, or 5′-IMP. Depending on enzyme source, the Michaelis constant values of 5PNase for ATP are usually one order of magnitude higher (10^{-3}–10^{-4} mol/L) than those determined with 5′-AMP (10^{-5} mol/L) as substrate (Barman 1969). Moreover, ATP in natural waters is also very effectively hydrolyzed mainly by APase (Karl and Craven 1980) and potentially may be degraded by ATP phosphohydrolase (E.C. 3.6.1.3), apyrase (E.C. 3.6.1.5), nucleotide pyrophosphatase (E.C. 3.6.1.9), and glucose-6-phosphatase (E.C. 3.1.3.9), which are the enzymes that are assayed with ATP as substrate (Barman 1969), and they are widely distributed throughout aquatic organisms. Thus, in light of the aforementioned information, [γ-^{32}P]ATP is hydrolyzed by a variety of phosphohydrolytic enzymes, and it should be used for determination of the total phosphohydrolytic activity rather than 5PNase activity in natural waters.

In studies of 5PNase activity in the Plußsee, the technique known as "enzyme diversion" (Belfield and Goldberg 1968) was applied to inhibit APase activity. Supplementation of water samples with β-glycerophosphate (final concentration 10 mmol/L) almost totally inhibited APase activity towards 5′-[^{32}P]monophosphate ([^{32}P]AMP), which was used as a substrate (final concentration 5 nmol/L) in the 5PNase assay. The competitive inhibition of APase activity, which was separately assayed with methylumbelliferyl phosphate (final concentration 100 nmol/L), by β-glycerophosphate (BGP) varied between 92% and 98% (average 96 $\pm$ 2%). The mechanism of the inhibition of APase hydrolysis of [^{32}P]AMP by β-glycerophosphate seems to involve diversion of the enzyme by an alternative substrate (BGP) present in excess (BGP-to-[^{32}P]AMP ratio in assay was 2×10^6 to 1). It was not possible to use orthophosphate to inhibit APase activity in 5PNase assays (Cotner and Wetzel 1991a,b) of the Plußsee water because previous studies have shown that bacterial (Chróst and Overbeck 1987) and extracellular, water-soluble (Chróst et al. 1986b) APase activity was not totally inhibited by orthophosphate supplementation, contrary to algal APase where activity was strongly inhibited by orthophosphate (Chróst 1991a). Moreover, high bacterial APase activity was found in water from the Plußsee even during high DIP content (Figs. 6.3A and 6.7A).

5′-Nucleotidase Activity in Lake Water

Rates of [^{32}P]AMP hydrolysis in water samples taken from the euphotic zone (0–2 m) of the Plußsee varied from 0.94 to 3.35 nmol L^{-1} h^{-1} and were the highest on April 19 (after the phytoplankton bloom collapsed) and August 14 (during the phytoplankton bloom) (Fig. 6.8A). When APase activity was suppressed by sup-

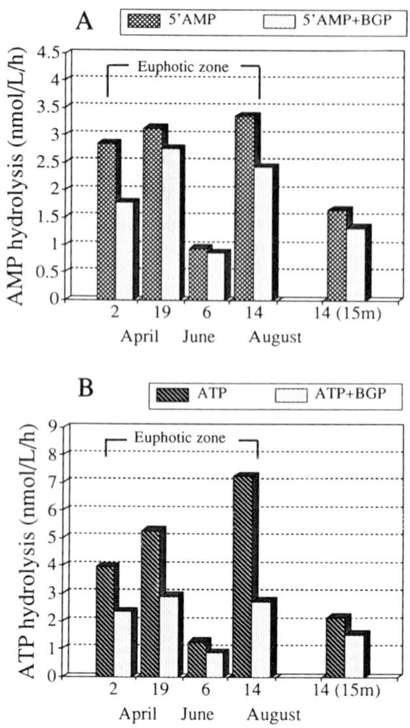

Figure 6.8. Rates of hydrolysis of [^{32}P]AMP (A) and [^{32}P]ATP (B) in Plußsee water samples collected from the euphotic zone (1 m) in April–August 1990. β-Glycerophosphate (BGP) was added to suppress the alkaline phosphatase activity toward radiolabeled 5′-nucleotides.

plementation of water samples with BGP, the rates of [^{32}P]AMP hydrolysis decreased, indicating that APase activity contributed from 9.6% to 37.4% of the total hydrolysis of the substrate. The highest contribution of APase activity to the hydrolysis of [^{32}P]AMP was found in water sampled on April 2 (37.4%) and August 14 (27.8%) during active development of algae forming spring and summer phytoplankton blooms (Table 6.2). 5PNase activity in water samples fluctuated between 0.85 and 2.89 nmol L^{-1} h^{-1}. Lowest rates of 5PNase activity were found in June, during the clear-water phase. The highest contribution of 5PNase to the total hydrolysis of AMP was found on April 19 during the breakdown of the spring phytoplankton bloom (88.4%) and during the clear-water phase (90.4%) at the beginning of June. 5PNase activity (1.32 nmol L^{-1} h^{-1}) hydrolyzed 80.5% of AMP in water sampled from the anoxic hypolimnion (15 m) on August 14, during summer stratification of the lake (Fig. 6.8A).

Phosphohydrolytic activity (i.e., sum of activity of 5PNase, APase, and other phosphohydrolases) in the euphotic zone of the Plußsee measured by means of [^{32}P]ATP varied from 1.25 to 7.25 nmol L^{-1} h^{-1} and was 32.9–116.4% higher than that measured by [^{32}P]AMP hydrolysis assay (Fig. 6.8B). The highest rates of [^{32}P]ATP hydrolysis, which were similar to hydrolysis of [^{32}P]AMP, were noted during the breakdown of the spring phytoplankton bloom and summer bloom. The hydrolysis of [^{32}P]ATP was 29.6–62.5% lower when APase activity was competi-

Table 6.2. Algal and Bacterial Biomass and the Concentration of Chlorophyll$_a$, Dissolved Inorganic Phosphorus (DIP), Dissolved Organic Phosphorus (DOP), and Dissolved Organic Carbon (DOC) in the Plußsee

	Biomass					
Sampling Date	Algae (μg C/L)	Bacteria (μg C/L)	Chlorophyll$_a$ (μg/L)	DIP (μg P/L)	DOP (μg P/L)	DOC (mg C/L)
April 2[a]	665.8	47.5	58.4	155	18.6	10.5
April 19[b]	72.2	93.7	4.5	92	26.8	15.2
June 6[c]	28.6	22.5	2.4	45	17.2	6.7
August 14[a]	386.4	115.2	31.2	0.5	22.1	11.4
August 14[d]	23.5	32.4	1.2	265	14.4	7.8

[a]Phytoplankton bloom.
[b]Phytoplankton bloom breakdown.
[c]Clear-water phase.
[d]Hypolimnion (15 m).

tively inhibited by BGP. Similar results on APase activity toward [^{32}P]ATP were reported for oligotrophic Lake Michigan and moderately eutrophic Third Sister Lake (Cotner and Wetzel 1991b). The significantly higher contribution of APase to [^{32}P]ATP hydrolysis, in comparison to its participation in [^{32}P]AMP dephosphorylation, indicated that ATP was a "better" substrate for APase than AMP. Activity of 5PNase determined by ATP hydrolysis varied from 0.88 to 2.89 nmol L^{-1} h^{-1} and was higher that 5PNase activity found in the [^{32}P]AMP assay. The ratio between [^{32}P]orthophosphate that was enzymatically released from [^{32}P]ATP and [^{32}P]AMP varied from 1.30 to 2.16, demonstrating that [^{32}P]ATP was also hydrolyzed by phosphohydrolases other than 5PNase.

Dephosphorylation of Nucleotides

Parallel determination of enzyme activity with the use of [^{32}P]AMP (5PNase + APase) and [^{32}P]ATP (5PNase + APase + other phosphohydrolases, e.g., ATPase, apyrase, nucleotide pyrophosphatase, glucose-6-phosphatase) and simultaneous suppression of APase activity by addition of β-glycerophosphate allowed us to estimate the contribution of these enzymes to ATP hydrolysis in water from the Plußsee (Table 6.3). The percentage of [^{32}P]ATP hydrolyzed by APase varied from 29% (in summer anoxic hypolimnion) to 62.5% (in the euphotic zone during summer phytoplankton bloom) and averaged 41.3 ± 12.2%. Assuming that 5PNase had identical affinity to ATP and AMP, 5′-nucleotidase hydrolyzed 33.4–68.0% (average 52.1 ± 12.2) of [^{32}P]ATP added to water samples. The contribution of phosphohydrolases other than 5PNase and APase to [^{32}P]ATP dephosphorylation varied between 2.4% and 14.5% (average 6.6 ± 4.7%). However, if 5PNase affinity for ATP was lower than its affinity for AMP, which seems to be the case when values of 5PNase Michaelis

Table 6.3. Contribution of Alkaline Phosphatase (APase), 5′-Nucleotidase (5PNase), and Other Phosphohydrolytic Enzymes (PE) to [γ-^{32}P]ATP Hydrolysis in the Euphotic Zone of the Plußsee

Sampling Date	Percentage of ATP Hydrolysis[a]		
	APase	5PNase	PE
April 2[b]	40.6	44.9	14.5
April 19[c]	44.9	52.6	2.5
June 6[d]	29.6	68.0	2.4
August 14[e]	62.5	33.4	4.1
August 14[f]	29.0	61.7	9.3

[a]Concentration of [γ-^{32}P]ATP was 5 nmol/L.
[b]Spring phytoplankton bloom.
[c]Collapse of phytoplankton bloom.
[e]Summer phytoplankton bloom.
[f]Anoxic hypolimnion (15 m).

constants are compared (see discussion above), then contribution of other phosphohydrolytic enzymes to [^{32}P]ATP dephosphorylation becomes greater than previously determined for the Plußsee.

Because [^{32}P]ATP was a nonspecific substrate for determination of 5PNase activity and was also hydrolyzed by a variety of phosphohydrolases, [^{32}P]ATP was used to measure an overall release of [^{32}P]orthophosphate by phosphohydrolytic activity present in the water samples. Figure 6.9 presents the contri-

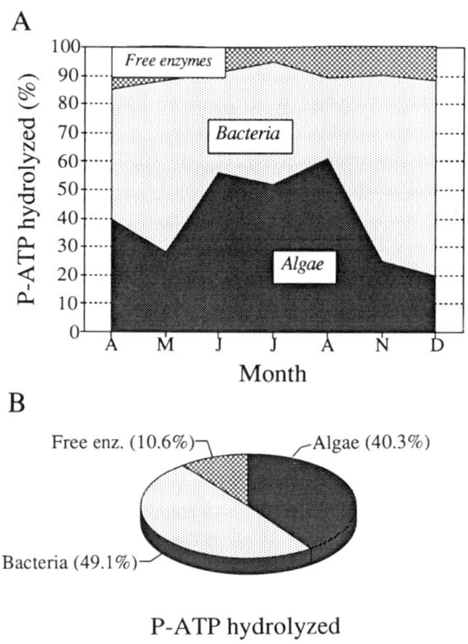

Figure 6.9. The contribution of activity of algal and bacterial size fractions of seston and extracellular free enzymes to [^{32}P]ATP hydrolysis (A) and average participation of seston fractions and free enzymes in [^{32}P]ATP dephosphorylation in the euphotic zone of the Plußsee (April–December 1989).

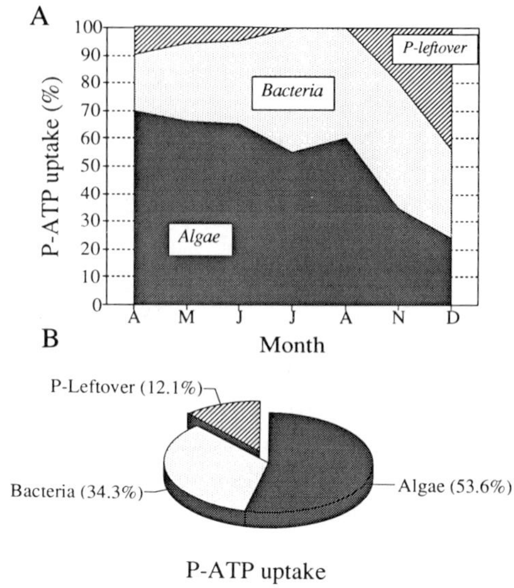

Figure 6.10. Seasonal (A) and average (B) percentage of uptake of [^{32}P]orthophosphate by algal and bacterial size fractions of seston during enzymatic dephosphorylation of [^{32}P]ATP and percentage of orthophosphate that was not taken up in the euphotic zone of the Plußsee (April–December 1989).

bution of phosphohydrolytic activity of bacterial and algal size fractions of microplankton and extracellular free enzymes to [^{32}P]ATP hydrolysis in the euphotic zone of the Plußsee. Phosphohydrolases associated with bacterial and algal size fractions liberated 28–68% and 20–61% of [^{32}P]orthophosphate, respectively. Highest percent of [^{32}P]ATP hydrolysis by bacteria was found in the spring (45–60%) and late autumn (65–68%). High participation of enzymes associated with the algal size fraction in [^{32}P]ATP dephosphorylation occurred in the summer, and algal phosphohydrolytic activity liberated 56–61% of [^{32}P]orthophosphate from the nucleotide (Fig. 6.9A). The contribution of extracellular free phosphohydrolases to the total hydrolysis of [^{32}P]ATP in water did not fluctuate markedly throughout the sampling period and varied from 9% to 15%. Enzymes of bacterial and algal size fractions and cell-free enzymes liberated on average 49.1 ± 15.2%, 40.3 ± 15.5%, and 10.6 ± 1.9% of [^{32}P]orthophosphate from [^{32}P]ATP (Fig. 6.9B).

From 88% to 100% of enzymatically released [^{32}P]orthophosphate from [^{32}P]ATP was simultaneously taken up and found in microplankton. Bacterial and algal size fractions assimilated 18–40% and 49–79% of [^{32}P]orthophosphate, respectively (Fig. 6.10A). Algae took more phosphorus (Fig. 6.10B) than they liberated from [^{32}P]ATP by their phosphohydrolases (Fig. 6.10B). On the other hand, phosphohydrolases associated with the bacterial size fraction released more [^{32}P]orthophosphate than bacteria could assimilate. These findings suggested that bacterial dephosphorylation of ATP could supply algae with additional amounts of inorganic phosphorus (Chróst 1986, 1988, Chróst and Overbeck 1987) and that bacteria were not P-limited and predominantly utilized the adenyl moiety rather than orthophosphate.

Significance of 5′-Nucleotidase to Phosphorus Mineralization

5PNase activity determined by the means of [^{32}P]AMP was proportional to the biomass of bacteria in water samples ($r^2 = 0.85$, $P \leq 0.005$), but the correlation between 5PNase and biomass of algae was statistically nonsignificant. This relationship suggested that bacteria, not algae, were the major producers of 5PNase activity in water from the Plußsee. Rates of 5PNase activity also positively correlated to the concentration of DOP ($r^2 = 0.78$, $P \leq 0.005$) and DOC ($r^2 = 0.92$, $P \leq 0.005$), suggesting that enzyme activity was dependent on the amount of 5′-nucleotides in the pool of organophosphoric compounds that was a part of the total DOC in lake water. Unfortunately, little is known about the variability of dissolved nucleotide concentrations in lakes and other freshwater environments. Few studies on the concentration of total dissolved adenosine (TDAN) nucleotides (AMP + ADP + ATP) were reported in marine ecosystems (McGrath and Sullivan 1981, Hodson et al. 1981a,b). In coastal seawater, the pool of adenylates determined periodically for one year varied between 1.2 and 5.6 nmol/L, (average 2.8 ± 1.7 nmol/L) and represented less than 0.1% of the total DOC.

From 15% to 85% of 5PNase-liberated [^{32}P]orthophosphate from [^{32}P]AMP and 45–95% of the [^{3}H]adenyl moiety of [^{3}H]AMP were simultaneously taken up by microplankton size fraction 0.2–2.0 μm, which was primarily occupied by bacteria (in this fraction 87–95% of the total bacterial number and only 2–8% of total chlorophyll$_a$ was found during the studies). The coupling of bacterial ortho-

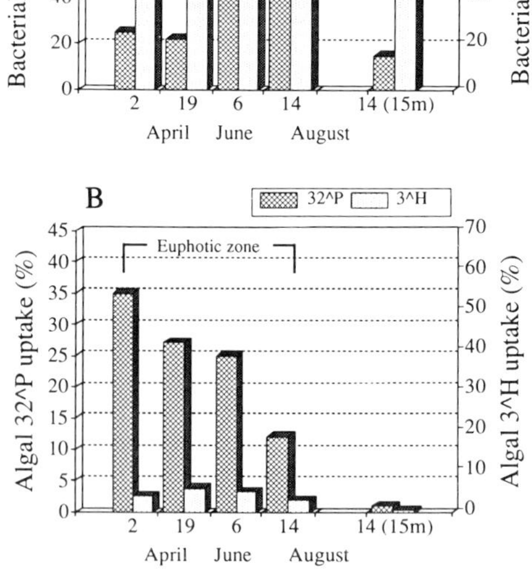

Figure 6.11. Percentage of uptake of [^{32}P]orthophosphate and [^{3}H]-adenyl moiety by bacterial (A) and algal (B) size fractions of seston during hydrolysis of [^{32}P]AMP and [^{3}H]AMP by 5′-nucleotidase in the euphotic zone of the Plußsee (April–August 1990).

phosphate uptake to 5PNase activity depended on the ambient concentration of DIP in water from the Plußsee (Fig. 6.11A, Table 6.2). The lowest bacterial uptake of [^{32}P]orthophosphate (corrected for dilution by naturally present nonradioactive orthophosphate) that was liberated from [^{32}P]AMP was found in the euphotic zone during spring and in the hypolimnion, i.e., in waters rich with DIP. This means that in non-P-limited environments most of the orthophosphate liberated by 5PNase hydrolysis was released into the water bulk phase and was available to phytoplankton (Fig. 6.11B). During DIP depletion in the euphotic zone in summer, the tightness of hydrolysis–uptake processes increased, and bacteria incorporated approximately 85% of [^{32}P]orthophosphate derived from [^{32}P]AMP hydrolysis by 5PNase. Similar dependence of the hydrolysis–uptake coupling system of 5PNase on P limitation in marine waters was recently reported by Ammerman and Azam (1991b). Turnover time of the TDAN pool in the euphotic zone of the Plußsee ranged from 5 h (August 14) to 3 days (April 2), which was in the range of TDAN turnover time found in marine ecosystems (McGrath and Sullivan 1981). Short turnover time and tight coupling of hydrolysis and P uptake during summer DIP depletion suggest that regeneration of inorganic phosphate by 5PNase is a very important process for P supply to bacteria. On the other hand, 5PNase hydrolysis of TDAN supplied bacteria with the adenyl moiety needed for nucleic acid synthesis. This study showed that bacteria simultaneously incorporated 60–95% (average 79.4 ± 11.4%) of [^{3}H]adenyl moiety liberated throughout the course of 5PNase hydrolysis of [^{3}H]AMP (Fig. 6.11A). Uptake of adenyl moiety by the algal size fraction was negligible and averaged 3.7 ± 1.9% of the total AMP hydrolysis (Fig. 6.11B).

The importance of 5PNase to organic phosphorus regeneration and overall P cycling in aquatic ecosystems depends on the concentration of dissolved nucleotides in the DOP pool that are available to the enzyme. It was suggested that the major inputs to the total dissolved nucleotide pool are from inefficient feeding of zooplankton on phytoplankton or grazing on bacteria by bacterivorous organisms, phage lysis, and decomposition of dead cells (McGrath and Sullivan 1981, Paul et al. 1991, Proctor and Fuhrman 1991, 1992, Turk et al. 1992, Maruyama et al. 1993). Release of nucleotides by healthy planktonic organisms (zooplankton, phytoplankton, protozoans, bacteria) probably contributes little to the dissolved pool because most organisms appear to conserve nucleotides, and many microorganisms possess metabolic salvage pathways for the recovery of free purines and pyrimidine that result from the hydrolytic degradation of nucleotides (Murray 1971). 5PNase hydrolysis of 5′-nucleotides seems to be coupled with hydrolysis of dissolved nucleic acids by nucleases that liberate nucleotides that serve as substrates for 5PNase. The presence of DNA-hydrolyzing bacteria and cell-surface bacterial DNase activity in seawater has been known for some time. Approximately 10–50% of the bacteria isolated from seawater could hydrolyze DNA, and 89% of those could also hydrolyze RNA (Maeda and Taga 1973). Rapid degradation of dissolved DNA and short turnover time of 6.5 h were recently reported in a marine environment (Paul et al. 1987, 1988). Concentrations of dissolved DNA and RNA in aquatic environments have ranges 0.05–81 μg/L and 7–51 μg/L, respectively (Karl and Bailiff (1989). This means that hydrolysis of combined maximal concentrations of both nucleic acids by nucleases may potentially liberate approximately 436 nmol/L of monophosphorylated nucleotides (assuming that average molecular weight of P nucleotide is about 300), and if they are further dephosphorylated by 5PNase, then the enzymatic hydrolysis of nucleotides in aquatic environments may yield about 13.5 μg P/L.

Much of the bacterial capacity for regenerating phosphorus from 5′-nucleotides is probably a consequence of phosphohydrolases such as 5PNase and APase that are regulated by the nutritional state of the bacterial cell with respect to organic carbon and nitrogen (purines and pyrimidine) but affect the supply of phosphorus to algae. This may stimulate metabolic activity of algae (and lead to higher primary production), which, through exudation of organic substrates into the water, can intensify bacterial secondary production in the euphotic zone of the lake during DIP depletion.

Decomposition of Polysaccharides: An Important Supply of Sugar Substrates to Heterotrophic Bacteria

Polysaccharides are common structural and storage polymers in both aquatic and terrestrial organisms and represent a major form of photosynthetically fixed carbon in the biosphere. In aquatic ecosystems carbohydrates may contribute 10–20% and more than 25% to the DOM and POM, respectively (Riley 1970, Ochiai and Ukiya 1981). Particulate and dissolved carbohydrates (DCHO) often constitute a major fraction of organic matter in waters and sediments (Münster 1984, Williams 1986; Chapter 2, this volume). DCHO are also major sources of carbon and energy for all bacteria (Fraenkel and Vinopal 1973), and they are of utmost importance for metabolism of aquatic heterotrophic bacteria (Overbeck 1979, Hagström et al. 1984). Only some of these compounds are ready for assimilation by bacteria (Billen et al. 1980, Burney 1986), and others must undergo further enzymatic hydrolysis prior to use (Chróst and Overbeck 1990). Therefore, the hydrolysis of polymeric carbohydrates is generally accepted as the first and rate-limiting step in their utilization by microheterotrophs (King 1986, Chróst 1989).

The great variety of dissolved carbohydrate compounds in freshwater lakes is a consequence of both autochthonous production and degradation of aquatic biota and flux of allochthonous carbohydrates from the surrounding area. Given their ubiquity and abundance, carbohydrates are potentially powerful tools in elucidating the sources, processes, and pathways of biologically important organic materials in natural waters. Based on the chemical structure and composition of carbohydrates it is possible to identify the origin of these compounds in aquatic environment (Cowie and Hedges 1984). Predomination of cellulose, hemicellulose, pectin, starch, and their complexes with humus materials in the carbohydrate pool indicates that it is mostly of terrestrial plant origin, whereas occurrence of glycogen may suggest that products of degradation of both animal and plant tissues contributed to the polysaccharide pool. Presence of chitin indicates that the major source of these carbohydrates were arthropod exoskeletons (Hackman and Goldberg 1981), molluscs (Jeuniaux et al. 1982), fungi and ascarides (Hackett and Chen 1978), or diatom products of photosynthesis (Smucker and Dawson 1986). Occurrence of peptidoglycans (i.e., glycan strands of β-1,4-linked amino sugars that are joined together with short peptides) and deoxysugars implies the contribution of bacterial cell wall components (Rogers 1979) and nucleic acid derivatives into the carbohydrate pool. In addition, carbohydrates also occur covalently bonded to other types of biomolecules, e.g., lipids (glycolipids) and proteins (glycoproteins).

β-Glucosidase

High diversity and ubiquity of polymeric carbohydrates are coupled to a variety of enzyme systems degrading these biopolymers in aquatic environments. Most of the hydrolytic enzymes acting on polysaccharides in waters and sediments are produced by heterotrophic microorganisms, and bacteria and fungi are certainly of the utmost importance in catalysis of these compounds (Chróst 1990, Chróst et al. 1989, Hoppe 1991, Smucker and Kim 1991, Boon 1991). The total enzymatic hydrolysis of polysaccharides, which predominate in the pool of organic matter in lakes, is extremely important for overall metabolism of microheterotrophs because it delivers to them significant amounts of readily utilizable monosaccharide substrates of high nutritional and energetic value (Chróst 1991c, Münster 1991). Total enzymatic hydrolysis of polysaccharides, however, is a complex process involving simultaneous action of different enzymes, both exo- and endoglucosidases, which consecutively divide a polymeric molecule into smaller subunits (Ljungdahl and Eriksson 1985, Coughlan 1991). The final step of the total hydrolysis of many polysaccharides (e.g., cellulose) is usually cleavage of the α- or β-glycosidic linkage of oligosaccharides (disaccharides), which results in liberation of monosaccharides (e.g., glucose).

β-Glucosidase (β-D-glucoside glucohydrolase, E.C. 3.2.1.21) is one of the most frequently found ectoenzymes hydrolyzing a variety of β-linked disaccharides (which are the byproducts of hydrolysis of many polisaccharides) in aquatic environments (Hoppe 1983, Somville 1984, King 1986, Meyer-Reil 1987, Chróst 1989, 1991a). β-Glucosidase (GLCase) is also one of the key enzymes involved in degradation of cellulose, which is the most abundant macromolecule in nature. The enzyme has high glycone specificity but low aglycone specificity. Aryl glycosides are in general better substrates than alkyl glycosides. GLCase is a broad-specificity enzyme that catalyzes the hydrolysis of β-linked (1→2, 1→3, 1→4, 1→6) disaccharides of glucose, celluhexose, and carboxymethylcellulose (Barman 1969).

Temperature of incubation had a pronounced effect on the kinetic parameters of GLCase in Plußsee water samples (Chróst 1989). At the optimal temperature of 28°C, the enzyme had lowest $^{a}K_{m}$ values, i.e., highest affinity for the substrate. GLCase also had high substrate affinity at temperatures between 15° and 30°C; however, both low (2–15°C) and high temperatures of incubation significantly decreased the affinity of an enzyme for substrate. The V_{max} of GLCase in lake water was strongly dependent on the pH of incubation; activity increased smoothly up to the optimal pH of 8.0 and rapidly decreased at pHs above 8.5. The pH optimum of GLCase tended to reflect the in situ value of the aquatic environment (King 1986, Chróst 1989).

β-Glucosidase Activity in the Euphotic Zone During Phytoplankton Bloom

During the spring phytoplankton bloom in April–May 1987, the kinetic parameters (V_{max}, ${}^{a}K_m$, and T_t) of GLCase showed a distinct temporal pattern in their fluctuation (Fig. 6.12A). V_{max} of β-glucosidase activity was low (4.83–10.38 nmol L^{-1} h^{-1}), and the apparent Michaelis constants did not vary markedly (105–110 μmol/L) during the first stage of bloom development, when algae grew rapidly (Fig. 6.12B). Algal biomass increased from 373.2 μg C/L at the beginning of the study to a maximum of 695.2 μg C/L after 2 weeks. V_{max} of GLCase began to increase rapidly during the bloom breakdown in the middle of April, and V_{max} reached the highest values (72.27–82.25 nmol L^{-1} h^{-1}) at the beginning of May, when chlorophyll$_a$ (4 μg/L, data not shown) and algal biomass (165.8 μg C/L) were very low. The increase in GLCase activity was proportional to the increase in biomass of bacteria in lake water (Fig. 6.12B). Bacterial biomass increased from 12.8 to 38.5 μg C/L during bloom development, and bacteria produced the highest biomass (68.4 μg C/L) after the bloom had broken down at the beginning of May.

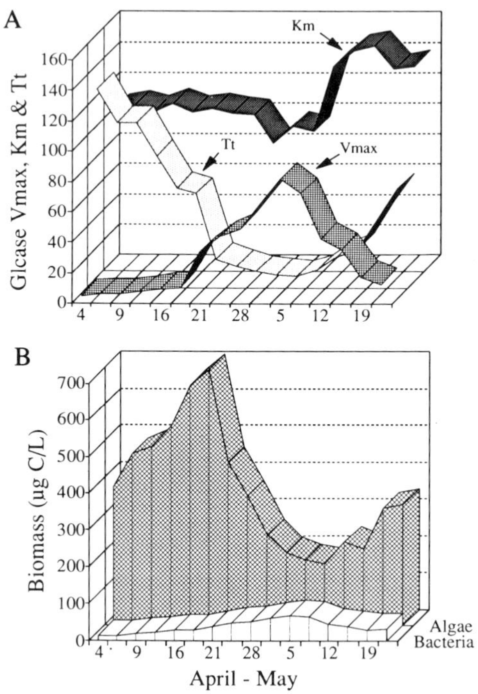

Figure 6.12. (A) β-Glucosidase V_{max} (nmol L^{-1} h^{-1}), turnover time (T_t, weeks), and apparent K_m (μmol/L), and (B) the biomass of algae and bacteria in the euphotic zone of the Plußsee during spring phytoplankton bloom (April–May 1987).

During this period the highest V_{max} values of GLCase were measured, and they were associated with the highest enzyme affinity for the substrate (lowest ${}^{a}K_m$ values). During the clear-water phase in the lake, which occurred in the second and third weeks of May, the V_{max} of GLCase gradually decreased (42.50–12.49 nmol L^{-1} h^{-1}), and apparent K_m values markedly increased from 84 to 148 µmol/L. The turnover time (T_t) of β-glucosidase varied between 7.3 and 130.5 weeks (Fig. 6.12A) and was the shortest during phytoplankton bloom collapse, when GLCase displayed the maximum activity. V_{max} of GLCase activity was proportional to the biomass of the bacteria ($r^2 = 0.88$, $P \leq 0.001$), contrary to algal biomass, which did not correlate significantly with activity of GLCase. The positive V_{max}–bacterial biomass relationship suggested that bacteria were the major producers of GLCase activity in the euphotic zone of the Plußsee.

This suggestion was further confirmed by measurements of GLCase activity in different size fractions of water samples. By differential filtration, the microplankton assemblages were crudely separated into the algal (larger than 3 µm) and bacterial (0.2–3.0 µm) size fractions. In the natural microplankton community in the Plußsee, the bulk of the photosynthetic microorganisms were retained by a 3.0-µm filter, i.e., in April, May, and September, on average, 97.7 ± 12.9% of total chlorophyll$_a$ was found in this seston fraction (Table 6.4). Photosynthetic picoplankton in the 0.2- to 3.0-µm size fraction contributed on average 2.3 ± 0.5% to the total chlorophyll$_a$. The highest chlorophyll$_a$ concentration (7–8% of the total) in the 0.2- to 3.0-µm size fraction was observed at the end of the phytoplankton bloom breakdown, which was accompanied by the development of small mixotrophic microflagellates (data not shown). The 0.2- to 3.0-µm size fraction was mainly associated with bacteria; i.e., on average 92% of the total number of bacteria in the unfiltered water sample passed through 3.0-µm filters. The highest bacterial abundance (12–22%) in the size fraction larger than 3.0 µm was noted at the end of the bloom, when a high content of suspended detritus colonized by bacteria was observed. Heterotrophic bacteria were the major producers of GLCase in the euphotic zone of the Plußsee because, on average, 88.9% of the total GLCase activity was found in the 0.2- to 3.0-µm size fraction, which also displayed the highest bacterial abundance and tritiated thymidine incorporation. Activity of free extracellular enzymes dissolved in the water was negligible and varied from 0 to 7% (mean 3.4%) of the total activity of GLCase. These results confirmed previous reports that most of the GLCase activity (>95%) in seawater and freshwater samples was associated with the bacterial size fraction (Hoppe 1983, Somville 1984, Chróst and Overbeck 1990).

Because bacteria were the major producers of β-glucosidase activity, it was possible to calculate the specific GLCase activity produced per unit of bacterial biomass, i.e., per microgram C biomass, during phytoplankton bloom development and breakdown. The specific activity of the enzyme began to increase gradually from 0.5 nmol µg C biomass^{-1} h^{-1} at the beginning of the bloom breakdown and reached the highest values of 4.2–4.5 nmol µg C biomass^{-1} h^{-1} during the late stage of phytoplankton collapse. The increase in β-glucosidase activity was proportional to the increase in secondary production of bacteria in the euphotic zone of the Plußsee (Chróst 1989).

Table 6.4. Percentage of β-Glucosidase Activity, Bacterial Number, Thymidine Incorporation, and Chlorophyll$_a$ Content in Size Fractions (μm) from the Euphotic Zone of the Plußsee Produced by Differential Filtration of Water through 3.0 μm and 0.2 μm Pore-Size Nuclepore Filters

Sampling	β-Glucosidase			Bacterial Number		Thymidine		Chlorophyll$_a$	
Date, 1987	0.2–3.0	>3.0	<0.2	0.2–3.0	>3.0	0.2–3.0	>3.0	0.2–3.0	>3.0
April 7	89	5	6	96	4	91	9	0	100
April 14	90	6	4	90	10	92	8	2	98
April 21[a]	80	13	7	88	12	94	6	7	93
April 28[a]	82	18	0	78	22	67	33	8	92
May 5	93	4	3	96	4	93	7	0	100
May 12	86	8	4	91	9	94	6	5	95
May 19	92	6	2	93	7	89	11	3	97
September 8	89	7	4	94	6	88	12	0	100
September 15	94	5	1	97	3	92	8	2	98
September 22	92	5	3	97	3	86	14	3	97
Average	88.9	7.7	3.4	92	8	88.6	11.4	2.3	97.7

[a]Phytoplankton bloom breakdown, high content of suspended particulate material in water.

β-*Glucosidase Activity in the Water Column*

Figure 6.13 presents GLCase activity and bacterial production during the late phase (May 12) of the bloom breakdown throughout the entire water column of the Plußsee. The lake was slightly stratified (temperature and oxygen) and had a thin metalimnetic layer between 4 and 6 m depth. The maximum of chlorophyll$_a$ (19.4 μg/L) occurred in the metalimnion at depth of 4 m. The highest rates of bacterial production (3.84 μg C L^{-1} h^{-1}) were also found in this layer of lake water. V_{max} of GLCase varied between 17.5 and 47.3 nmol L^{-1} h^{-1}, and there was distinct variation of the enzyme activity throughout the water column (Fig. 6.13A). Low ^aK_m values and high hydrolytic rates occurred in the metalimnion of the lake. GLCase V_{max} was almost uniformly distributed throughout the deep water, however, and had an average activity of 31 nmol L^{-1} h^{-1}. The distribution pattern of specific activity of GLCase in the water column was totally different in comparison to its V_{max} (Fig. 6.13B). Low specific activity of GLCase (1.47–1.71 nmol μg C biomass^{-1} h^{-1}) occurred in the epilimnion and metalimnion, and high specific activity (2.40–2.59 nmol μg C biomass^{-1} h^{-1}) of the enzyme was found in the hypolimnion of the lake. Michaelis constants of GLCase varied from 67 to 142 μmol/L, and the enzyme had the highest substrate affinity (lowest ^aK_m values) in the water layer between 0 and 8 m. Turnover time of substrate hydrolysis varied from 11.8 to 28.5 weeks and was shortest in the metalimnetic layer of the lake.

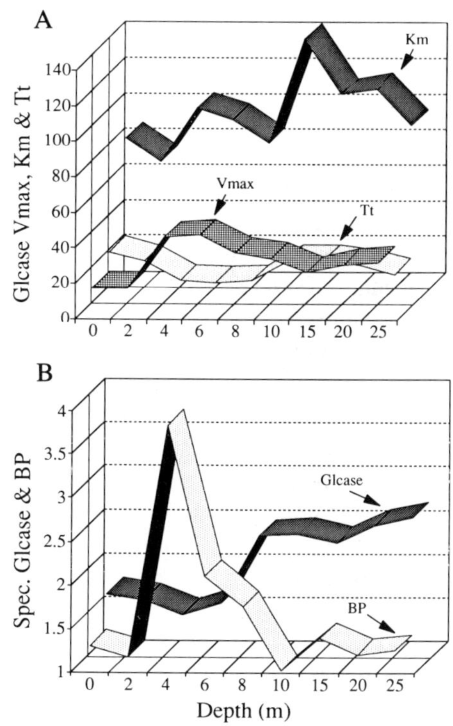

Figure 6.13. (A) β-Glucosidase V_{max} (nmol L^{-1} h^{-1}), apparent K_m (μmol/L), and turnover time (T_t, weeks), and (B) the specific activity of β-glucosidase (GLCase, nmol μg C biomass^{-1} h^{-1}) and bacterial secondary production (BP, μg C L^{-1} h^{-1}) in the entire water column of the Plußsee during late phase of phytoplankton bloom breakdown (May 1987).

Environmental Control of β-Glucosidase Synthesis and Activity

The specific activity of GLCase and the rates of uptake of glucose (which was the final product of substrate hydrolysis) and bacterial secondary production showed a strong positive correlation in Plußsee water sampled in euphotic zone during spring phytoplankton blooms in March–April 1984 (Chróst et al. 1989) and April–May 1987 (Chróst 1989). These studies showed that carbohydrate hydrolysis and the uptake of glucose were tightly coupled processes ($r^2 = 0.87$, $P \leq 0.001$) that could stimulate bacterial secondary production in lake water. The dependence of bacterial production on GLCase activity ($r^2 = 0.83$, $P \leq 0.001$) indicates that bacteria efficiently utilized polymeric carbohydrates, but it must be noted that the relationship between these two processes primarily depends on circumstances that regulate the rate of enzyme synthesis and activity in the microbial cell. Laboratory studies indicated that GLCase synthesis is induced in the presence of disaccharides in the ambient environment (Berry and Dekker 1984, Freer and Detroy 1985) or is regulated by the growth-rate-dependent derepression mechanism (Strobel and Russell 1987). Supplementation of lake water with cellobiose (a byproduct of cellulose degradation that is a disaccharide composed of two subunits of β-linked D-glucose) or glucose (final product of cellobiose hydrolysis by GLCase) induced GLCase synthesis or caused strong catabolite repression of enzyme synthesis, respectively (Chróst 1991a). The specific GLCase activity in Plußsee water during spring phytoplankton bloom in 1984 correlated well ($r^2 = 0.65$, $P \leq 0.005$) with the amount of the dissolved combined glucose, i.e., pool of natural substrates for the enzyme. The specific GLCase activity in spring 1987 also displayed a significant relationship ($r^2 = 0.62$, $P \leq 0.005$) with the concentration of the dissolved organic carbon, indicating indirectly that DC-glucose constituted a weighty portion of the total DOC pool in lake water (unfortunately, DC-glucose was not measured in 1987). Comparison of rates of phytoplankton exudation of organic carbon (EOC) and the specific activity of GLCase (spring 1984) revealed the strong negative correlation ($r^2 = -0.56$, $P \leq 0.005$) between these variables.

The above results of field studies suggested that β-glucosidase in lake water was under the control of complex regulatory mechanisms. These regulatory mechanisms may include (1) induction/derepression of synthesis, (2) repression of synthesis, and/or (3) inhibition of activity (Chróst 1989, 1990, 1991a). To verify some of the postulated regulatory mechanisms of GLCase synthesis and activity, Plußsee water samples were supplemented with glucose (final product of GLCase hydrolysis of disaccharides) and DOC extracted from spring phytoplankton (Siuda et al. 1991) composed of a variety of organic compounds, naturally present in the water, which are potential products of phytoplankton decomposition. Depending on the compound added, results of these experiments revealed the stimulation or inhibition of GLCase synthesis and activity in natural bacterial assemblages in Plußsee water (Fig. 6.14). The supplementation of lake water with organic substrates markedly increased bacterial cell numbers and biomass. The highest biomass of bacteria was produced in samples where algal DOC was added. Contrary to control samples, lake water supplemented with organic substrates did not

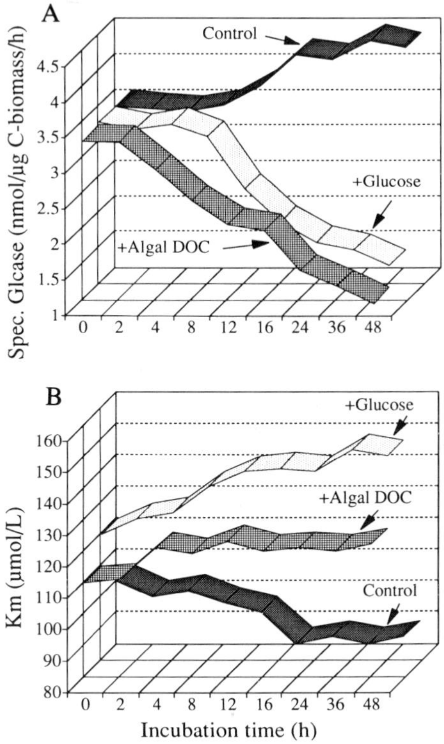

Figure 6.14. Effect of enrichment of water from the Plußsee with glucose and DOC extracted from phytoplankton (5 mg C/L of both substrates was added) on (A) specific activity and (B) apparent K_m of β-glucosidase (GLCase). Samples were collected from the euphotic zone (a depth of 0–1 m) in August 1988.

display a significant increase in V_{max} of GLCase. V_{max} of β-glucosidase in control samples increased markedly from 85 nmol L^{-1} h^{-1} at the beginning of experiment to 268 nmol L^{-1} h^{-1} after 48 h of incubation, whereas V_{max} in enriched samples increased to a range of 85–129 nmol L^{-1} h^{-1}. The specific activity of GLCase gradually decreased within 48 h from 3.49 to 1.47 nmol µg C biomass^{-1} h^{-1} and from 3.46 to 1.17 nmol µg C biomass^{-1} h^{-1} in samples supplemented with glucose and algal extract, respectively (Fig. 6.14A). A decrease in the specific activity of GLCase in enriched water samples suggested the repression of enzyme synthesis when bacteria were grown on readily utilizable substrates such as glucose and algal DOC (Chróst 1984, Chróst and Overbeck 1990, Chróst 1991a, Siuda et al. 1991). The specific activity of GLCase in control samples remained almost constant (3.41–3.49 nmol µg C biomass^{-1} h^{-1}) within the first 12 h of the experiment. However, within the next 36 h of incubation, the specific activity of the enzyme gradually increased from 3.78 to 4.42 nmol µg C biomass^{-1} h^{-1}, suggesting that within the first 12 h of incubation the readily utilizable substrates were removed, then bacteria began to synthesize GLCase at a higher specific activity to hydrolyze and utilize combined carbohydrates naturally present in the water. This assumption seems to be correct because in the control samples the affinity of GLCase to substrate increased during the same time period; i.e., the apparent K_m decreased from 105 to 85 µmol/L (Fig. 6.14B). The apparent Michaelis constants in samples supplemented with glucose increased markedly from 125 to 152 μmol/L, indicating repression of GLCase synthesis and/or inhibition of its activity (Fig. 6.14B).

In enrichment experiments reported previously (Chróst 1991a), readily utilizable bacterial substrates added to water samples acted as repressors, and they caused strong catabolic repression of enzyme synthesis. Moreover, in parallel to the catabolic repression of synthesis of "new enzyme protein," glucose inhibited the initial velocity of enzyme reaction because it drastically decreased the affinity of GLCase for the substrate. Competitive inhibition of GLCase activity in lake water by glucose has been well documented in previous reports (Chróst 1989, 1990, Chróst and Overbeck 1990).

To summarize the field studies on β-glucosidase activity described above during phytoplankton bloom development and breakdown in the Plußsee, the hypothetical model of GLCase synthesis and activity regulation can be postulated (Chróst 1989). During active growth of phytoplankton, algal populations excreted into the water photosynthetic products, including readily assimilable small molecules (Chróst 1981, Chróst and Faust 1983) that supported bacterial growth and metabolism (Cole et al. 1982, Chróst 1986, Chróst and Rai 1993). They (e.g., glucose; Chróst et al. 1989) inhibited the activity and repressed the synthesis of β-glucosidase in bacterial cells. During breakdown of the phytoplankton bloom senescent algae released a high amount of polymeric material (e.g., polysaccharides). These compounds induced the synthesis of β-glucosidase in bacterial cells and increased the enzyme specific activity. Another "switch-on" mechanism of GLCase could be the derepression (i.e., return to an active state) of enzyme synthesis by low concentrations of readily utilizable substrates (glucose, amino acids, etc.) in the water because of their rapid uptake by microheterotrophs. Catabolic repression of β-glucosidase by glucose is very well documented in pure cultures of many microbial strains (Smith and Gold 1979, Berry and Dekker 1984, Freer and Detroy 1985, Strobel and Russell 1987). Catabolic repression is, in its most basic sense, the inhibition of enzyme synthesis by better substrates, which are available for microbial uptake. It appeared to be an important factor controlling β-glucosidase synthesis in the Plußsee. Thus, when EOC, glucose, or other simple monomers were used as carbon sources, a high GLCase activity did not appear until the level of easily assimilable substrates in the water was nearly depleted.

Decomposition of Proteinaceous Compounds: Important Source of Amino Acids for Microbial Metabolism

Mode of Protein Hydrolysis and Utilization

Besides carbohydrates, proteins and their derivatives are widely distributed in nature, and in eukaryotic and prokaryotic organisms they are the most abundant class of biopolymers in cells. As is true of many biomolecules, proteins exhibit functional versatility and are therefore utilized in a variety of biological roles. Proteins are not only important functional regulators (such as enzymes and hormones) or protection agents (antibodies) but are also structural and storage components of living organisms. Obviously every cell contains thousands of different proteins that range in molecular weight from about 6000 (single protein chain) to several millions (protein complexes); all are polymers composed of covalently linked amino acids. Proteins are classified as fibrous and globular. The fibrous proteins are composed of polypeptides and are often cross-linked by various types of bonding to yield tough and water-insoluble structural proteins, e.g., α-keratin, collagen. Globular proteins are water soluble and have spheroidal, compact structures, and they are the biologically active proteins of living organisms.

Because all aquatic biota are protein-rich organisms, it is not surprisingly that proteins in the course of grazing, lysis, and death of aquatic micro- and macroorganisms are released into water and contribute greatly to the POM and DOM in aquatic environments (Jumars et al. 1989, Nagata and Kirchman 1991, 1992). Proteins of POM and DOM pools of organic matter in natural waters constitute excellent substrates that, after degradation to amino acids, can be utilized by a variety of both autotrophic and heterotrophic microorganisms. They serve as carbon, nitrogen, and energy sources and also are building subunits for synthesis of novel cellular proteins. Enzymatic degradation of proteins is a complex process involving the simultaneous action of various endo- and exopeptidases, which finally yield free amino acids or small peptides, i.e., products that can be transported into a microbial cell and then utilized in metabolic pathways (Keil and Kirchman 1992).

Ideas on proteolytic enzymes have changed considerably in recent years. It was previously thought that some enzymes, the proteinases, were adapted to attack only large molecules, whereas others only hydrolyzed small molecules. It is now realized that it is not a matter of the size of the substrate molecule at all; it is really a matter of specificity. Not all peptide links are hydrolyzed by all peptidases; whether a given enzyme will act or not depends on the nature of the other chemical groups in the neighborhood of the peptide bond (Dixon and Webb 1979). Endopeptidases do not require an adjacent terminal carboxyl or amino group and may even be prevented from acting by such groups, whereas exopeptidases require a free terminal group. The endopeptidases can thus act in the middle of a peptide chain, breaking protein molecules into smaller fragments. The exopeptidases cannot hydrolyze bonds in the middle of the protein molecule. Some act from the carboxyl (carboxypeptidases) and others from the amino end of the chain (aminopeptidases) and remove the terminal amino acids successively, one at a time. The endo- and exopeptidases cooperate very effectively in protein degradation; it may be said that the main function of the former is to produce a large number of free ends on which the latter can act.

Protein degradation in water from the Plußsee was studied by means of leucine aminopeptidase (AMPase), the model enzyme that hydrolyzes proteins and peptides to yield free amino acids (Jones and Lock 1991). Leucine aminopeptidase (L-leucyl-peptide hydrolase, E.C. 3.4.1.1) is widely distributed in aquatic habitats (Halemejko and Chróst 1986, Chróst et al. 1986a, 1989, Hoppe et al. 1988, Mayer 1989, Hoppe 1991, Billen 1991, Jacobsen and Rai 1991, Martinez and Azam

1993). Most studies of aquatic environments have indicated that ectoenzymatic activity of AMPase is always associated with heterotrophic bacteria (Rego et al. 1985, Rosso and Azam 1987, Jacobsen and Rai 1988, Cunningham and Wetzel 1989, Keil and Kirchman 1992). AMPase is a zinc-containing metalloenzyme (molecular weight approximately 300,000) requiring Mg^{2+} and Mn^{2+} ions for activation. Ions of heavy metals (Cu^{+}, Cd^{2+}, Hg^{2+}, and Pb^{2+}) are completely inhibitory (Smith and Hill 1960). The enzyme hydrolyzes a large number of peptides and amino acid amides of the L-configuration; however, L-leucyl peptides and L-leucyl amides are the best substrates. AMPase from lake water samples had an optimum temperature of 20°C. The enzyme displayed a rapid increase of activity between 15 and 20°C and a strong drop of activity at temperatures above 20°C. The activity of AMPase increased with an increase of pH from 5 to 7.5 and decreased rapidly at a pH of 8 to 9 (Hałemejko and Chróst 1986).

In natural environments, aminopeptidases coexist with other proteolytic enzymes and create the microbial enzyme systems required for utilization of exogenous proteinaceous materials. The utilization of exogenous peptides and proteins by bacteria requires their degradation to small peptides and amino acids before cellular uptake, and the direct uptake of utilizable proteins and large peptides probably does not occur as such in bacteria (Law 1980). The role of aminopeptidases in protein and peptide utilization by bacteria is clearly seen in the context of advances in our understanding of peptide and amino acid transport. Most of the known aminopeptidases act as exoectoenzymes and presumably contribute to the protein-utilizing abilities of bacteria by liberating free amino acids directly from peptides that are too large for direct transport into the organism. The scheme for the use of proteinaceous materials by bacteria producing proteolytic enzymes would be expected to follow a pattern of extracellular protein and polypeptide degradation to amino acids and peptides, whose maximum transportable size appears to be limited to about seven amino acid residues. Before further utilization, transported peptides are hydrolyzed to their constituent amino acids by intracellular peptidases of wide specificity (Law 1980). Amino acids not only serve as components for intracellular protein synthesis but are also catabolized as carbon and nitrogen sources.

Temporal Changes of Aminopeptidase Activity in the Euphotic Zone

Environmental studies of AMPase activity carried out from April to October 1989 in the Plußsee indicated that the enzyme had the highest V_{max} during the collapse of spring (3275 nmol L^{-1} h^{-1}, end of April) and summer (2297 nmol L^{-1} h^{-1}, end of July) phytoplankton blooms (Fig. 6.15B). AMPase was actively produced when the concentration (36–426 μg C/L) of dissolved combined amino acids (DCAA), products of algal cells degradation, increased markedly in the water, indicating the induction of the enzyme synthesis. The enzyme activity in surface water (0–3 m) correlated significantly ($r^2 = 0.86$, $P \leq 0.0005$) to bacterial biomass (Fig. 6.15A), and on average 86% of the total activity was found in the 0.2 to 2.0-μm size fraction of microplankton, which was predominantly occupied by bacteria (on

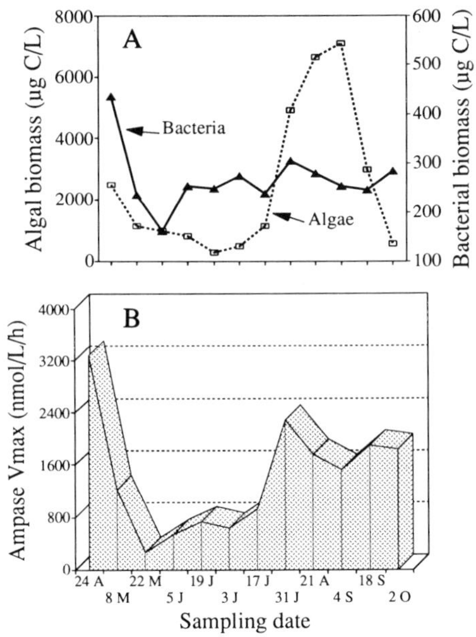

Figure 6.15. (A) Algal and bacterial biomass and (B) V_{max} of aminopeptidase (AMPase) in the euphotic zone of the Plußsee (April 24 to October 2, 1989).

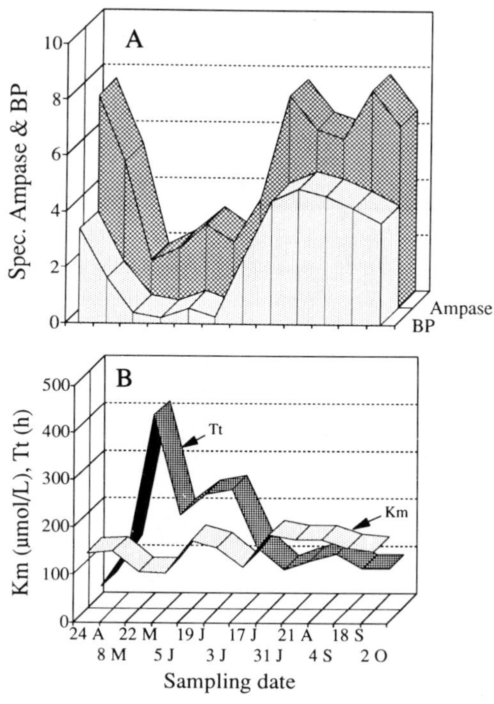

Figure 6.16. (A) The specific activity of aminopeptidase (AMPase) per unit of bacterial biomass (nmol µg C biomass^{-1} h^{-1}) and bacterial secondary production (BP, µg C L^{-1} h^{-1}), and (B) apparent Michaelis constant (K_m) and turnover time (T_t) of peptide degradation in the euphotic zone of the Plußsee (April 24 to October 2, 1989).

Table 6.5. Percentage of Aminopeptidase Activity, Bacterial Biomass, and Bacterial Secondary Production in Size Fractions (µm) from the Euphotic Zone of the Plußsee Produced by Differential Filtration of Water through 2.0-µm- and 0.2-µm-Pore-Size Nuclepore Filters

	Aminopeptidase			Bacterial Biomass		Bacterial Production	
Sampling Date	0.2–2.0	>2.0	<0.2	0.2–2.0	>2.0	0.2–2.0	>2.0
April 24[a]	85	11	4	82	18	88	12
May 22	91	6	3	90	10	92	8
June 19[b]	92	2	6	93	7	91	9
July 31[c]	86	6	8	87	13	90	10
September 4[c]	88	8	4	86	14	88	12
October 2[a]	76	18	6	81	19	83	17
Average	86.3 ± 5.2	8.5 ± 5.0	5.2 ± 1.7	86.5 ± 4.2	13.5 ± 4.2	88.7 ± 2.9	11.3 ± 2.9

[a]Phytoplankton bloom breakdown.
[b]Clear-water phase.
[c]Actively growing algae.

average 86.5–88.7% of the total bacterial biomass and production was measured in this fraction of seston). Activity of free extracellular AMPase dissolved in the water was negligible and constituted only 8.5% of the total AMPase activity of unfiltered samples (Table 6.5).

The specific activity of AMPase (per unit of bacterial biomass) displayed a similar temporal distribution pattern as found for its V_{max} and varied from 1.63 to 7.75 nmol μg C biomass^{-1} h^{-1}. Lowest specific AMPase activity was noted between phytoplankton blooms, during the clear-water phase in the lake. High specific activity of the enzyme in the euphotic zone corresponded to high rates of bacterial secondary production (Fig. 6.16A). The apparent Michaelis constants (K_m) of AMPase fluctuated between 116 and 184 μmol/L and did not show a correlation to the enzyme V_{max} or specific activity (Fig. 6.16B). The turnover time of protein hydrolysis varied from 44 (during spring bloom collapse) to 405 h (clear-water phase in the lake) and was relatively short (80–114 h) during the summer period, when epilimnetic water had 18–21°C, temperature close to the optimal for AMPase activity (Fig. 6.16B).

Depth Distribution of Aminopeptidase Activity

V_{max} of AMPase activity was always higher in the euphotic zone than in the profundal of the lake during the study period. Activity of AMPase in deep lake waters was strongly dependent on the sedimentation rates of detritus produced in the euphotic zone. Usually lag periods were observed between the maxima of AMPase activity in the surface and deep waters of the lake. An example of distribution of AMPase kinetics and activity in the summer stratified water column (August 21, 1989) is presented in Figures 6.17 and 6.18.

Lake water was sampled during the phase of active growth of summer phytoplankton bloom. Algal biomass production decreased dramatically from 6957 to 390 μg C/L between the depth of 0 and 4 m and was very low in deep waters of the lake. The distribution of bacterial biomass showed three maxima: at the surface (260 μg C/L), at the boundary (6 m depth) between epi- and metalimnion (202 μg C/L), and at a depth of 25 m (184 μg C/L), i.e., 1 m above the sediments (Fig. 6.17A). V_{max} of AMPase activity gradually decreased from 2328 nmol L^{-1} h^{-1} at the surface to 440–478 nmol L^{-1} h^{-1} at 10–15 m and increased again to 807–1226 nmol L^{-1} h^{-1} in the deepest waters of the Plußsee (Fig. 6.17B).

The specific activity of AMPase did not covary strongly with the enzyme V_{max} in the water column. There were three sampling depths where the enzyme had distinct high specific activity: 9.93 nmol μg C biomass^{-1} h^{-1} (at 2 m), 6.41 nmol μg C biomass^{-1} h^{-1} (at 6 m), and 6.66 nmol μg C biomass^{-1} h^{-1} (at 25 m; Fig. 6.18A). Bacterial secondary production within the sampling profile varied from 0.37 to 4.56 μg C L^{-1} h^{-1}, and the surface water (0–2 m) displayed the highest rates; it decreased continuously in deeper waters (2–20 m), approaching a higher value at 25 m (Fig. 6.18A). Apparent K_m values of AMPase fluctuated between 91 and 226 μmol/L and were proportional to the concentration of DCAA (170–280 μg C/L; data not shown) in the Plußsee water column (Fig. 6.18B). Short turnover

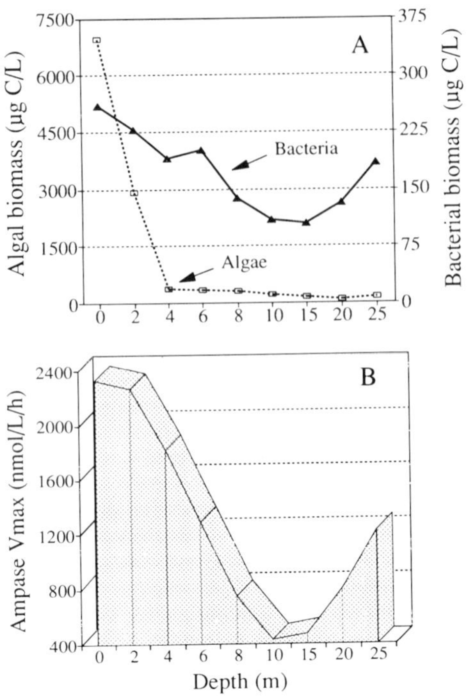

Figure 6.17. (A) Algal and bacterial biomass and (B) V_{max} of aminopeptidase (Ampase) activity throughout the entire water column of the Plußsee (sampled on August 21, 1989).

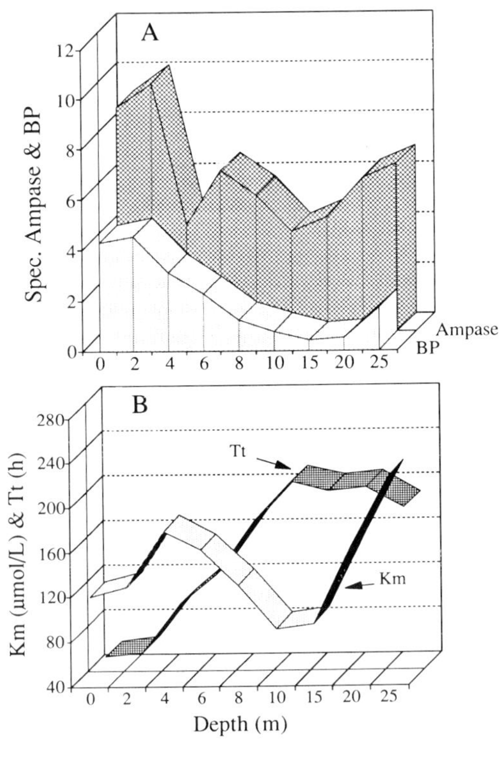

Figure 6.18. (A) The specific activity of aminopeptidase (AMPase) per unit of bacterial biomass (nmol µg C biomass^{-1} h^{-1}) and bacterial secondary production (BP, µg C L^{-1} h^{-1}), and (B) apparent Michaelis constant (K_m) and turnover time (T_t) of peptide degradation in the water column of the Plußsee (sampled on August 21, 1989).

time of protein degradation was observed in surface waters (52–56 h), and the longest T_t (184–208 h) was found in the profundal zone (Fig. 6.18B).

Aminopeptidase, Dissolved Organic Constituents, and Bacterial Production in Lake Water

The specific activity of AMPase from the euphotic zone of the Plußsee was inversely proportional ($r^2 = -0.62$, $P \leq 0.005$) to the rates of organic carbon exudation (EOC) by algae. It can be hypothesized that synthesis of bacterial AMPase was repressed, and/or the enzyme activity was inhibited, because a variety of readily utilizable constituents were exuded by photosynthesizing algae (Münster and Chróst 1990), which directly supported bacterial metabolism during the active growth phase of phytoplankton (Chróst 1986; Chapters 4 and 5, this volume). Bacteria readily utilized low-molecular-weight organic substrates (e.g., amino acids); therefore, AMPase synthesis for enzymatic degradation and catabolism of peptides and proteins in bacterial cells was unnecessary.

Dissolved combined amino acids (DCAA) constituted from 0.41% to 4.07% (average 2.43 ± 0.75%) of the total DOC pool in the Plußsee water samples. Thus, under steady-state conditions only a small portion of the total DOC could be degraded by peptidases. Such a situation, however, does not exist in the lake because proteins are continuously supplied into the water and are quickly enzymatically degraded; therefore, the ambient concentration of DCAA represents only the dynamic equilibrium between protein flux and its degradation and utilization. The specific activity of AMPase correlated significantly to DCAA concentration ($r^2 = 0.83$, $P \leq 0.0001$) in the profundal zone of the Plußsee, suggesting that synthesis of the enzyme was directly dependent on the presence of protein substrates in lake water. Induction and synthesis of AMPase in the euphotic zone of the lake were not instantly influenced by DCAA concentration, and other environmental parameters regulated AMPase production in bacterial cells. There was a nonsignificant relationship between the specific AMPase activity and DCAA content in the euphotic zone of the Plußsee.

The correlation between the specific activity of AMPase and DOC concentration was highly significant for the profundal samples ($r^2 = 0.77$, $P \leq 0.0005$), indicating that DCAA were important contributors to the DOC pool. This relationship was statistically less significant in the euphotic zone ($r^2 = 0.34$, $P \leq 0.05$) of the lake, which suggests that regulation of AMPase synthesis was more complex, and perhaps both induction by DCAA and repression of the enzyme by other constituents of the DOC pool (e.g., free amino acids, EOC) were responsible for the enzyme production in bacteria.

Bacteria had a very efficient peptide hydrolysis–uptake coupling system because the relationship between V_{max} of AMPase and uptake of leucine (a model amino acid studied) was very tight in the euphotic ($r^2 = 0.82$, $P \leq 0.0001$) and profundal ($r^2 = 0.94$, $P \leq 0.0001$) zones of the Plußsee. Degradation of peptides by AMPase produced by bacteria supplied their metabolism with free amino acids, which were immediately taken up by microorganisms. Continuous supply of bacterial metabolism with amino acids via AMPase hydrolysis of peptides was very important to bacterial secondary production in the profundal zone of the lake, where labile organic substrates exuded by phytoplankton were not present in the water. Bacterial production correlated significantly to V_{max} of AMPase ($r^2 = 0.76$, $P \leq 0.0005$) in the profundal zone of the Plußsee. There was no tight relationship between bacterial secondary production and V_{max} of AMPase in the euphotic zone of the lake, which means that, contrary to the profundal zone, bacteria living in the presence of photosynthesizing algae mostly utilized algal EOC, and the degradation of peptides by AMPase was unnecessary. This hypothesis was also suggested by the inverse relationship between the specific activity of AMPase and EOC in the lake.

Control of Aminopeptidase Synthesis and Activity in Aquatic Bacteria

Previous laboratory studies suggested that AMPase synthesis in aquatic bacteria is subject to control by catabolic repression (Chróst 1991a). The specific activity of AMPase drastically decreased when natural bacterial assemblages grew in the Plußsee water samples supplemented with readily transportable and utilizable low-molecular-weight substrates (acetate, glucose, amino acids). AMPase specific activity of bacteria grown in samples supplemented with acetate and glucose decreased markedly only after 20 h and 44 h of incubation. Samples to which a mixture of amino acids was added displayed a rapid decrease in AMPase activity just after 4 h. Further lowering of activity was observed after 20 and 44 h. The affinity of AMPase for its substrate in samples supplemented with acetate and glucose was almost unchanged in comparison to the control, as indicated by the $^{a}K_m$ values. Augmentation of water samples with amino acids not only repressed "new enzyme protein" synthesis but also decreased AMPase affinity for the substrate (significant increase of $^{a}K_m$ values was found) through inhibition of the initial velocity of the enzyme reaction. The inhibition had a competitive nature. A complex pattern of bacterial AMPase response to dipeptides and protein was observed when Plußsee water samples were supplemented with leucyl-leucine, alanyl-serine, or albumin. The addition of dipeptides, which are directly transported through the bacterial cytoplasmic membrane (Payne 1980), slightly decreased specific activity of the enzyme synthesized by bacteria in comparison to control samples, where specific (per-cell) activity increased slightly after 24 and 48 h of incubation. When bacteria grew in samples enriched with albumin (approximate molecular weight 66,000), which is not readily transportable, rapid induction of AMPase synthesis in bacterial cells was observed. Newly synthesized AMPase also had a higher affinity for the substrate, i.e., $^{a}K_m$ values decreased after 24 h of incubation. At this stage of bacterial growth and AMPase synthesis, when water samples were fortified with readily utilizable glucose, the enzyme specific activity gradually decreased after 48, 60, and 72 h. This catabolic repression of AMPase synthesis was coupled with the inhibition of the enzyme's initial velocity because of the decline in AMPase affinity for the substrate; i.e., a strong increase of $^{a}K_m$ values after glucose supplementation was found (Chróst 1991a).

The results of the experiments on AMPase kinetics in lake water demonstrated that the enzyme synthesis and activity were under different control mechanisms, which were dependent on the physico-chemical conditions of the studied habitat. There is ample evidence for general catabolite repression of proteolytic enzyme synthesis by readily utilizable sources of carbon (Boethling 1975, Litchfield and Prescott 1976) as well as more specific end-product repression by amino acids (Glenn 1976). However, control of the aminopeptidases appears to be distinct and more complex. In some bacteria, amino acids, peptides, and/or proteins seem to induce aminopeptidase synthesis (Daatselaar and Harder 1974, Litchfield and Prescott 1976). It is not known specifically how aminopeptidase induction operates, especially since amino acids were reported to act as inducers in some bacteria, rather than acting in their more predictable role as end-product inhibitors.

The ability of bacteria living in the euphotic zone of the lake to produce AMPase seems to be strongly affected by the availability of the low-molecular-weight readily utilizable substrates that are exuded by algae and are known to be excellent substrates for bacteria (Cole et al. 1982, Coveney 1982). Chróst and Rai (1993) found that the rates of AMPase and GLCase production by aquatic bacteria strongly depend on the bacterial organic carbon demand. When the amount of readily utilizable substrates of algal EOC covers the bacterial organic carbon requirement, microorganisms do not synthesize enzymes for hydrolysis of the polymeric substrates because their utilization is unnecessary. Such a situation might occur in the Plußsee euphotic zone, where the specific activity of AMPase and bacterial production did not correlate to DCAA and AMPase V_{max}, respectively. Moreover, the specific activity of AMPase correlated negatively to the rates of algal EOC. The above relationships confirmed that bacterial production in the euphotic zone was dependent mainly on algal exudation of EOC (Chapter 5, this volume). On the other hand, when bacterial growth and metabolism are limited by readily utilizable substrates, bacteria produce enzymes with high specific activity to degrade and utilize proteins and peptides (and other directly nonutilizable compounds). Bacteria living in the profundal zone of the lake are often substrate limited (Chróst et al. 1988; see Chapter 5, this volume) because the amount of substrate in deep waters primarily depends on the sedimentation rates of organic matter produced in the euphotic zone. There is no direct supply of labile organic compounds exuded by algae. Moreover, during sedimentation processes within the water column, the labile organic compounds are utilized, and sinking constituents that approach the profundal zone are mostly polymers, which are not readily utilizable compounds. Under such environmental circumstances bacterial metabolism and production are strongly dependent on the activity of synthesized ectoenzymes and the presence of their polymeric substrates. Therefore, it is not surprising that the specific activity of AMPase and bacterial production in the profundal zone of the Plußsee correlated significantly to DCAA and V_{max} of AMPase, respectively.

Enzymatic Solubilization of POM

One of the important sources of dissolved organic matter in lake waters is ectoenzymatic solubilization of detrital particulate organic matter (Halemejko and Chróst 1984, 1986, Hoppe 1986). Different microorganisms (bacteria, fungi, protozoa) are known to be closely associated with organic particles, and through stepwise ectoenzymatic hydrolysis of insoluble particulate material, they may liberate a substantial portion of solutes (Hoppe et al. 1988, Chróst 1989, 1992, Chróst et al. 1989). An interesting but still controversial issue is whether microorganisms hydrolyze particulate organic matter in a biochemically controlled process and assimilate most of the hydrolysis products, "or whether much of the hydrolysate diffuses into the environment as DOM" (Azam and Cho 1987).

There is some evidence that attached bacteria are highly effective in ectoenzyme production. The specific activity of some ectoenzymes, calculated per bacterium, may be 2 to 20 times higher in attached cells than in free-living bacteria (Hoppe 1991, Chróst 1990). Our studies indicate that the specific activity of β-glucosidase and aminopeptidase produced by bacteria attached to particles (>2 μm) was on average 3.4 and 4.9 times higher, respectively, than the enzyme activity associated with free-living cells of the 0.2- to 2.0-μm size fraction (Table 6.6, Fig. 6.19A and C). Moreover, the specific uptake rates of glucose (end

Table 6.6. Ratio between the Specific Activity[a] of β-Glucosidase (GLCase) and Aminopeptidase (AMPase) Produced by Attached and Free-Living Bacteria, and Ratio Between the Specific Uptake of Glucose and Leucine of Free-Living and Attached Bacteria in the Euphotic Zone of the Plußsee

	Enzyme Activity Attached/Free-Living		Product Uptake Free-Living/Attached	
Sampling Date	GLCase[b]	AMPase[c]	Glucose[b]	Leucine[c]
April 9	2.95	nd[d]	3.33	nd
April 18	2.13	nd	2.85	nd
May 5	3.07	nd	3.76	nd
May 8	4.36	nd	3.70	nd
April 24	nd	1.61	nd	3.04
May 8	nd	3.36	nd	4.26
July 3	nd	3.79	nd	4.25
July 31	nd	11.34	nd	4.32
September 18	nd	5.76	nd	5.21

[a]Specific enzyme activity and uptake were calculated per bacterium.
[b]Data from 1987.
[c]Data from 1989.

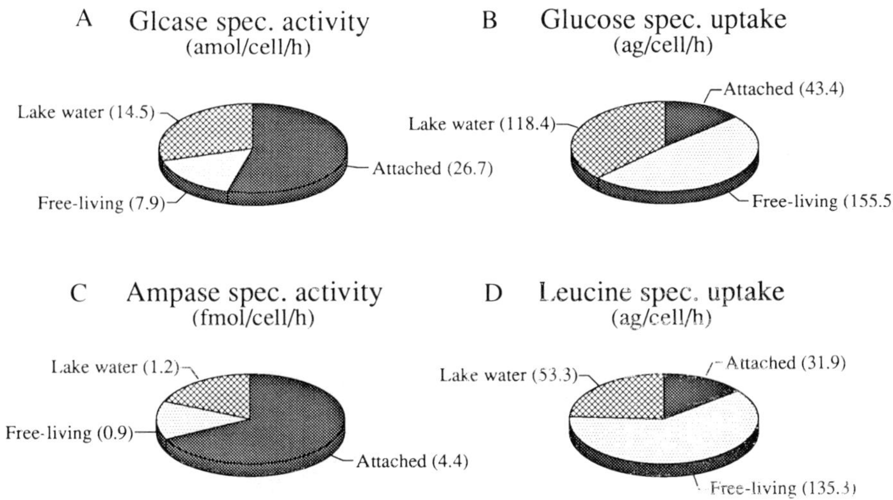

Figure 6.19. Comparison of the specific activity of (A) β-glucosidase (GLCase) and (C) aminopeptidase (AMPase), and the specific uptake of (B) glucose and (D) leucine in free-living (0.2- to 2.0-μm seston fraction) and attached bacteria (size fraction larger than 2 μm) and in unfiltered lakewater of the Plußsee. Numbers in parentheses show average values estimated for triplicates taken on six sampling dates in the euphotic zone. GLCase and glucose uptake data are from April–June 1987; AMPase and leucine uptake were measured in April–September 1989.

product of β-glucosidase hydrolysis) and leucine (end product of aminopeptidase catalysis) of attached bacteria were on average 3.6 and 4.2 times lower, respectively, than that of free-living bacteria (Table 6.6, Fig. 6.19B and D). Similar observations were reported by Jacobsen and Azam (1985), who found that bacteria colonizing ^{14}C-labeled fecal pellets released into the water two or three times as much ^{14}C as they assimilated.

Concluding Remarks

Microbial Processing of Organic Matter

The stepwise degradation of polymeric organic matter is obviously mediated by heterotrophic bacteria, and their hydrolytic enzymatic activities are of the utmost importance in this process. Thus, polymers, in this sense, are biologically the most important nutritional sources, and compounds such as proteins, polysaccharides (cellulose, chitin, starch), and organophosphoric esters (sugar phosphates, nucleic acids, and nucleotides) predominate in aquatic nonpolluted environments. Heterotrophic bacteria are excellent producers of these hydrolytic enzymes, and their hydrolysis of polymers is an acknowledged rate-limiting step in the microbial utilization of DOM and POM in aquatic environments. Most of the bacterial

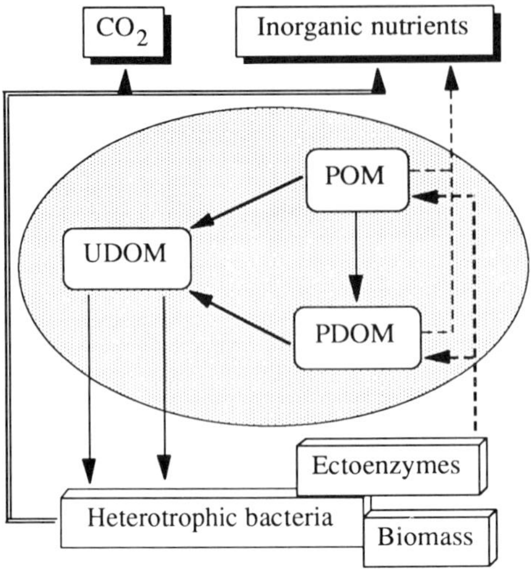

Figure 6.20. Scheme of enzymatic transformation of organic matter (oval shaded area) in a lake showing the role of ectoenzymes in bacterial biomass production and regeneration of inorganic nutrients and carbon dioxide. Thick broken lines show ectoenzymatic solubilization and hydrolysis of the particulate organic matter (POM) and the polymeric dissolved organic matter (PDOM), respectively; thin broken lines display regeneration of inorganic nutrients, and thick solid lines demonstrate the creation of utilizable dissolved organic matter (UDOM) in the course of POM and PDOM ectoenzymatic hydrolysis; thin solid lines and stripes exhibit UDOM assimilation and its mineralization by heterotrophic bacteria.

enzymes responsible for degradation of polymeric organic matter are inducible catabolic enzymes, which are associated with the bacterial cell surface or located in the periplasmic space of gram-negative bacteria. They are called "ectoenzymes" and are involved in the degradation of nonpenetrating cell membrane substrates and polymers, which are not continuously available in the water.

The microorganisms that produce ectoenzymes are presumably superior competitors for organic and inorganic nutrients in aquatic environments. Through the production of ectoenzymes, they are capable of utilizing a variety of polymeric compounds that are not otherwise utilizable. Ectoenzymes react on polymeric substrates outside the cell. They catalyze hydrolytic conversion of high-molecular-weight compounds to low-molecular-weight subunits, the final products of ectoenzyme reaction. Ectoenzymes solubilize organic particles, convert POM to DOM, and transform nonutilizable compounds of the organic matter pool to readily transportable and utilizable substrates (UDOM) for microorganisms (Fig. 6.20). Because these very efficient degradative enzyme systems are located at the cell surface, they make UDOM directly available for substrate-transporting systems. This gives an increased chance of survival to bacteria in aquatic environments when easily available energy and nutrient sources become limiting and also enables them to increase their growth and biomass production. Obviously the production of ectoenzymes close to the cell that degrade substrates resistant to attack by other microbial species may allow the producer to dominate an ecosystem through its nutritional advantage. Moreover, bacteria producing the ectoenzymes form a vital part of aquatic ecosystems in which they not only make carbon, phosphorus, and nitrogen available to themselves by degradation but also interact with algae supplying them with nutrients.

The nutritional status of the environment and physiological state of the microbial cell regulate the rates of ectoenzyme synthesis and activity in the Plußsee. Ectoenzyme synthesis is under a repression/derepression control mechanism. The production of studied ectoenzymes in aquatic bacteria and alkaline phosphatase in algae is repressed when concentrations of readily utilizable substrates fulfill bacterial (or algal) requirements for metabolism and growth. The synthesis of ectoenzymes only becomes derepressed once the concentration of readily utilizable substrates in the water falls below a critically low level. Using the repression strategy for ectoenzyme synthesis, microorganisms could avoid the wasteful production of inducible enzymes that are not useful when their growth is not limited by carbon, nitrogen, or phosphorus sources. It was also documented that synthesis of studied ectoenzymes in aquatic microorganisms was repressed by the end product that was derived from the substrate and accumulated in the cell or in the surrounding environment. Thus, ectoenzyme repression was tightly coupled to the availability of DOM in lake water.

Both spatial and seasonal ectoenzymatic activities fluctuated markedly in the Plußsee environment. The production of ectoenzymes by microorganisms was strongly correlated to the influx of polymeric organic matter and/or the depletion of readily utilizable UDOM in the environment. The highest ectoenzyme activities were observed in the euphotic zone of the lake, indicating that both the phytoplankton production of organic matter and its enzymatic decomposition occurred simultaneously, both spatially and temporally. There was also a strong covariation of enzymatic decomposition of organic matter with bacterial numbers and bacterial secondary production.

Ectoenzyme production and activity also showed marked seasonal variation in both surface and deep waters of the lake. In surface waters, the maxima of ectoenzymatic activities occurred during the late stage of phytoplankton bloom development and after its breakdown, and the minima of activities were recorded during the clear-water phase in the lake. Ectoenzymatic activity was lower in deep lake waters than in the surface layers, especially during the summer thermal stratification, and the activity was strongly dependent on the sedimentation rates of detritus produced in the euphotic zone. Usually a lag period was observed between the maximum ectoenzymatic activity in surface and deep waters of a lake. The seasonal and spatial fluctuations of ectoenzymatic activity clearly indicated that ectoenzymes were coupled to the decomposition of both particulate and dissolved organic matter in the Plußsee and that they played a crucial role in its mineralization by microorganisms.

Significance of Bacterial Enzymatic Activity for Functioning of Lake Ecosystem

The concept of the "microbial loop" (Azam et al. 1983, Hagstrom et al. 1988) has brought new light and attention to studies of composition and mechanisms of utilization of organic matter in aquatic environments. From 10% to 50% of the total organic carbon formed by phytoplankton primary production is transformed into bacterial biomass (Bell and Kuparinen 1984, Lovell and Konopka 1985; Chapter 5, this volume), channeled to the microbial loop, and subsequently transferred to higher trophic levels through the grazing food chains (Chapters 11 and 12, this volume). The metabolic regulations within the trophic levels and their interspecific relationships are the most important mechanisms steering the flow of organic matter in aquatic food webs.

Heterotrophic bacteria operating on every level in aquatic food webs are responsible for the regeneration of nutrients and the transfer of carbon and energy within the whole lake ecosystem. Above all, they are the only single component of aquatic biota that converts the DOM to the particulate phase (bacterial biomass). Moreover, heterotrophic bacteria (and fungi) are also known to colonize the detrital POM, and, through stepwise ectoenzymatic hydrolysis of insoluble polymeric material, they liberate a substantial portion of solutes. Bacteria associated with organic detrital particles channel energy into hyperproduction of hydrolytic ectoenzymes, which rapidly solubilize POM and support the DOM pool in a microenvironment. Attached bacteria become "baby machines" and release their progeny in the particle's vicinity, which is enriched by DOM liberated from POM. Thus, it appears that colonized POM, through the ectoenzymatic action of attached bacteria, can become a source of both DOM and free-living bacteria. Moreover, bacteria by colonizing detrital POM, enrich the particles with cell components (proteins, polysaccharides, nucleic acids, lipids, etc.) and significantly increase the nutritional value of the POM when it is consumed by particle-feeding animals (zooplankton, benthic fauna, fish, etc.). On the other hand, attached bacteria operating on different levels of the aquatic food webs compete with these animals

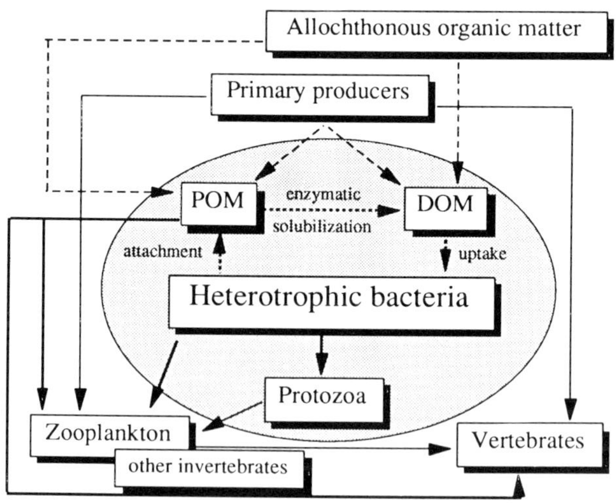

Figure 6.21. Schematic presentation of the interaction of heterotrophic bacteria with POM and DOM in the microbial loop (oval shaded area) and its importance for partitioning and overall flux of organic matter in the lake ecosystem.

by rapidly solubilizing POM. Thus, the particles become a source of DOM that is not accessible for animals.

These unique functions of heterotrophic bacteria are of the utmost significance for the flow and partitioning of organic matter that transmits energy to the higher trophic levels of the lake ecosystem. Figure 6.21 presents the flow of organic matter in a lake ecosystem and emphasizes the pathways where the activity of heterotrophic bacteria directly or indirectly influences these processes. The central issue of the DOM and POM processing are bacterial interactions within the microbial loop. They are now regarded as a crucial step in organic matter processing in aquatic environments. The major step in the microbial loop is the conversion of DOM into bacterial biomass and production. Through the microbial loop bacteria contribute energy to higher trophic levels when they are grazed by protozoans and zooplankton. DOM-utilizing bacteria are now regarded as an important trophic link between the detrital and grazing food chains for energy and nutrient transfer. Heterotrophic bacteria have a very high respiratory potential for oxidation of organic matter, thereby releasing CO_2, PO_4^{3-}, NH_4^-, and other nutrients that are required for phytoplankton. Thus, heterotrophic bacteria, through their enzymatic activity, are important contributors to the steady supply of algal nutrients and promote a steady state of algal biomass and primary production in the absence of other sources of nutrients.

References

Aaronson S (ed) (1981) *Chemical Communication at the Microbial Level, vol 1*. CRC Press, Boca Raton, Florida

Ammerman JW (1991) Role of ecto-phosphohydrolases in phosphorus regeneration in estuarine and coastal ecosystems. In Chróst RJ (ed) *Microbial Enzymes in Aquatic Environments*. Springer-Verlag, New York, pp 165–186

Ammerman JW, Azam F (1985) Bacterial 5′-nucleotidase in aquatic ecosystems: A novel mechanism of phosphorus regeneration. Science 227:1338–1340

Ammerman JW, Azam F (1991a) Bacterial 5′-nucleotidase activity in estuarine and coastal marine waters: Characterization of enzyme activity. Limnol Oceanogr 36: 1427–1436

Ammerman JW, Azam F (1991b) Bacterial 5′-nucleotidase activity in estuarine and coastal marine waters: Role in phosphorus regeneration. Limnol Oceanogr 36:1437–1447

Azam F, Cho BC (1987) Bacterial utilization of organic matter in the sea. In Fletcher M, Gray TRG, Jones JG (eds) *Ecology of Microbial Communities*. Cambridge University Press, Cambridge, pp 261–281

Azam F, Fenchel T, Field JG, Gray JS, Meyer-Reil LA, Thingstad F (1983) The ecological role of water-column microbes in the sea. Mar Ecol Prog Ser 10:257–263

Barman TE (ed) (1969) *Enzyme Handbook, vol II*. Springer-Verlag, Heidelberg

Belfield A, Goldberg DM (1968) Inhibition of the nucleotidase effect of alkaline phosphatase by β-glycerophosphate. Nature 219:73–75

Bell RT, Kuparinen J (1984) Assessing phytoplankton and bacterioplankton during early spring in Lake Erken, Sweden. Appl Environ Microbiol 48:1221–1231

Bengis-Garber C (1985) Membrane-bound 5′-nucleotidase in marine luminous bacteria: Biochemical and immunological properties. Can J Microbiol 31:543–548

Bengtsson G (1988) The impact of dissolved amino acids on protein and cellulose degradation in stream waters. Hydrobiologia 164:97–102

Bergbauer M, Newell SY (1992) Contribution to lignocellulose degradation and DOC formation from a salt marsh macrophyte by the ascomycete *Phaeosphaeria spartinicola*. FEMS Microbiol Ecol 86:341–348

Berry RK, Dekker RFH (1984) Induction studies showing evidence of the similarities between an inducible intracellular and extracellular β-D-glucosidase produced by a species of *Monilia*. FEMS Microbiol Lett 21:309–312

Billen G (1991) Protein degradation in aquatic environments. In Chróst RJ (ed) *Microbial Enzymes in Aquatic Environments*. Springer-Verlag, New York, pp 123–143

Billen G, Joiris C, Wijnant J, Gillain G (1980) Concentration and microbial utilization of small organic molecules in the Scheldt estuary, the Belgian coastal zone of the North Sea and the English Channel. Estuar Coast Mar Sci 11:279–294

Boethling RS (1975) Regulation of extracellular protease secretion in *Pseudomonas maltophilia*. J Bacteriol 123:954–961

Boon PI (1991) Enzyme activities in billabongs of Southeastern Australia. In Chróst RJ (ed) *Microbial Enzymes in Aquatic Environments*. Springer-Verlag, New York, pp 286–297

Burney CM (1986) Bacterial utilization of total in situ dissolved carbohydrate in offshore waters. Limnol Oceanogr 31:427–431

Burney CM, Sieburth JMcN (1977) Dissolved carbohydrates in seawater. II. A spectrophotometric procedure for total carbohydrate analysis and polysaccharide estimation. Mar Chem 5:15–28

Campbell DM (1962) Determination of 5′-nucleotidase in blood serum. Biochem J 82:34P

Chróst RJ (1981) The composition and bacterial utilization of DOC released by phytoplankton. Kiel Meeresforsch Sonderh 5:325–331

Chróst RJ (1984) Use of ^{14}C-dissolved organic carbon (RDOC) released by algae as a realistic tracer of heterotrophic activity measurements for aquatic bacteria. Arch Hydrobiol Beih Ergeb Limnol 19:207–214

Chróst RJ (1986) Algal–bacterial metabolic coupling in the carbon and phosphorus cycle in lakes. In Megusar F, Gantar M (eds) *Perspectives in Microbial Ecology*. Slovene Society for Microbiology, Ljubljana, pp 360–366

Chróst RJ (1988) Phosphorus and microplankton development in an eutrophic lake. Acta Microbiol Polon 37:205–225

Chróst RJ (1989) Characterization and significance of β-glucosidase activity in lake water. Limnol Oceanogr 34:660–672

Chróst RJ (1990) Microbial ectoenzymes in aquatic environments. In Overbeck J, Chróst RJ (eds) *Aquatic Microbial Ecology; Biochemical and Molecular Approaches*. Springer-Verlag, New York, pp 47–78

Chróst RJ (1991a) Environmental control of the synthesis and activity of aquatic microbial ectoenzymes. In Chróst RJ (ed) *Microbial Enzymes in Aquatic Environments*. Springer-Verlag, New York, pp 29–59

Chróst RJ (ed) (1991b) *Microbial Enzymes in Aquatic Environments*. Springer-Verlag, New York

Chróst RJ (1991c) Ectoenzymes in aquatic environments: Microbial strategy for substrate supply. Int Ver Theor Angew Limnol Verh 24:2597–2600

Chróst RJ (1992) Significance of bacterial ectoenzymes in aquatic environments. Hydrobiologia 243/244:61–70

Chróst RJ, Faust MA (1983) Organic carbon release by phytoplankton: Its composition and utilization by bacterioplankton. J Plankton Res 5:477–493

Chróst RJ, Krambeck HJ (1986) Fluorescence correction for measurements of enzyme activity in natural waters using methylumbelliferyl-substrates. Arch Hydrobiol 106:79–90

Chróst RJ, Overbeck J (1987) Kinetics of alkaline phosphatase activity and phosphorus availability for phytoplankton and bacterioplankton in Lake Plußsee (north German eutrophic lake). Microb Ecol 13:229–248

Chróst RJ, Overbeck J (1990) Substrate–ectoenzyme interaction: Significance of β-glucosidase activity for glucose metabolism by aquatic bacteria. Arch Hydrobiol Beih Ergeb Limnol 34:93–98

Chróst RJ, Rai H (1993) Ectoenzyme activity and bacterial secondary production in nutrient impoverished and enriched freshwater mesocosms. Microb Ecol 25:131–150

Chróst RJ, Siuda W, Halemejko GZ (1984) Longterm studies on alkaline phosphatase activity (APA) in a lake with fish-aquaculture in relation to lake eutrophication and phosphorus cycle. Arch Hydrobiol 70:1–32

Chróst RJ, Wcislo R, Halemejko GZ (1986a) Enzymatic decomposition of organic matter by bacteria in an eutrophic lake. Arch Hydrobiol 197:145–165

Chróst RJ, Siuda W, Albrecht D, Overbeck J (1986b) A method for determining of enzymatically hydrolyzable phosphate (EHP) in natural waters. Limnol Oceanogr 32:662–667

Chróst RJ, Overbeck J, Wcislo R (1988) Evaluation of the [^{3}H]thymidine method for estimating bacterial growth rates and production in lake water: Re-examination and methodological comments. Acta Microbiol Pol 37:95–112

Chróst RJ, Münster U, Rai, H, Albrecht D, Witzel KP, Overbeck J (1989) Photosynthetic production and exoenzymatic degradation of organic matter in euphotic zone of an eutrophic lake. J Plankton Res 11:223–242

Cole JJ, Likens GE, Strayer DL (1982) Photosynthetically produced dissolved organic carbon: An important carbon source for planktonic bacteria. Limnol Oceanogr 27:1080–1090

Cotner JB, Wetzel RG (1991a) Bacterial phosphatases from different habitats in a small, hardwater lake. In Chróst RJ (ed) *Microbial Enzymes in Aquatic Environments.* Springer-Verlag, New York, pp 187–205

Cotner JB, Wetzel RG (1991b) 5′-Nucleotidase activity in a eutrophic lake and an oligotrophic lake. Appl Environ Microbiol 57:1306–1312

Coughlan MP (1991) Mechanisms of cellulose degradation by fungi and bacteria. Animal Feed Sci Technol 32:77–100

Coveney MF (1982) Bacterial uptake of photosynthetic carbon from fresh water phytoplankton. Oikos 38:8–29

Coveney MF, Wetzel RG (1992) Effects of nutrients on specific growth rate of bacterioplankton in oligotrophic lake water cultures. Appl Environ Microbiol 58:150–156

Cowie GL, Hedges JI (1984) Carbohydrate sources in a coastal marine environment. Geochim Cosmochim Acta 48:2075–2087

Cunningham HW, Wetzel RG (1989) Kinetic analysis of protein degradation by a freshwater wetland sediment community. Appl Environ Microbiol 55:1963–1967

Currie DJ, Kalff J (1984a) A comparison of the abilities of freshwater algae and bacteria to acquire and retain phosphorus. Limnol Oceanogr 29:298–310

Currie DJ, Kalff J (1984b) The relative importance of bacterioplankton and phytoplankton in phosphorus uptake in freshwater. Limnol Oceanogr 29:311–315

Daatselaar MCC, Harder W (1974) Some aspects of the regulation of the production of extracellular proteolytic enzymes by a marine bacterium. Arch Hydrobiol 101:21–34

Dixon WJ, Massey FJ (1969) *Introduction to Statistical Analysis.* McGraw-Hill, London

Dixon M, Webb EC (ed) (1979) *Enzymes.* Longman Group, London

Ejsmont-Karabin J (1984) Phosphorus and nitrogen excretion by zooplankton (rotifers and crustaceans) in relation to individual bodyweights of animals, ambient temperature and presence or absence of food. Ekol Pol 32:3–42

Fenchel T (1987) *Ecology of Protozoa.* Springer-Verlag, New York

Fraenkel DG, Vinopal RT (1973) Carbohydrate metabolism in bacteria. Annu Rev Microbiol 27:69–100

Francko DA, Heath RT (1979) Functionally distinct classes of complex phosphorus compounds in lake water. Limnol Oceanogr 24:463–473

Freer SN, Detroy RW (1985) Regulation of β-1–4-glucosidase expression by *Candida wickerhamii.* Appl Environ Microbiol 50:152–159

Gage MA, Gorham E (1985) Alkaline phosphatase activity and cellular phosphorus as an index of the phosphorus status of phytoplankton in Minnesota lakes. Freshwater Biology 15:227–233

Gajewski A, Chróst RJ, Siuda W (1993) Bacterial lipolytic activity in a eutrophic lake. Arch Hydrobiol 128:107–126

Gardner WS, Chandler JF, Laird GA, Scavia D (1986) Microbial response to amino acid additions in Lake Michigan: Grazer control and substrate limitation of bacterial populations. J Great Lakes Res 12:161–174

Glenn AR (1976) Production of extracellular proteins by bacteria. Annu Rev Microbiol 30:41–62

Golterman HL (ed) (1977) *Interactions between Sediments and Fresh Water.* Junk, The Hague

Hackett CJ, Chen KC (1978) Quantitative isolation of native chitin from resistant structures of *Sordaria* and *Ascaris* species. Anal Biochem 89:487–500

Hackman RH, Goldberg M (1981) A method for determination of microgram amounts of chitin in arthropod cuticles. Anal Biochem 110:277–280

Hagström A, Ammerman JW, Henrichs S, Azam F (1984) Bacterioplankton growth in seawater. 2. Organic matter utilization during steady-state growth in seawater cultures. Mar Ecol Prog Ser 18:41–48

Hagström A, Azam F, Andersson A, Wikner J, Rassoulzadegan F (1988) Microbial loop in an oligotrophic pelagic marine ecosystem: Possible roles of cyanobacteria and nannoflagellates in the organic fluxes. Mar Ecol Prog Ser 49:171–178

Hałemejko GZ, Chróst RJ (1984) The role of phosphatases in phosphorus mineralization during decomposition of lake phytoplankton blooms. Arch Hydrobiol 101:489–502

Hałemejko GZ, Chróst RJ (1986) Enzymatic hydrolysis of proteinaceous and dissolved material in an eutrophic lake. Arch Hydrobiol 107:1–21

Healey FP, Hendzel LL (1980) Physiological indicators of nutrient deficiency in lake phytoplankton. Can J Fish Aqua Sci 37:442–453

Heath RT, Cooke GD (1975) The significance of alkaline phosphatase in a eutrophic lake. Int Ver Theor Angew Limnol Verh 19:959–965

Heinz F, Haeckel R (1984) 5′-Nucleotidase. UV-method. In Bergmeyer J, Graßl M (eds) *Methods of Enzymatic Analysis, vol IV.* Verlag Chemie, Weinheim, pp 106–119

Hodson RE, Maccubbin AE, Pomeroy LR (1981a) Dissolved adenosine triphosphate utilization by free-living and attached bacterioplankton. Mar Biol 64:43–51

Hodson, RE, Azam, F, Carlucci AF, Fuhrman JA, Karl DM, Holm-Hansen O (1981b) Microbial uptake of dissolved organic matter in McMurdo Sound, Antarctica. Mar Biol 61:89–94

Holm-Hansen O (1984) Composition and nutritional mode of nanoplankton. Arch Hydrobiol Beih Ergeb Limnol 19:125–129

Hooper FF (1973) Origin and fate of organic phosphorus compounds in aquatic systems. In Griffith EJ, Beeton A, Spencer JM, Mitchell DT (eds) *Environmental Phosphorus Handbook*. John Wiley & Sons, New York, pp 834–865

Hoppe HG (1983) Significance of exoenzymatic activities in the ecology of brackish water: Measurements by means of methylumbelliferyl-substrates. Mar Ecol Prog Ser 11:299–308

Hoppe HG (1986) Degradation in sea water. In Rehm HJ, Reed G (eds) *Biotechnology, vol 8*. VCH Verlagsgesellschaft, Weinheim, pp 453–474

Hoppe HG (1991) Microbial extracellular enzyme activity: A new key parameter in aquatic ecology. In Chróst RJ (ed) *Microbial Enzymes in Aquatic Environments.* Springer-Verlag, New York, pp 60–83

Hoppe HG, Kim SJ, Gocke K (1988) Microbial decomposition in aquatic environments: Combined processes of extracellular enzyme activity and substrate uptake. Appl Environ Microbiol 54:784–790

Hoppe HG, Ducklow H, Karrasch B (1993) Evidence for dependency of bacterial growth on enzymatic hydrolysis of particulate organic matter in the mesopelagic ocean. Mar Ecol Prog Ser 93:277–283

Hutchinson GE (1957) *A Treatise on Limnology, vol 1*. John Wiley & Sons, New York

Jacobsen TR, Azam F (1985) Role of bacteria in copepod fecal pellet decomposition: Colonization, growth rates and mineralization. Bull Amar Sci 35:495–502

Jacobsen TR, Rai H (1988) Determination of aminopeptidase activity in lakewater by a short term kinetic assay and its application in two lakes of differing eutrophication. Arch Hydrobiol 113:359–370

Jacobsen TR, Rai H (1991) Aminopeptidase activity in lakes of differing eutrophication. In Chróst RJ (ed) *Microbial Enzymes in Aquatic Environments*. Springer-Verlag, New York, pp 155–164

Jeuniaux C, Voss-Foucart MF, Bussers JC (1982) Preliminary results of chitin biomass in some benthic marine biocenoses. In Hirano S, Tokura S (eds) *Chitin and Chitosan*. Tottori University Press, Tottori, pp 200–204

Jones SE, Lock MA (1991) Peptidase activity in river biofilms by product analysis. In Chróst RJ (ed) *Microbial Enzymes in Aquatic Environments*. Springer-Verlag, New York, pp 144–154

Jørgensen NOG, Søndergaard M (1984) Are dissolved free amino acids free? Microb Ecol 10:301–316

Jumars PA, Penry DL, Baross JA, Perry MJ, Frost BW (1989) Closing the microbial loop: Dissolved carbon pathway to heterotrophic bacteria from incomplete ingestion, digestion and absorption in animals. Deep Sea Res 36:483–495

Jürgens K, Güde H (1990) Incorporation and release of phosphorus by planktonic bacteria and phagotrophic flagellates. Mar Ecol Prog Ser 59:271–284

Karl DM, Bailiff MD (1989) The measurement and distribution of dissolved nucleic acids in aquatic environments. Limnol Oceanogr 34:543–558

Karl DM, Craven DB (1980) Effects of alkaline phosphatase activity on nucleotide measurements in aquatic microbial communities. Appl Environ Microbiol 40:549–560

Keil RG, Kirchman DL (1992) Bacterial hydrolysis of protein and methylated protein and its implications for studies of protein degradation in aquatic systems. Appl Environ Microbiol 58:1374–1375

King GM (1986) Characterization of β-glucosidase activity in intertidal marine sediments. Appl Environ Microbiol 51:373–380

Kobori H, Taga N (1979) Phosphatase activity and its role in the mineralization of organic phosphorus in coastal seawater. J Exp Mar Biol Ecol 36:23–39

Kreutzberg GW, Reddington M, Zimmermann H (eds) (1986) *Cellular Biology of Ectoenzymes*. Springer-Verlag, Berlin

Law BA (1980) Transport and utilization of proteins by bacteria. In Payne JW (ed) *Microorganisms and Nitrogen Sources*. John Wiley & Sons, New York, pp 381–409

Lean DRS, Nalewajko C (1976) Phosphorus uptake and excretion by freshwater algae. J Fish Res Board Can 33:1312–1323

Lee S, Fuhrman JA (1987) Relationships between biovolume and biomass of naturally derived marine bacterioplankton. Appl Environ Microbiol 53:1298–1303

Litchfield CD, Prescott JM (1976) Regulation of proteolytic enzyme production by *Aeromonas proteolytica*. II. Extracellular aminopeptidase. Can J Microbiol 16:23–27

Ljungdahl LG, Eriksson KE (1985) Ecology of microbial cellulose degradation. In Marshall KC (ed) *Advances in Microbial Ecology*. Plenum Press, New York, pp 237–299

Lovell CR, Konopka A (1985) Primary and bacterial production in two dimictic Indiana lakes. Appl Environ Microbiol 49:485–492

Maeda M, Taga N (1973) Deoxiribonuclease activity in seawater and sediment. Mar Biol 20:58–63

Martinez J, Azam F (1993) Periplasmic aminopeptidase and alkaline phosphatase activities in a marine bacterium: Implications for substrate processing in the sea. Mar Ecol Prog Ser 92:89–97

Maruyama A, Oda M, Higashihara T (1993) Abundance of virus-sized non-DNase-digestible DNA (coated DNA) in eutrophic seawater. Appl Environ Microbiol 59:712–717

Mayer LM (1989) Extracellular proteolytic enzyme activity in sediments of an intertidal mudflat. Limnol Oceanogr 34:973–981

McGrath SM, Sullivan CW (1981) Community metabolism of adenylates by microheterotrophs from the Los Angeles Harbor and Southern California coastal waters. Mar Biol 62:217–226

Meyer-Reil LA (1987) Seasonal and spatial distribution of extracellular enzymatic activities and microbial incorporation of dissolved organic substrates in marine sediments. Appl Environ Microbiol 53:1748–1755

Meyer-Reil LA, Köster M (1992) Microbial life in pelagic sediments: The impact of environmental parameters on enzymatic degradation of organic material. Mar Ecol Prog Ser 81:65–72

Münster U (1984) Distribution, dynamics and structure of free dissolved carbohydrates in Plußsee, a north German eutrophic lake. Int Ver Theor Angew Limnol Verh 22:929–935

Münster U (1991) Extracellular enzyme activity in eutrophic and polyhumic lakes. In Chróst RJ (ed) *Microbial Enzymes in Aquatic Environments.* Springer-Verlag, New York, pp 96–122

Münster U, Chróst RJ (1990) Origin, composition and microbial utilization of dissolved organic matter. In Overbeck J, Chróst RJ (eds) *Aquatic Microbial Ecology: Biochemical and Molecular Approaches.* Springer-Verlag, New York, pp 8–46

Murray AW (1971) The biological significance of purine salvage. Annu Rev Biochem 50:811–826

Nagata T, Kirchman DL (1991) Release of dissolved free and combined amino acids by bacteriovorous marine flagellates. Limnol Oceanogr 36:433–443

Nagata T, Kirchman DL (1992) Release of macromolecular organic complexes by heterotrophic marine flagellates. Mar Ecol Prog Ser 83:233–240

Ochiai M, Ukiya T (1981) Seasonal variations of dissolved organic constituents in Lake Nakanuma during March and November 1979. Int Ver Theor Angew Limnol Verh 21:682–687

Overbeck J (1961) Die Phosphatasen von *Scenedesmus quadricauda* und ihre ökologische Bedeutung. Int Ver Theor Angew Limnol Verh 14:226–231

Overbeck J (1962) Untersuchungen zum Phosphathaushalt von Grünalgen. II. Die Verwertung von Pyrophosphat und organisch gebundenen Phosphaten und ihre Beziehung zu den Phosphatasen von *Scenedesmus quardicauda* (Turp.) Breb Arch Hydrobiol 58:281–308

Overbeck J (1979) Studies on heterotrophic functions and glucose metabolism of microplankton in Plußsee. Arch Hydrobiol Beih Ergeb Limnol 13:56–76

Overbeck J, Babenzien HD (1963) Nachweis von freien Phosphatasen, Amylase und Saccharase im Wasser eines Teiches. Naturwissenschaften 50:571–572

Paul JH, Jeffrey WH, DeFlaun MF (1987) Dynamics of extracellular DNA in the marine environment. Appl Environ Microbiol 53:170–179

Paul JH, DeFlaun MF, Jeffrey WH (1988) Mechanisms of DNA utilization by estuarine microbial populations. Appl Environ Microbiol 54:1682–1688

Paul JH, Jiang SC, Rose JB (1991) Concentration of viruses and dissolved DNA from aquatic environments by vortex flow filtration. Appl Environ Microbiol 57:2197–2204

Payne JW (1980) Transport and utilization of peptides by bacteria. In Payne JW (ed) *Microorganisms and Nitrogen Sources.* John Wiley & Sons, New York, pp 211–256

Perry MJ (1972) Alkaline phosphatase activity in subtropical Central North Pacific waters using a sensitive fluorometric method. Mar Biol 15:113–119

Pettersson K (1980) Alkaline phosphatase activity and algal surplus phosphorus as phosphorus-deficiency indicators in Lake Erken. Arch Hydrobiol 89:54–87

Pieczyńska E (1976) *Selected Problems of Lake Littoral Ecology.* Warsaw University Press, Warsaw

Pomeroy LR, Wiebe WJ (1988) Energetics of microbial food webs. Hydrobiologia 159:7–18

Proctor LM, Fuhrman JA (1991) Roles of viral infection in organic particle flux. Mar Ecol Prog Ser 69:133–142

Proctor LM, Fuhrman JA (1992) Mortality of marine bacteria in response to enrichments of the virus size fraction from seawater. Mar Ecol Prog Ser 87:283–293

Rego VJ, Billen G, Fontigny A, Somville M (1985) Free and attached proteolytic activity in water environments. Mar Ecol Prog Ser 21:245–249

Reichardt W, Overbeck J, Steubing L (1967) Free dissolved enzymes in lake waters. Nature 216:1345–1347

Riley GA (1970) Particulate matter in seawater. Adv Mar Biol 8:1–118

Rogers HJ (1961) The dissimilation of high molecular weight organic substrates. In Gunsalus IC, Stanier RY (eds) *The Bacteria, vol 2*. Academic Press, New York, pp 261–318

Rogers HJ (1979) The function of bacterial autolysins. In Berkeley RCW (ed) *Microbial Polysaccharides and Polysaccharases*. Academic Press, London, pp 237–268

Rosso AL, Azam F (1987) Proteolytic activity in coastal oceanic waters: Depth distribution and relationship to bacterial populations. Mar Ecol Prog Ser 41:231–240

Sanders RW, Porter KG (1988) Phagotrophic phytoflagellates. In Marshall KC (ed) *Advances in Microbial Ecology, vol 10*. Plenum Press, New York, pp 167–192

Siuda W (1984) Phosphatases and their role in organic phosphorus transformation in natural waters. A review. Pol Arch Hydrobiol 31:207–233

Siuda W, Chróst RJ (1987) The relationship between alkaline phosphatase (APA) activity and phosphate availability for phytoplankton and bacteria in eutrophic lakes. Acta Microbiol Pol 36:247–257

Siuda W, Wcisło R, Chróst RJ (1991) Composition and bacterial utilization of photosynthetically produced organic matter in an eutrophic lake. Arch Hydrobiol 121:473–484

Smith MH, Gold MH (1979) *Phanerochaete chrysosporium* β-glucosidases: Induction cellular location, and physical characterization. Appl Environ Microbiol 37:938–942

Smith EL, Hill RL (1960) Leucine aminopeptidase. In Boyer PD, Lardy H, Myrback K (eds) *The Enzymes, vol 4*. Academic Press, London, pp 37–62

Smith RE, Kalff J (1981) The effect of phosphorus limitation on algal growth rates: Evidence from alkaline phosphatase. Can J Fish Aqua Sci 38:1421–1427

Smucker RA, Dawson R (1986) Products of photosynthesis by marine phytoplankton: Chitin in TCA "protein" precipitates. J Exp Mar Biol Ecol 104:143–152

Smucker RA, Kim CK (1991) Chitinase activity in estuarine waters. In Chróst RJ (ed) *Microbial Enzymes in Aquatic Environments*. Springer-Verlag, New York, pp 249–269

Solorzano L (1978) Soluble fractions of phosphorus compounds and alkaline phosphatase activity in Loch Creran and Loch Etire, Scotland. J Exp Mar Biol Ecol 34:227–232

Somville M (1984) Measurement and study of substrate specificity of exoglucosidase activity in eutrophic water. Appl Environ Microbiol 48:1181–1185

Steinberg C, Münster U (1985) Geochemistry and ecological role of humic substances in lakewater. In Aiken GR, McKnight DM, Wershaw RL, McCarthy P (eds) *Humic Substances in Soil, Sediment, and Water. Geochemistry, Isolation and Characterization*. John Wiley & Sons, New York, pp 105–145

Sterner RW (1989) The role of grazers in phytoplankton succession. In Sommer U (ed) *Plankton Ecology*. Springer-Verlag, New York, pp 124–136

Stewart AL, Wetzel RG (1982) Phytoplankton contribution to alkaline phosphatase activity. Arch Hydrobiol 93:265–271

Strickland JDH, Parsons TR (1972) *A Practical Handbook of Seawater Analysis*. Bull Fish Res Bd Can 167:1–462

Strobel HJ, Russell JB (1987) Regulation of β-glucosidase in *Bacteroides ruminicola* by a different mechanism: Growth rate-dependent derepression. Appl Environ Microbiol 53:2505–2510

Tamminen T (1989) Dissolved organic phosphorus regeneration by bacterioplankton: 5′-Nucleotidase activity and subsequent phosphate uptake in a mesocosm enrichment experiment. Mar Ecol Prog Ser 58:89–100

Turk V, Rehnstam AS, Lundberg E, Hagström Å (1992) Release of bacterial DNA by marine nanoflagellates, an intermediate step in phosphorus regeneration. Appl Environ Microbiol 58:3744–3750

Udenfriens S, Stein S, Böhlen P, Dairman W, Leimgruber W, Weigele M (1972) Fluorescamine: A reagent for assay of amino acids, peptides, proteins, and primary amines in the picomole range. Science 178:871–872.

Van Husen N, Gerlach, U (1984) 5′-Nucleotidase. Colorimetric method. In Bergmeyer J, Graßl M (eds) *Methods of Enzymatic Analysis, vol IV.* Verlag Chemie, Weinheim, pp 113–119

Vincent WV (1981) Rapid physiological assays for nutrient demand by the plankton. II. Phosphorus. J Plankton Res 3:699–710

Wetzel RG, Likens GE (1979) *Limnological Analyses.* WB Saunders, Philadelphia

Williams PJLeB (1981) Incorporation of microheterotrophic processes into the classical paradigm of the planktonic food web. Kieler Meeresforsch Sonder 5:1–28

Willimas PM (1986) Chemistry of the dissolved and particulate phases in the water column. In Eppley RW (ed) *Plankton Dynamics of the Southern California Bight.* Springer-Verlag, New York, pp 645–682

Wynne D, Gophen M (1981) Phosphatase activity in freshwater zooplankton. Oikos 37:369–376

7. Saprophytic and Oligotrophic Bacteria in the Plußsee

Jürgen Overbeck

Introduction

Conventionally the bacterial microflora is classified into two groups, saprophytic (zymogenous, copiotrophic) and oligotrophic (autochthonous) bacteria (Winogradsky 1949). Saprophytic bacteria dominate when dissolved organic nutrients are supplied in high concentrations. Usually they grow very well on nutrient agar, from which they can be isolated also. They are widespread throughout the whole biosphere: water, soil, and air. It is doubtful whether they comprise typical aquatic bacteria (Ruttner 1932). The fraction of gram-positive bacteria in the aquatic environment is apparently rather small, and gram-negative bacteria predominate, in contrast to the high percentage of gram-positive bacteria in soil (Kusnetzov 1959). Furthermore, the fraction of spore-forming bacteria in aquatic environments corresponds with the trophic state of the water: the more eutrophic a lake, the smaller the part of spore-forming bacteria.

In contrast to saprophytes, the oligotrophic bacteria are adapted to the low quantities of dissolved organic material (DOM) generally present in an unpolluted habitat. They grow only on highly diluted nutrient agar, if they grow at all. Their numbers (direct counts, AODC, or DAPI) are much higher than the plate counts and range from 10^4 to 10^6 bacteria/ml in the Plußsee, and the numbers of CFUs (colony-forming units) usually amount to far less than 1% of the total counts.

Plate counts of the saprophytic and oligotrophic bacteria of the Plußsee are described in this chapter. From these data some basic rules about structure and function of heterotrophic microbial life in limnic systems can be generalized.

Abundance of Saprophytic and Oligotrophic Bacteria During the Annual Cycle

For the determination of saprophytic bacteria, Difco Nutrient Agar B1 and the CPS medium of Collins (1963), which contains a variety of organic substrates, were used. The number of oligotrophic bacteria was estimated using dilutions of 10^{-3} and 10^{-6} of the same medium. In addition KAZ mineral agar (Moaledj 1978) and 1.1% ion agar in bidistilled water was used. A 0.1-ml sample of lake water (or a suitable dilution), calculated to get 20 to 150 colonies, was spread in triplicate on the dried agar surface of a petri dish by means of a Drygalsky spreader. The incubation temperature was 26°C. Colonies were grown under air or under a defined atmosphere (79% N_2, 1% CO_2, 20% O_2); for further details see Moaledj and Overbeck (1982).

Figure 7.1 shows the number of colonies grown on nutrient agar (Difco Nutrient Agar B1). Bacteria were sampled from August 1971 until February 1973 at a depth of 1 m. Remarkably, there was a pronounced maximum from April until July 1972 during the main development of phytoplankton (Chapter 1, this volume). From October until December in both years (1971 and 1972), during the lake turnover, higher numbers of bacteria were observed, especially on diluted media, followed by a minimum in winter 1971–1972. The numbers of saprophytic bacte-

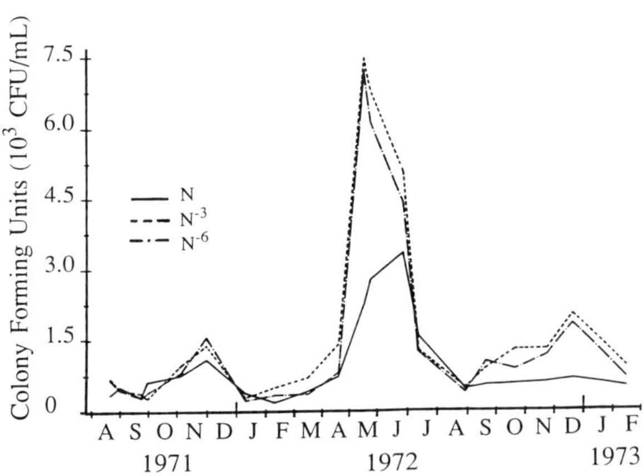

Figure 7.1. Annual changes in the number of the colony-forming units (CFU) grown on undiluted and diluted nutrient agar (Difco) in the Plußsee water sampled at the depth of 1 m. N, Difco nutrient agar B1; N^{-3}, diluted 1000 times; N^{-6}, nutrient agar diluted 1 million times (modified according to Moaledj and Overbeck 1982).

ria on undiluted nutrient agar ranged between 180 CFU/ml in February and 3350 CFU/ml in June 1972 with an average of 950 CFU/ml during the sampling period.

After dilution of the media, the numbers of colonies increased strongly (Fig. 7.1). Similar results were reported by numerous studies (Babenzien 1964, Overbeck 1968a,b, Sorokin and Kadota 1972, Buck 1974, Bölter 1977). During the maximum of bacterial development, the numbers of colonies on the diluted media were doubled in comparison to undiluted nutrient agar. The highest number of bacteria during the study period was observed on diluted nutrient agar N^{-3} in May 1972 (7500 CFU/ml), the average of 20 samples on diluted nutrient agar N^{-3} amounted to 1700 CFU/ml. The largest part of the colonies on diluted nutrient agar had diameters less than 1 mm ("*pin points*"), even after an incubation time of 4 weeks. The colonies were rather uniform morphologically and hard to differentiate.

The number of saprophytic bacteria growing on CPS medium was remarkably higher than the corresponding numbers on nutrient agar (up to 20,000 CFU/ml in May 1972; Fig. 7.2). However, the number of colonies decreased as dilution increased; i.e., the highest numbers were found on undiluted CPS medium. The average number of bacterial colonies was 5620 CFU/ml on undiluted CPS medium and 1340 CFU/ml and 1140 CFU/ml on CPS medium diluted 10^3 and 10^6 times, respectively. On CPS medium a summer maximum is very pronounced with smaller maxima in autumn, as is the case on nutrient agar.

The number of oligotrophic bacteria from a depth of 1 m on three different media, 1000-fold diluted nutrient agar (DA), mineral agar (MA), and agar agar plus bidistilled water (AAW), followed a similar pattern. Distinct maxima of bacteria in spring 1972 (April–June) coincided with the main development of phytoplankton. Lower bacterial abundance was found in winter and during the summer stratification of the lake. However, during the fall turnover the numbers increased again. Actually the highest numbers (up to 7000 CFU/ml) of the whole series are found on pure mineral agar (Kaz) in May 1972, followed by highly diluted nutrient agar (DA). But on AAW medium about 6000 CFU/ml was counted. Most of the colonies were very small "*pin points*." The bacteria growing

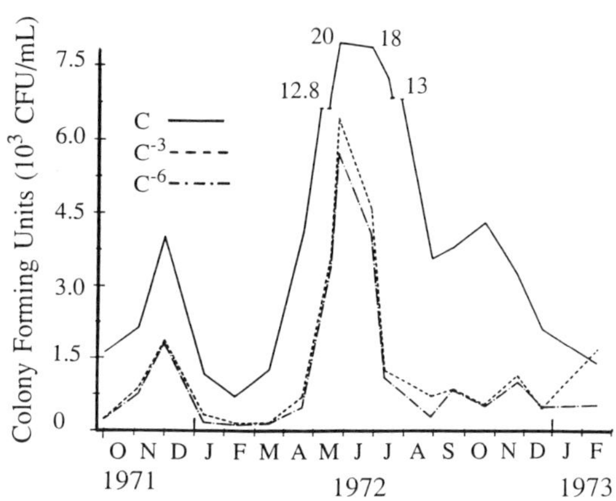

Figure 7.2. Annual variation of the number of the colony-forming units grown on undiluted and diluted CPS medium in water sampled from the Plußsee at the depth of 1 m. C, CPS medium; C^{-3}, CPS medium diluted 1000 times; C^{-6}, CPS medium diluted 1 million times (modified according to Moaledj and Overbeck 1982).

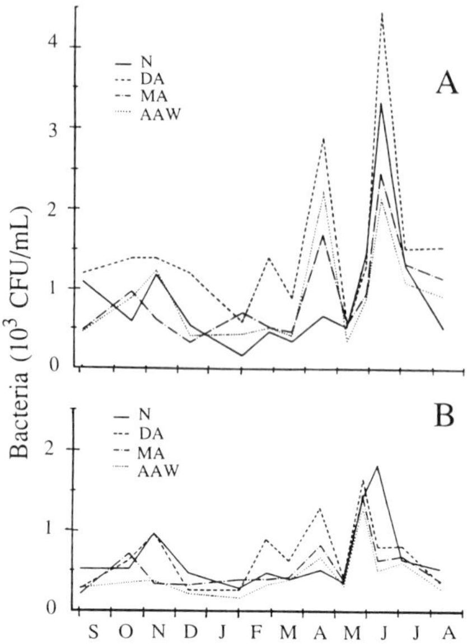

Figure 7.3. Annual variation of the number of bacterial colony forming units grown on different media in the Plußsee water sampled at the depth of 5 m. Media were incubated (A) under air and (B) under a defined atmosphere (20% O_2, 79% N_2, 1% CO_2). N, Difco nutrient agar B1; DA, nutrient agar diluted 1000 times; MA, mineral agar; AAW, agar agar plus bidistilled water (modified according to Moaledj and Overbeck 1982).

on agar that was free of organic material (AAW) probably took volatile organic substances from the air. The highest numbers of bacteria from all three media were found in June 1973, but the actual numbers are much smaller, about one half, compared with 1972.

Figure 7.3 compares the annual cycle of the growth of bacteria incubated in air and under a defined atmosphere sampled at the depth of 5 m. The pattern of the bacterial development in deeper waters of the lake was similar to that at the surface. All fractions of bacteria (saprophytic, oligotrophic, total bacteria) displayed a distinct growth during periods of phytoplankton bloom and breakdown. Under a defined atmosphere (20% O_2, 79% N_2, 1% CO_2), growth was smaller than in the natural air. This was also found in samples from other depths.

Enhanced growth on diluted media or mineral agar without addition of organic substrate pointed out the ability of oligotrophic growth for many aquatic bacteria. Oligotrophic bacteria were not restricted to the aerobic part of the lake; they were present during the year throughout the whole water column. Many of them were apparently facultatively anaerobic (Moaledj 1978), although their numbers were decreased slightly as depth increased. Average numbers of 13 monthly samples during the period 1971–1973 amounted to 980 CFU of oligotrophic bacteria/ml at 1 m, 940 CFU/ml at 5 m, 890 CFU/ml at 10 m, and 780 CFU/ml at 20 m depth.

Stratification of Bacterial Communities

The distribution pattern of phytoplankton and bacteria in vertical profiles of the stratified water column of the Plußsee clearly reflected the distinct metabolic

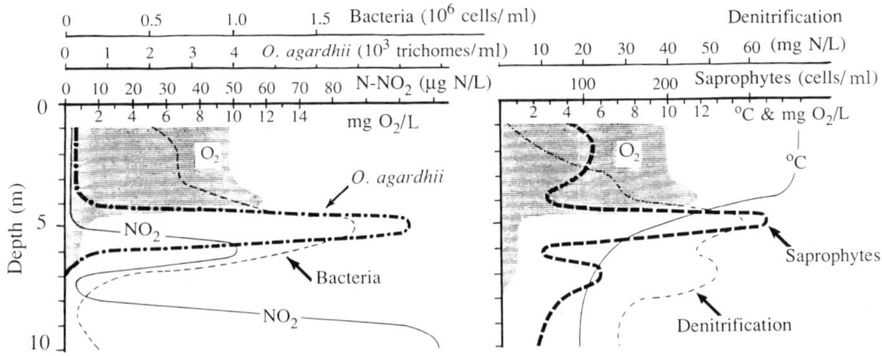

Figure 7.4. Distribution of the microorganisms in the upper 10 m of the water column of the Plußsee on August 30, 1966. Saprophytic bacteria were enumerated as the number of colony-forming units grown on Difco nutrient agar B1. For further explanations see text (modified according to Overbeck 1968b).

coupling between the autotrophic and heterotrophic compartments of the ecosystem. One example may refer to a vertical profile of the water column sampled on August 30, 1966 (Overbeck 1968a). One of the dominant features of this profile was the correspondence between phytoplankton and bacteria at the same depth. Figure 7.4 shows the total number of bacteria (direct counts; left side) and the number of saprophytic bacteria (plate counts on Difco Nutrient Agar B1; right side). The fraction of saprophytic bacteria amounted to only 0.02% of the total number of bacteria.

In the thermocline (a depth of 5 m), oxygen was totally depleted. It was noteworthy that at this anaerobic depth, a high abundance of *Oscillatoria agardhii* was found. The reason for this was the depletion of phosphorus in the epilimnion. Within the layer of the thermocline there was enough light for photosynthesis, and inorganic phosphorus was supplied from the hypolimnion, which was very rich in this element (Chapter 1, this volume). This resulted in a sharp stratification of *Oscillatoria*. A maximum of bacteria and high concentrations of NO_2 from high rates of denitrification were present at the same depth. *Oscillatoria agardhii* supplied the aerobic heterotrophic bacteria with excreted organic carbon. However, to be metabolically active in an anaerobic environment, these aerobic heterotrophic bacteria probably switch their respiratory metabolism to nitrate respiration.

The following typical annual cycle of the vertical stratification of heterotrophic bacteria in the Plußsee was regularly observed. The main part of the bacterial biomass was found in the epilimnion, which coincided with the spring bloom of phytoplankton. At the end of the phytoplankton bloom, as a result of nutrient depletion and high grazing pressure by zooplankton and heterotrophic nanoflagellates (Chapters 11 and 12, this volume), the peaks of the bacterial biomass were sinking down to the thermocline. With the beginning of destratification of the lake, the heterotrophic bacterial biomass started to mineralize the accumulated DOM in the aerated parts of the hypolimnion (Overbeck 1974).

Table 7.1. Primary Production and Numbers of Bacteria in Four Lakes of East Holstein[a]

Lake	Primary Production[b] (mg C m^{-3} d^{-1})	Saprophytic Bacteria (CFU/ml)	Total Number[c] of Bacteria (10^6 cells/ml)
Plußsee	57	1000	1.1
Großer Plöner See	35	650	1.0
Schluensee	35	350	0.5
Schöhsee	16	300	0.5
Ratio[d]			
Plußsee	1	1	1
Großer Plöner See	0.61	0.65	0.90
Schluensee	0.30	0.35	0.45
Schöhsee	0.28	0.30	0.45

[a]Average numbers from 1961–1964 (1 m depth). Saprophytic bacteria were grown on Difco nutrient agar. (From Overbeck 1965).
[b]According to Ohle (1962).
[c]Direct count method.
[d]Ratio between studied lake and the Plußsee.

This basic distribution pattern of phytoplankton and heterotrophic bacteria has been documented from almost the beginning of aquatic microbiology (Henrici 1937, Waksman 1941, ZoBell 1946, Kusnetzov 1959, Kriss 1961, Czeczuga 1962, Overbeck and Babenzien 1964, Schegg 1971) in East Holstein lakes (Overbeck 1965, 1968a,b). Kusnetzov (1958) presented a vertical profile of the Dalny Lake that was similar in many respects to the results from the Plußsee. An example of these early studies is shown in Table 7.1. Primary production and number of bacteria (saprophytic bacteria grown on nutrient agar and total number of bacteria) from four lakes of the Plußsee area are compared. There is a succession from the Plußsee to the Schöhsee that is parallel in number of bacteria and rate of primary production.

The biochemical basis for the metabolic coupling between phytoplankton and bacteria lies in the extracellular organic carbon (EOC) released by phytoplankton (Nalewajko 1977, Cole et al. 1982, Chróst and Faust 1983, Baines and Pace 1991, Sell and Overbeck 1992; Chapters 4, 5, and 10, this volume). Because of the variety of metabolic types of algae and bacteria and the complexity of the environmental parameters, no uniform metabolic coupling may be expected, a fact that is often referred to in early studies (Gusewa 1951, Collins 1963, Goldman et al. 1968, Jones 1971). The relationship between photosynthesis and extracellular release is not simple and is additionally complicated by the effects of changing light intensity during the day. The excretion products range from simple organic molecules to large polymers, and the pattern of excretion varies throughout the day (Chróst and Faust 1980, 1983, Münster and Chróst 1990, Sundh 1992). The aquatic ecosystem reveals a set of oscillating systems. The heterotrophic bacteria

Table 7.2. Excretion of Extracellular Organic Carbon (EOC) by *Limnothrix redekei* and Maximal Uptake Rates (V_{max}) of EOC by Bacteria Isolated from *Limnothrix redekei* Cultures[a]

Experiment	Incubation Time (h)	*Limnothrix redekei* (10^7 cells/ml)	EOC (mg C $L^{-1}h^{-1}$)	Bacteria (10^7 cells/ml)	V_{max} (mg C $L^{-1}h^{-1}$)
Limnothrix redekei	18.5	2.80	0.205	8.52	19.46
+ *Lactobacillus*	44	7.35	0.450	15.5	35.40
sp.	67.5	10.05	0.825	20.0	45.68
Limnothrix redekei	25	2.83	0.300	12.69	40.79
+ *Flavobacterium*	45	12.0	0.675	21.0	67.50
sp.	73	17.75	1.250	30.32	97.46

[a]Modified from Herbst (1984).

are involved in short-time oscillations including day–night cycles, which exhibit synchronous activity patterns of whole groups of organisms in the ecosystem.

Herbst (1984) studied the metabolic coupling between *Limnothrix redekei* and bacteria that were isolated from *Limnothrix* cultures (Chapter 10, this volume). After labeling of *Limnothrix* with $^{14}CO_2$, the excretion rate of [^{14}C]EOC by *Limnothrix* and the uptake rate of bacteria were measured. Table 7.2 shows that the uptake rate of bacteria was much higher than the EOC excretion rate. The amount of excreted EOC limited the growth of the accompanying bacteria. On the other hand, the high V_{max} values of the bacteria enabled a fast response to a sudden increase of DOM, e.g., after the breakdown of a phytoplankton bloom. The utilization of *Limnothrix* EOC by *Lactobacillus* sp., one of the accompanying bacteria that was isolated from *Limnothrix* cultures, was studied by gel chromatography. A rapid utilization of the EOC (fractions of molecular weight 100–1500) was measured.

Analysis of Microbial Communities: Numerical Taxonomy of Isolates

We can assume that aquatic bacteria belong to different groups of microorganisms with distinct physiological and morphological capabilities. Up to now, most studies of aquatic bacteria dealt with the numbers or biomass of microbes as a function of changing environmental conditions with time and space. Little is known about the qualitative changes within bacterial communities. This may be primarily because the identification of bacterial genera is extremely laborious (Dundas 1974) and often very uncertain because of the great variability of species.

In the past, there have been several approaches to describe natural bacterial populations. Plate counts using different media (Jannasch and Jones 1959, Oláh and Vasarhelyi 1970, Jones 1970) have been used as well as physiological tests of isolated strains (Fondén 1969, Straskrabova 1974, Reichardt 1974, Strzelczyk et al. 1975) or morphological observations of natural samples (Hirsch 1980). Generally most of the studies were limited by laboratory methods, as only a small part of the total natural bacterial community could be isolated for taxonomic studies.

There is an approach (that is still in use) that is a numerical treatment of taxonomic data from laboratory cultures that had been isolated from different natural habitats. Most of these studies used strains isolated from soil (Goodfellow 1968, Sundman 1970), from marine habitats (Pfister and Burkholder 1965, Bölter 1977), or from estuaries (Mallory et al. 1977, Austin et al. 1977). Only a few studies characterized freshwater microbial populations by means of numerical taxonomy (Lighthardt 1974, Rosswall and Persson 1980).

The observations show that the number of bacteria in lakes coincides with the development of phytoplankton during an annual cycle and the well-known changes in phytoplankton community structure (Chapter 1, this volume). Therefore, it can be concluded that there should also be a significant qualitative change within the composition of the bacterial community in connection with phytoplankton successions and the variable supply of EOC. To study the expected successions of bacterial communities, bacteria were isolated from water samples of the Plußsee on four sampling dates: June 23, 1967 (group A), December 11, 1967 (group B), March 4, 1968 (group C) and September 30, 1968 at 1 m depth (group D) and 7 m depth (group E). Nine strains (group F), isolated from enrichments with methanol as the sole source of carbon, were included in the numerical analysis. Water samples were collected aseptically, using a Sorokin sampler with 500-ml glass bottles. Difco nutrient agar plates were inoculated with appropriate dilutions of lake water samples with sterile tap water and incubated aerobically at 25°C for 7 days. Visible colonies were counted and randomly picked from the plates for further purification by streaking on Difco nutrient agar plates. Purity of the isolated strains was verified by streaking and microscopic examination.

Shortly after isolation, each strain was examined for morphological, physiological, and biochemical characteristics (total set of 105 variables). If not otherwise stated, the methods employed were conventional techniques of microbial taxonomy (Skerman 1967, Hallmann 1961). For further details see Witzel et al. (1982). The data were analyzed using a simple matching coefficient, which includes positive and negative matches. Clustering was achieved by the single-linkage procedure (Sneath and Sokal 1973), constructing a sorted similarity matrix that included all strains. Clusters of strains were separated manually at a minimum average similarity level of 75%. Subsequently, the intra- and interspecific similarity indices, the real median strain of each cluster, and the relative frequency of characters within a cluster were determined with the aid of a computer.

Table 7.3 shows the vertical profile of the number of bacterial colonies and their pigment production determined on ZoBell agar without NaCl (ZoBell and Conn 1940), Difco nutrient agar, and Kaz agar (Kaserer's mineral agar; Overbeck 1965) at the time of sampling for the present study. The variation in the number and pigmentation of colonies on the different media, over time and depths, already indicates that there could be significant variations in the composition of the microbial populations and the accompanying phytoplankton at different times and different depths.

Table 7.3. Bacterial Colony-Forming Units Grown on Different Media and Distribution of Bacteria Throughout the Water Column (1–10 m) of the Plußsee (June 28, 1967)

	Colony-Forming Units (CFU)												
	ZoBell Agar				KAZ Agar				Nutrient Agar				
Depth (m)	White (%)	Yellow (%)	Pink (%)	Total (CFU/ml)	White (%)	Yellow (%)	Pink (%)	Total (CFU/ml)	White (%)	Yellow (%)	Pink (%)	Total (CFU/ml)	Direct Counts (10^7 cells/ml)
1[a]	100	0	0	1912	100	0	0	1499	100	0	0	307	3.7
2[a]	100	0	0	2443	100	0	0	1616	100	0	0	973	3.1
3[b]	65.3	33.1	1.5	1183	100	0	0	1600	100	0	0	1195	2.6
4[b]	82.1	12.1	5.8	207	100	0	0	1229	92.2	0.9	6.9	847	2.3
5[b]	89.2	8.5	2.3	798	100	0	0	1588	90.6	0.8	8.6	980	2.4
6[b]	67.5	13.4	19.1	231	100	0	0	1624	90.8	0	9.2	1059	1.7
7	80	15.1	4.4	451	100	0	0	1140	88.5	4.1	7.4	773	1.3
8	85	13.4	1.8	112	100	0	0	1662	91.5	1.7	6.8	709	1.2
9	76.1	21.7	2.2	92	100	0	0	1053	96.3	1.1	2.6	568	1.1
10	81.4	15.9	2.7	113	100	0	0	1280	87.9	2.1	10.1	535	0.9

[a] *Limnothrix redekei*, *Aphanizomenon* sp.
[b] *Limnothrix redekei*.

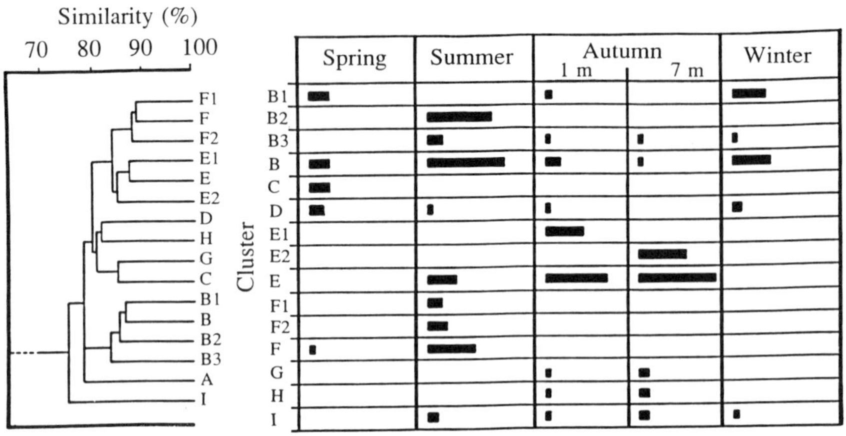

Figure 7.5. Dendrogram prepared on the basis of the mean intercluster similarity coefficients (left), and frequency of bacterial strains isolated in different seasons, within the clusters B1 to I (right). Length of the bar is proportional to the number of strains (modified from Witzel et al. 1982).

By numerical analysis 108 of the total set of 124 strains (87%) isolated from six different samples were arranged in nine clusters at a similarity level of ≥ 75%. In some of these groups, subclusters could be distinguished at a much higher similarity. The two largest clusters, B and E, contained more than half of the strains grouped. The rest were distributed over seven clusters, which contained between 3 and 11 strains each. The relationships between the clusters are shown by a dendrogram, which is constructed on the basis of the mean intercluster similarity values (Fig. 7.5).

As expected, the bacterial isolates from different sampling dates and water depths were not randomly distributed among the clusters. There were clear preferences; in most cases strains isolated from the same sample were grouped together into the same cluster. For each season, except winter, at least one typical cluster could be identified. The strains from the winter sample were distributed among the unspecified clusters. The results show that the microbial flora in the Plußsee consisted of different components, and the composition was variable over time and at different depths. With the numerical taxonomic approach applied here, it was possible to recognize these changes. Studies of cluster-analytical methods require the determination of a broad spectrum of characters (usually more than 100) and require much more information regarding each organism than does conventional classification or factor analysis. It seems that this approach is still useful in ecological studies, since it considers each organism as having a set of adaptations to environmental conditions that may be very variable and rapidly changing (Chapter 9, this volume).

The identification to the genus level of freshly isolated bacteria from aquatic biota usually causes problems because of weak or inconclusive reactions in the tests or contradictory characteristics in the assignment to certain taxa (Austin et al.

1977, Mallory et al. 1977, Overbeck 1974). Therefore, in this study the clusters of strains were not assigned to certain taxonomic groups. Moreover, it was suggested that generic groupings of bacteria were less important with respect to ecological questions than groups of organisms with similar physiological possibilities. This seems to be a principal difference between a taxonomy under ecological aspects and one with the aim of a generic classification. The latter must be based on more stable characters such as DNA base composition, cell wall components, etc. (Witzel 1990). The ecological importance of these characters is not clear, although some others have distinct advantages under certain environmental conditions.

Oligotrophic Bacteria

The oligotrophic autochthonous bacteria are adapted to very small amounts of DOM. Presumably they predominantly utilize substances of high molecular weight such as humic compounds or cellulose. They might be responsible for the turnover of large amounts of the dissolved organic matter and play an important role in the carbon cycle of lakes. Up to now little is known of the taxonomic status of the oligotrophic aquatic bacteria. One of the reasons for this is that oligotrophic bacteria grow very slowly, if at all, on the culture media. In a comprehensive qualitative analysis, Moaledj (1978) studied the oligotrophic bacterial community of the Plußsee for the first time.

In order to enrich oligotrophic bacteria selectively, water samples from the Plußsee were inoculated into liquid inorganic nutrient media or media that contained only small amounts of organic substrates (Moaledj 1978). Aquatic bacteria obviously show a preference for liquid media, since these more closely resemble the natural conditions in the lake than do solid media (Melchiorri-Santolini 1966). In these media the oligotrophic bacteria can develop in free competition with any other microorganisms and, if necessary, also adapt to new compounds. After being inoculated in air for 98 days, and 120 days after inoculation in a benzene or petroleum atmosphere, the enrichment process was terminated in order to isolate pure cultures. It could be assumed that after such a long period all the carbon sources initially available had been exhausted. The enriched bacteria were isolated by plating out in duplicate on the relevant media with 1.1% ion agar.

Sixty-one pure cultures of bacteria were obtained. A few days after the final isolation step, they were able to grow reasonably well aerobically at 26°C on mineral agar without addition of organic compounds. The gaseous secretions of growing and decaying cultures (amines, ketones, organic acids) are important sources of volatile carbon and energy for the growth of the cultures (Hirsch 1963). On the basis of the definitions of oligotrophic (oligocarbophilic) bacteria of Hirsch (1963) and Overbeck (1968a,b), all the strains isolated here belong to this group of bacteria. Ishida et al. (1980) compared critically the different definitions of the term "oligotrophic bacteria" used by various authors.

Taxonomic Classification of the Strains

The isolated strains were classified according to *Bergey's Manual of Determinative Bacteriology* (eighth edition). Since the features we studied did not always correspond with those given in the manual, the classification into genera is preliminary. The strains of oligotrophic bacteria that were isolated from the Plußsee were arranged according to *Bergey's Manual* as shown in Table 7.4.

Although the oligotrophic bacteria are widely distributed in nature, to date there are only a few studies on their morphology and taxonomy. With the exception of the *Hyphomicrobium* sp. and *Caulobacter* sp. strains isolated from various water bodies, only a few of these have been cultured and identified. With selective methods and the use of various dilute nutrient media under defined conditions, a great number of oligotrophic strains could be isolated from the Plußsee. The oligotrophic community consists of species of various taxonomic status. Besides the microorganisms so far known to be oligotrophic (*Actinomycetes, Mycobacter, Nocardia, Caulobacter,* and *Hyphomicrobium* spp.), other oligotrophic bacteria play an important role in the aquatic ecosystem. Eleven strains of *Vibrio* species predominated, as Henrici and Johnson (1935) also observed in Lake Alexander. A high proportion of the isolated bacterial strains (67%) were gram-negative, as is normally the case with aquatic bacteria. The taxonomy, distribution patterns, and their composition show that the group of oligotrophic microorganisms comprises quantitatively and qualitatively the largest fraction of bacterioplankton, and their number exceeds that of the saprophytes cultured on nutrient agar (Moaledj 1975). Because of their high substrate affinity (Chapter 8, this volume), many of these bacteria are very competitive over saprophytic bacteria. They are adapted to utilize rapidly the small resources of DOM present in the unpolluted Plußsee. Thus, they are extremely important for understanding the carbon cycle in the lake.

Table 7.4. Oligotrophic Bacteria That Were Isolated from the Plußsee

Part	Group	Species (number of strains)
Part 4	Budding and/or appendaged bacteria	*Hyphomicrobium* sp. (4); *Caulobacter* sp. (4); undetermined (1)
Part 6	Spiral and curved bacteria	*Microcyclus aquaticus* (1); *Pseudomonas* sp. (6); undetermined (4)
Part 7	Gram-negative aerobic rods and cocci	*Agrobactericum* sp. (2); undetermined (1)
Part 8	Gram-negative facultatively anaerobic rods	*Photobacterium* sp. (3); *Vibrio* sp. (11); *Aeromonas* sp. (1); *Flavobacterium* sp. (2)
Part 10	Gram-negative cocci and coccobacilli	*Paracoccus* sp. (1)
Part 14	Gram-positive cocci	*Micrococcus roseus* (1); *Micrococcus luteus* (3); *Staphylococcus* sp. (2); *Staphylococcus saprophyticus* (2)
Part 17	Actinomycetes and related organisms	*Corynebacterium* sp. or *Arthrobacter* sp. (4); *Brevibacterium* sp. (1)

Functions of Oligotrophic Bacteria in the Plußsee

Very little research has been done on the degradation of refractory DOC in lake waters, and the mechanisms involved in the breakdown processes of the different classes of molecules still remains unknown (Chapter 6, this volume). In laboratory experiments Steinberg (1977) and Stabel (1977) measured the uptake of 20-fold-concentrated DOC from lake water by inoculation with a natural bacterial community. They found that after 13 weeks of incubation 37% of the initial concentration of the total DOC remained unaffected. De Haan (1974, 1977) published details of the metabolism of fulvic acids from Tjeukemeer by oligotrophic bacteria, showing that the microbial processes affect molecular size distribution of naturally occurring DOC.

In studies of Stabel et al. (1979) the degradation and uptake of different fractions of the pool of DOC from the Plußsee by different strains of the autochthonous microflora was examined in order to get more insight into mechanisms of the consumption of naturally occurring organic substances by aquatic bacteria. A vacuum evaporation method was used to concentrate 200-fold the amount of natural DOC from the Plußsee. After discharge of precipitated calcium carbonate, the solution was applied stepwise to a Sephadex G-15 column and eluted using the technique described by Stabel (1977). At least three different fractions of molecules were separated by their molecular weight and chromatographic behavior. Fraction I was composed of compounds of molecular weight larger than 1500 (macromolecules); fraction II consisted of substances with molecular weight between 400 and 600 (oligomers); and fraction III included retarded substances.

Three different media were prepared by mixing the fractions of different molecular-weight DOC (fractions I–III) with Kaz mineral nutrient solution (Overbeck 1965). After sterile filtration these three solutions (each containing tenfold the natural concentration of its organic constituents) were inoculated with eight oligocarbophilic and one saprophytic bacterial strain. The microorganisms were isolated and described morphologically and physiologically by Moaledj (1978) and Overbeck and Toth (1978). Optical density (at a wavelength of 260 nm) and the amount of DOC were monitored to measure the bacterial growth and DOC decomposition. The lag phases were very long (for example, in the case of *Pseudomonas* sp., strain P_2 took more than 30 days in the presence of fractions I and II). Only oligotrophic bacteria were able to degrade oligomeric and retarded fractions. Contrary to oligotrophic strains, the saprophytic *Klebsiella* strain isolated from the Plußsee (Overbeck and Toth 1978) did not grow under the same conditions on any of the three media.

Glucose appeared to stimulate the utilization of macromolecules (fraction I) and retarded substances (fraction III) by *Actinomycetes* sp., strain N1. A rapid increase in turbidity in medium containing fraction I was seen within 12 days, which continued more slowly until the end of the experiment. In medium containing DOC fraction III a similar pattern was recorded but on a smaller scale. A rapid increase in turbidity within the first 24 h was followed by a rapid decrease in medium with DOC fraction II. From the latter curve it can be concluded that there was only a rapid uptake of glucose and no consumption of oligomeric molecules.

In a review, Horvarth (1972) defined the terms "cometabolism and cooxidation." The phenomena occur under two different assumptions. Some authors report that structural similarity between the substrate and the cometabolized compound is needed (Horvarth 1972, De Haan 1977), although in the paper of Lehtomäki and Niemelä (1975) different classes of substances are cometabolized. In the Plußsee experiments co-metabolism of glucose and macromolecules as well as retarded substances (DOC fractions I and III) could occur during the first period of the experiment. The bacteria were able to induce specific enzymes enabling them to degrade the macromolecules (Chapter 6, this volume). Some of these macromolecules are easily degradable constituents of DOC such as proteins and polysaccharides (Stabel 1977, Steinberg 1977, Stabel and Steinberg 1976; Chapter 6, this volume).

In spite of the fact that results obtained from experiments with cultivated organisms should be interpreted with great care, some conclusions may be drawn. There was a rapid uptake of DOC until a steady-state balance between utilizable and refractory fractions of organic compounds was achieved. The remaining refractory DOC could only be degraded by the oligotrophic (autochthonous) bacteria. The following conclusions from these decomposition studies can be drawn. (1) Oligocarbophilic bacteria were able to utilize refractory molecules, they may play an important role in the carbon cycle in fresh waters. (2) The degradation processes occurred very slowly. The autochthonous bacteria preferably utilized oligomeric molecules and compounds retarded on Sephadex columns. (3) The pool of macromolecular DOC, although consisting of easily degradable substances such as proteins and polysaccharides, were more stable against microbial degradation. (4) If smaller molecules that could be readily utilized were present (in this investigation glucose was added), a cometabolism of refractory molecules occurred. (5) The utilization of refractory substances by oligotrophic bacteria was indicated by a simultaneous decrease of DOC concentration and increased density of the bacterial cultures.

These results support important, generally valid mechanisms of degradation of refractory substances. In the presence of readily utilizable low-molecular-weight substrates, oligotrophic bacteria are able to develop enzymes for degradation of macromolecules. For example, after addition of glucose to typical humic lake water (Lakes Stråken and Småland, Sweden), the degradation of humic acids occurred much faster than without glucose (J. Overbeck, unpublished). The addition of glucose apparently simulated the supply of readily utilizable DOM (e.g., phytoplankton EOC in these lakes). Tranvik (1988) demonstrated that bacterial assemblages utilize on average 9.5% of the total organic carbon pool, hydrolyzing the dissolved organic matter–dissolved humic matter (DOM-DHM) complex. They utilize the labile organic substrates (carbohydrates, amino acids, proteins) that are bound to the humic material. DHM acts as a matrix for the binding of DOM. The refractory humic material is important for a continuous supply of low-molecular-weight substrate from autochthonous DOM production.

References

Austin B, Allen DA, Mills AL, Colwell RR (1977) Numerical taxonomy of heavy metal-tolerant bacteria isolated from an estuary. Can J Microbiol 23:1433–1447

Babenzien HD (1964) Hydrobiologische Untersuchungen im Stechlinsee. Limnologica 2:9–34

Baines SB, Pace ML (1991) The production of dissolved organic matter by phytoplankton and its importance to bacteria: Patterns across marine and freshwater systems. Limnol Oceanogr 36:1078–1090

Bölter M (1977) Numerical taxonomy and character analysis of saprophytic bacteria isolated from the Kiel Fjord and Kiel Bight. In Rheinheimer G (ed) *Microbial Ecology of a Brackish Water Environment.* Springer-Verlag, Berlin, pp. 148–176

Buck JD (1974) Effects of medium composition of the recovery of bacteria from seawater. J Exp Mar Biol Ecol 15:25–34

Chróst RJ, Faust MA (1980) Molecular weight fractionation of dissolved organic matter (DOM) released by phytoplankton. Acta Microbiol Pol 29:79–88

Chróst RJ, Faust MA (1983) Organic carbon release by phytoplankton: Its composition and utilization by bacterioplankton. J Plankton Res 5:477–493

Cole JJ, Likens GE, Strayer DL (1982) Photosynthetically produced dissolved organic carbon: An important carbon source for planktonic bacteria. Limnol Oceanogr 27:1080–1090

Collins VG (1963) The distribution and ecology of bacteria in freshwater. Proc Soc Wat Treat Examin 12:40–73

Czeczuga B (1962) An attempt of establishing the production and numerical relations of bacterioplankton biomass. Acta Hydrobiol 4:1–20

Dundas J (1974) Needs and uses of automation in environmental microbiology. In *Proceedings of a Symposium on Automation in Microbiology,* Nordforsk, The Scandinavian Council for Applied Research, Copenhagen, pp 61–68

Fondén R (1969) Nutritional classification of some limnic bacterial populations. Oikos 20:373–383

Goldman CR, Gerletti M, Javornicky P, Melchiorri-Santolini U, De Amezaga E. (1968) Primary production, bacteria, phyto- and zooplankton in lake Maggiore. Correlations and relationships with ecological factors. Mem Ist Ital Idrobiol 23:49–127

Goodfellow M (1968) Properties and composition of the bacterial flora of a pine forest soil. J Soil Sci 19:154–167

Gusewa KA (1951) Relation between phytoplankton and saprophytic bacteria in waters. Trudy Zool Inst AN SSSR 1:34–38 [in Russian]

De Haan H (1974) Effect of a fulvic acid fraction on the growth of a *Pseudomonas* from Tjeukemeer (The Netherlands). Freshwater Biol 4:301–310

De Haan H (1977) Effect of benzoate on microbial decomposition of fulvic acids in Tjeukemeer (The Netherlands). Limnol Oceanogr 22:38–44

Hallmann L (1961) *Bakteriologie und Serologie. Ausgewählte Untersuchungs-Methoden für das bakteriologische und serologische Laboratorium.* Thieme-Verlag, Stuttgart

Henrici AT (1937) Studies of freshwater bacteria IV. Seasonal fluctuations of lake bacteria in relation to plankton production. J Bacteriol 35:129–139

Henrici AT, Johnson DE (1935) Studies on freshwater bacteria. II. Stalked bacteria, a new order of *Schizomycetes.* J Bacteriol 30:61–93

Herbst V (1984) Physiologische Untersuchungen zur Kopplung des Stoffwechsels von *Oscillatoria redekei* van Goor und Begleitbakterien. Arch Hydrobiol Suppl 69:525–594

Hirsch P (1958) Stoffwechselphysiologische Untersuchungen an *Nocardia petroleophila* n. sp. Arch Mikrobiol 29:368–393

Hirsch P (1963) Oligocarbophilie (Wachstum auf Kosten von Luftverunreinigungen) bei Mycobakterien und einigen ihnen nahestehenden Actinomyceten. Zbl Bakt 194:70–82

Hirsch P (1980) Distribution and pure culture studies of morphologically distinct solar lake microorganisms. In Nissenbaum A (ed) *Hypersaline Brines and Evaporitic Environments*. Elsevier, Amsterdam, pp 41–60
Horvarth RS (1972) Microbial co-metabolism and the degradation of organic compounds in nature. Bacteriol Rev 36:146–155
Ishida Y, Shibahara K, Uchida H, Kadota H (1980) Distribution of obligately oligotrophic bacteria in Lake Biwa. Bull Jpn Soc Sci Fish 46:1151–1158
Jannasch HW, Jones GE (1959) Bacterial populations in sea water as determined by different methods of enumeration. Limnol Oceanogr 4:128–139
Jones JG (1970) Studies on freshwater bacteria: Effect of medium composition and method on estimates of bacterial population. J Appl Bacteriol 33:679–686
Jones JG (1971) Studies on freshwater bacteria: Factors which influence the population and its activity. J Ecol 59:593–613
Kriss AE (1961) *Meeresmikrobiologie*. VEB Gustav Fischer Verlag, Jena
Kusnetzov SI (1958) Number of bacteria and formation of organic matter. Verh Int Ver Limnol 13:156–169
Kusnetzov SI (1959) *Die Rolle der Mikroorganismen im Stoffkreislauf der Seen*. VEB Deutscher Verlag der Wissenschaften, Berlin
Lehtomäki M, Niemelä S (1975) Improving microbial degradation of oil in soil. Ambio 4:126–129
Lighthardt B (1974) A cluster analysis of some bacteria in the water column of Green Lake, Washington. Can J Microbiol 21:392–394
Mallory LM, Austin B, Colwell RR (1977) Numerical taxonomy and ecology of oligotrophic bacteria isolated from the estuarine environment. Can J Microbiol 23:733–750
Melchiorri-Santolini U (1966) Pelagic heterotrophic bacteria in the Ligurian Sea and Lago Maggiore. Mem Ist Ital Idrobiol 20:261–287
Moaledj K (1975) Untersuchungen über oligocarbophile Bakterien im Plußsee. PhD Thesis, University of Kiel, Germany
Moaledj K (1978) Qualitative analysis of an oligocarbophilic microflora in the Plußsee. Arch Hydrobiol 82:98–113
Moaledj K, Overbeck J (1982) Verteilung der oligocarbophilen und saprophytischen Bakterien im Plußsee. Arch Hydrobiol 93:287–302
Münster U, Chróst RJ (1990) Origin, composition and microbial utilization of dissolved organic matter. In Overbeck J, Chróst RJ (eds) *Aquatic Microbial Ecology: Biochemical and Molecular Approaches*. Springer-Verlag, New York, pp 8–46
Nalewajko C (1977) Extracellular release in freshwater algae and bacteria: Extracellular products of algae as a source of carbon for heterotrophs. In Cairns J (ed) *Aquatic Microbial Communities*. Garland, New York, pp 589–626
Ohle W (1962) Der Stoffhaushalt der Seen als Grundlage einer allgemeinen Stoffwechseldynamik der Gewässer. Kiel Meeresforsch 18:107–120
Oláh, J, Vasarhelyi R (1970) Comparative bacteriological investigation of three shallow Hungarian lakes with different trophic levels. Annal Biol Tihany 37:223–234
Overbeck J (1965) Primärproduktion und Gewässerbakterien. Naturwissenschaften 51:145
Overbeck J (1968a) Prinzipielles zum Vorkommen der Bakterien im See. Mitt Int Ver Limnol 14:134–144
Overbeck J (1968b) Bakterien im Gewässer—ein Beispiel für die gegenwärtige Entwicklung der Limnologie. Mitt aus der MPG 3:148–156
Overbeck J (1974) Microbiology and biochemistry. Mitt Int Verein Limnol 20:198–228
Overbeck J, Babenzien HD (1964) Bakterien und Phytoplankton eines Kleingewässers im Jahreszyklus. Z Allgem Mikrobiol 4:59–76
Overbeck J, Toth D (1978) Einfluß des Phosphatgehalts auf die Glucoseaufnahme bei Bakterien. Arch Hydrobiol 82:114–122
Pfister RM, Burkholder PR (1965) Numerical taxonomy of some bactreria isolated from antarctic and tropical seawaters. J Bacteriol 90:863–872

Reichardt W (1974) Zur Ökophysiologie einiger Gewässerbakterien aus der *Flavobacterium-Cytophaga*-Gruppe. Zbl Bakt Hyg I Abt Orig A 227:85–93

Rosswall T, Persson IB (1980) Functional description of bacterial populations from seven Swedish lakes. Limnologica 14:1–16

Ruttner F (1932) Bericht über ältere, bisher unveröffentlichte bakteriologische Untersuchungen an den Lunzer Seen. Int Rev Ges Hydrobiol 26:438–444

Schegg E (1971) Produktion und Destruktion in der trophogenen Schicht. Untersuchungen ökologischer Parameter im polytrophen Rotsee und in der mesotrophen Horwer Bucht (Vierwaldstätter See). Schweiz Z Hydrol 33:425–532

Sell A, Overbeck J (1992) Exudates: Phytoplankton–bacterioplankton interactions in Plußsee. J Plankton Res 14:1199–1215

Skerman VDB (1967) *A Guide to the Identification of the Genera of Bacteria*, 2nd ed. Williams & Wilkins, Baltimore

Sneath HA, Sokal RR (1973) *Numerical Taxonomy*. WH Freeman, San Francisco

Sorokin YI, Kadota H (ed) (1972) *Techniques for the Assessment of Microbial Production and Decomposition in Freshwaters. IBP Handbook No 23*. Blackwell, London

Stabel HH (1977) Gebundene Kohlenhydrate als stabile Komponenten im Schöhsee und in *Scenedesmus*-Kulturen. Arch Hydrobiol Suppl 53:159–254

Stabel HH, Moaledj K, Overbeck J (1979) On the degradation of dissolved organic molecules from Plußsee by oligocarbophilic bacteria. Arch Hydrobiol Beih Ergeb Limnol 12:95–104

Stabel HH, Steinberg C (1976) Cleavage of macromolecular allochthonous soluble organic matter. Naturwissenschaften 63:533

Steinberg C (1977) Schwer abbaubare, stickstoffhaltige gelöste Substanzen im Schöhsee und in Algenkulturen. Arch Hydrobiol Suppl 53:48–158

Straskrabova V (1974) Seasonal occurrence of several groups of heterotrophic bacteria in two reservoirs. Int Rev Ges Hydrobiol 59:9–16

Strzelczyk E, Stopiacu-nski M, Donderski W (1975) Studies on metabolic activity of planktonic bacteria isolated from three lakes. Acta Microbiol Pol Ser B 7:177–183

Sundh I (1992) Biochemical composition of dissolved organic carbon derived from phytoplankton and used by heterotrophic bacteria. Appl Environ Microbiol 58:2938–2947

Sundman V (1970) Four bacterial soil populations characterized and compared by a factor analytical method. Can J Microbiol 16:455–464

Tranvik L (1988) Availability of dissolved organic carbon for planktonic bacteria in oligotrophic lakes of differing humic content. Microb Ecol 16:311–322

Waksman SA (1941) Aquatic bacteria in relation to the cycle of organic matter in lakes. In *Symposium on Hydrobiology*. University of Wisconsin Press, Madison, pp 86–105

Winogradsky S (1949) *Microbiologie du Sol—Problèmes et Méthodes*. Masson, Paris

Witzel KP (1990) Approaches to bacterial population dynamics. In Overbeck J, Chróst RJ (eds) *Aquatic Microbial Ecology: Biochemical and Molecular Approaches*. Springer-Verlag, New York, pp 96–128

Witzel KP, Overbeck J, Moaledj K (1982) Microbial communities in Lake Plußsee—an analysis with numerical taxonomy of isolates. Arch Hydrobiol 94:38–52

ZoBell CE (1946) *Marine Microbiology. A Monograph on Hydrobacteriology*. Chronica Botanica, Waltham

ZoBell CE, Conn JE (1940) Studies on the thermal sensitivity of marine bacteria. J Bacteriol 40:223–238

8. Heterotrophic Potential of Bacteria

Jürgen Overbeck

This chapter is concerned with the study of the heterotrophic potential of bacterial assemblages, an approach that has been widely used since its introduction by Parsons and Strickland (1962), Hobbie and Wright (1965a,b), and Wright and Hobbie (1965, 1966). Three parameters relating to the function of the aquatic microflora can be computed from uptake kinetics: (1) the maximum theoretical rate of uptake (V_{max}), (2) the sum of the substrate affinity and the substrate concentration present in situ ($K_t + S_n$), and (3) the time required by the population for the transformation of the substrate (i.e., turnover time T_t).

V_{max} and ($K_t + S_n$) are affected by the applied range of glucose concentration. The substrate transport constant (K_t, µg/L) is a physiological parameter that is dependent on the specific type of uptake mechanisms for a given substrate, inherent in the different physiological types of microorganisms. The maximum rate of uptake V_{max} (µg L^{-1} h^{-1}) and turnover time (hours) depend on the number of bacteria that are utilizing the ^{14}C-labeled substrate. According to Wright and Hobbie (1966), V_{max} should provide real information about the size and functioning of a bacterial community. As with all procedures currently used, this method has certain drawbacks. Three fundamental ones are:

1. It is questionable whether the labeled substrate used (normally glucose or acetate) and its turnover are representative of the substrate turnover in the water body. The in situ rates of the natural, more complex substrates are not measured. Of 115 strains isolated from different depths of the Plußsee (Difco

Nutrient Broth B1 in dilutions), 72 (63%) could utilize glucose. Of the strains isolated by Düsing (1973) from the Fulda river, 72% were glucose positive. In these figures the oligotrophic bacteria are not included, and they form a major part of the total bacterial biomass. The respiration percentage of incorporated glucose ranges from 25% to 50% and should be considered in calculation of the total uptake rate (Hobbie and Crawford 1969).

2. The substrate concentrations used in the experiments are usually higher than those found in the field. Large amounts of added substrates in comparison to their ambient concentration may lead to an overestimation of turnover times (Gocke 1978).
3. It is often very difficult to interpret the uptake kinetics of bacterial assemblages.

Uptake of [^{14}C]Glucose in the Water Column

An example of this kind of study is the vertical profile of the upper 10 m from April 14, 1969 (Overbeck 1975). The physical and chemical properties of the Plußsee are shown in Table 8.1. The thermocline at a depth of 3–7 m at that time of the year was not very distinct. There is a marked O_2 depletion at 5 m, and below 10 m H_2S is present. The levels of P and N concentrations are still high. The conspicuous fluctuations in COD suggest a stratification of the metabolically active biomass. The epilimnic phytoplankton assemblage was dominated as usual in spring by a diatom population, mostly *Stephanodiscus hantzschii* together with flagellates (various species of the genus *Cryptomonas*) and unicellular nanoplankton. Primary production reaches the highest values of the annual cycle (1162 mg C m^{-2} d^{-1}).

The pattern of glucose uptake is consistent with Michaelis–Menten kinetics only at depths of 4–8 m, i.e., below the zone of active phytoplankton (Fig. 8.1). Irregular or S-shaped curves are recorded at lesser and greater depths and do not follow simple Michaelis–Menten kinetics. This result shows that there are physiologically distinct communities in the vertical strata of the lake. Glucose uptake by highly diverse bacterial assemblages should not follow Michaelis-Menten kinetics since the constants for the uptake kinetics for individual species are probably very different (Vaccaro and Jannasch 1967). The great number of bacteria associated with the phytoplankton have apparently not adapted to glucose as the main substrate. Using a mathematical model Williams (1973) was able to show that heterogeneous populations do not adhere to Michaelis–Menten kinetics if the transport constant (K_t) varies over a wide range within the natural population. The author stated that the greater the species diversity, the less successful the kinetic analysis would be. Such deviations have also been observed by Azam and Hodson (1981). A gradient exists between 4 and 8 m. The highest rate of glucose uptake is at 4 m, and it decreases with increasing depth. The quantity of glucose uptake was independent of the kind of response and amounted to 0.1–1.0 µg glucose L^{-1} h^{-1}. This is in agreement with published data for eutrophic lakes (Morgan and Kalff 1972, Overbeck 1974).

Table 8.1. Physical and Chemical Parameters of the Water Column (0–10 m) of the Plußsee (April 14, 1969)

Depth (m)	Temperature (°C)	Oxygen (mg O_2/L)	P-PO_4^{3-} (µg P/L)	P-total (µg P/L)	N-NH_4^+ (µg N/L)	N-NO_2^- (µg N/L)	N-NO_3^- (µg N/L)	S-H_2S (µg S/L)	COD (mg C/L)
0	7.5	13.1	15	69	60	5	10	0	111.2
1	7.5	13.4	15	62	72	4	2	0	10.5
2	7.5	13.4	10	56	29	3	24	0	8.2
3	7.1	13.1	10	59	15	5	0	0	140.3
4	5.1	10.5	15	69	12	6	2	0	12.7
5	5.1	5.1	27	88	45	6	50	0	6.6
6	5.1	2.0	60	127	20	7	113	0	133.2
7	4.5	2.2	77	133	28	5	114	0	9.1
8	4.4	0.2	137	196	180	12	24	0	10.1
9	4.4	0	165	236	300	9	0	0	105.1
10	4.4	0	162	254	302	9	0	120	14.4

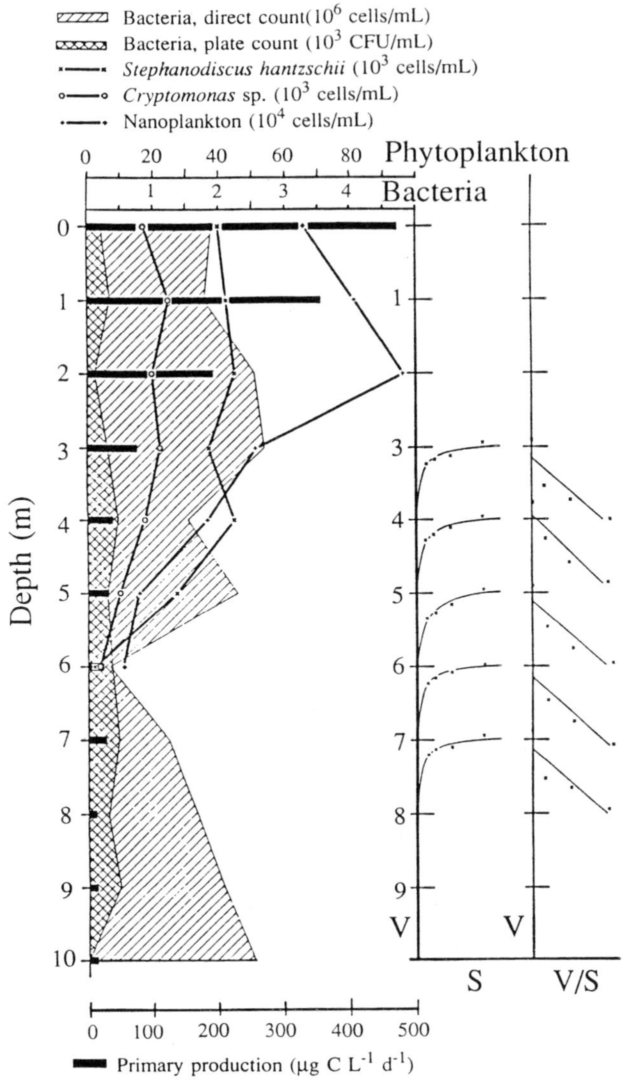

Figure 8.1. Vertical profile of the Plußsee (April 14, 1969) with saturation glucose uptake response and Eadie–Hofstee transformation.

As examples of the complex situation for measurement and interpretation of uptake kinetics in a stratified lake, some vertical profiles may be compared. In February, there are few phytoplankton, low primary production, and low values for V_{max}. In April, there are diatom bloom, high primary production, and high values for V_{max} (Fig. 8.1). The uptake kinetics of bacteria below the highly productive epilimnion follow Michaelis–Menten kinetics. In June, there are few phytoplankton and low primary production; the values for V_{max} are depressed. In

July, the profile is similar to that of April (Fig. 8.1). There is a bloom of *Anabaena flos-aquae* with a high production density. Immediately below the euphotic zone, beginning from a depth of 4 m, high numbers of bacteria (more than 2×10^6 cells/ml) show Michaelis–Menten kinetics. In August, a population of *Cryptomonads* developed at 6 m, accompanied by high numbers of bacteria with an enhanced rate of glucose uptake, showing Michaelis–Menten kinetics.

In summary, the rate of uptake (v and V_{max}) is determined by the production level, i.e., by the amount of available dissolved organic carbon (Table 8.2). Primary production and rate of uptake are well correlated. The type of uptake, whether Michaelis–Menten kinetics or deviations, has no appreciable influence on the amount of uptake of [^{14}C]glucose. The form of the uptake curve is an index of the physiological state of the bacterial community. As with all generalizations, this summary tends to simplify a very complex situation. Michaelis–Menten kinetics as well as irregular or sigmoid responses are distributed in a convoluted way. The reasons for these differences are not yet clear. Chapter 7 (this volume) shows that the taxonomic composition of bacterial assemblages is highly dynamic, with distinct successions. The saturation-type response results largely from a more or less uniform bacterial community with a similar affinity for the substrate. If glucose is taken up by a large number of different metabolic types, constants for the uptake kinetics should differ for each species, thus resulting in an irregular or sigmoid uptake response.

The distribution pattern of different uptake kinetics of [^{14}C]glucose in vertical profiles of the stratified Plußsee can be summarized as follows. Bacteria accompanying primary producers are following an irregular or sigmoid-type response; below the zone of the primary producers a bacterial microflora with Michaelis–Menten kinetics dominates, and bacteria accompanying distinct populations of flagellates also follow Michaelis–Menten kinetics.

The distribution pattern of different glucose uptake responses reflects the metabolic coupling between autotrophic and heterotrophic structures of the ecosystem. The bacteria accompanying phytoplankton assemblages are adapted to EOC and not to the one labeled substrate used in uptake kinetics measurements. Similar results are reported by Francisco (1971): "The values of V_{max} tended to increase with total number of bacteria, but the nonuniformity of the proportionality suggested qualitative as well as quantitative changes in the bacterial population." The whole limnetic ecosystem seems to be subdivided into physiological compartments in which, in a highly dynamic way, the best adapted metabolic types are always active—a prerequisite of the fast running turnover processes.

The coupling between autotrophic and heterotrophic processes is also demonstrated by the correlation between primary production and V_{max} that many authors have reported (e.g., Hobbie 1966; Hobbie and Wright 1968). Morgan and Kalff (1972) compared kinetic data from several lakes, and found a relationship between the bacterial uptake rate of glucose and phytoplankton production. Our data from a stratified lake, however, reveal that production and decomposition can be spatially separated (Fig. 8.1). For example, Olah et al. (1973) observed a distinct uptake of acetate in the water column by phototrophic bacteria only. Such hetero-

Table 8.2. Primary Production and Glucose Uptake Throughout the Water Column (0.2–10 m) of the Plußsee (1969)

	Feb	Apr	May	Jun	Jul	Aug	Sept	Oct	Nov	Average
Primary production (mg C m^{-2} d^{-1})	10.3	1161	672	341	1086	368	225	171	140	464
Glucose uptake (mg C m^{-2} d^{-1})	19.1	32.4	52.3	22.5	24.5	19.2	19.2	16.3	17.2	24.7
Glucose uptake (% of primary production)	—	2.8	7.8	6.3	2.3	5.2	8.5	9.5	12.3	6.8
V_{max} of glucose uptake[a] (μg C L^{-1} d^{-1})	0.09	0.15	0.24	0.11	0.12	0.12	0.09	0.09	0.08	0.12

[a]Mean value.

geneous uptake rates provide information on in situ active metabolic pathways. In Lake Kinneret there was no relationship between glucose cycling and primary production or between DOM and phytoplankton biomass (Berman et al. 1979).

The dimensions of the kinetic parameters in the Plußsee (water layer of 0.2–10 m) are as follows:

$$V_{max}\ (\mu g\ \text{glucose}\ L^{-1}\ h^{-1}) = 0.2\text{–}1.2,\ \bar{x}\ 0.45 \pm 0.27$$
$$K_t + S_n\ (\mu g\ \text{glucose}\ L^{-1}) = 3.8\text{–}46.9,\ \bar{x}\ 19.26 \pm 13.31$$
$$T_t\ (\text{hours}) = 6.0\text{–}201.8,\ \bar{x}\ 51.28 \pm 39.66$$

The heterotrophic potential calculated on the basis of uptake kinetics comprises only a small part of the actual degradation processes in situ. The percentage of glucose uptake from the overall heterotrophic activity, measured by sediment traps, amounted to 10% in Plußsee, and to less than 5% in other eutrophic lakes of the Plön area. With glucose as substrate the percentage decreases with less primary production (Overbeck 1979).

In pure culture studies on oligocarbophilic bacteria, four of nine isolated strains (two strains of *Pseudomonads,* two *Ascinomycetes*) followed Michaelis–Menten kinetics for glucose and acetate (Moaledj and Overbeck 1980). When uptake velocity of the oligocarbophilic bacteria was plotted against substrate concentration, a hyperbolic curve was obtained which indicated saturation of the transport system at a given substrate concentration. The Lineweaver–Burk transformations show a clear relationship between V_{max} and the bacterial biomass, but the kinetic parameters of the oligocarbophilic strains are different under comparable conditions, especially when substrate concentrations were equal.

The amount of available dissolved phosphate can be important for the bacterial uptake kinetics. Overbeck and Toth (1978) showed that changes in kinetic parameters in a vertical profile with a phosphorus gradient may not reflect qualitative changes of the distribution of bacterial assemblages throughout the water column. P-dependent transport systems of the cell were strongly influenced by P availability.

Uptake of [^{14}C]Glucose in the Sediments of Plußsee

Blauw (1988) studied nitrogen fixation as well as uptake kinetics of [^{14}C]glucose by heterotrophic bacteria in the sediments of the Plußsee. In contrast to the V_{max} values of the water column (0.2–1.2 μg L^{-1} h^{-1}, Overbeck 1975), in surface sediments the V_{max} values were as high as 350 μg L^{-1} h^{-1}. The values demonstrate the well-known high bacterial activity at the sediment surface (Wetzel 1975). Similar differences between water column and sediment were found by Wood (1973), Wood and Chua (1973), Meyer-Reil (1978), and Novitsky (1983).

V_{max} and (K_t + S_n) are affected by the applied glucose concentration range. Hence, a comparison of the observed data and literature studies must be taken with care. Under comparable concentration ranges of glucose, the values of the Plußsee sediments are slightly lower than those reported by Toerien and Cavari (1982) for

Lake Kinneret sediments and can be compared with values from Hall et al. (1972) in the shallow oligotrophic Marion Lake. Decomposition in the water column of the Plußsee is very effective (Ohle 1962). Eight percent of the primary production reached the bottom at the deepest parts of the lake. Therefore, a relatively low input of organic matter into the sediments of the Plußsee should be the reason for the unexpectedly low V_{max} values. The T_t values can be compared with measurements by Hall et al. (1972) in Marion Lake sediments (0.06–0.4 h). From the relatively low values in the Plußsee sediments one can conclude that the heterotrophic bacteria are taking up glucose very efficiently and that glucose is a common substrate in these sediments. T_t increased as water depth increased, demonstrating the more intensive glucose metabolism of the littoral zone.

Final Remarks

The kinetic parameters reveal that the lake ecosystem is subdivided into manifold compartments. The in situ kinetic approach is not representative of the total bacterial assemblages. In spite of many difficulties in the metabolic interpretation of the data, it can serve for us as an important tool to get a better understanding of the strategies of biochemical adaptations to changing substrate supply conditions.

References

Azam F, Hodson RE (1981) Multiphasic kinetics for D-glucose uptake by assemblages of natural marine bacteria. Mar Ecol Prog Ser 6:213–222

Berman T, Hadas O, Marchaim U (1979) Heterotrophic glucose uptake and respiration in Lake Kinneret. Hydrobiologia 62:275–282

Blauw TS (1988) Nitrogen fixation (acetylene reduction) in the sediments of the Pluss-See with special attention to the role of sedimentation. Arch Hydrobiol Suppl 81:359–489

Düsing F (1973) Zur Stoffwechseldynamik der fließenden Welle: Bakterieller Abbau gelöster organischer Komponenten. Thesis, University of Kiel, Germany, 154 pp

Francisco DE (1971) Glucose and acetate utilization by the natural microbial community in a stratified reservoir. PhD. Thesis, North Carolina State Univsity, Chapel Hill, 83 pp.

Gocke K (1978) Comparison of methods for determining the turnover time of dissolved organic compounds. Mar Biol 42:131–140

Hall KJ, Kleiber PM, Yesaki I (1972) Heterotrophic uptake of organic solutes by microorganisms in the sediments. Mem Ist Ital Idrobiol 29:441–471

Hobbie JE (1966) Glucose and acetate in freshwater: Concentrations and turnover rates. In Goltermann HL, Clymo RS (eds) *Chemical Environment in the Aquatic Habitat.* NV Noord-Hollandsche Uitgevers Maatschappij, Amsterdam, pp 245–251

Hobbie JE, Crawford CC (1969) Respiration corrections for bacterial uptake of dissolved organic compounds in natural waters. Limnol Oceanogr 13: 361–363

Hobbie JE, Wright RT (1965a) Bioassay with bacterial uptake kinetics: Glucose in freshwater. Limnol Oceanogr 10:471–474

Hobbie JE, Wright RT (1965b) Competition between planktonic bacteria and algae for organic solutes. Mem Ist Ital Idrobiol 18:175–185

Hobbie JE, Wright RT (1968) A new method for the study of bacteria in lakes: Description and results. Mitt Int Ver Limnol 14:64–71

Meyer-Reil LA (1978) Uptake of glucose by bacteria in the sediment. Mar Biol 44:293–298

Moaledj K, Overbeck J (1980) Studies on uptake kinetics of oligocarbophilic bacteria. Arch Hydrobiol 89:303–312

Morgan KC, Kalff J (1972) Bacterial dynamics in two high-arctic lakes. Freshwat Biol 2:217–228

Novitsky JA (1983) Microbial activity at the sediment–water interface in Halifax Harbor, Canada. Appl Environ Microbiol 45:1761–1766

Ohle W (1962) Der Stoffhaushalt der Seen als Grundlage einer allgemeinen Stoffwechseldynamik der Gewässer. Kieler Meeresforsch 18:107–120

Olah, J, Biebl H, Overbeck J (1973) Foto-organotróf acetát hasnositás egy rétegzett eutróf tóban. Hidrologiai Közlöny 1973:21–27

Overbeck J (1974) Über die Kompartimentierung der stehenden Gewässer—ein Beitrag zur Struktur und Funktion des limnischen Ökosystems. Verh Ges Ökol 1973:211–223

Overbeck J (1975) Distribution pattern of uptake kinetic responses in a stratified eutrophic lake (Plußsee ecosystem study IV). Verh Int Ver Limnol 19:2600–2615

Overbeck J (1979) Studies on heterotrophic functions and glucose metabolism of microplankton in Plußsee. Arch Hydrobiol Beih Ergeb Limnol 13:56–76

Overbeck J, Tóth D (1978) The influence of phosphate on the uptake rate of glucose in bacteria. Arch Hydrobiol 82:114–122

Parsons TR, Strickland JDH (1962) On the production of particulate organic carbon by heterotrophic processes in sea water. Deep Sea Res 8:211–222

Toerien DF, Cavari B (1982) Effect of temperature on heterotrophic glucose uptake, mineralization and turnover rates in lake sediments. Appl Environ Microbiol 43:1–5

Vaccaro RF, Jannasch HW (1967) Variations in uptake kinetics for glucose by natural populations in seawater. Limnol Oceanogr 12:540–542

Wetzel RG (1975) *Limnology.* WB Saunders, Philadelphia

Williams PJLeB (1973) The validity of the application of simple kinetic analysis to heterogeneous microbial populations. Limnol Oceanogr 18:159–165

Wood LW (1973) Pollution stress and respiration of glucose by natural microbial communities of the sediment and water of Toronto Harbour. In *Proceedings of the 16th Conference on Great Lakes Research.* Ann Arbor, Michigan, pp 204–213

Wood LW, Chua KE (1973) Glucose flux at the sediment–water interface of Toronto Harbour, Lake Ontario, with reference to pollution stress. Can J Microbiol 19:413–420

Wright RT, Hobbie JE (1965) The uptake of organic solutes in lake water. Limnol Oceanogr 10:22–28

Wright RT, Hobbie JE (1966) Use of glucose and acetate by bacteria in aquatic ecosystems. Ecology 47:447–464

9. Phosphoenolpyruvate Carboxylase: Allosteric Regulation at the Ecosystem Level by Dissolved Organic Matter

Jürgen Overbeck

Introduction

In natural unpolluted aquatic ecosystems the carbon sources that are available for heterotrophic microbial life are limited. Readily utilizable dissolved organic matter (UDOM) is present only in very small amounts with a high substrate diversity, and frequently the supply is discontinuous (Münster and Chróst 1990; Chapter 2, this volume). The qualitative and quantitative composition and metabolic activity of heterotrophic bacterial assemblages in aquatic ecosystems is to a great extent determined by the availability of DOM, although abiotic factors are also important. Because of the fundamental significance of dissolved organic substrates, very effective and rapid regulatory mechanisms that enable bacteria within a genetically determined range to adapt continuously to different and often growth-limiting levels of DOM are to be expected. Compared with the very beginning of aquatic microbiology in the early 1960s, adequate methods for measurements of activity and production of aquatic bacteria are now available. The different "activities" we are measuring are the expression of catabolic (production of low-molecular-weight intermediates and energy) and anabolic (biosynthetic) pathways (production of biomass). Behind the data stands the bacterial metabolism with its adaptive capabilities for life and survival in continuously changing environments (e.g., substrate, temperature, diurnal cycles, pressure). However, our knowledge of regulatory properties of enzymes of ecological importance in aquatic bacteria is very limited.

Only in the last decade did the importance of different ways of steering enzyme activities as control mechanisms of cell metabolism become apparent. Allosterically regulated enzymes are one of the major possible means for metabolic control of heterotrophic bacteria under natural conditions. We have therefore tried to transfer into aquatic microbial ecology biochemical methods for studying the regulation of adaptive mechanisms in situ, going from the whole-cell level down to the enzyme level.

Basic Strategies of Biochemical Adaptation to the Environment

Regulation of Enzyme Synthesis at the Molecular Level

Transcriptional regulation is thought to be the primary and most important step in gene expression. "Gene expression" refers to the appearance of a protein as a final gene product that may be physiologically active. This type of regulation is rather slow, requiring hours to days. For example, in *Mesembryanthemum crystallinum* (Dicotyledonae, *Mesembryanthemaceae*), phosphoenolpyruvate carboxylase (PEPC) protein can be induced with NaCl, drought, and other factors. Within 24 to 36 h of irrigation with 500 mM NaCl, novel mRNAs for early stress proteins of unknown function can be detected. PEPC mRNA increases after 2 days. After that the amounts of PEPC and other crassulacean acid metabolism (CAM) enzymes increase. CAM metabolism itself does not commence until about 7 days later (Bohnert et al. 1992).

Regulation of the Enzyme Activity at the Metabolic Level

The specific activity of enzymes can change as a result of interaction with a class of metabolites termed "enzyme modulators" or "enzyme effectors," as receptors of these effector molecules serve allosteric proteins. Enzymes that are regulated in a qualitative way by effectors are termed regulatory enzymes. They are mostly key enzymes of metabolic control mechanisms. This kind of regulation is very fast, allowing an instantaneous fine-tuning of enzyme functions. Effectors, as "cellular signals" (Hecker and Babel 1988), provide the metabolism with a "most rapid, accurate, and sensitive level of regulation" (Hochachka and Somero 1973). Positive effectors increase enzyme activity; negative effectors inhibit it. Because environmental changes may occur within seconds or minutes, survival may depend on biochemical adaptations that can occur at the same rapid rates. In this way qualitative enzyme regulation is the most effective metabolic regulation of ecological significance.

If intermediates of the TCA cycle are drained off for biosynthetic purposes, the pool of C_4 dicarboxylic acids of the TCA cycle must be refilled. These anaplerotic reactions (Kornberg 1966; anaplerotic means filling up the TCA cycle with C_4 compounds that have been lost for biosyntheses) are so vital that it is not surprising that the enzymes catalyzing anaplerotic reactions are widely distributed.

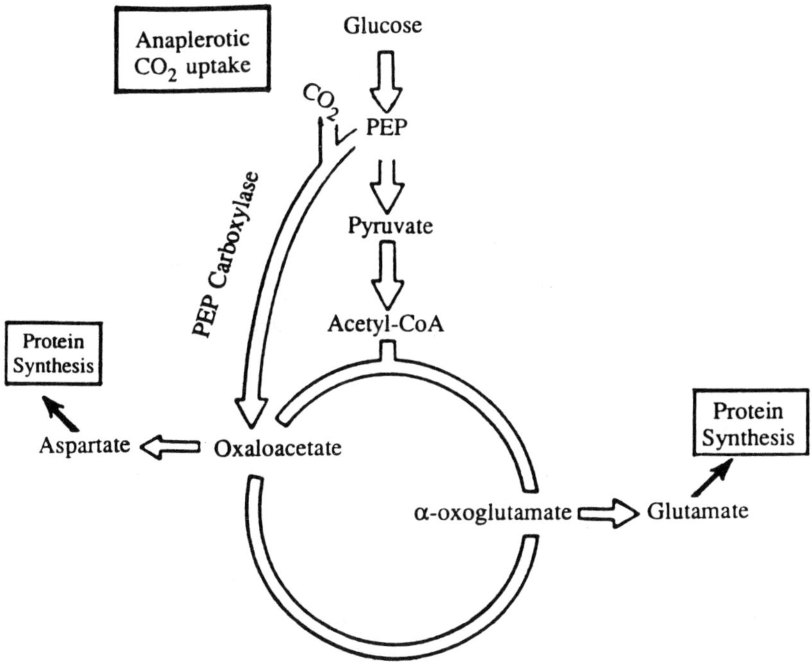

Figure 9.1. The function of PEP carboxylase as an anaplerotic enzyme during growth of *E. coli* on glucose (modified after Gottschalk 1988).

One of the most important enzymes involved in these reactions is PEP carboxylase. CO_2 dark uptake is a basic metabolic process related to the biosynthetic processes that start from the pool of tricarboxylic acid cycle intermediates. In bacteria, algae, and higher plants, PEP carboxylase is one of the key enzymes with anaplerotic functions carboxylating C_3 compounds under formation of oxaloacetate and other manifold functions (Latzko and Kelly 1983, Kluge 1983). The enzyme plays a fundamental role each time the cells have to reorganize their central metabolism in order to adapt or improve metabolic activity as a response to environmental changes. This role can be explained partly by the anaplerotic ability of the enzyme and the possibility of its regulation by effectors. Figure 9.1 shows the anaplerotic functions of this enzyme during growth of *E. coli* on glucose.

CO_2 Uptake as Indicator of Biosynthetic Activity

Sorokin (1961, 1964), Romanenko (1964), and Kusnetsov and Romanenko (1966) introduced a method for the estimation of bacterial growth from the incorporation of $^{14}CO_2$ into the bacterial biomass for 24 h. They suggested that the rate of CO_2 uptake proceeds at a rate strictly proportional to heterotrophic bacterial carbon increase. Many studies show that there is a functional relationship between CO_2

uptake and the growth rate of heterotrophic bacteria (e.g., Overbeck 1972, 1979), but the assumption of a constant relationship of 6% between $^{14}CO_2$ uptake and total heterotrophic microbial carbon uptake is contradictory (Overbeck and Daley 1973). However, the general conclusion of Splittstoesser (1966) that dark CO_2 uptake can be considered an indicator of overall biosynthetic activity is valid. Under defined conditions the correspondence between CO_2 uptake and metabolic activity is satisfactory. Independent determinations of heterotrophic bacterial production by means of a direct measurement of bacterial biomass with a particle analyzer agreed rather well with growth rates derived from $^{14}CO_2$ uptake (anaplerotic sequences) in mixed cultures of *Scenedesmus quadricauda* and *Enterobacter aerogenes* in a chemostat where the exudate pool of the alga was the sole carbon source for the bacteria (Overbeck 1984).

In contrast to experimental conditions, we have to assume that dissolved organic material (DOM) is qualitatively quite variable in the different compartments of the lake (Chapter 2, this volume). On the other hand, we are confronted with a complex metabolic situation in situ. The changing relationships of CO_2 and substrate uptake by whole cells apparently reflect bacterial metabolic pathways under in situ conditions (Overbeck 1979). Only with a study of defined enzyme systems under controlled conditions with few variables will precise measurements and a closer approximation to understanding of the regulation of heterotrophic microbial pathways of carbon metabolism at the ecosystem level be possible. The following section will, therefore, deal mainly with the structure, physiological functions, kinetics, and regulatory properties of phosphoenolpyruvate carboxylase as one of the most important carboxylating enzymes with ecological relevance that stands behind CO_2 uptake.

Phosphoenolpyruvate Carboxylase

Phosphoenolpyruvate carboxylase (PEP carboxylase, PEPC, E.C. 4.1.1.31) represents an enzyme universally distributed with a number of regulatory functions. The enzyme has attracted great interest and was studied thoroughly during the past two decades because it plays a major role in C_4 photosynthesis and crassulacean acid metabolism (CAM). Several reviews of PEP carboxylase properties and functions appeared (e.g., O'Leary 1982, Latzko and Kelly 1983, Kluge 1983, Stiborova 1988).

The molecular weight of the PEP carboxylase of *Enterobacter aerogenes* is about 360,000. The enzyme is a tetramer composed of identical subunits of about 90,000 molecular weight. The regulatory properties of the enzyme molecule are very flexible with respect to various ligands (PEP, bicarbonate, divalent metal ions), and differences in catalytic activity result from complex interactions among ligands, metabolites, pH, temperature, etc. Therefore, it is questionable whether in vitro studies with purified enzymes under defined conditions with only one substrate as a ligand are directly comparable with in vivo experiments, not to mention the study of the complex in situ conditions. For further details see Overbeck (1990).

Regulation of Phosphoenolpyruvate Carboxylase by Effectors

PEP carboxylase is allosterically regulated. Effectors are bound to allosteric enzymes in a very cooperative manner. In most cases, they are not identical with the substrate of the enzyme they act on. Allosteric enzymes contain, therefore, special binding sites for effectors (inhibiting or activating). Effectors that are bound to these sites alter the activities of allosteric enzymes, which may be characterized by sigmoidal saturation curves. Under the aspect of metabolic regulation a sigmoidal curve implies that only a relatively small difference in the supply of the substrate (S) is required to change the activity of the enzyme. This makes regulation very fast and sensitive (fine-tuning of enzyme regulation).

To study the enzyme spectrophotometric as well as radiochemical assays are usually applied. We used PEP carboxylase from *Enterobacter aerogenes* (DSM 30053) purified to homogeneity (Fig. 9.2) and commercially available PEP carboxylase from *Escherichia coli* (Sigma P 7150) in a spectrophotometric assay (Fig. 9.3). The formation of oxaloacetate was measured in a malate-dehydrogenase-coupled system: NADH and malate dehydrogenase were added to the assay, and the change in absorbance (at 340 nm) as a result of reduction of oxaloacetate to malate was measured.

PEP carboxylase was strongly activated by acetyl-CoA as well as by phosphorylated compounds of the glycolytic pathway (glucose-6-phosphate, fructose-6-phosphate, and fructose-1,6-bisphosphate; Fig. 9.3). The enzyme was inhibited by aspartate and malate. Fumarate also showed a remarkable inhibition, but other intermediates of the TCA cycle did not (Nishikido et al. 1968; J. Overbeck, unpublished).

From experiments with purified PEP carboxylase it became evident that fructose and glucose are not positive activators of the enzyme, but only phosphory-

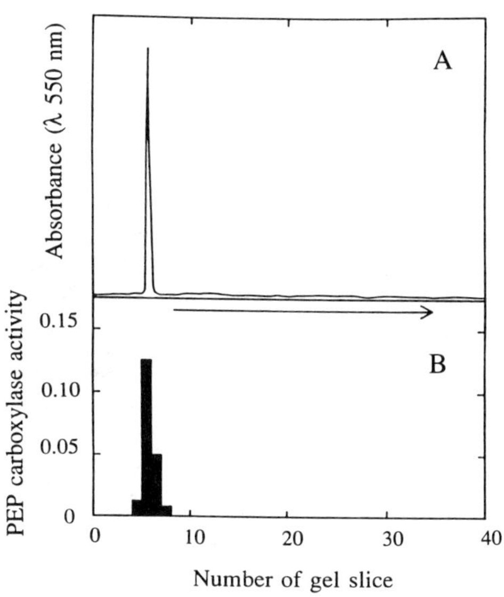

Figure 9.2. Purification of PEP carboxylase of *Enterobacter aerogenes* (DSM 30053) to homogeneity. Polyacrylamide gel electrophoresis of preparation VI of PEP carboxylase. Two gels were carried out in parallel. One (lower panel B) was cut into 2.5-mm slices, which were assayed for PEP carboxylase activity. The other (upper panel A) was stained with Coomassie blue and scanned at 550 nm (Sako unpublished).

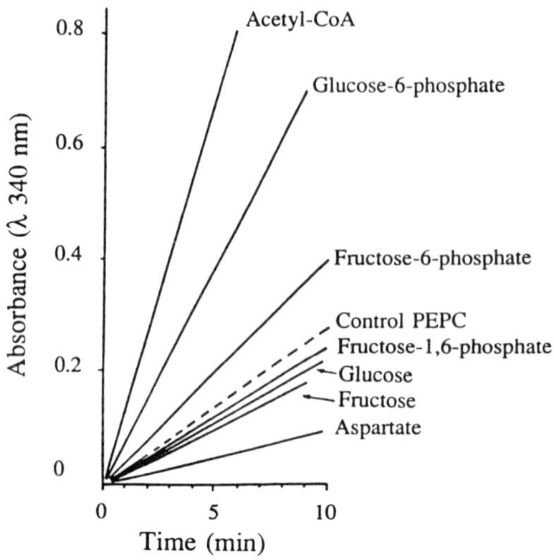

Figure 9.3. Spectrophotometric measurement of activators [+] and inhibitors [–] of PEP carboxylase (purified PEPC from *E. coli,* Sigma P 7150), 0.2 units/ml. One unit forms 1 μmol of oxaloacetate from PEP and HCO_3^- at pH 8.5 at 25°C. Concentration of effectors 0.25 mM (acetyl-CoA 0.10 mM). Results with the purified enzyme of *Enterobacter aerogenes* are identical (from Overbeck 1990).

lated derivatives. The removal of ATP from crude enzyme extracts by desalting with a Sephadex G25 column resulted in inactivation of glucose and fructose as effectors (J. Overbeck, unpublished). Apparently the reason for this was lack of phosphorylation of carbohydrates in the absence of ATP. As in our experiments with PEP carboxylase from *Enterobacter aerogenes* and *Escherichia coli,* in purified maize leaf PEP carboxylase, glucose-6-phosphate was the most active sugar phosphate in stimulating enzyme activity; the next most active sugar phosphate was fructose-6-phosphate. The activation by diphosphorylated sugars in maize leaf PEP carboxylase is only small, which also corresponds with bacterial PEP carboxylase.

Because of substantial decarboxylation, measurements with $H^{14}CO_3^-$ as the substrate could be imprecise because of instability of oxaloacetate, which is the product of carboxylation of PEP via PEP carboxylase. Therefore, ^{14}C-labeled oxaloacetate has to be converted to malate during carboxylation by addition of NADH and malate dehydrogenase to the assay.

Figure 9.4 exhibits a model of PEP carboxylase regulation in the bacterial cell. PEPC is activated by the substrates of the glycolytic channel through a complex cooperativity between the enzyme and the substrate as a positive effector. This activation means adaptation of the specific activity of the enzyme and, in a broader sense, of the central carbon metabolism to changing substrate concentrations. Acetyl-CoA acts as a feedforward control promoting its own oxidation by increasing the amount of oxaloacetate and consequently promoting the reactions of the tricarboxylic acid cycle. Aspartate regulates the protein biosynthesis by feedback inhibition.

With the CO_2 dark uptake we are measuring basically activities of the tricarboxylic acid cycle. But the model does not consider the complex internal metabolism of PEP and CO_2 that was discussed above. The feedback of PEP carboxylase by aspartate and malate means that these compounds are controlling the CO_2

Fine Tuning System

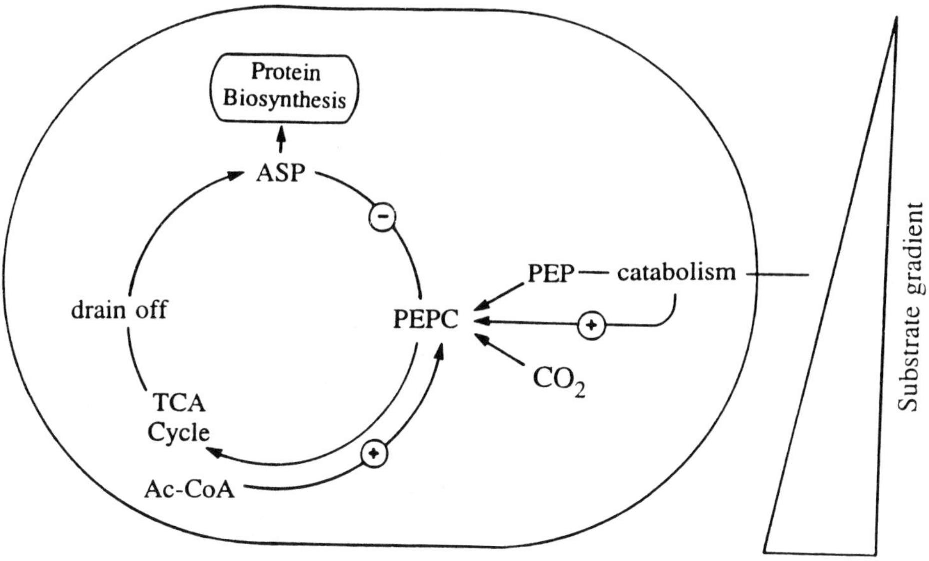

ASP - aspartate
TCA cycle - tricarboxylic acid cycle
Ac-CoA - acetyl-coenzymA
PEPC - phosphoenolpyruvate carboxylase
PEP - phosphoenolpyruvate

Figure 9.4. Regulation of PEP carboxylase activity and protein biosynthesis in a substrate gradient. Synergistic activation of PEP carboxylase by phosphorylated substrates of the glycolytic pathway and acetyl-CoA. Regulation of protein biosynthesis by feedback inhibition of aspartate (from Overbeck and Sako 1989).

uptake, with the restriction that there is no internal pool of CO_2. This is apparently true in general for measurements at the ecosystem level with substrate limitation.

Phosphoenolpyruvate Carboxylase and Phosphoenolpyruvate Carboxykinase

Using crude or only partially purified enzyme extracts for PEP carboxylase assays, we get confusing results. In the presence of phosphoenolpyruvate carboxykinase with aspartate, instead of inhibition a strong activation is observed. This enzyme catalyzes PEP carboxylase CO_2 fixation with conversion of phosphoenolpyruvate to oxaloacetate. PEP carboxykinase (PEPCK, E.C. 4.1.1.32) is widely distributed in bacteria, plants, and animals. The primary function of PEPCK is gluconeogenic, i.e., formation of fructose-1,6-bisphosphate from PEP. The PEP carboxykinase-catalyzed reaction is readily reversible, quite in contrast to the formation of oxaloacetate by the PEP carboxylase reaction. In bacteria, ADP is the preferred

Table 9.1. Comparison of Phosphoenolpyruvate Carboxylase (PEPC) and Phosphoenolpyruvate Carboxykinase (PEPCK)

	PEPC	PEPCK
Main functions	Formation of oxaloacetate	Gluconeogenesis
	Anaplerotic reactions	Formation of carbohydrates
Divalent metal inons	Mg^{2+}, Mn^{2+}	Mg^{2+}, Mn^{2+}
Molecular weight	360,000 (tetramer)	70,000
Effect of aspartate	Strong inhibition	Strong activation

phosphate acceptor of PEP carboxykinase (Utter and Kolenbrander 1972). The study of PEP carboxylase in crude systems that contain PEP carboxykinase with differing regulatory properties, can therefore lead to erroneous results. PEP carboxykinase is induced and activated by C_4 compounds, including TCA-cycle intermediates and amino acids such as aspartate and glutamate, whereas PEP carboxylase is strongly inhibited by aspartate.

PEP carboxykinase can be assayed in crude or only partially purified enzyme extracts other than PEP carboxylase by using an isotope exchange method (Hsie and Rickenberg 1966). Quite in contrast to PEP carboxylase, the equilibrium of the PEP carboxykinase reaction is in favor of PEP formation. The method is based on the incorporation of $^{14}CO_2$ into a pool of nonradioactive oxaloacetate in an exchange reaction resulting a highly labeled oxaloacetate. In Table 9.1 these two enzymes are compared. The presence of the two enzymes utilizing the same intermediates for opposing regulatory systems is a strong hint that studies on enzyme regulation in any case must be supplemented by experiments with pure enzymes to avoid erroneous interpretations.

Control of PEP Carboxylase at the Ecosystem Level

The metabolic complexity of bacterioplankton is not comparable to pure culture studies, and the regulation of carbon metabolism in situ can hardly be understood by measuring the CO_2 uptake of whole cells of bacterioplankton. In the Plußsee and in several mesotrophic/eutrophic lakes in East Holstein, the uptake of glucose comprises only about 10% of the total heterotrophic activity estimated from heterotrophic potential studies (Overbeck 1981; Chapter 8, this volume). Renewed studies on CO_2 uptake in 1990 showed that after addition of glucose to natural water samples the CO_2 uptake by aquatic bacteria increased only occasionally.

For the study of the regulation of PEP carboxylase at the ecosystem level, we developed the following preparation protocol.

The Preparation Protocol

Plußsee lakewater (50 L)

1. Prefiltration (10-µm pore-size nylon net)
2. Prefiltration (3-µm pore-size membrane filters)

3. Continuous centrifugation (10°C, 15,000 rpm)
4. Washing of the pellet in Tris buffer, pH 7.5

Enzyme purification (0–4°C)

5. Grinding of pellet in sand (Merck 7712, precooled mortar)
6. Addition of Tris buffer (pH 7.5, 2 ml)
7. Centrifugation
8. Removal of cell membranes (pellet)
9. Desalting of protein with NAP-5 Sephadex G-25 DNA-Grade (prepacked column, sample volume 0.5 ml)
10. Elution (1 ml Tris buffer pH 7.5)
11. Concentration to total volume of 500 μl (Sartorius Centrisart SM 13249 E, molecular weight cutoff 20,000)

Miniassay 200 μl:

12. 50-μl assay (Overbeck 1990); 50 μl cell extract; 100 μl $NaH^{14}CO_3$ 1 μCi, supplemented with NADH/malate dehydrogenase for conversion of [^{14}C]oxaloacetate to [^{14}C]malate

To obtain a better yield of cells, in addition to continuous centrifugation of untreated lake water, 24-h preincubation at 22°C after prefiltration (3-μm pore-size membrane filters) with addition of glucose (500 mg/L) or yeast extract (500 mg/L) was useful. Additionally, the results of field work were compared with studies of purified enzymes of *Enterobacter aerogenes* (DSM 30053) (Fig. 9.3). For concentration of the enzyme protein, the Centrisart ultrafiltration system (Sartorius, Göttingen) was used to cut off protein with a molecular weight below 20,000 (Table 9.2).

Table 9.3 shows yield, protein content, and specific activity of PEP carboxylase of lake bacteria after preincubation with glucose. The values are classified by increasing specific activity showing independence from the sampling date. Yield and protein content of the bacterial biomass are nearly equal, in contrast to the specific activity, which covers a broad range in spite of uniform handling and preincubation. This is an expression of the undefined metabolic state of natural bacterial assemblages and depends on the actual situation in the lake. After addition of 1 mmol glucose to the assay as effector, a distinct activation in nearly

Table 9.2. Application of Centrisart for Purification of *Enterobacter aerogenes*[a] PEP Carboxylase

Assay	Protein (mg/ml)	DPM/mg Protein	Malate (DPM)
104/1	2.43	49,915	710
104/2 (Centrisart)	3.75	125,034	2714

[a]*Enterobacter aerogenes* 24-h batch culture, mineral solution + 0.25% yeast extract, 22°C. Sephadex G-25 (desalting), Centrisart (Sartorius Göttingen, molecular weight cutoff <20,000), miniassay PEP + [^{14}C]HCO_3^-.

Table 9.3. Growth Yield, Protein Content, and PEP Carboxylase Specific Activity in Preincubated[a] Lakewater Bacteria and the Effect of Glucose on Specific Activity of PEP Carboxylase

				Glucose Added			
				0.5 mmol		1 mmol	
Sampling Date (1990)	Yield (g/L)	Protein (mg/ml)	Specific Activity (dpm/mg protein)	Specific Activity (dpm/mg protein)	Activation (%)	Specific Activity (dpm/mg protein)	Activation (%)
January 15	4.2	1.88	3,744	5,106	136	4,638	124
January 21	1.40	2.78	5,232	5,064	95	5,467	193
August 21	1.08	3.33	7,111	5,741	81	5,429	76
September 3	1.50	1.01	11,485	10,376	90	12,594	110
February 5	8.4	2.05	28,058	34,809	124	34,497	123
February 19	1.64	3.50	32,274	36,662	114	38,468	119
June 19	1.36	0.94	57,957	68,857	119	67,404	116

[a]5 L lake water (Plußsee, Schöhsee) was filtered through 3-μm pore-size membrane filter and preincubated with 500 mg/L (2.8 mmol/L) glucose in batch culture for 48 hours at 22°C. Glucose was added as effector to the assay. Yield is wet weight (g/L). The values are classified by increasing specific activities. Enzyme preparation, see Preparation Protocol.

Table 9.4. Comparison of the Results of Preincubation[a] of Plußsee Water Bacteria with Yeast Extract and Glucose

	Preincubation	
	Yeast Extract	Glucose
Yield (g wet weight/L)	0.44–1.12 (after 24 h)	1.08–8.4 (after 48 h)
Protein (mg/ml)	2.28–13.57	0.94–3.33
Specific activity (dpm/mg protein)	3,286–44,846	3,744–57,957
Activation (%)		
0.5 mmol glucose	71–111 (91)[b]	81–124 (108)
1.0 mmol glucose	66–116 (93)	76–124 (110)
0.5 mmol asparate	77–167 (134)	No activation
1.0 mmol aspartate	80–216 (155)	No activation

[a]Five liters of lake water was prefiltered through 3-μm pore-size membrane filters and preincubated in batch culture with 500 mg/L yeast extract (24 h) or 500 mg/L glucose (48 h) at 22°C.
[b]Average value

all series up to 124% could be observed; the activation was less pronounced after addition of 500 μmol glucose. After addition of aspartate the specific activity remained unchanged.

In Table 9.4 results of preincubation of lake bacteria (Plußsee) with glucose and yeast extract are compared. The experiments were performed from January to December 1991, and seasonal differences did not exist; the yields were nearly equal (mean 0.7 g wet weight/L after 24 h with yeast extract and 1.4 g wet weight/L after 48 h incubation with glucose). The protein content of the bacterial biomass is much higher after preincubation with yeast extract. Unexpectedly, the biomass with a lower protein content after glucose preincubation has a much higher specific activity of PEP carboxylase. Also, strong differences exist in the activation of the enzyme in dependence on preincubation. Glucose acts as a distinct positive effector after preincubation with glucose. There was no activation with glucose after preincubation with yeast extract. Aspartate, which strongly inhibits PEP carboxylase, now acts as a strong activator. A high activation by aspartate is indicative of the presence of an active PEP carboxykinase (PEPCK) in the assay (Table 9.1). PEPCK is a key enzyme in gluconeogenesis with anabolic functions. The growth of bacteria on noncarbohydrate carbon sources apparently requires gluconeogenesis. Aspartate thus can be estimated as an indicator of the metabolic shift of the bacterial assemblage of the lake after preincubation with yeast extract.

From the increase in specific activity with glucose as a positive effector one cannot conclude that we are following an allosteric activation of PEP carboxylase of bacterioplankton. Moreover, the product of the PEP carboxylase reaction has to be identified. For this purpose we used a thin-layer chromatography (TLC) technique (Berthold Tracemaster Linear Analyzer) with a computer-processed three-

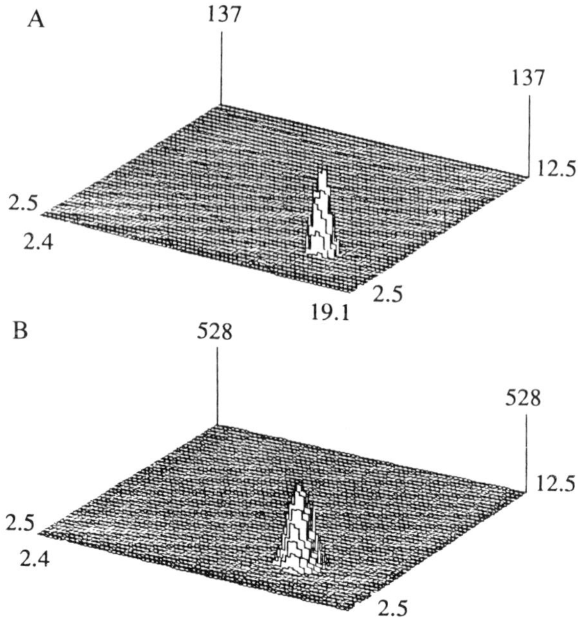

Figure 9.5. Three-dimensional evaluation of ^{14}C-labeled spot on the TLC plate with computer-processed Berthold Tracemaster Linear Analyzer. Fingerprint of [^{14}C]malate (A) without and (B) with addition of 0.001 μCi [^{14}C]malate to the assay. Two-dimensional TLC system according to Metzner (1962), cellulose plate (Merck 5716). Enzyme source: *Enterobacter aerogenes* (DSM 30053), batch culture 24 h (rotary shaker), 22°C, mineral medium Kaz (Kaserer + Hoagland A–Z) + yeast extract (0.25%).

Table 9.5. Effect of Glucose-6-Phosphate and Glucose on [^{14}C]Malate Formation as a Result of Activity of PEP Carboxylase Harvested from Plußsee Bacteria Preincubated with Yeast Extract or Glucose[a]

	Radioactivity of [^{14}C]Malate (dpm)	
Effectors (1 mmol/L)	Yeast Extract	Glucose
Glucose-6-phosphate	801	1124
Glucose	729	1283

[a]Five liters of Plußsee water were prefiltered through 3.0 μm pore-size membrane filters and preincubated with yeast extract (500 mg/L) or glucose (500 mg/L) in aerated batch culture at 20°C for 24 h. [^{14}C]Malate radioactivity was determined by thin-layer chromatography. For enzyme preparation, see Preparation Protocol.

Table 9.6. Effect of Glucose and Glucose-6-Phosphate on the Specific Activity and Activation of PEP Carboxylase in Plußsee Bacteria without Preincubation[a]

Variant	Specific Activity (dpm/mg protein)	Activation (%)
Control	6,162	100
Glucose (50 μmol/L)	11,297	183
Glucose-6-phosphate (50 μmol/L)	6,000	97

[a]Yield 0.17 g bacterial biomass (wet weight) from 405 L lake water; six samples were taken from 27 March until 23 May 1990. Protein content was 1.48 mg/ml. For enrichment and preparation of the enzyme, see Preparation Protocol.

dimensional evaluation of the radioactive spots. Because of the instability of oxaloacetate as product of carboxylation of PEP via PEP carboxylase, [^{14}C]oxaloacetate had to be converted to [^{14}C]malate by addition of NADH and malate dehydrogenase to the assay (see Preparation Protocol). Malate is therefore the first product of PEP carboxylase activity that has to be identified by fingerprints on the thin-layer plates (Fig. 9.5).

Table 9.5 shows that after preincubation with glucose the amount of [^{14}C]malate as a result of phosphoenolpyruvate carboxylation is nearly twice as high as after preincubation with yeast extract, which is in accordance with results from lake studies (Tables 9.3 and 9.4).

But to get real information on PEP carboxylase activity and regulation in lake bacteria, study of the enzyme in bacterioplankton without preincubation was essential. According to the Preparation Protocol bacteria from the Plußsee were enriched by continuous centrifugation; the results are shown in Tables 9.6 and 9.7. The specific activity of both series is equal and within the range of specific activity

Table 9.7. Effect of Different Effectors on the Specific Activity and Activation of PEP Carboxylase in Plußsee Bacteria without Preincubation[a]

Effector	Specific Activity (dpm/mg protein)	Activation (%)
Control	5,815	100
Glucose (50 μmol/L)	7,226	124
Glucose (500 μmol/L)	8,806	151
Glucose-6-phosphate (50 μmol/L)	6,823	117
Glucose-6-phosphate (500 μmol/L)	5,301	91
Asparate (50 μmol/L)	6,352	109
Aspartate (500 μmol/L)	6,655	114

[a]Yield 1.6 g bacterial biomass (wet weight) from 1465 L lake water; 27 samples taken from 30 May 1990 until 11 December 1991. Protein content 2.38 mg/ml. For enrichment and preparation of the enzyme, see Preparation Protocol.

after preincubation (Table 9.4). In the spring series (Table 9.6) a high activation of the enzyme by glucose is observed; G-6-P has no effect. In the summer series of 1990 (Table 9.7), the enzyme is distinctly activated by glucose again, and not by glucose-6-P and aspartate, respectively.

Conclusions

One of the main results of the experiments on PEP carboxylase regulation at the ecosystem level is the fact that the assemblage of lake bacteria follows metabolic pathways that were studied in detail by means of purified enzymes. Starting from the question of whether allosteric regulation of PEP carboxylase may be of importance for instantaneous adaption of this key enzyme to rapid changes of substrate supply in situ, we get the following results. First, with biochemical methods using concentration of enzyme protein by gel filtration, PEP carboxylase as a key allosteric enzyme can be measured qualitatively and quantitatively in bacterioplankton. The specific activity of PEP carboxylase in bacterioplankton without preincubation is continuously rather high, of the same order of magnitude as in bacterioplankton after preincubation with glucose. Furthermore, activation of PEP carboxylase in bacterioplankton by effectors as an expression of the adaptive capabilities of bacterial metabolism can be demonstrated. Low-molecular-weight dissolved organic material acts as effectors, providing metabolism with a "most rapid, accurate, and sensitive level of regulation" (Hochachka and Somero 1973). This type of regulation offers the chance to address the crucial question of whether allosteric regulation is of ecological importance. From the results it can be concluded that this type of regulation actually plays an important role in aquatic bacteria. The optimum concentration of the effector lies within an order of magnitude of the actual concentration in situ. The final point is that preincubation with yeast extract, i.e., a high supply of nitrogen-rich DOM, modifies the metabolic situation of bacterioplankton in a rather consistent way. Aspartate instead of glucose or glucose-6-P acts as a strong activator, indicating a shift to gluconeogenic (anabolic) pathways. Thus, the efficiency of different effectors can be indicative of the metabolic status of bacterioplankton.

PEP carboxylase as a key allosteric enzyme is not a representative enzyme for the whole carbon metabolism in bacteria. But detailed comparative studies of one enzyme with highly purified enzyme protein and partially purified bacterioplankton protein can give a lot of information on metabolic pathways and adaptative capabilities of bacteria. The general importance of allosteric regulation from the point of view that effectors act as signals to convey information from outside to key enzymes of cellular metabolism is summarized in Chapter 17 (this volume).

References

Bohnert HJ, Vornon DM, DeRocher EJ, Michalowski CB, Cushman JC (1992) Biochemistry and molecular biology of CAM. In Wray JL (ed) *Inducible Plant Proteins: Their*

Biochemistry and Molecular Biology. Cambridge University Press, Cambridge, pp 113–137.

Gottschalk G (1988) *Bacterial Metabolism*. Springer-Verlag, New York

Hecker M, Babel W (eds) (1988) *Physiologie der Mikroorganismen*. Gustav Fischer Verlag, Jena

Hochachka PW, Somero GN (1973) *Strategies of Biochemical Adaptation*. WB Saunders, Philadelphia

Hsie AW, Rickenberg HV (1966) A mutant of *Escherichia coli* deficient in phosphoenolpyruvate carboxykinase activity. Biochim Biophys Res Commun 25:676–683

Kluge M (1983) The role of phosphoenolpyruvate carboxylase in C_4-photosynthesis and crassulacean acid metabolism. Physiol Veget 21:817–882

Kornberg HL (1966) Anaplerotic sequences and their role in metabolism. In Campbell PN, Greville GD (eds) *Essays in Biochemistry, vol 2*. Academic Press, New York, pp 1–31

Kusnetsov SJ, Romanenko WJ (1966) Produktion der Biomasse heterotropher Bakterien und die Geschwindigkeit ihrer Vermehrung im Rybinsk-Stausee. Verh Int Ver Limnol 16:1493–1500

Latzko E, Kelly GJ (1983) The many-faceted function of phosphoenolpyruvate carboxylase in C_3 plants. Physiol Veget 21:805–815

Metzner H (1962) Papierchromatographische Trennung der Photosynthese-Intermediärprodukte. Naturwissenschaften 49:183

Münster U, Chróst RJ (1990) Origin, composition and utilization of dissolved organic matter. In Overbeck J, Chróst RJ (eds) *Aquatic Microbial Ecology: Biochemical and Molecular Approaches*. Springer-Verlag, New York, pp 8–46

Nishikido T, Izui A, Iwatani H, Katsuki H, Tanaka S (1968) Control of carbon dioxide fixation in *Escherichia coli* by the compounds related to TCA cycle. J Biochem 63:532–541

O'Leary MH (1982) Phosphoenolpyruvate carboxylase: An enzymologist's view. Annu Rev Plant Physiol 33:297–315

Overbeck J (1972) Experimentelle Untersuchungen zur Bestimmung der bakteriellen Produktion im See. Verh Int Ver Limnol 18:176–187

Overbeck J (1979) Dark CO_2 uptake—biochemical background and its relevance to in situ bacterial production. Arch Hydrobiol Beih Ergeb Limnol 12:38–47

Overbeck J (1981) A new approach for estimating the overall heterotrophic activity in aquatic ecosystems. Verh Int Ver Limnol 21:1355–1358

Overbeck J (1984) Application of TCA cycle metabolism for growth estimates of heterotrophic bacterioplankton. Arch Hydrobiol Beih Ergeb Limnol 19:23–36

Overbeck J (1990) Aspects of aquatic microbial carbon metabolism: Regulation of phosphoenolpyruvate carboxylase. In Overbeck J, Chróst RJ (eds) *Aquatic Microbial Ecology: Biochemical and Molecular Approaches*. Springer-Verlag, New York, pp 78–95

Overbeck J, Daley RJ (1973) Some precautionary comments on the Romanenko technique for estimating heterotrophic bacterial production. Bull Ecol Res Commun 17:342–344

Overbeck J, Sako Y (1989) Ecological aspects of enzyme regulation in aquatic bacteria. In *Proceeding 5th International Symposium on Microbial Ecology (ISME 5)*. Japan Scientific Societies Press, Kyoto, pp 378–382

Romanenko VI (1964) Heterotrophic assimilation of CO_2 by the bacterial aquatic flora. Mikrobiologiya 30:679–683

Sorokin YI (1961) Role of chemosynthesis in production of organic substance in water reservoir. Investigation of chemosynthesis production in Kuibyshev Water Reservoir in 1958–1959. Mikrobiologiya 30:928–937

Sorokin YI (1964) On the trophic role of chemosynthesis in water bodies. Int Rev Ges Hydrobiol 49:307–324

Splittstoesser WE (1966) Dark fixation and its role in the growth of plant tissue. Plant Physiol 41:755–759

Stiborova M (1988) Phosphoenolpyruvate carboxylase: The key enzyme of C_4-photosynthesis. Photosynthetica 22:240–263

Utter MF, Kolenbrander HM (1972) Formation of oxaloacetate by CO_2 fixation on phosphoenolpyruvate. In Boyer PD (ed) *The Enzymes, vol 6*. Academic Press, New York, pp 117–168

10. Relations between *Limnothrix redekei* (Cyanophyta/Cyanobacteria) and Bacteria: Mutual Control of Growth by Extracellular Algal Products

Maria-Elisabeth Meffert

Introduction

Populations of *Limnothrix redekei* (Meffert 1987), which were accompanied by *Planktothrix agardhii* (Anagnostidis and Komarek 1988), *Aphanizomenon gracile,* and sometimes *Limnothrix rosea,* appeared in the Plußsee several times between 1960 and 1978 (Overbeck 1972, 1973, Zaiss 1976, Hickel 1978, 1988; Chapter 1, this volume). The reasons for their occurrence and long-term blooms were unknown, but parallel growth of cyanophytes and heterotrophic bacteria was observed several times and described by Overbeck (1965, 1968, 1972; Chapter 7, this volume). Maxima and minima of algae and bacteria corresponded temporally and spatially. The isolation of *L. redekei* clones from Plußsee populations in nonaxenic as well as axenic cultures (Meffert 1972, Meffert and Chang 1978) gave us the opportunity to investigate the relationship between *L. redekei* and heterotrophic bacteria. Extracellular algal products represented by their type, nitrogenous compounds, the mode of release, and their support of bacterial growth were of special interest, as were bacterial activities with respect to growth of *L. redekei* (Meffert 1973a,b, 1989, Meffert and Zimmermann-Telschow 1979, Meffert and Overbeck 1979, 1981).

Materials and Methods

Details are given by Meffert (1989), Meffert and Overbeck (1979, 1981), and Meffert and Zimmermann-Telschow (1979). Axenic and nonaxenic clones of

Limnothrix redekei originally isolated from the Plußsee (Meffert 1972) were cultivated in Erlenmeyer and Roux flasks (50–4000 ml) at 20°C at various photon flux densities (Osram L 58 W/36 Natura) and measured with a Li-192S quantum sensor connected to a Li-185A Q/R/P meter (Lambda Instruments, USA). The light and dark cycles were 15 and 9 h. Fast-growing suspensions were cultivated under continuous illumination in aerated gas washing bottles (modified for our purposes) at 25°C and pH 9. Air was supplied at a rate of 2.5 L/h per 300 ml suspension. The pH values of the culture and the periodic supply of CO_2 added to the air were controlled by a pH meter and regulator with magnetic valve and time relay (precision 0.1 pH) (Meffert 1973a, 1975). Air and CO_2 were filtered through glass wool. For studies involving CO_2 deficiency, air without CO_2 was supplied. The light intensity within the gas washing bottles amounted to 0.88 mW/cm^2. Determinations of number of cells, trichomes, cells per trichome, dry weight, and extinction at 436 nm were carried out as described by Meffert and Zimmermann-Telschow (1979), as were determinations of extra- and intracellular total nitrogen, protein-N, NO_3-N, NO_2-N, NH_3-N and free and soluble peptide amino acids.

Results

Algal and Bacterial Growth

Fast-growing *Limnothrix redekei* cultures had more than four divisions in 4 days (Figs. 10.1 and 10.3; Meffert and Zimmermann-Telschow 1979). In axenic cultures the algal cell number increased from 8.3 to 145 × 10^6 cells/ml, but nonaxenic cultures increased to 199 × 10^6 cells/ml. The accompanying bacteria had a similar growth rate with approximately four doublings in 4 days and corresponding cell numbers. Algal and bacterial growth were within the same range and were strongly correlated ($r^2 = 0.98$, $P \leq 0.005$). Stimulation of *L. redekei* growth in the presence of bacteria was apparent after 48 h, when about 50% of the nitrate was consumed (Fig. 10.2). The increase in intracellular nitrogen compounds corresponded to the increase in the number of cells and was distinctly higher in nonaxenic suspensions.

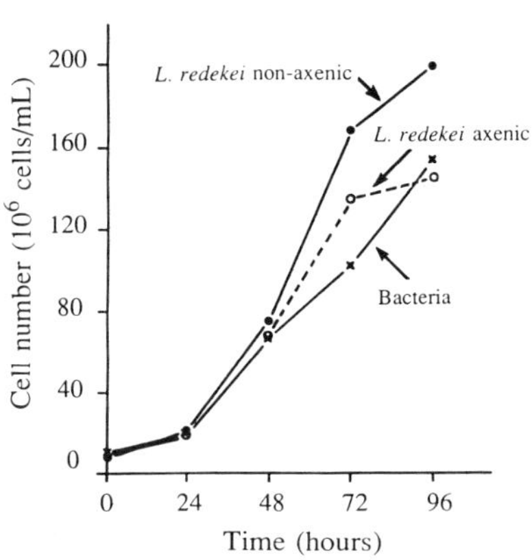

Figure 10.1. Number of cells of *Limnothrix redekei* and bacteria in axenic and nonaxenic cultures.

Figure 10.2. Nitrate uptake and change of nitrogen compounds in intracellular nitrogen pool of *Limnothrix redekei* axenic and nonaxenic cultures.

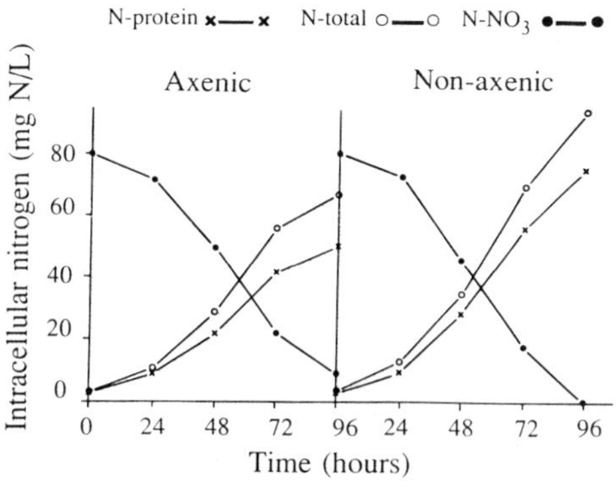

Release of Nitrogen Compounds by *Limnothrix redekei*

During the first 24 h after incubation, the release rates of protein nitrogen and ammonia nitrogen were relatively high in the axenic and nonaxenic cultures of *Limnothrix redekei* (Fig. 10.3; Meffert and Zimmermann-Telschow 1979). Subsequently the amount of protein nitrogen and ammonia nitrogen increased at similar rates in axenic suspensions, but both stayed relatively constant in bacteria-containing cultures that were started at a low initial concentration of protein and ammonia-nitrogen (40 μg N/L in each, compared to 200 μg N/L in the axenic cultures). Nitrite was released in relatively small amounts (maximum 300 μg N/L) by both types of cultures and decreased when the concentration of nitrate nitrogen was about 20 mg/L or 25% of the initial amount, indicating the beginning of nitrogen limitation.

In axenic cultures, increases in cell densities and intracellular nitrogen compounds of *L. redekei* correlated to the release rates of protein and ammonia nitrogen ($r_2 = 0.97$, $P \leq 0.005$). The number of doublings within 4 days was similar. Bacterial growth followed the production of extracellular algal substances

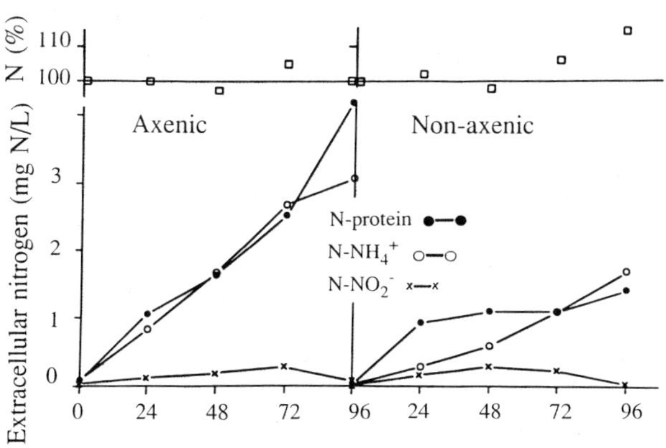

Figure 10.3. Extracellular nitrogen compounds and nitrogen balance in *Limnothrix redekei* axenic and nonaxenic cultures. □, Sum of the nitrogen values as a percentage of the initial value (84 mg N/L).

with corresponding rates as revealed by changes in release rates and the subsequent bacterial growth rates (Fig. 10.4). Bacterial growth was controlled by extracellular *Limnothrix* products, which were proportional to the increasing biomass of cyanophytes (measured as cell numbers or intracellular nitrogen compounds).

Free extracellular amino acids and peptide amino acids from soluble peptides (Table 10.1) were present in relatively small amounts. This finding was expected from the nitrogen balance and was also found to be true of other algae (Fogg 1952, Gocke 1969). After 68 h of algal culture, the concentration of amino acid nitrogen in filtrates of axenic suspensions was 304 µg N/L, approximately 12% of the extracellular protein nitrogen. There were relatively high concentrations of glycine, proline, alanine, and glutamine and relatively low concentrations of threonine, serine, and phenylalanine. Tyrosine was not detected. In filtrates of bacteria-containing cultures, after 68 h of incubation there was only 64 µg N/L of amino acid nitrogen, which was equivalent to approximately 500 µg/L of amino acids. Similar values were obtained from the Plußsee, with an average of 70 µg N/L amino nitrogen from peptides (Brehm 1966) and 331 µg/L free amino acids plus amino acids from soluble peptides (Gocke 1969). Tyrosine, threonine, phenylalanine, and arginine were not detected in filtrates of bacteria-containing cultures, and the concentrations of the remaining amino acids in these cultures were approximately 20% of those in axenic cultures with two exceptions: serine comprised 53% and glycine 28%.

Mode of Release and Cell Number per Trichome

Two modes of trichome breakage occurred in *L. redekei* (Meffert et al. 1981, Meffert 1987): intercellular and transcellular breakage. The first was never observed in natural populations and was only observed once in one culture, but with numerous filaments fragmenting in this first mode. Transcellular trichome breakage in *Limnothrix,* which is not identical to the process investigated by Lamont (1969), can be described as a type of lysis involving the destruction of cell walls and release of the complete cell contents (Fig. 10.5). Two or more shortened fragments with a reduced number of cells are the result of lysis. This process, which accompanies algal growth, was frequently observed in natural populations and in cultures of *L. redekei* and may reflect an apparent lack of metabolic control, as observed in other cyanophytes (Carr et al. 1969, Fogg et al. 1973). Lysis can increase rapidly and can destroy single cells as well as the majority of filaments within minutes. The number of cells per trichome indicates the degree of fragmentation and lysis, describes the metabolic state of the algae, and characterizes the increase of extracellular substances.

Under optimal conditions the fragmentation rate of filaments and doubling rate of cells resulted in a constant mean number of cells per trichome. Increase of fragmentation rate and number of shortened trichomes with a decreasing mean cell number indicates suboptimal conditions. In axenic fast-growing cultures with continuous illumination (Fig. 10.1) the number of cells per trichome decreased

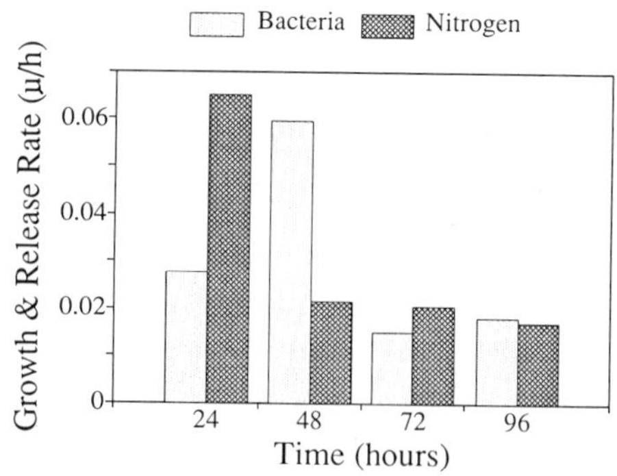

Figure 10.4. Rates of release of extracellular nitrogen compounds (protein and ammonia nitrogen) and the growth rate of bacteria in *Limnothrix redekei* culture.

Table 10.1. Extracellular Amino Acids Released in Axenic and Nonaxenic Cultures of *Limnothrix redekei*[a]

	Concentration of Amino Acid (μg N/L)		
Amino Acid	Axenic	Nonaxenic	Percentage of Axenic Suspension
Cysteic acid	13.4	2.7	20.1
Aspartic acid	22.4	4.8	21.4
Threonine	3.5	<0.01[b]	—
Serine	4.7	2.5	53.2
Glutamic acid	32.9	6.6	20.0
Proline	31.8	6.4	20.1
Glycine	58.8	16.6	28.2
Alanine	30.6	5.4	17.6
Valine	20.0	5.3	26.5
Cystine	<0.01	<0.01	—
Methionine	4.7	1.2	25.5
Isoleucine	15.3	3.6	23.5
Leucine	27.0	4.9	18.1
Tyrosine	<0.01	<0.01	—
Phenylalanine	3.5	<0.01	—
Lysine	23.5	4.0	17.0
Arginine	11.8	<0.01	—
Ammonia	174.1	109.4	62.3

[a]Values are sums of free amino acids of soluble peptides. Modified from Meffert and Zimmermann-Telschow (1979).
[b]Below the detection limit of 0.01 μg N/L.

during the first and last 24 h. Cell density decreased in the nonaxenic parallel cultures during the first 24 h only. The extracellular nitrogen compounds, mainly protein and ammonia nitrogen, increased simultaneously from about 10% to 20% of the intracellular nitrogen in axenic cultures and from 3% to 12% in bacteria-containing cultures. Contrary to this increase of number of cells per trichome, there was a decrease in percentage of extracellular nitrogen compounds.

Similar processes were regularly observed in natural populations (Gibson 1975, Meffert 1989). The mean cell number per trichome of *L. redekei* assemblages in the Edebergsee decreased from 30–40 to 5–10 between January and June (Fig. 10.6). When algal growth started in February, the fragmentation rate increased as growth-stimulating light intensities increased and decreasing nutrient concentrations reached their minimum in June–July. During the beginning of circulation in September, the number of cells per trichome increased again, reaching their initial maximal values in October–November. The process was accompanied by distinct changes of biomass and chlorophyll$_a$ concentrations. The chlorophyll$_a$ concentration decreased from 0.2 to 0.06 pg/cell and increased again in September. The increase of fragmentation and lysis of cells during growth of *L. redekei* is connected with a release of cell contents, supporting the parallel growth of accompanying heterotrophic bacteria in situ. Thus the corresponding distribution pattern of *L. redekei* and bacterioplankton, e.g., in vertical profiles of the lake, can be explained (Overbeck 1972).

Factors Inducing Lysis

The tendency of the *Limnothrix* species to increase their lysis rate under suboptimal and variable conditions is a particularly interesting feature observed several times and under different conditions (Meffert 1973a,b, 1975, 1989, Gibson 1975, Meffert and Zimmermann-Telschow 1979, Zaiss 1985). Under nutrient-deficient conditions the lysis rate rapidly increased, with simultaneous depletion of cellular reserves, and intensified when growth-stimulating radiation and temperatures increased (Table 10.2). Cellular requirements and storage capacity were inversely related to the availability of nutrients. Inorganic carbon, the most abundant cellular and extracellular element, was hardly stored at all. Bicarbonate deprivation induced extensive lysis, although the cultures were aerated during their light–dark cycle. During the 4 days of survival, the number of cells per trichome decreased from about 30 to 6. The carbon content of dry substance (Table 10.2) varied over a narrow range (Reynolds 1984) while, according to availability, nitrogen and phosphorus were reduced from about 10% to 5% and from about 1% to 0.1%, respectively, indicating the different storage capacities for nitrogen and phosphorus compounds. The survival time also differed between nitrogen deficiency (20 days) and phosphorus deficiency (30 days), when the number of cells per trichome decreased from 31 to 5 and from 36 to 12, respectively. Thus survival time of *L. redekei* was dependent largely on initial cellular reserves.

Under growth-stimulating light intensities the lysis-inducing effect of CO_2 deficiency increased distinctly. Irreversible damage was caused to 48-h-old axenic

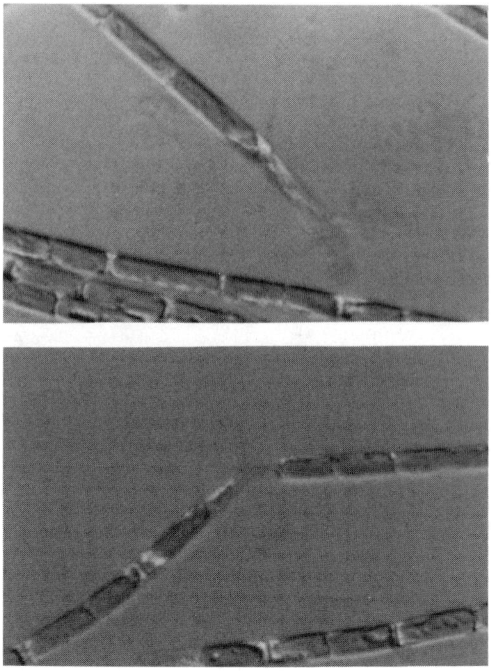

Figure 10.5. Lysing cells of *Limnothrix redekei*.

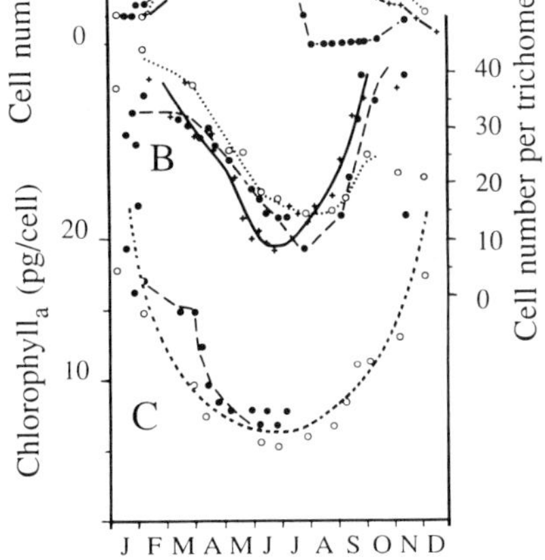

Figure 10.6. Total number of cells (A), chlorophyll$_a$ content per cell (B), and number of cells per trichome (C) during *Limnothrix redekei* development in lake in 1974, 1983, and 1984.

Table 10.2. Lysis, Growth, and Composition of *Limnothrix redekei* Grown under Deficient Concentrations of Carbon, Nitrogen, or Phosphorus in Batch Cultures[a]

	Deficiency of					
	Carbon		Nitrogen		Phosphorus	
Limnothrix redekei	Beginning	End	Beginning	End	Beginning	End
Cellular content (% dry substances)	48	48	15–10	4.8	2.1	0.1
$Chlorophyll_a$ (% of dry substance)			2.4	1.0	2.1	1.6
Number of cells per trichome	30	6.4	31	4.8	36	12
Doublings	0.3		1.5–1		3–4	
Survival time (days)	4		20		30	

[a]Nonaerated, 1.1 E m^{-2} d^{-1}, light/dark cycle 15:9 h, 20°C, deficiency: no supply of bicarbonate, nitrate, or phosphate. Modified from Meffert (1989).

cultures of *L. redekei* grown under continuous illumination (Fig. 10.7) when CO_2 concentration was reduced by supplying air instead of the air–CO_2 mixture for 6 h, and when the pH increased from 9 to 10. In the following 48 h, 90% of the cells lysed, and the mean cell number per trichome fell from 27 to 3. Considerable quantities of organic compounds were released into solution. The concentration of extracellular protein nitrogen increased from 8% to 45% of the intracellular protein nitrogen. Some of the nitrogen compounds measured as intracellular could have come from particulate cell debris.

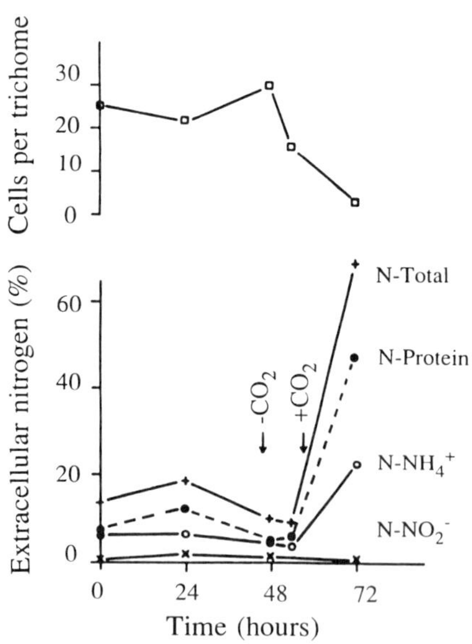

Figure 10.7. Release of extracellular nitrogen compounds and number of cells per trichome after 6 h of CO_2 deficiency in *Limnothrix redekei* culture.

In lakes with *L. redekei* blooms, the nitrate, phosphate, and inorganic carbon minima were reached in April, May, and June, respectively (Meffert 1989). The number of cells per trichome of *L. redekei* decreased from about 30 in March–April to 10 in June (Fig. 10.6) when, simultaneously, incident radiation increased.

Effect of Bacteria on Algal Growth in Long-Term Cultures

Long-term studies (longer than 5 years) of axenic and nonaxenic batch cultures of *L. redekei* again revealed a complete degradation of the algae in axenic cultures, which began 3 months after inoculation (Meffert 1989). During the following months and years neither filaments nor chlorophyll reappeared, whereas in nonaxenic cultures *L. redekei* filaments and 5–10% of the maximal chlorophyll$_a$ amount were present after 3 years. After 5.5 years these algae could still be recultivated. In both culture types (axenic and nonaxenic batch cultures), the growth maximum was passed when 80% of the nitrate and 90% of the phosphate were assimilated. Lysis with release of the complete cell contents could occur in several growth stages, as could be seen by the presence of extracellular dissolved organic nitrogen (N_{org}) and phosphorus (P_{org}) compounds. The largest amounts of extracellular organic substances were measured during later degradation stages and decreased only slightly. Constantly high concentrations of extracellular products were measured throughout the following months. Bacterial cell density increased with algal biomass measured as chlorophyll. The maximal bacterial cell density appeared during early algal degradation stages, and the subsequent substance release did not support bacterial growth. Similar results were obtained for numerous bacterial strains that were tested (K. Moaledj, unpublished results). A large fraction of spheroblasts, cell walls, and debris accumulated over time; the dry weight was about 60% of the maximum, whereas chlorophyll was %. During the later stages of degradation of *L. redekei,* inorganic phosphate and organic phosphorus compounds were measured in distinct amounts. About 50% of the assimilated phosphorus reappeared as extracellular SRP, and its concentration remained constant throughout the following months. On the other hand, the assimilated nitrogen was largely released as organic nitrogen compounds that were scarcely metabolized by *L. redekei* and only partially metabolized by bacteria. Nitrite and ammonia nitrogen appeared in relatively small amounts. Inorganic metabolizable nitrogen compounds were thus scarcely produced in the course of algal degradation.

After 5.5 years the recultivation of algae was checked by addition of nutrients in the original amounts (Table 10.3). Only nitrogen compounds, i.e., inorganic and organic substances, induced *L. redekei* and bacterial growth with reassimilation of the released inorganic SRP, indicating the previous nitrogen deficiency. Phosphate deficiency inhibited nitrate uptake (Meffert 1989). Optimal growth when nitrogen was added further revealed that accumulated lysis products had no growth-inhibiting effects.

The heterotrophic capacities of axenic *L. redekei* were investigated with numerous combinations of low-molecular-weight organic substances. None of the

Table 10.3. Growth of *Limnothrix redekei* and Bacteria in Nonaxenic, Long-Term Cultures (5.5 Years Old) That Were Supplemented with Nutrients[a]

Supply	Cell Number (10^6 cells/ml)		Chlorophyll$_a$ (mg/L)	SRP uptake (mg P/L)
	Algae	Bacteria		
Control	3	7	0.1	0
N-NO_3^-	101	26	5.2	0.96
P-PO_4^{3-}	3	4	0.2	0.12
N-NO_3^- + P-PO_4^{3-}	95	23	5.1	2.00
Nutrient broth	87	220	3.2	2.78
Glucose	3	26	0.1	0.25

[a]Batch culture, 1.1 E m^{-2} d^{-1}, light/dark cycle 15:9 h, 20°C. Variables were determined after 4 weeks of incubation. From Meffert (1989).

clones tested could grow in the presence of glucose with photosynthesis inhibition by DCMU (3[3,4-dichlorophenyl]-1,1-dimethylurea). Carbohydrates had no growth-supporting effect in the absence of DCMU. Attempts to grow axenic *L. redekei* heterotrophically at different concentrations of amino acids, either individually or variously combined, and in the presence of carbohydrates were unsuccessful. Addition of amino acids to the complete nutrient solution was also unsuccessful. These findings suggested obligate autotrophy in *L. redekei*.

Experiments with several nitrogen compounds showed optimal algal growth when ammonium compounds were added in lower concentrations and pH was in the optimal range. When nutrient broth was the only nitrogen source for axenic *L. redekei*, there was only limited growth correlated with a low ammonium nitrogen content, whereas growth of *L. redekei* was clearly supported in the presence of bacteria (Fig. 10.8).

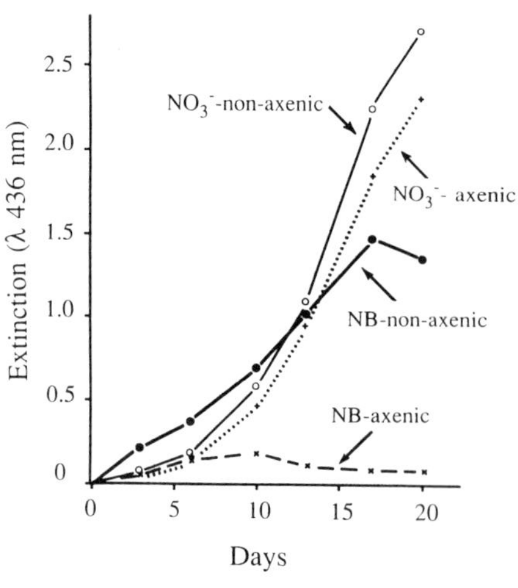

Figure 10.8. Growth of *Limnothrix redekei* on nutrient broth (NB) and NO_3^- as the only nitrogen source. Initial extinction was 0.01, light intensity 20 μE m^{-2} s^{-1}, light and dark cycle 15:9 h, temperature 20°C.

Conclusions

The disposition of *Limnothrix redekei* to lysis supports the metabolic coupling with heterotrophic bacteria, a presupposition of the long-term survival of the population. The lysis was examined in connection with the obligate autotrophy of *L. redekei* and its storage capacity for nitrogen and phosphorus compounds as well as its ability to grow at low light intensities.

Lysis with the release of the complete cell content seems to be a normal process during growth of *L. redekei* (Fig. 10.5). Growth and lysis controlled the size of the *Limnothrix* population, adapting it to changing environmental conditions. The rate of lysis increases with decreasing nutrient concentrations and is high under nutrient-deficient conditions. Furthermore, it depends on cellular reserves and the growth rate. The storage capacity varies with elements and their abundance (Table 10.2, Fig. 10.8). Carbon was hardly stored at all, in contrast to phosphorus. Under growth-stimulating light intensities and temperatures a nutrient deficiency causes a collapse of the population in culture and in nature (Fig. 10.8). Under low light intensities and at low temperatures the nutrient deficiency had no effect on lysis. Constant numbers of bacteria reveal the minor release of *L. redekei* products (Meffert and Overbeck 1981).

Extracellular organic nitrogen compounds released as a result of *L. redekei* lysis support the growth of bacteria, which assimilate about 80% of the released nitrogen compounds in fast-growing as well as long-term batch cultures. About 20% of the extracellular dissolved organic nitrogen compounds seem to be refractory and cannot be metabolized when nitrogen is limited.

Organic nitrogen compounds allowed growth of *L. redekei* in the presence of heterotrophic bacteria, thus confirming the obligate autotrophy of the alga. This and the previous findings give reasons to assume that inorganic nitrogen compounds released during bacterial degradation of organic substrates are assimilated by *Limnothrix*. This short-circuited cycle, a term that was introduced by Ohle (1962), ensures the survival of the population and explains the optimal development of *L. redekei* in eutrophic waters with organic pollution. The relationship between *L. redekei* and the accompanying bacteria has a symbiotic character.

References

Anagnostidis K, Komarek J (1988) Modern approach to the classification system of cyanophytes. 3 Oscillatoriales. Arch Hydrobiol, Suppl 80, Algol Stud 50–53:327–472.

Brehm J (1966) Untersuchungen über den Aminosäure-Haushalt holsteinischer Gewässer, insbesonderes des Plußsees. PhD thesis, Christian-Albrechts-Universität, Kiel, Germany

Carr NG, Hood W, Pearce J (1969) Control and intermediary metabolism of blue-green algae. Prog Photosynth Res 3:1565–1569

Fogg GE (1952) The production of extracellular nitrogenous substances by a blue-green alga. Proc R Soc B 139:372–397

Fogg GE, Fay P, Stewart WDP, Walsby AE (eds) (1973) *The Blue-Green Algae*. Academic Press, London

Gibson CE (1975) Cyclomorphosis in natural populations of *Oscillatoria redekei* Van Goor. Freshwat Biol 5:279–286

Gocke K (1969) Untersuchungen über Abgabe und Aufnahme von Aminosäuren und Polypeptiden durch Planktonorganismen. PhD thesis, Christian-Albrechts-Universität, Kiel, Germany

Hickel B (1978) Phytoplankton population dynamics in Plußsee (Eastholstein, Germany). Verh Ges Ökol 1977:119–126

Hickel B (1988) Unexpected disappearance of cyanophyte blooms in Plußsee (North Germany). Arch Hydrobiol, Suppl 80, Algol Stud 50–53:545–554

Lamont HC (1969) Sacrificial cell death and trichome breakage in an oscillatoriacean blue-green alga: The role of murein. Arch Mikrobiol 69:237–259

Meffert ME (1972) Zur Isolierung von *Oscillatoria redekei* Van Goor. Arch Hydrobiol, Suppl 41, Algol Stud 7:235–241

Meffert ME (1973a) Einfluß von pH, CO_2-Konzentration und Bakterien auf das Wachstum der Blaualge *Oscillatoria redekei* Van Goor. Arch Hydrobiol 72:186–201

Meffert ME (1973b) Kultur und Wachstum von *Oscillatoria redekei* Van Goor. Zur Bedeutung der Bakterien. Verh Int Ver Limnol 18:1359–1366

Meffert ME (1975) Analysis of the population dynamics of *Oscillatoria redekei* Van Goor in Lake Edeberg. Verh Int Ver Limnol 19:2682–2688

Meffert ME (1987) Planktic unsheathed filaments (Cyanophyceae) with polar and central gas vacuoles. I. Their morphology and taxonomy. Arch Hydrobiol Suppl 76:315–346

Meffert ME (1989) Planktic unsheathed filaments (Cyanophyceae) with polar and central gas vacuoles. II. Biology, population dynamics and biotopes of *Limnothrix redekei* (Van Goor) Meffert. Arch Hydrobiol 116:257–282

Meffert ME, Chang TP (1978) The isolation of planktonic blue green algae (*Oscillatoria* species). Arch Hydrobiol 82:231–239

Meffert ME, Overbeck J (1979) Regulation of bacterial growth by algal release products. Arch Hydrobiol 87:118–121

Meffert ME, Overbeck J (1981) Interactions between *Oscillatoria redekei* (Cyanophyta) and bacteria. Verh Int Ver Limnol 21:1432–1435

Meffert ME, Zimmermann-Telschow H (1979) Net release of nitrogenous compounds by axenic and bacteria-containing cultures of *Oscillatoria redekei* (Cyanophyta). Arch Hydrobiol 87:125–138

Meffert ME, Oberhäuser R, Overbeck J (1981) Morphology and taxonomy of *Oscillatoria redekei* (Cyanophyta). Br Phycol J 16:107–114

Ohle W (1962) Der Stoffhaushalt der Seen als Grundlage einer allgemeinen Stoffwechseldynamik der Gewässer. Kiel Meeresforsch 18:107–120

Overbeck J (1965) Primärproduktion und Gewässerbakterien. Naturwissenschaften 51:145

Overbeck J (1968) Prinzipielles zum Vorkommen der Bakterien im See. Mitt Int Ver Limnol 14:134–144

Overbeck J (1972) Zur Struktur und Funktion des aquatischen Ökosystems. Ber Deutsch Bot Ges 85:553–577

Overbeck J (1973) Über die Kompartimentierung der stehenden Gewässer—Ein Beitrag zur Struktur und Funktion des limnischen Ökosystems. Verh Ges Ökol 1973:211–223

Reynolds CS (1984) *The Ecology of Freshwater Phytoplankton*. Cambridge University Press, Cambridge

Zaiss U (1976) Physiologische und ökologische Untersuchungen zur Regulation der Phosphatspeicherung bei *Oscillatoria redekei*. PhD thesis, Christian-Albrechts-Universität, Kiel, Germany

Zaiss U (1985) Physiologische und ökologische Untersuchungen zur Regulation der Phosphatspeicherung bei *Oscillatoria redekei*. II. Der Einfluß ökologischer Parameter auf die Regulation des Polyphosphatstoffwechsels. Arch Hydrobiol Suppl 72:166–219

11. Zooplankton Grazing on Bacteria

Andrea Bemmer and Jürgen Overbeck

Introduction

The interrelationship of zooplankton and bacteria in the limnic food chain has become of increasing interest. Numerous authors have investigated the grazing of different groups of zooplankton on freshwater bacteria: ciliates (Taylor and Berger 1976, Taylor 1978, 1979), rotifers (Dougherty et al. 1960, Pilarska 1977, Starkweather et al. 1979, Bogdan and Gilbert 1982, Pourriot 1980, Starkweather 1980), cladocerans (McMahon and Rigler 1963, 1965, Fenchel 1980, Haney 1973, Gophen et al. 1974, Lampert 1974, Kerfoot and DeMott 1980, Hadas et al. 1982, 1983, Pace et al. 1983), and copepods (Malovitskaya and Sorokin 1961, Rieper 1978). The predatory pressure on natural assemblages of bacteria caused by zooplankton communities has also been investigated (Gliwicz 1969, Haney 1973, Crisman et al. 1981, Pace and Orcutt 1981, Pedrós-Alió and Brock 1983). Of special interest is the impact of microzooplankton (heterotrophic nanoflagellates, protozoa) grazing on bacteria (Sherr et al. 1987), which is described in Chapter 12 (this volume).

In this chapter we discuss grazing studies of macrozooplankton on bacterioplankton from the Plußsee. Grazing experiments using ^{14}C-labeled bacteria were carried out during the late summer of 1983 in the course of the limnological study of the Plußsee. The main objectives of these experiments were the measurement of filtration rates and feeding rates of different zooplankton organisms and the examination of the grazing impact of zooplankton on cultivated bacteria as well as on the natural bacterial community of the lake as a whole.

The grazing activity of zooplankton animals is dependent on a variety of biotic and abiotic factors. The following factors are of special interests:

1. The growth stadium of the animals (Webster and Peters 1978, Gliwicz 1980).
2. The age of the animals (Schindler 1968, Pilarska 1977, Bogdan and Gilbert 1982).
3. Temperature (Burns 1969a, Geller 1975, Lampert 1977).
4. Illumination (Buikema 1973).
5. Time of day (Nauwerck 1959, Starkweather 1975).
6. Density of food (Burns 1968a, Lampert 1977, Pilarska 1977).
7. Size of the food particles (Burns 1968b, 1969b, Gliwicz and Siedlar 1980).
8. Morphology of the food (Lampert 1977, McNaught et al. 1980).
9. Toxic substances in the food (Ryther 1954, Lampert 1981, 1982).

Peters and Downing (1984) provided a review of the literature published through 1979 on the subject. Various findings have already been made demonstrating that the nutritional value and the acceptability of the food vary remarkably, depending on the species of the consumer and of the food organism. Another important finding is that several zooplankton species show symptoms of malnutrition when they feed only on axenic algal cultures (Manuilova 1958, Dougherty and Solberg 1959, Taub and Dollar 1968).

Materials and Methods

Bacteria

Cultivation and ^{14}C-Labeling of Laboratory Strains

In situ as well as in vitro grazing experiments were carried out on strains of culture collections and natural lake water bacteria. The following laboratory strains were used: *Bacillus subtilis* (strain BC 019) and *Pseudomonas putida,* both kindly made available by the Institute of General Botany (University of Hamburg, Department of Microbiology, Prof. Fortnagel), and *Klebsiella* sp. (Culture Collection of the Max-Planck-Institute for Limnology, Department Microbial Ecology, Plön). The *Klebsiella* strain was originally isolated from Plußsee water and used (as Strain 8) in studies on uptake kinetics, metabolic regulations, and other numerous studies (Chapter 9, this volume). *Klebsiella* sp. are gram-negative bacteria, short rods, immobile, with polysaccharide macrocapsules.

The bacteria were cultivated in 10-fold-diluted nutrient broth medium (Difco Lab.) at 27°C (*Bacillus subtilis* and *Klebsiella* sp.) or at 30°C (*Pseudomonas putida*). After 18 h of cultivation, the log-phase cells were labeled with [^{14}C]protein hydrolysate (algal protein hydrolysate, Amersham Buchler, specific activity 56 mCi/matom carbon, radioactive concentration 50 μCi/ml). The procedure for radiolabeling of *Klebsiella* sp. was shortened as follows: 3 ml of the bacterial suspension was transferred directly from the nutrient medium (10-fold-diluted nutrient broth) to Kaserer medium containing Hoagland's solution and 50 μCi of protein hydrolysate, and incubated for 1 h at 27°C. The labeled bacteria were kept deep frozen, and they were thawed in an ice bath for half an hour before grazing experiment. All experiments were performed in triplicate. Unlabeled bacteria (which served as food during the adaptation phase or as "dilution" of the labeled bacteria) were transferred to 100 ml of sterile 10-fold-diluted nutrient broth medium 18 h before the experiment. After washing they were transferred to the experimental medium (0.2-μm pore size filtered lake water for the cladocerans) or synthetic freshwater.

Concentration and ^{14}C Labeling of Lake Bacteria

During the study period the numbers of bacteria in the littoral water of Plußsee amounted to 2–4 × 10^6 cells/ml. To achieve sufficient concentration of limnic bacteria for ^{14}C labeling, it became necessary to concentrate the cells down from a larger volume to a few milliliters. Two systems were applied: a Pellicon cassette system (Millipore, 0.22-μm pore size) and an Amicon hollow fiber concentrator CH4 with a diaflow hollow fiber cartridge (type H1-P100–43, lower limit 100,000 daltons). The most convenient for concentration of bacteria was the Pellicon cassette system. Immediately after reaching a sufficient concentration bacteria were labeled as described above but without being transferred into a minimal medium. After labeling with [^{14}C]protein hydrolysate the lake bacteria were washed by filtering through Nuclepore filter (0.2-μm pore size, diameter 25 mm) and shock frozen. It was found that after slowly defrosting about 80–90% of bacterial cells had survived.

The average volume and length of cultivated bacteria and natural bacterioplankton from Plußsee are compared in Table 11.1. Both volume and length of bacteria from cultures were about twice those with lake bacteria. The statistics show that the standard deviation of measurements of lake bacteria was much greater. Specific activity of labeled bacterioplankton varied from only 10^{-4} dpm/cell to 1.8 × 10^{-3} dpm/cell. One reason for the low

Table 11.1. Average Values of Morphometric Characteristics of Bacteria Used in Grazing Experiments

Bacteria	Cell Volume (μm^3)	Ratio of Cell Length to Width (%)	Cell Length (μm)	N[a]
Bacillus subtilis	0.2302 ± 0.0903	293.5 ± 80.2	1.5 ± 0.41	115
Pseudomonas putida	0.3179 ± 0.2633	637.0 ± 423.6	2.5 ± 1.66	92
Klebsiella sp.	0.1772 ± 0.0858	182.0 ± 64.0	1.0 ± 0.35	145
Natural bacteria from the Plußsee	0.0863 ± 0.0902	260.4 ± 132.1	0.8 ± 0.41	141

[a]Number of measurements.
[b]Numbers given ± S.D.

activity may be that the cells had been damaged slightly during the concentration process. Otherwise the physiological state of the natural bacterial assemblages is not comparable to that of cultivated bacteria in the log phase. It could be stated, on the other hand, that during the incubation period of 1 h a number of cell divisions took place, indicating a rapid response of natural bacteria to the algal protein hydrolysate as suitable substrate. Therefore, it is doubtful that the labeled lake bacteria are still identical to the natural community before labeling. Morphologically Plußsee bacteria did not differ from the concentrated, labeled bacteria.

Zooplankton

Laboratory Strains

The following species were used for grazing experiments:

1. *Daphnia magna,* isolated from a pond (at the University Hospital, Eppendorf, Hamburg), cultivated in filtered lake water, fed with baker's yeast or *Scenedesmus quadricauda.*
2. *Daphnia longispina,* isolated from Plußsee, cultivated in filtered lake water, fed initially with *Scenedesmus quadricauda* and later with *Chlorella pyrenoidosa* (because of difficulties in keeping isolated newborn individuals alive).
3. *Chydorus sphaericus,* isolated from Plußsee and cultivated on *Chlorella pyrenoidosa* in filtered lake water.
4. *Brachionus leydigi* (culture from the Zoological Institute of the University of Frankfurt) grown in synthetic lake water and fed with *Chlorella pyrenoidosa.*

In Vitro Grazing Experiments

The experimental conditions were maintained for the zooplankton according to the following criteria: adaptation to temperature and medium is not required, the concentration of food should approximate that encountered in the Plußsee, the abundance of the animals in the experiments should be similar to density of the cultures.

Standard conditions of in vitro experiments on zooplankton grazing on bacteria are listed in Table 11.2. The time of the day at which the incubation started depended on the time required to complete the preparation of the experiment. In practice the incubation began at about 1 p.m. *Daphnia magna* or *Daphnia longispina* were selected from cloned cultures 14 days before the start of the experiment, and 100 individuals of the same age were distributed to each of four petri dishes. The period of culture was lengthened or shortened as required for the experiments with *D. magna* of different ages. *Chydorus sphaericus* and *Brachionus leydigi* were taken directly from the main cultures.

After incubation with ^{14}C-labeled bacteria the animals were preserved in formaldehyde. Further analysis of the samples required 3–4 h each. According

Table 11.2. Standard Conditions of in Vitro Experiments of Zooplankton Grazing on Radiolabeled Bacteria

Daphnia magna and *Daphnia longispina*	Age 14 days
Chydorus sphaericus and *Brachionus leydigi*	Mixed ages
Number of individuals in one experiment	100
Abundance of bacteria	2×10^6 cells/ml
Volume of medium during experiment:	
D. magna and *D. longispina*	500 ml
C. sphaericus and *B. leydigi*	20 ml
Period of adaptation	1 h
Temperature of incubation	20 ± 1°C
Time of incubation	10 min

to Holtby and Knoechel (1981) and Lampert and Taylor (1985), the samples lose radioactivity when preserved in the formaldehyde solution. This loss may account for up to 49% of the ^{14}C. In addition to time and the type of radioisotope the loss of radioactivity may depend on the species and the size of the preserved animal as well as the previous nutrition of the animal. These factors make it difficult to evaluate the magnitude of loss of radioactivity during storage.

The ingestion rates of zooplankton were measured as the amount of labeled bacteria ingested within 10 min. Hence, the incubation time was smaller than the duration of gut passage. The feeding rate was calculated as the ratio of radioactivity of the food cells to the radioactivity of the animals after incubation. So the data could be compared directly, the dry weights and carbon content of the bacteria and the consumers were determined (Krambeck et al. 1981). Also, the biovolume of the bacteria was measured according to Krambeck and Krambeck (1984).

Field Experiments

Grazing experiments in the field were carried out between July and September 1983 in the littoral zone of the Plußsee. Zooplankton from the center of Plußsee were collected (with a 2.7-L Ruttner sampler) in the morning on the days of the experiments. The samples were stored in 10-L PVC buckets with lids. The sampling depth was 2 m and at the upper metalimnion (4–8 m depth). The sample volumes were 27 L and 18 L, respectively. From each sample a 1-L subsample was taken for identification of the zooplankton species. The main sample was carefully concentrated with a 20-μm pore-size gauze bag to a final volume of 500–600 ml. For grazing experiments, two samples of concentrated zooplankton from two different depths (epilimnion, metalimnion) were available, according to the working scheme presented in Figure 11.1. Table 11.3 compiles a survey of the field grazing experiments.

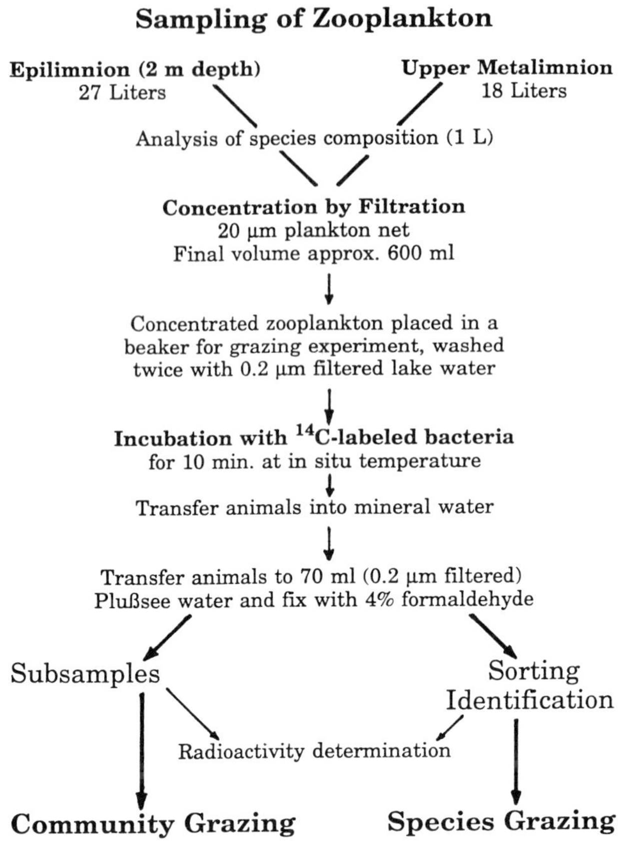

Figure 11.1. General protocol of field experiments of zooplankton grazing on bacteria.

Table 11.3. Some Characteristics of Field Grazing Experiments on Bacteria

Date (1983)	Bacteria	Abundance ($\times 10^6$ cells/ml)	Sampling Depth (m)	Water Temperature (°C)	Degree of Labeling (cells/dpm)	Counts
September 16	*Bacillus subtilis*	1.81	2	16	23	68
			8	9		49
August 4	*Pseudomonas putida*	1.12	2	20	14	27
			5	16		43
September 8	*Klebsiella* sp.	1.87	2	17	17	68
			8	9		47
July 28[a]	Concentrated bacteria from the Plußsee	2.73	2	22	2,100	24
			4.5	18		18
August 18	Concentrated bacteria from the Plußsee	1.71	2	20	2,380	60
			7	11		53
September 22	Concentrated bacteria from the Plußsee	9.83	2	15	11,200	64
			8	9		51
September 30	Concentrated bacteria from the Plußsee	1.71	2	15	1,950	71
			8	9		55

[a]Test only.

Composition of Plußsee Plankton During the Experiments

Table 11.4 gives an overview of zooplankton species composition during the study period. Cladocerans were most abundant at the beginning. Later rotifers became increasingly more frequent. *Daphnia cucullata, Diaphanosoma brachyurum,* and *Bosmina coregoni* were regularly found in all 1-L samples. *Daphnia longispina* and *Chydorus sphaericus,* on the other hand, were only present during the first half of the experimental period. The calanoid copepod *Eudiaptomus gracilis* appeared mainly in its copepodid stage in all samples. The most abundant rotifers were *Keratella cochlearis* and *Keratella quadrata.*

Noticeable mass development of *Ceratium* sp. was observed throughout the entire experimental period. A mass development of *Oscillatoria* sp. in the beginning of September 1983 had a disturbing influence on the grazing experiments by clogging bacteria during concentration and rinsing. *Pediastrum, Staurastrum, Nitzschia, Fragillaria, Asterionella,* and *Pandorina* species were less frequent or rare.

Results and Discussion

The grazing rates of the selected zooplankton species of Plußsee (Table 11.5) were measured under the following varying conditions: zooplankton species, food quality and concentration, temperature, biomass, and body size of the consumers.

Dependence of Grazing Rates on the Zooplankton Species

There were strong differences in filtration rates and feeding rates and in food selection between the zooplankton species. *Diaphanosoma brachyurum* was the strongest predator and the only one with preference for *Klebsiella.* Throughout all experiments *Eudiaptomos gracilis* was the species with the smallest ingestion rate (Fig. 11.2). Zooplankton grazing rates on natural bacteria from Plußsee water were generally higher than on cultivated laboratory bacterial strains (Fig. 11.3). The filtration rates on *Bacillus subtilis* and *Klebsiella* sp. were found to be the smallest. Figure 11.2 presents the ingestion rates per individual related to the carbon content of the bacteria. Besides the preference for lake bacteria, the dependence of the grazing rates based on the concentration of bacteria is quite clear, especially for ingestion of bacteria from the Plußsee. The increase of the ingestion rates and food concentration indicated that food concentrations in the experiments were clearly in the range of concentration dependence. This fact was also described by Porter et al. (1983).

Low feeding rates, found for all five zooplankton species, can be interpreted as rejection of *Bacillus subtilis* and *Klebsiella,* which was observed by Meyer-Hinrichsen (1980). If feeding only depends on biovolume or carbon content of bacteria, higher feeding rates of these two bacteria had to be expected. The feeding rates may have been influenced by the *Oscillatoria* bloom at time of the experiment (Burns 1968a, Gliwicz 1977, Lampert 1981, 1982). In our experiments

Table 11.4. Zooplankton Species Composition in Plußsee during the Study Period

	28 July		4 August		18 August		8 September		16 September		22 September		30 September	
Depth (m)	2	4.5	2	5	2	7	2	8	2	8	2	8	2	8
Cladocera														
Daphnia cucullata	○	◆	◆	★	●	○	◆	○	○	○	◆	○	★	○
Daphnia longispina	○	▲	○	★	○	○	○	○	○	○	○	○	○	○
Ceriodaphnia quadrangula	○	◆	○	●	○	▲	◆	○	○	○	○	○	○	○
Diaphanosoma brachyurum	○	◆	★	▲	○	○	◆	○	○	○	◆	○	○	○
Bosmina coregoni	★	◆	●	●	★	▲	●	○	○	○	★	○	○	○
Bosmina longirostris	○	○	○	◆	○	○	○	○	○	○	○	○	○	○
Chydorus sphaericus	○	●	★	■	▲	○	▲	○	○	○	○	○	○	○
Copepoda														
Eudiaptomus gracilis	○	◆	○	○	○	○	○	○	○	○	★	○	○	★
Mesocyclops leuckarti	○	★	★	●	○	○	★	○	○	○	○	○	○	○
Thermocyclops oithonoides	○	○	★	●	★	○	★	○	○	○	★	○	▲	○
Diacyclops bicuspidatus	○	▲	○	●	▲	▲	◆	○	○	○	★	○	▲	○
Nauplii	○	★	★	●	◆	▲	●	▲	○	▲	●	▲	◆	○
Rotatoria														
Keratella cochlearis	○	○	▲	■	★	○	◆	▲	★	▲	●	★	▲	○
Keratella quadrata	○	●	○	◆	★	○	▲	○	○	○	★	○	◆	○
Trichocerca similis	○	○	○	▲	◆	○	★	○	★	○	★	○	○	○
Trichocerca capuzina	○	○	▲	○	▲	○	○	○	○	○	◆	○	○	○
Filinia sp.	○	○	○	◆	○	○	○	○	○	○	○	○	★	○
Kellikottia longispina	○	○	○	○	○	○	○	○	○	○	○	○	○	○
Synchaeta pectinata	○	○	○	○	○	○	○	○	○	○	★	○	◆	○
Polyarthra sp.	▲	○	○	■	○	○	○	○	○	○	○	○	○	○
Ascomorpha sp.	○	○	○	○	○	○	○	○	○	○	○	○	▲	○
Asplanchna sp.	○	○	○	○	★	○	○	○	○	○	◆	○	○	◆

[a]○, 0–5 ind/L; ▲, 6–10 ind/L; ★, 11–20 ind/L; ◆, 21–100 ind/L; ●, 101–1000 ind/L; ■, >1000 ind/L.

Table 11.5. Zooplankton Species from the Plußsee Used in Grazing Experiments

Species	Number of Individuals	Number of Measurements
Eudiaptomus gracilis	360	34
Daphnia cucullata	557	52
Diaphanosoma brachyurum	644	57
Ceriodaphnia quadrangula	448	45
Bosmina coregoni	372	34

clogging of the cells may have caused disturbances of grazing. These findings are generally in accord with results of other authors. Malovitskaya and Sorokin (1961) and Shcerbak et al. (1983) found that calanoid copepods were only very weak bacteria grazers. Malovitskaya and Sorokin (1961) even state that calanoid copepods are only able to filter bacteria in forms of bigger cell aggregates. However, our Plußsee experiments showed that *Eudiaptomus gracilis* could also ingest single bacteria cells.

Geller and Müller (1981) noted that *Diaphanosoma brachyurum* was an effective bacteria filter feeder, as a result of the fine mesh size of its filter apparatus. According to the same authors, *Ceriodaphnia quadrangula* and *Daphnia cucullata* also belong to the group of fine-particle filter feeders. Our results confirm these findings. *Bosmina coregoni* has a medium filter-feeding efficiency

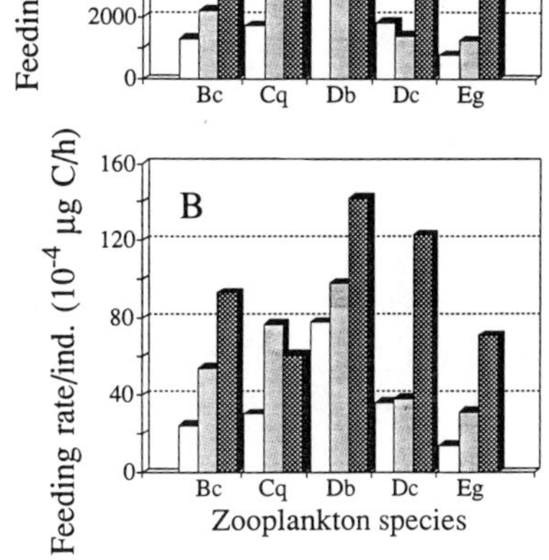

Figure 11.2. Average feeding rates of selected zooplankton species from the Plußsee on ^{14}C-labeled laboratory strains of different bacteria. Bc, *Bosmina coregoni;* Cq, *Ceriodaphnia quadrangula;* Db, *Diaphanosoma brachyurum;* Dc, *Daphnia cucullata;* Eg, *Eudiaptomus gracilis.* Water temperature was 15–17°C.

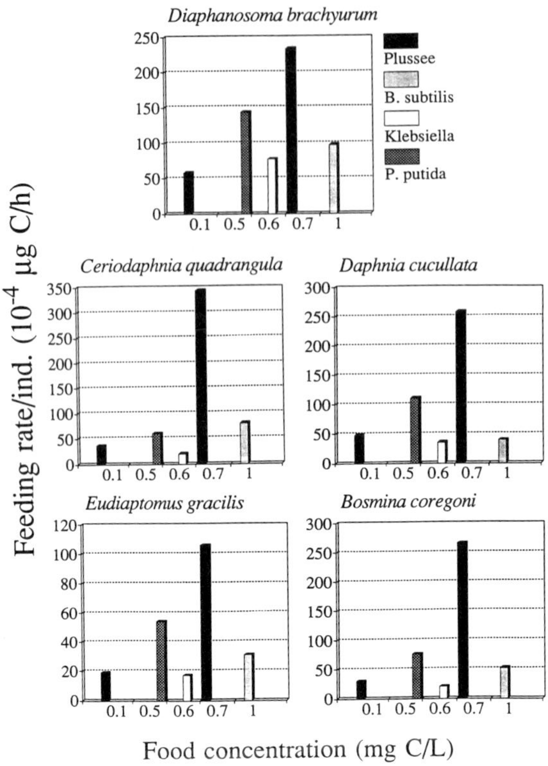

Figure 11.3. Feeding rates (average values) of five zooplankton species from Plußsee on ^{14}C-labeled laboratory strains and ^{14}C-labeled natural bacteria from the Plußsee water. Note that the scale of the *Y* axis varies for different species.

(Geller and Müller 1981). The Plußsee experiments did not exactly confirm these results, because *B. coregoni* exhibited the same filtration rates and feeding rates as the other cladocerans. *Bosmina coregoni* proved to be a very effective consumer when feeding on the small natural bacteria.

The filtration rates of *Daphnia cucullata* on limnic bacteria from the Plußsee varied between 0.03 and 0.04 ml ind^{-1} h^{-1}. These values correspond with data published by Riemann and Bosselmann (1984), even though their experiments were carried out at much lower temperatures. They found feeding rates from 0.9 to 4.6 ng C ind^{-1} h^{-1}, using a formula by Watson et al. (1977) for assessing the carbon content of the bacteria. Their data are comparable to the results of bacteria from the Plußsee.

Food Quality and Food Concentration

Table 11.6 presents estimates of the filtration rates and feeding rates of five zooplankton species of the Plußsee used in these experiments. The feeding rates apparently depended on the food concentration. In terms of filtration rates, the natural bacteria concentrations in the lakes seem to be in the range of concentration dependence. Only for *Diaphanosoma brachyurum* did the filtration rate decrease at higher food concentrations in the range of 9.83×10^6 cells/ml, which indicated that *D. brachyurum* fed independently of the food concentration. Gener-

Table 11.6. Relative Rates of Feeding and Filtration on Different Bacteria by Five Selected Zooplankton Species from the Plußsee

Bacteria (cells/ml)	Zooplankton Species	Feeding Rate (cells/min)	Specific Feeding Rate (10^{-5} μg C/min)	Biovolume (μm^3/min)	Specific Filtration Rate (μg d.w. μl^{-1} min^{-1})
Bacteria from the Plußsee 1.71×10^6	*Eudiaptomus gracilis*	56	0.42	4.82	0.033
	Diaphanosoma brachyurum	248	1.87	21.41	0.145
	Ceriodaphnia quadrangula	289	2.19	25.00	0.170
	Daphnia cucullata	287	2.16	24.73	0.158
	Bosmina coregoni	200	1.51	17.24	0.117
Bacteria from the Plußsee 9.83×10^6	*Eudiaptomus gracilis*	327	2.47	28.23	0.033
	Diaphanosoma brachyurum	1,000	7.54	86.34	0.102
	Ceriodaphnia quadrangula	2,559	19.30	220.88	0.260
	Daphnia cucullata	1,508	11.37	130.11	0.153
	Bosmina coregoni	1,786	13.47	154.14	0.182
Bacillus subtilis 1.81×10^6	*Eudiaptomus gracilis*	14	0.76	3.12	0.007
	Diaphanosoma brachyurum	58	3.21	13.26	0.032
	Ceriodaphnia quadrangula	80	4.46	18.32	0.044
	Daphnia cucullata	33	1.85	7.61	0.018
	Bosmina coregoni	49	2.72	11.24	0.027
Klebsiella sp. 1.87×10^6	*Eudiaptomus gracilis*	11	0.37	2.00	0.006
	Diaphanosoma brachyurum	75	2.49	13.30	0.040
	Ceriodaphnia quadrangula	34	1.13	6.01	0.018
	Daphnia cucullata	50	1.66	8.86	0.027
	Bosmina coregoni	31	1.02	5.44	0.016
Pseudomonas putida[a] 1.12×10^6	*Eudiaptomus gracilis*	34	1.62	10.81	0.030
	Diaphanosoma brachyurum	97	4.62	30.86	0.087
	Ceriodaphnia quadrangula	74	3.50	23.34	0.066
	Daphnia cucullata	116	5.51	36.80	0.103
	Bosmina coregoni	100	4.77	31.88	0.090
Pseudomonas putida[b] 1.12×10^6	*Eudiaptomus gracilis*	16	0.76	5.03	0.014
	Diaphanosoma brachyurum	149	7.08	47.31	0.133
	Ceriodaphnia quadrangula	57	2.72	18.17	0.051
	Daphnia cucullata	98	4.68	31.29	0.088
	Bosmina coregoni	62	2.97	19.86	0.056

[a]Temperature 16°C.
[b]Temperature 20°C.

ally the cultivated bacteria were less frequently ingested than lake bacteria, as shown in Figures 11.2 and 11.3. Lowest filtration rates and feeding rates were found for *Klebsiella* sp. Except for *Diaphanosoma brachyurum,* all zooplankton rejected *Klebsiella* sp. This fact is supported by findings of Meyer-Hinrichsen (1980) and Meyer (1985).

Grazing rates on *Bacillus subtilis* were rather low. Highest rates for all laboratory bacterial strains were found for *Pseudomonas putida.* But even grazing on this laboratory strain was lower than grazing on natural bacteria. It is not clear how far bloom of *Oscillatoria* sp. in the Plußsee influenced the filtration activity of cladocerans (Burns 1968a, Gliwicz 1977, Gliwicz and Siedlar 1980). Because of clogging, microalgae could remain in the incubation medium and could have been preferred by zooplankton. It is, in principle, impossible to exclude the possibility that low feeding rates on *Bacillus subtilis* may be caused by factors other than *B. subtilis* itself.

Based on the results presented, it can be concluded that bacteria of different species are ingested differently by zooplankton. Table 11.6 shows that between the zooplankton species there were clear differences in the grazing rates. The cell size was not the only criterion for zooplankton feeding behavior. The experiments showed that lake bacteria, which were much smaller than the cells of cultivated strains (Table 11.1) and could hardly be retained by the cladoceran filters, were predominantly ingested.

Influence of Temperature on Grazing

The water temperature was not a major factor for zooplankton grazing activity, at least not in the range of 10–20°C (Table 11.7). At 20°C higher ingestion rates (statistically significant) were measured only for *Ceriodaphnia quadrangula* and *Diaphanosoma brachyurum.* Between 9° and 15°C no differences existed in these two species. The decline of the filtration rates and feeding rates of *Bosmina coregoni* at higher temperatures was not statistically significant. During field experiments in the Plußsee the water temperature was apparently of only minor importance. Hayward and Gallup (1976) demonstrated in vitro that the impact of changing temperature conditions upon *Daphnia schoedleri* grazing on *Ankistrodesmus* was compensated by simultaneous variations of other factors such as food concentration or light.

Biomass and Body Size of Bacterivorous Zooplankton

The dependence of grazing on size or biomass of the filter feeders has been demonstrated by a number of authors. Burns (1968b) found a correlation between body size and feeding rate of daphnids. Porter et al. (1983) concluded that the specific filtration rates per microgram dry weight decreased with increasing body size of the individual. A regression analysis in connection with a literature review was given by Peters and Downing (1984). They correlated the filtration and feeding rates of cladocerans, marine calanoid copepods, and natural zooplankton community as dependent on some biotic and nonbiotic factors. They found that the

Table 11.7. Influence of Temperature on Feeding Rates (cells $ind^{-1} h^{-1}$) and Filtration Rates (µl $ind^{-1} h^{-1}$) of Five Selected Zooplankton Species from the Plußsee on Bacteria of the Plußsee

	9–11°C		15°C		20°C		
Species	Feeding Rate	Filtration Rate	Feeding Rate	Filtration Rate	Feeding Rate	Filtration Rate	*N*[a] (*n*)[b]
Eudiaptomus gracilis	30,972 ± 3,441	18.12 ± 1.98	20,430 ± 549	11.94 ± 0.30	22,908 ± 6,108	13.44 ± 3.60	10 (84)
Diaphanosoma brachyurum	73,716 ± 3,711	43.20 ± 2.16	79,530 ± 9,000	46.62 ± 5.28	185,922 ± 25,620	108.90 ± 15.00	14 (169)
Ceriodaphnia quadrangula	50,694 ± 2,772	29.70 ± 1.62	55,056 ± 11,181	32.22 ± 6.54	113,022 ± 513	66.24 ± 0.30	19 (182)
Daphnia cucullata	84,138 ± 21,156	49.26 ± 12.42	56,592 ± 7,928	33.18 ± 4.62	69,174 ± 10,713	40.50 ± 6.30	13 (168)
Bosmina coregoni	72,072 ± 5,211	42.24 ± 3.06	39,546 ± 3,171	23.16 ± 1.86	23,598 ± 4,698	13.80 ± 2.76	10 (94)

[a]Number of vials.
[b]Number of individuals per vial.

grazing rates generally rise with the biomass of the consumers. The deviations of the data presented suggest, on the other hand, that additional factors can also determine the filtration rates.

In contrast, our data do not allow for a strong and uniform correlation between biomass of the filter feeders and the relative grazing rates on different bacteria. However, these deviations may be explained by different filter mechanisms of calanoid copepods and cladocerans, in particular *Daphnia, Bosmina,* and *Chydorus* species (Storch 1924, Storch and Pfisterer 1925, Jørgensen 1966, DeMott and Kerfoot 1982, DeMott 1982); different ingestion rates of the heterogeneous lake bacteria; and influence of the time of day. Most of our experiments were performed at noon. According to Haney and Hall (1975), the filtration rate of *Daphnia pulex* at that time of the day was independent of body size. Otherwise our data set used for regression analysis is relatively small, in comparison with Peters and Downing (1984).

Bacterial Grazing of the Total Zooplankton Community

Table 11.8 shows that the grazing pressure of the total zooplankton community on the bacteria was remarkably higher than the grazing of the five zooplankton species alone. Even though the biomass of the five zooplankton used in the grazing experiments contributed a major part to the total zooplankton biomass, it represented only a relatively small fraction of the total grazing activity on bacteria. This result emphasizes the importance of microzooplankton, mostly flagellates and ciliates, as grazers (Chapter 12, this volume). Ciliates and microflagellates were often found in the Plußsee at concentrations of 10^3–10^4 ind/ml (Fukami et al. 1991). It is very well known that grazing on bacteria by "microbial loop" organisms is a very important factor controlling bacterial biomass in aquatic ecosystems (Azam et al. 1983, Sherr and Sherr 1987, 1988, McManus and Fuhrman 1988, Pace et al. 1990, Bernard and Rassoulzadegan 1990, Fukami et al. 1991; Chapters 5 and 12, this volume).

Clear differences existed between samples from the epilimnion and the metalimnion, and between lake bacteria and laboratory strains (Table 11.8). The grazing rates of the zooplankton community from the epilimnion were always higher than the hypolimnic one. The microzooplankton seemed to prefer lake bacteria, because the percentage of community grazing was always remarkably higher on lake bacteria than on laboratory strains. Otherwise *Pseudomonas putida,* for example, was nearly exclusively fed on by the five selected zooplankton species.

Studies on feeding of cladocerans described by Haney (1971, 1973) confirm our results. In the upper metalimnion layer the five most abundant zooplankton species represented on average only 4.9% of total community grazing, and 8.5% at most. However, 33.3–92.6% of the grazing was found in the epilimnion. These vertical differences, according to Lampert and Taylor (1984, 1985), may be caused by diurnal vertical migration of the zooplankton and may effect quantitative and qualitative changes in the zooplankton abundance and community structure. These

Table 11.8. Percentage Contribution of Grazing of Five Selected Zooplankton Species (*E. gracilis, D. brachyurum, C. quadrangula, D. cucullata, B. coregoni*) of the Total Zooplankton Community Grazing on Bacteria

Date of Experiment, Biovolume of Bacteria (μm^3/ml)	Depth (m)	Feeding Rate (10^3 μm^3 L^{-1} h^{-1})			Filtration Rate (ml L^{-1} h^{-1})			Biomass (µg dry weight/L)		
		A[a]	B[b]	%[c]	A	B	%	A	B	%
18 August										
Bacteria[d]	2	1,114	203	18.2	7.55	1.38	18.2	270.4	164.4	60.8
147,000	8	838	71	8.5	5.68	0.49	8.5	94.8	51.6	54.4
30 September										
Bacteria[d]	2	2,080	119	5.7	14.09	0.81	5.7	295.8	102.0	34.5
147,000	8	571	27	4.6	3.87	0.17	4.3	125.3	99.2	79.1
22 September										
Bacteria[d]	2	5,582	1,986	35.6	6.58	2.34	35.6	537.1	385.2	71.7
849,000	8	3,418	166	4.9	4.74	0.20	4.9	63.7	35.4	55.6
16 September										
B. subtilis	2	613	40	6.5	1.47	0.10	6.5	67.0	47.6	71.0
417,000	8	2,483	55	2.2	5.96	0.13	2.2	34.3	21.4	62.4
4 August										
Ps. putida	2	1,354	1,254	92.6	3.80	3.52	92.6	928.3	813.8	87.7
356,000										
8 September										
Klebsiella sp.	2	1,019	419	41.1	3.07	1.27	41.1	763.1	493.5	64.7
331.000	8	347	16	4.5	1.02	0.05	4.5	36.0	12.6	35.0

[a]Total zooplankton community.

[b]Selected zooplankton species.

[c]Percentage contribution of selected species to total community.

[d]Bacteria from lake water.

Table 11.9. Rates of Feeding and Filtration of Different Bacteria by Nauplii and Rotifers

Bacteria	Animal	Feeding Rate (μm^3 ind^{-1} min^{-1})	Filtration Rate (μl ind^{-1} min^{-1})	N^a $(n)^b$
Bacillus subtilis	Nauplii	60.4	0.14	3(22)
	Asplanchna sp.	131.2	0.31	3 (17)
Klebsiella sp.	Nauplii	57.8	0.17	2 (17)
Pseudomonas	Nauplii	5.6	0.01	3 (29)
putida	*Asplanchna* sp.	55.9	0.15	2 (2)
Bacteria from	*Asplanchna* sp.	180.7	1.23	3 (42)
the Plußsee	*Synchaeta* sp.	65.8	0.45	1 (8)

[a]Number of set measurements.
[b]Number of measured individuals.

variations can directly influence zooplankton grazing pressure on bacteria and algae (Porter et al. 1983, Kankaala 1988).

In the upper metalimnion, the grazing rates of the five selected zooplankton species had dropped considerably. Rotifers and protozoans, another group of bacteriovores, are now regarded as the major bacterial grazers in aquatic environments (Starkweather et al. 1979, Pace and Orcutt 1981, Porter et al. 1985, Sherr et al. 1987). In the Plußsee, high grazing rates found in the metalimnion could be caused by *Asplanchna* sp., *Synchaeta* sp., and nauplii (Table 11.9).

Zooplankton Grazing Pressure on the Plußsee Bacteria

We used the results of community grazing on natural bacteria to estimate the grazing pressure by zooplankton on bacterioplankton from the Plußsee. Filtration rates of the zooplankton community in 1 L of water from the Plußsee varied between 3.87 and 14.09 ml/h; the average was 7.09 ml/h (Table 11.8). This means that potentially from 9.3% to 33.8% (mean 17%) of the pelagic bacteria were ingested daily. Consequently, from 71 to 258 hours (mean 141 hours, i.e., 5 days) would be necessary for complete clearance of bacterioplankton by zooplankton in the Plußsee.

In experiments using lake water (in August–October 1979), generation times of bacteria were between 41 and 222 h, and division rates varied from 0.004 to 0.01 h^{-1} (Merck 1980, Meyer-Hinrichsen 1980). After filtration (through 20-µm pore-size plankton net) of the lake water, the generation times of bacteria decreased to 18–84 h, and division rates were 0.01–0.05 h^{-1}, which shows the impact of zooplankton grazing on the bacterial production (see Chapter 5, this volume). Moreover, Fukami et al. (1991) determined in autumn 1988 (using Whatman GF/F filtered water from the Plußsee) the doubling times of bacteria in the range 32 to 160 h (average 73.9 h), which agree very well with our and Chróst and Rai's (Chapter 5, this volume) data. The bacterial production-to-consumption ratio estimated from the above studies is roughly about 2. If the grazing of zooplankton larger than 140 μm amounts to only 20% of the grazing community, then the bacterial production-to-consumption ratio increases to about 10.

Riemann and Bosselmann (1984) calculated that *Daphnia cucullata* ingested 27% of the bacterial production in an eutrophic lake (in November). For another eutrophic lake Bell and Bern (1984) estimated that a total of 5–15% of the daily bacterial production was consumed by the cladoceran community. Pedrós-Alió and Brock (1983) measured daily grazing rates of 1–60% of the bacterial biomass in Lake Mendota and an annual grazing rate of 1–10% of the bacterial production. An additional control for bacteria, besides zooplankton grazing, is discussed in Chapter 5, this volume.

In enclosure experiments in the Frederiksborg Slotsø, feeding rates of zooplankton larger than 140 μm in size in the absence of fish increased to about 3 μg bacterial carbon L^{-1} h^{-1} (Riemann 1985). The major zooplankton species were *Daphnia cucullata, D. galeata,* and *Eudiaptomus graciloides.* In enclosures containing fish the feeding rates values dropped to 0.3 μg bacterial carbon L^{-1} h^{-1}. The grazing of the total zooplankton community amounted to 48–51% of the bacterial net production in the enclosures without fish. The grazing of zooplankton in the enclosures with fish and in the open lake removed 4% and 21% of bacterial production, respectively. Comparison of the grazing rates of zooplankton larger than 140 μm and flagellates points out that in enclosures without fish 80–90% of ingestion was made out by macrozooplankton; with fish, bacterivory grazing was almost exclusively by flagellates.

Our data for grazing in the Plußsee agree well with grazing rates in other eutrophic lakes (Pedrós-Alió and Brock 1983, Porter et al. 1983, Scavia 1988). They imply that only during periods of high filter feeder abundance, probably together with low bacterial production, a "top-down" control of bacterioplankton by grazing macrozooplankton is possible. Especially in spring, during the first phytoplankton bloom, a "bottom-up" control of the bacterial dynamics by the newly produced extracellular organic compounds as a limiting factor should be of major importance (Chapter 5, this volume).

Long-Term Feeding Experiments in Vitro

The aim of additional in vitro studies was to find out whether natural bacteria from the Plußsee can serve as sufficient food for cultivated *Daphnia magna, D. longispina,* and *Chydorus sphaericus.* For this purpose animals of each species were kept in petri dishes at 20°C (± 2°C) and washed daily with 0.2-μm filtered lake water. ^{14}C-labeled bacteria from the Plußsee were added at a concentration of 2×10^6 cells/ml. This concentration resembled the natural bacteria abundance in the Plußsee (Chapter 5, this volume). A control group of animals was exclusively fed with *Scenedesmus quadricauda* and *Chlorella pyrenoidosa,* and another control subgroup of zooplankton was not fed at all.

Bacterial density of 2×10^6 cells/ml was found to be sufficient for maintaining the metabolic functions of the three species of animals. Cladocerans of the Plußsee were not able to reproduce grazing solely on lake bacteria containing 0.15 μg carbon/ml. This may demonstrate the importance of bacteria as an additional food source for cladocerans during starvation. According to Lampert (1977), the carbon content of bacteria should suffice for reduced reproduction of *Daphnia pulex*. Similar experiments were carried out by Pace et al. (1983). In another in vitro experiment *Chydorus sphaericus* ingested *Bacillus subtilis* at a rate of up to 8.5×10^6 cells/ml (ca. 4.7 μg C/ml).

Filtration of Bacterioplankton

Scanning electron micrographs of cladoceran filter apparatus showed that bacteria are not only caught by the setulae: small limnic bacteria were also found in the coarser bristles and at the basis of the setulae. Exocrine glands may effectively support the catchability of the filters (Jørgensen 1966). In order to visualize the position of filtered bacteria in the filter apparatus of cladocerans, *Daphnia magna, D. longispina,* and *Chydorus sphaericus* were incubated for approximately 1 h with a suspension of 0.2-μm filtered lake water and bacteria from the Plußsee. The bacterial density was about 10^6 cells/ml. Scanning electron micrographs of in situ and in vitro experiments showed that the mesh size of the zooplankton filters was not the only criterion for their catchability. Cladocerans preferred feeding on smaller bacteria and to a lesser extent ingested bigger laboratory cultivated bacteria. In contrast, *Daphnia magna* preyed preferably on bigger *Bacillus subtilis* cells, even though according to its fine filter meshes it is classified as a "small-particle filter feeder" (Meyer 1985, Meyer-Hinrichsen 1985). *Chydorus sphaericus* (also classified as a fine-particle filter feeder) clearly preferred the smaller natural bacteria.

All our results contradict general opinion (Geller and Müller 1981, DeMott 1982, Gophen and Geller 1984). These authors believe that only mechanical processes determine the efficiency of zooplankton filtering. Rubinstein and Koehl (1977) and Porter et al. (1983) described some physical properties of the filter action, such as electrostatic absorption or adhesive effects. However, different opinions about the efficiency of zooplankton filtration exist. Børsheim and Olsen (1984) showed that *Daphnia pulex* can filter microalgae and bacteria with the same efficiency. Peterson et al. (1978) demonstrated in contrast that bacterial grazing was about 30% less effective than grazing on algae. Recently Hart and Jarvis (1993) published extensive studies of bacterial selectivity and filtration rates by cladocerans (*Daphnia, Diaphanosoma, Bosmina, Ceriodaphnia,* and *Moina* species) in a hypereutrophic lake.

It is difficult to estimate the actual filtration rates and efficiency of the Plußsee zooplankton from these results. The long-term feeding experiments revealed, however, that *Daphnia magna, Daphnia longispina,* and *Chydorus sphaericus* ingested lake bacteria with sufficient efficiency to maintain their metabolic requirements.

Acknowledgments. We are grateful to Prof. R. J. Chróst for his comments and suggestions, which greatly improved the final version of our manuscript.

References

Azam F, Fenchel T, Field JG, Gray JS, Meyer-Reil LA, Thingstand F (1983) The ecological role of water column microbes in the sea. Mar Ecol Prog Ser 10:257–263

Bell Rt, Bern L (1984) Bacterioplankton dynamics in eutrophic Lake Norrviken. Verh Int Ver Limnol 22:735

Bernard C, Rassoulzadegan F (1990) Bacteria or microflagellates as a major food source for marine ciliates: Possible implications for the microzooplankton. Mar Ecol Prog Ser 64:147–155

Bogdan KG, Gilbert JJ (1982) Seasonal patterns of feeding by natural populations of *Keratella, Polyarthra,* and *Bosmina:* Clearance rates, selectivities, and contribution to community grazing. Limnol Oceanogr 27:918–934

Børsheim KY, Olsen Y (1984) Grazing activities by *Daphnia pulex* on natural populations of bacteria and algae. Verh Int Ver Limnol 22:644–648

Buikema A (1973) Filtering rate of the cladocera, *Daphnia pulex,* as a function of body size, light and acclimation. Hydrobiologia 41:515–527

Burns CW (1968a) Direct observations of mechanisms regulating feeding behavior of *Daphnia* in lakewater. Int Rev Ges Hydrobiol 53:83–100

Burns CW (1968b) The relationship between body size of filter-feeding cladocera and the maximum size of particle ingested. Limnol Oceanogr 13:675–678

Burns CW (1969a) Relation between filtering rate, temperature, and body size in four species of *Daphnia.* Limnol Oceanogr 14:693–700

Burns CW (1969b) Particle size and sedimentation in the feeding behavior of two species of *Daphnia.* Limnol Oceanogr 14:392–402

Crisman TL, Beaver JR, Bays JS (1981) Examination of the relative impact of microzooplankton and macrozooplankton on bacteria and algae in Florida lakes. Verh Int Ver Limnol 21:359–362

DeMott WR (1982) Feeding selectivities and relative ingestion rates of *Daphnia* and *Bosmina.* Limnol Oceanogr 23:518–527

DeMott WR, Kerfoot WC (1982) Competition among cladocerans: Nature of the interaction between *Bosmina* and *Daphnia.* Ecology 63:1949–1966

Dougherty EC, Solberg B (1959) Laboratory culture of rotifers and gastrotrichs. II. Dixenic, monomixenic, and attempted axenic cultures. Anat Rec 134:555–556

Dougherty EC, Solberg B, Harris LG (1960) Synxenic and attempted axenic cultivation of rotifers. Anat Rec 137:350–351

Fenchel T (1980) Relation between particle size selection and clearance in suspension-feeding ciliates. Limnol Oceanogr 25:733–738

Fukami K, Meier B, Overbeck J (1991) Vertical and temporal changes in bacterial production and its consumption by heterotrophic nanoflagellates in a north German eutrophic lake. Arch Hydrobiol 122:129–145

Geller W (1975) Die Nahrungsaufnahme von *Daphnia pulex* in Abhängigkeit von der Futterkonzentration, der Temperatur, der Körpergröße und dem Hungerzustand der Tiere. Arch Hydrobiol Suppl 48:47–107

Geller W, Müller H (1981) The filtration apparatus of cladocera: Filter mesh-sizes and their implications on food selectively. Oecologia 49:316–321

Gliwicz ZM (1969) The share of algae, bacteria and trypton in the food of the pelagic zooplankton of lakes with various trophic characteristics. Bull Acad Pol Sci 17:159–165

Gliwicz ZM (1977) Food size selection and seasonal secession of filter feeding zooplankton in an eutrophic lake. Ekol Pol 25:179–225

Gliwicz ZM (1980) Filtering rates, food size selection, and feeding rates in cladocerans—another aspect of interspecific competition in filter-feeding zooplankton. In Kerfoot WC (ed) *Evolution and Ecology of Zooplankton Communities.* University Press, New England, pp 282–291

Gliwicz ZM, Siedlar E (1980) Food size limitation and algae interfering with food collection in *Daphnia.* Arch Hydrobiol 88:155–177

Gophen M, Geller W (1984) Filter mesh size and food particle uptake by *Daphnia.* Oecologia 64:408–412

Gophen M, Cavari BZ, Berman T (1974) Zooplankton feeding on differentially labeled algae and bacteria. Nature 247:393–394

Hadas O, Cavari BZ, Kott Y, Bachrach U (1982) Preferential feeding behavior of *Daphnia magna*. Hydrobiologia 89:49–52

Hadas O, Bachrach U, Kott Y, Cavari BZ (1983) Assimilation of *E. coli* cells by *Daphnia magna* on the whole organism level. Hydrobiologia 102:163–169

Haney JF (1971) An in situ method for the measurement of zooplankton grazing rates. Limnol Oceanogr 16:970–977

Haney JF (1973) An in situ examination of the grazing activities of natural zooplankton communities. Arch Hydrobiol 72:87–132

Haney JF, Hall DJ (1975) Diel vertical migration and filter-feeding activities of *Daphnia*. Arch Hydrobiol 75:413–441

Hart RC, Jarvis AC (1993) In situ determinations of bacterial selectivity and filtration rates by five cladoceran zooplankters in a hypereutrophic subtropical reservoir. J Plankton Res 15:295–315

Hayward RS, Gallup DN (1976) Feeding, filtering and assimilation in *D. schoedleri* Sars as affected by environmental conditions. Arch Hydrobiol 77:139–164

Holtby LB, Knoechel R (1981) Zooplankton filtering rates: Error due to loss of radioisotopic label in chemically preserved samples. Limnol Oceanogr 26:774–780

Jørgensen CB (ed) (1966) *Biology of Suspension Feeding*. Pergamon Press, Oxford

Kankaala P (1988) The relative importance of algae and bacteria as food for *Daphnia longispina* (Cladocera) in a polyhumic lake. Freshwater Biol 19:285–296

Kerfoot WC, Demott WR (1980) Foundations for evaluating community interactions: The use of enclosures to investigate coexistence of *Daphnia* and *Bosmina*. In Kerfoot WC (ed) *Evolution and Ecology of Zooplankton Communities*. University Press, New England, pp 725–741

Krambeck C, Krambeck HJ (1984) Morphometric analysis of cell-cycle responses in bacterioplankton. Arch Hydrobiol Ergeb Limnol 19:111–118

Krambeck HJ, Lampert W, Brede H (1981) Messung geringer Mengen von partikulärem Kohlenstoff in natürlichen Gewässern. GIT 25:1009–1012

Lampert W (1974) A method for determining food selection by zooplankton. Limnol Oceanogr 19:995–998

Lampert W (1977) Studies on the carbon balance of *Daphnia pulex* De Geer as related to environmental conditions. II. The dependence of carbon assimilation on animal size, temperature, food concentration and diet species. Arch Hydrobiol Suppl 48:310–335

Lampert W (1981) Inhibitory and toxic effects of blue-green algae on *Daphnia*. Int Rev Ges Hydrobiol 66:285–298

Lampert W (1982) Further studies on the inhibitory effect of the toxic blue-green *Microcystis aeruginosa* on the filtering rate of zooplankton. Arch Hydrobiol 95:207–220

Lampert W, Taylor BE (1984) In situ grazing rates and particle selection by zooplankton: Effects of vertical migration. Verh Int Ver Limnol 22:943–946

Lampert W, Taylor BE (1985) Zooplankton grazing in an eutrophic lake: Quantitative and qualitative implications of diel vertical migration. Ecology 66:68–82

Malovitskaya LM, Sorokin YJ (1961) On the feeding of some species of Diaptomids (Copepoda, Calanoida) on bacteria. Dokl Akad Nauk SSSR 136/137:94–95

Manuilova EF (1958) The question of the role of bacterial numbers in the development of cladocera in natural condition. Dokl Acad Sci USSR Biol Sci 120:438–441

McMahon JW, Rigler FH (1963) Mechanisms regulating the feeding rate of *Daphnia magna* Straus. Can J Zool 41:321–332

McMahon JW, Rigler FH (1965) Feeding rate of *Daphnia magna* Straus in different foods labeled with radioactive phosphorus. Limnol Oceanogr 10:105–113

McManus GB, Fuhrman J (1988) Control of marine bacterioplankton populations: Measurement and significance of grazing. Hydrobiologia 159:51–62

McNaught DC, Griesmer D, Kennedy M (1980) Resource characteristics modifying selective grazing by copepods. In Kerfoot WC (ed) *Evolution and Ecology of Zooplankton Communities*. University Press, New England, pp 292–298

Merck S (1980) Zur Methodik von Untersuchungen über das Grazing von Zooplankton auf Bakterien unter spezieller Berücksichtigung der Expositionsdauer. Dissertation, Universität Hamburg, Germany

Meyer S (1985) Experimentelle Untersuchungen zum Bakterien-Grazing ausgewählter Zooplanktonarten. Dissertation, Universität Hamburg

Meyer-Hinrichsen A (1980) Zur Methodik von Untersuchungen über das Grazing von Zooplankton auf Bakterien unter spezieller Berücksichtigung von verschiedenen Maschenweiten zum planktonfrei Filtrieren. Dissertation, Universität Hamburg

Meyer-Hinrichsen A (1985) Untersuchung zur Regulation Zooplankton—Bakterien in der Nahrungskette des Plußsees. Dissertation, Universität Hamburg

Nauwerck A (1959) Zur Bestimmung der Filtrierrate limnischer Planktontiere. Arch Hydrobiol Suppl 25:83–101

Pace ML, Orcutt JD (1981) The relative importance of protozoans, rotifers, and crustaceans in fresh water zooplankton communities. Limnol Oceanogr 26:822–830

Pace ML, Porter KG, Feig YS (1983) Species- and age-specific differences in bacterial resource utilization by two co-occurring cladocerans. Ecology 64:1145–1156

Pace ML, McManaus GB, Findlay SEG (1990) Planktonic community structure determines the fate of bacterial production in a temperate lake. Limnol Oceanogr 35:795–808

Pedrós-Alió C, Brock TD (1983) The impact of zooplankton feeding on the epilimnetic bacteria of an eutrophic lake. Freshwater Biology 13:227–239

Peters RH, Downing JA (1984) Empirical analysis of zooplankton filtering and feeding rates. Limnol Oceanogr 29:763–784

Peterson BJ, Hobbie JE, Haney JF (1978) *Daphnia* grazing on natural bacteria. Limnol Oceanogr 23:1039–1044

Pilarska J (1977) Eco-physiological studies on *Brachionus rubens* Ehrb. (Rotatoria) I. Food selectivity and feeding rate. Pol Arch Hydrobiol 24:319–328

Porter KG, Feig YS, Vetter EF (1983) Morphology, flow regime, and filterung rates of *Daphnia, Ceriodaphnia,* and *Bosmina* fed natural bacteria. Oecologia 58:156–163

Porter KG, Sherr EB, Sherr BF, Pace M, Sanders RS (1985) Protozoa in planktonic food webs. J Protozool 32:409–415

Pourriot R (1980) Workshop on culture techniques of rotifers. Hydrobiologia 73:33–35

Riemann B (1985) Potential importance of fish predation and zooplankton grazing on natural populations of freshwater bacteria. Appl Environ Microbiol 50:187–193

Riemann B, Bosselmann S (1984) *Daphnia* grazing on natural populations of bacteria. Verh Int Ver Limnol 22:795–799

Rieper M (1978) Bacteria as food for marine harpacticoid copepods. Mar Biol 45:337–345

Rubinstein DI, Koehl MAR (1977) The mechanisms of filter feeding: Some theoretical considerations. Am Nat 111:981–994

Ryther JH (1954) Inhibitory effects of phytoplankton upon the feeding of *Daphnia magna* with reference to growth, reproduction and survival. Ecology 35:522–533

Scavia D (1988) On the role of bacteria in secondary production. Limnol Oceanogr 33:1220–1224

Schindler DW (1968) Feeding, assimilation and respiration rates of *Daphnia magna* under various environmental conditions and their relation to production estimates. J Anim Ecol 37:369–385

Shcerbak VI, Golovko TV, Zhdanova GA (1983) Consumption of bacteria and algae by planktonic crustaceans in the Kiev Reservoir. Hydrobiol J 19:24–29

Sherr EB, Sherr BF (1987) High rates of consumption of bacteria by pelagic ciliates. Nature 325:710–711

Sherr E, Sherr B (1988) Role of microbes in pelagic food webs: A revised concept. Limnol Oceanogr 33:1225–1227

Sherr BF, Sherr EB, Fallon RD (1987) Use of monodispersed, fluorescently labeled bacteria to estimate in situ protozoan bacterivory. Appl Environ Microbiol 53:958–965

Starkweather PL (1975) Diel patterns of grazing in *Daphnia pulex* Leydig. Verh Int Ver Limnol 19:2851–2857

Starkweather PL (1980) Behavioral determinants of diet quantity and diet quality in *Brachionus calyciflorus*. In Kerfoot WC (ed) *Evolution and Ecology of Zooplankton Communities*. University Press, New England, pp 151–157

Starkweather PL, Gilbert JJ, Frost TM (1979) Bacterial feeding by the rotifer *Brachionus calyciflorus:* Clearance and ingestion rates, behavior and population dynamics. Oecologia 44:26–30

Storch O (1924) Morphologie und Physiologie des Fangapparates der Daphniden. Ergeb Fortschr Zool 6:125–234

Storch O, Pfisterer O (1925) Der Fangapparat von *Diaptomus*. Z Vergl Physiol 3:330–376

Taub FB, Dollar AM (1968) The nutritional inadequacy of *Chlorella* and *Chlamydomonas* as food for *Daphnia pulex*. Limnol Oceanogr 13:607–617

Taylor WD (1978) Growth responses of ciliate protozoa to the abundance of their bacterial prey. Microb Ecol 4:207–214

Taylor WD (1979) Overlap among cohabiting ciliates in their growth responses to various prey bacteria. Can J Zool 57:949–951

Taylor WD, Berger J (1976) Growth responses of cohabiting ciliate protozoa to various prey bacteria. Can J Zool 54:1111–1114

Watson SW, Novitsky TJ, Quinby HL, Valois FW (1977) Determination of bacterial number and biomass in the marine environment. Appl Environ Microbiol 33:940–946

Webster KE, Peters RH (1978) Some size-dependent inhibitions of larger cladoceran filters in filamentous suspensions. Limnol Oceanogr 23:1238–1245

12. Nanoflagellate and Ciliate Grazing on Bacteria

Bodo G. Meier and Edith Reck

Heterotrophic and Mixotrophic Nanoflagellate Grazing on Bacteria

In the early 1950s, Ohle (1952) reported the algal photoautotrophic refixation of CO_2 released by bacteria during decomposition of organic matter in the epilimnion. He summarized his observations by the term "short-circuit cycle of organic matter in the epilimnion" ("kurzgeschlossener Stoffkreislauf des Epilimnions"). In 1962, he showed that the Plußsee's mean depth of productivity was 5 m and that 70–93% of primary production was utilized within that compartment of the pelagial. This means that the carbon passes through the "short-circuit" about 10 times in the upper layer of the water column, theoretically. These early investigations pointed out the nonlinear turnover of organic matter, but there was still little information about the organisms involved.

Fenchel (1982c) was among the first who recognized the important role of small heterotrophic flagellates (diameter 2–20 μm) as bacterial consumers. They are called either heterotrophic microflagellates or, according to Sieburth (1979), nanoflagellates (HNF). He concluded that in marine systems, the bacterial biomass is controlled mainly by HNF grazing on bacteria. Azam et al. (1983) introduced the concept of the microbial loop, comprising release of DOM by

---

The section on Heterotrophic Nanoflagellate Grazing on Bacteria was written by Bodo G. Meier. The section on Pelagic Ciliates was written by Edith Reck.

phytoplankton and uptake and decomposition of DOM by pelagic bacteria grazed on by nanoflagellates. HNF are then preyed on by microzooplankton, e.g., ciliates. The authors considered the microbial loop to be "rather inefficient" in terms of returning energy lost by phytoplankton to the main (grazing) food chain. Nanoflagellates should therefore act as remineralizers in the sea. This was later confirmed for lake ecosystems by Güde (1986) and Bloem et al. (1988), who addressed the major effect of the microbial loop in the regeneration of nitrogen and phosphorus.

It became evident that despite extensive investigations of the microbial ecology of the Plußsee the function and role of nanoflagellates as part of the microbial loop were still unknown. When Sherr et al. (1987) presented the use of monodispersed, fluorescently labeled bacteria (FLB) to estimate in situ protozoan bacterivory, it proved to be a reliable method when connected with the [^{3}H]thymidine ([^{3}H]TdR) incorporation method for in situ measurement of heterotrophic bacterial production (Fuhrman and Azam 1980, 1982; Chapter 5, this volume). These methods give us an overview of the relationship between bacteria and nanoflagellates in the pelagial. The aim of the following examinations was to demonstrate seasonal changes and differences in function of the microbial loop in the Plußsee.

Experimental Design: Estimation of Nanoflagellate Grazing Rates Using the FLB Technique

As described by Sherr et al. (1987), fluorescently labeled bacteria (FLB) were employed to determine in situ grazing rates of heterotrophic nanoflagellates (HNF). For this purpose, bacteria previously stained by 5-(4,6-dichloro-triazinyl)aminofluorescein (DTAF) were directly added to 100-ml water samples in concentrations of approximately 20% of the actual bacteria numbers. Sherr et al. (1987, 1989) demonstrated that FLB are not toxic to protozoa, and HNF do not discriminate between natural bacteria and FLB—both are digested.

After an incubation period of 30 or 60 min, the samples were stained by 4,6-diamidino-2-phenylindol (DAPI). Bacteria and HNF were counted directly under an epifluorescence microscope (Porter and Feig 1980) equipped with four filter sets in a quick-change housing. Phototrophic and mixotrophic nanoflagellates were distinguished from HNF by their chlorophyll and phycobilin (Cryptophyceae) autofluorescence and are referred to as autofluorescent nanoflagellates (ANF). At an excitation wavelength of 450–490 nm, ingested FLB could be directly observed inside HNF and counted. From the ratio of FLB to natural bacteria in the samples it was then possible to calculate the ingestion rates (bacteria HNF^{-1} h^{-1}) by multiplying by the number of the HNF community grazing rate (bacteria ml^{-1} h^{-1}).

Several authors proved that HNF and ciliates have the tendency preferentially to ingest larger bacteria (Andersson et al. 1986, Chrzanowski and Šimek 1990, Gonzalez et al. 1990, Kuuppo-Leinikki 1990). To obtain realistic grazing rates, the size distribution of FLB should correspond to the distribution of the bacteria in the system investigated. In this study, two different approaches to collect bacteria suitable for FLB preparation were used. For the first approach, approximately 10 L of Plußsee surface water was poured through sterile glass fiber filters (Whatman GF/F) for removal of larger particles and microzooplankton. The water was then incubated with 50 mg/L yeast extract at 20°C for 2 days, and bacteria cells were harvested by centrifugation at midlogarithmic growth phase. After washing twice with buffer 1 solution (0.05 M Na_2HPO_4 in 0.85% NaCl solution, pH 9; Sherr et al. 1987) the cells were resuspended in 100 ml of the same buffer. This concentrated suspension was kept overnight in a refrigerator to induce a size reduction of large, nutrient-rich cells by fragmentation on starvation conditions (Kjelleberg and Hermansson 1984). FLB prepared from these bacteria were used in the investigations during autumn–winter 1988–1989.

For the second approach, approximately 10 L of the Plußsee water from a depth of 1 m was prefiltered through glass fiber filters (Whatman GF/C) and then concentrated to 130–180 ml by an Amicon hollow fiber concentrator CH4A, cartridge type H1P100–15 or H1MP01–43. This turbid and opalescent suspension contained about 10^9 bacteria/ml, which were harvested by centrifugation and resuspended in 15 ml of buffer 1. In both cases, the staining of the bacteria suspensions by DTAF was in accordance with Sherr et al. (1987).

Results

Samples of natural Plußsee bacteria and of both FLB types were scanned by digital image analysis (Schröder and Krambeck 1991). Table 12.1 compares their cell volumes divided into five size classes. Obviously, both FLB preparations had shifted the mean cell volumes toward larger values, which was most significant in the case of yeast extract incubation. Therefore, at first glance, there appeared to be an overestimation of the determined HNF and ANF ingestion rates. In fact, the use of "larger" yeast extract FLB in the autumn–winter experiments revealed ingestion rates for a wider range (0.78–15.1 bacteria HNF^{-1} h^{-1}) than the hollow-fiber FLB during spring (0.44–5.06 bacteria HNF^{-1} h^{-1}) and summer (0.74–5.03 bacteria HNF^{-1} h^{-1}), but this could be because of variable seasonal community structures. If, on the other hand, the ingestion rates of this study are compared to the range of data (1–254 bacteria HNF^{-1} h^{-1}) published by several authors and obtained by different methods (Bennett et al. 1990, Bloem et al. 1989a,b, Fenchel

Table 12.1. Percentage of Size Distribution of Bacteria Used in Grazing Experiments

	Percentage of Cell Volume (μm^3)				
Bacteria	≤0.1	≤0.2	≤0.3	≤0.4	≤0.5
Natural Plußsee bacteria	98.2	1.6	0.1	0	0
FLB (Plußsee bacteria concentrated by cartridge)	93.5	5.2	1.1	2.2	1.1
FLB (Plußsee bacteria enriched by yeast extract)	84.8	10.5	2.9	1.0	1.0

1982b, McManus and Fuhrman 1988, and others), they rather appear to be conservative.

Autumn and Winter Period

The lake was still strongly stratified in September; both thermocline and oxycline were becoming less pronounced and, at the same time, going deeper as the water cooled down in October and November. On December 13, the straight temperature and oxygen profiles indicated that autumn mixis had occurred.

To detect any differences, the sampling depth was at 2 m for the epilimnion and 12 m for the hypolimnion, but for the metalimnion, sampling followed the sinking thermocline from 6 to 9 m. These depths were still used after the autumn turnover, when the division into epilimnion, metalimnion, and hypolimnion no longer existed in the strict sense. Data for number of bacteria and HNF (DAPI direct count), bacterial production ([^{3}H]TdR technique), and HNF grazing on bacteria (FLB method) are shown in Figure 12.1. The uniform scale for all three graphs (upper, epilimnion; middle, metalimnion; lower, hypolimnion) allows direct comparison of the parameters. Investigations were carried out from mid-September to early January at different intervals.

Epilimnion. Decreasing concentrations of chlorophyll$_a$ in the 0- to 1-m layer (September 15 μg/L, November 5 μg/L) indicated that the decay of phytoplankton lasted until mid-November. Released EOC (Sell and Overbeck 1992; Chapter 4, this volume) and detritus supported high bacterial production (3–8.4 × 10^4 cells ml^{-1} h^{-1}) in the epilimnion up to November (Fig. 12.1, open bars). High numbers of HNF (4.8–8.5 × 10^3 cells/ml) were found in the beginning of the period. Their grazing rates (Fig. 12.1, solid bars) determined by FLB grazing exceeded the bacterial production measured by the [^{3}H]TdR incorporation method at certain times without any apparent negative effect on the bacterial numbers.

Other authors (Šimek and Straskrabova 1992) reported similar findings of protozoan grazing exceeding bacterial production. In the Plußsee, this paradoxical situation of FLB grazing exceeding bacterial production rates was observed during the autumn–winter and spring periods and at depths of relatively high concentrations of dissolved oxygen and DOC, which suggests actively metabolizing and growing bacteria. It might be that the method used neglected to address diurnal variations of production (Münster and Chróst 1990) and grazing (Wikner et al. 1990), having used a "one-time-point measurement." Moreover, the applied concentration of [^{3}H]TdR (9.8, 12.1, and 17.8 nM in this study; Bell 1990), the conversion factor for the [^{3}H]TdR technique used (2 × 10^{18} cells/mol; Fuhrman and Azam 1982, Bell et al. 1983), and ingestion preferences of nanoflagellates may have been sources of methodological inaccuracies and must be considered in the interpretation of the data.

However, the increasing numbers of bacteria indicated that HNF could not completely remove the bacterial increase. Finally, it resulted in a maximum number of bacteria (10 × 10^6 cells/ml) at the end of October, since the HNF had been strongly reduced. This culmination was shifted about 18 days against the

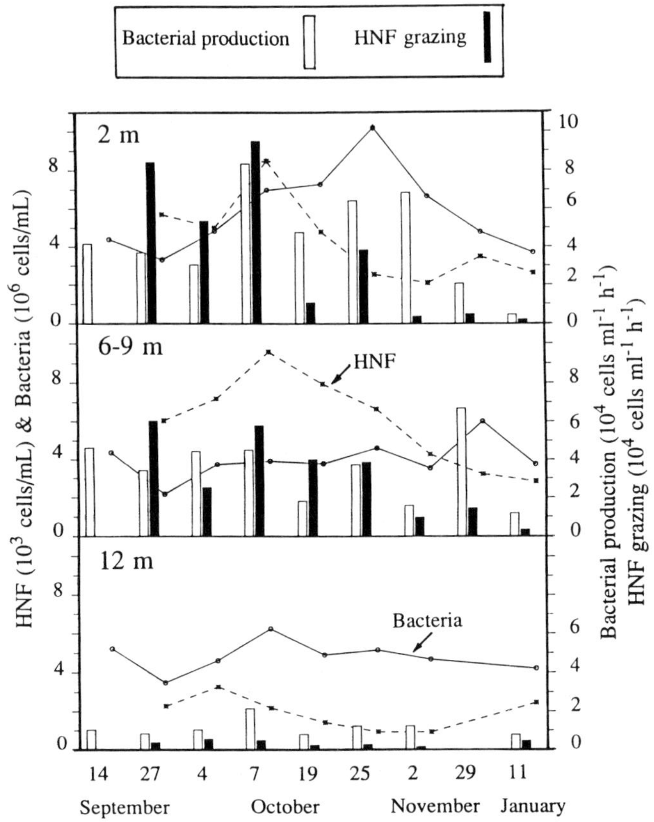

Figure 12.1. Abundance of bacteria and heterotrophic nanoflagellates (HNF) and the rates of bacterial production and HNF grazing on bacteria in the Plußsee during autumn and early winter.

maxima of HNF (October 7; Fig. 12.1). Although data before September are missing, the tendencies of the bacteria and HNF densities indicated that a predator–prey relationship between nanoflagellates and pelagic bacteria may exist in the epilimnion during the autumn period. Comparable oscillations were reported for marine systems by Fenchel (1982c) and Andersen and Fenchel (1985).

Metalimnion. It is important to note that over the entire period observed, HNF in the metalimnion were most abundant out of the three strata (mean 6.0×10^3 HNF/ml) and reached their absolute highest number (9.6×10^3 HNF/ml) at the same time as HNF in the epilimnion. On the other hand, the bacteria cell counts were the lowest (mean 4.0×10^6 cells/ml) and revealed remarkably small alterations from mid-September to the beginning of November (Fig. 12.1). Production and grazing rates varied and were lower than in the upper layer. The determined grazing rates were often higher than the measured rates of production of bacteria in the 6- to 9-m water layer, as observed in the epilimnion.

It is evident that until November, the HNF effectively controlled the number of bacteria by taking up most of the bacterial production. The course of the bacteria counts gives the impression that a density of approximately 4×10^6 cells/ml represented a critical threshold for grazing efficiency and growth of the HNF community. Assuming a mean bacterial cell volume of 0.065 μm^3 (Krambeck et al. 1981) and a volume-to-carbon factor of 0.56 pg C/μm^3 (Bratbak 1985), this corresponds to 146 μg/L of bacterial particulate organic carbon. This value is within the range of 50–200 μg C/L found by Fenchel (1982a) for isolated flagellates and cultured bacteria. However, it is higher than the reported 40 μg C/L in Lake Biwa (Nagata 1988) and 66 μg C/L in Lake Vechten (Bloem and Bär-Gilissen 1989), which may be explained by physiological differences between the respective HNF communities and also by different volume-to-carbon factors employed.

HNF preferentially remove larger bacteria by size-selective uptake (Andersson et al. 1986, Šimek and Chrzanowski 1992), which also means actively growing cells that are ready to divide (Gonzalez et al. 1990). In that way, not only is the number of bacteria reduced but also a considerable part of their production is "prevented," and the mean cell size is reduced. It is obvious that the bacteria may drop below the threshold value, as was the case on September 27 (2×10^6 cells/ml; Fig. 12.1). As a consequence, from September 27 to October 4, the HNF ingestion rate changed from 10 to 3.5 bacteria HNF^{-1} h^{-1}, and the clearance rate from 0.94 to 4.5 nl HNF^{-1} h^{-1}. During this time, bacteria could recover to a number of 3.7×10^6 cells/ml (Fig. 12.1). This also illustrates that control mechanisms between HNF and bacteria may work in a bidirectional mode rather than as a simple top-down control (see Chapter 5, this volume).

The decrease of HNF in the metalimnion from the end of October was not as pronounced as in the epilimnion, but their grazing rates were strongly reduced, which also may have been the effect of lower water temperature. The thermocline fell below 9 m around November 9, which caused higher oxygen concentrations, and lake turnover probably caused input of DOC from the surface water layers. At this time, bacteria relieved from HNF grazing pressure reached the highest production and density values of the investigation period.

Hypolimnion. The situation at a depth of 12 m was characterized by bacteria densities (3.5–6×10^6 cells/ml) that were higher overall than in the 6- to 9-m water layer, but production rates were very low. The HNF numbers were the lowest of all three depths, and their estimated grazing rates were always lower than the determined bacterial production (Fig. 12.1). It appears reasonable that low water temperatures of 4.4–5.6°C and low oxygen concentrations (below 1 mg O_2/L until the beginning of November) restricted the HNF activity. This is emphasized by the HNF specific clearance rates, which were on average 1.36 and 1.41 nl HNF^{-1} h^{-1} in the epilimnion and in the 6- to 9-m water layer, respectively, but 0.40 nl HNF^{-1} h^{-1} in the deep water (12 m). The mean bacteria numbers of 5.8, 4.0, and 4.8×10^6 cells/ml, respectively, showed no such marked differences. On the other hand, a significant increase in the clearance activity after the oxygen input to a depth of

12 m caused by turnover at the beginning of November could not be detected. During the observation period, as water temperatures sank, the activity of HNF, expressed as specific clearance rate, decreased from 4.48 to 0.21 nl HNF^{-1} h^{-1} in the epilimnion and from 4.53 to 0.31 nl HNF^{-1} h^{-1} in the metalimnion but remained unchanged at 0.41–0.44 nl HNF^{-1} h^{-1} in the hypolimnion of the lake.

Hypolimnion communities are adapted to very low oxygen concentrations or even a strictly anaerobic environment, as described for protozoa by Fenchel and Finlay (1990) and Fenchel et al. (1990). No special precautions were made in this study to avoid aeration of the samples, and a bias in the measured grazing rates cannot be excluded. However, the figures are within the range of values reported for the hypolimnion of Lake Vechten by Bloem et al. (1989b), who prevented sample aeration. In conclusion, plotting bacteria numbers and production against HNF grazing demonstrates that HNF produced no significant effect on bacteria during the autumn–winter period in the hypolimnion of the Plußsee.

Spring

The spring observation period lasted from March 13 until June 5 and covered the spring phytoplankton bloom as well as the subsequent clear-water phase. At the beginning, thermal stratification was not yet pronounced. However, distribution of dissolved oxygen, 13.6 mg O_2/L at the surface and 10.7 mg O_2/L at 10 m, and $chlorophyll_a$ contents from 14.2 µg/L (0- to 1-m layer) to 4.9 µg/L (8- to 9-m depth) indicated that the development of the spring bloom had already set in. The phytoplankton community consisted mainly of Bacillariophyceae (diatoms, e.g., *Stephanodiscus* sp.) and Cryptophyceae (cryptomonads, e.g., *Rhodomonas* sp.; Chapter 1, this volume). The maximum $chlorophyll_a$ concentration was recorded on April 10 (56.2 µg/L at the depth of 0–1 m) when water transparency (Secchi disk visibility) reached its lowest value of 1.5 m. On May 8, as a result of both oxygen-evolving phytoplankton photosynthesis and oxygen-consuming bacterial decomposition, a particular heterograde oxygen curve could be observed, with a metalimnetic oxygen maximum at 3 m and a minimum at 5 m. The following dramatic reduction of algae from intensive cladoceran feeding led to the clear-water phase (Lampert et al. 1986) with Secchi disk visibility of more than 7 m and 0.7–1.5 µg/L of $chlorophyll_a$ on May 22. At the end of the observations, the lake was markedly stratified, and the oxygen minimum was still present at 5 m.

For distinct investigation of epilimnion, metalimnion, and hypolimnion, lake water was sampled at depths of 2, 4–5, and 9 m, respectively. Sampling was performed weekly. The data shown in Figure 12.2 are arranged in accordance with Figure 12.1 but complemented by the number of autofluorescent nanoflagellates (ANF) and their grazing rate, which was generally of the same order of magnitude as HNF grazing.

Epilimnion. On March 13, the general activity of the ecosystem was indicated by high numbers of bacteria (6×10^6 cells/ml), heterotrophic nanoflagellates (12×10^3 HNF/ml), and autofluorescent nanoflagellates (18×10^3 ANF/ml). Until April 10, the bacteria increased sharply to the first spring peak (23×10^6 cells/ml)

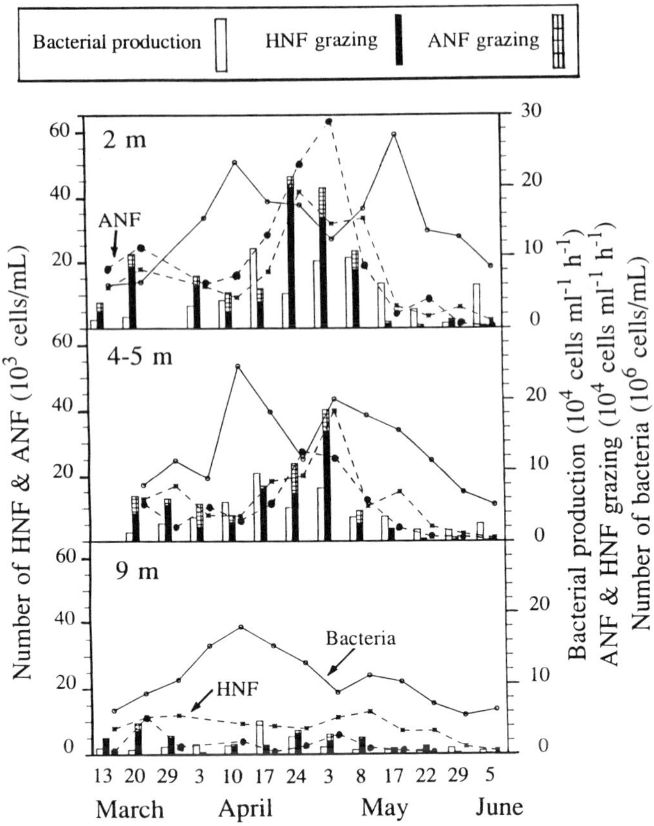

Figure 12.2. Abundance of bacteria and autotrophic (ANF) and heterotrophic nanoflagellates (HNF) and the rates of bacterial production and grazing on bacteria in the Plußsee during spring.

simultaneously with chlorophyll$_a$. This indicates that the bacterial community was intensively metabolizing readily utilizable dissolved organic matter released by living phytoplankton (Chapters 2, 4, and 5, this volume). The mass development of HNF and ANF was delayed by 2 and 3 weeks, respectively. Heterotrophic nanoflagellates increased to 41×10^3 HNF/ml on April 24, and autofluorescent nanoflagellates, predominantly *Chrysochromulina* sp., increased to 63×10^3 ANF/ml on May 3 (Fig. 12.2). These numbers were very high indeed, but they are around the same levels reported for eutrophic Lake Oglethorpe by Bennett et al. (1990), 15×10^3 HNF/ml in April and 90×10^3 ANF/ml in February, or found in mesoeutrophic Lake Constance by Jürgens and Güde (1991), 25×10^3 HNF/ml in May. From immense grazing pressure bacteria were diminished by a factor of 2 until May 3 (Fig. 12.2). At first sight, the delay in population growth of HNF and ANF when bacteria were abundant suggests a similar predator–prey relationship as was outlined for the epilimnion during autumn (see above). Reck (1987) described high numbers of mainly phytophagous ciliates in spring in the Plußsee, which may feed on nanoflagellates as well. Moreover, the dominant ciliate species *Tintinnidium pusillum* preferentially ingested *Rhodomonas* sp. or algae 3–5 μm in diameter. These findings explain the temporary decrease of HNF and ANF from March 20 to the beginning of April in spite of the excellent prey supply (Fig. 12.2). The observed change of the predominant ANF species from *Rhodomonas* sp. in March to the much smaller *Chrysochromulina* sp. on May 3 also fits this theory. It was evident that in the Plußsee during the spring, bacteria and ciliates grew parallel to the phytoplankton, and the increase of nanoflagellates was retarded as a result of ciliate grazing pressure. A comparable situation was noticed in Lake Constance by Weisse (1990, 1991).

At the turning point from spring phytoplankton bloom to clear-water phase around May 8, the nanoflagellates were reduced dramatically (Fig. 12.2). This event was also observed in other lakes (Jürgens and Güde 1991, Šimek et al. 1990, Weisse 1991). The common explanation is that cladocerans feed directly on nanoflagellates as they do on algae. Reck (1987), on the other hand, states that the decrease of nanoflagellates at the beginning of May coincides with the ciliates' spring maximum (see below). Using the reported feeding rate and density of *Tintinnidium pusillum* one can roughly calculate the observed loss of ANF and HNF that could have been taken up by ciliates. Since ciliates themselves were reduced markedly from the middle of May (Reck 1987; see below), it is possible that cladocerans had grazing impact on nanoflagellates.

Favored by DOM liberated from senescent algae and macrozooplankton "sloppy feeding" (Lampert 1978; Chapter 2, this volume), the bacteria had high production rates of 6–9 $\times$ 10^4 cells ml^{-1} h^{-1}. Relieved from nanoflagellate (NF) grazing pressure, they attained a second spring peak on May 17 (27 $\times$ 10^6 NF/ml; Fig. 12.2). The decline of the bacterial density toward the end of May, which was not as strong as the NF, truly was the effect of cladoceran size-selective feeding on bacteria (Güde 1988, Šimek et al. 1990).

Metalimnion. In general, at the depth of 4–5 m no severe differences appeared between the epilimnion and the metalimnion during spring bloom (Fig. 12.2, middle panel). Bacteria reached their highest density on April 10 (24 $\times$ 10^6 cells/ml), which was as high as that in the surface water. HNF numbers were comparable, whereas ANF were only roughly half as dense (27 $\times$ 10^3 ANF/ml). Interestingly, ANF showed grazing rates that were within the same range as in the epilimnion, which at certain times could even exceed those of the HNF. The measured Secchi disk visibility of less than 2 m during spring bloom indicated that a photoautotrophic metabolism at the investigated depths of 4–5 m was impossible because of insufficient light energy (Chapter 3, this volume). This may explain the pronounced lower number of ANF. If the mean ANF clearance rates in the metalimnion (0.19 nl ANF^{-1} h^{-1}) and epilimnion (0.07 nl ANF^{-1} h^{-1}) are compared, the impression is that ANF in a depth of 4–5 m tried to compensate for low irradiance by intensifying their phagotrophy. In the case of HNF, the difference observed between the clearance rates (0.24 at depth of 2 m and 0.19 nl HNF^{-1} h^{-1}

in the water layer between 4 and 5 m) was not as significant and was just inverted. However, the results in the 4- to 5-m water layer as well as in the epilimnion demonstrate the basic ability of the mixotrophs among the ANF to be bacterial consumers as important as HNF, and this has been confirmed by other authors (Sanders et al. 1989, Bennett et al. 1990).

During the onset of the clear-water phase, HNF and ANF numbers decreased in the water layer between 4 and 5 m, and in the beginning of June they were as low as in the other strata, which was around 10^3 HNF/ml and 0.4×10^3 ANF/ml, which was near the detection limit. The bacteria reached no second peak during the clear-water phase, but the bacterial density curve exhibited smoother, more continuous descent until June. This suggests that there was a minor influence of macrozooplankton grazing effects on bacteria in this layer.

Hypolimnion. Both low water temperature (4.1–5.7°C) and low input of photosynthetically available radiation (PhAR) because of shading caused by the high algal density in the strata above made a comparable mass development of phytoplankton impossible in the hypolimnion. The highest chlorophyll$_a$ concentrations in the 8- to 9-m layer (12 and 18 μg/L) were recorded on March 20 and 29. Absence of PhAR may also explain the low density of ANF, which reached a maximum of 11.4×10^3 ANF/ml on March 20 (Fig. 12.2, lower panel). From the end of May, ANF were no longer detectable at a depth of 9 m. The highest abundance of HNF was $12–13 \times 10^3$ HNF/ml, and they tended to disappear at the end of the observation period. As Reck (1987) reported, ciliates could be found in low numbers at this depth (see below). The development of the nanoflagellates, however, indicated that there was no major grazing impact caused by ciliates. Since photosynthetically active algae were missing, bacteria were dependent on DOM, which had not been metabolized above, and DOC released by sinking decaying phytoplankton. Therefore, their production remained low. As they did in the higher layers, bacteria peaked on April 10 (18×10^6 cells/ml) and then decreased steadily. Comparing bacterial production and grazing of nanoflagellates (Fig. 12.2, lower panel) suggests that a certain fraction of bacteria was consumed. The development of bacteria did not show grazing effects as in the epilimnion. One exception may have been the decrease on May 3. It is evident that the supply of DOM from upper layers had the biggest impact on bacteria in the hypolimnion during spring.

Summer

Investigations were carried out from July 9 to August 13. Besides the pronounced stratification and strong oxygen depletion at depths greater than 5 m, the summer phytoplankton bloom set the tone of the period. Chlorophyll$_a$ concentration reached its maximum of 62 μg/L on July 30, when intensive photosynthesis led to oxygen concentrations of 25 mg O_2/L at the surface and 23 mg O_2/L at 1 m depth, which meant oxygen supersaturation between 110% and 290% in the epilimnion. Water transparency (Secchi disk visibility) varied from 3.2 m to 1.0 m. Because of the dense algal community light was limited, so photosynthetic activity was

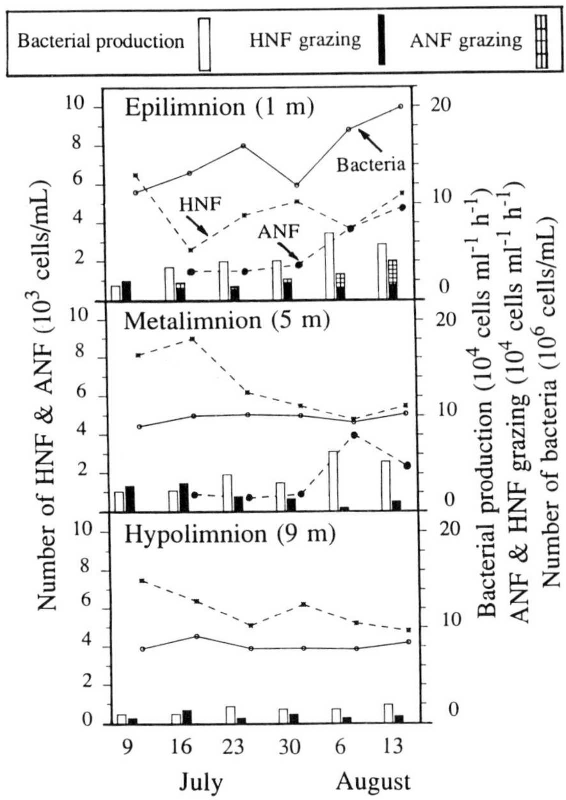

Figure 12.3. Abundance of bacteria and autotrophic (ANF) and heterotrophic nanoflagellates (HNF) and the rates of bacterial production and grazing on bacteria in the epilimnion, metalimnion, and hypolimnion of the Plußsee during summer stratification.

certainly restricted to the epilimnion down to 3-m depth. PhAR at 5 m was 0–1.3% of the surface value, leading to light limitation of the phytoplankton primary production as shown previously (Meffert and Overbeck 1985, Sell 1990). With respect to the physical conditions, the sampling depth of 1 m represented the epilimnion, 5 m the metalimnion, and 9 m the hypolimnion (Fig. 12.3).

Epilimnion. Primary production was the main factor characterizing this layer. The bacterial production (1.5–6.9 × 10^4 cells ml^{-1} h^{-1}) was directly supplied by phytoplankton EOC (Chapter 5, this volume). Since nanoflagellate grazing was lower than bacterial production, numbers of bacteria increased steadily from 11 × 10^6 cells/ml to 20 × 10^6 cells/ml. The only drop in density, on July 30, could not be explained by protozoan grazing, and also an observed community of the ciliate *Astylozoon fallax* in the layer at 0–4 m depth (Reck 1987) could hardly account for a loss of that magnitude. It was probably a technical problem during filtration. It could not be clarified whether macrozooplankton feeding or eventual vertical migrations were responsible for the changes in HNF numbers. Feeding could also have influenced the larger ANF; however, their density was stable until the end of July and then increased from 1.8 to 4.8×10^4 ANF/ml (Fig. 12.3), which was slower than the majority of phytoplankton by about 2 weeks. The bulk of ANF grazing on August 13, which exceeded the HNF grazing, was by *Cryptomonas* sp. (Fig. 12.3). At this time, the ANF ingestion rate was 5.1 bacteria ANF^{-1} h^{-1}; the clearance rate was 0.26 nl ANF^{-1} h^{-1}, which was in the range reported for *Cryptomonas* sp. by Tranvik et al. (1989). The consumption of 42% (HNF 27%) of the measured bacterial production by ANF indicates the basic importance of these organisms to food webs in the lake.

Metalimnion. As a result of methane oxidation and degradation of sinking matter by bacteria (Anagnostidis and Overbeck 1966; Chapter 16, this volume), the detected concentration of dissolved oxygen at this depth was as low as 0.2–2.6 mg O_2/L. Bacterial production (average 4.4×10^4 cells ml^{-1} h^{-1}) in the metalimnion was equal to that in the epilimnion (average 4.5×10^4 cells ml^{-1} h^{-1}). Mean clearance rates of 0.21 nl HNF^{-1} h^{-1} in the metalimnion as in the epilimnion indicate that HNF in this microaerobic environment were as active as in the upper layers. Moreover, the relationship between grazing and bacterial production was comparable to that in the epilimnion (Fig. 12.3). Bacterial density remained almost unchanged throughout 6 weeks (0.9–1.0×10^7 cells/ml), which obviously could not be because of HNF grazing control (Fig. 12.3). In addition to cladocerans feeding on bacteria attached to detritus, the main impact on bacteria probably came from ciliates. According to Reck (1987), *Coleps* sp. could be found at high densities of up to 19 ind/ml at the beginning of August in the metalimnion. Also, Sanders et al. (1989) and Šimek et al. (1990) reported on a major ciliate grazing impact on bacteria during summer. Ciliates may also have reduced the HNF and stopped the increase of ANF (Fig. 12.3) as they controlled the HNF development in Lake Constance during the summer (Weisse 1991). Since ANF grazing on bacteria was rarely observed and was insignificant in the metalimnion, it remains an open question whether ANF were really able to survive at insufficient irradiation (see above) or if the cells were inactive or dying. However, an increase of ANF abundance could be detected, as in the epilimnion on June 30. Perhaps ANF performed grazing exclusively during the day or the nighttime, totally different from the sampling time (9–10 a.m.), or they took up DOM osmotrophically (Sanders and Porter 1988).

Hypolimnion. With the exception of July 30, when intensive photosynthetic oxygen production occasionally raised O_2 concentration to 2.4 mg O_2/L, the hypolimnion was anoxic (Fig. 12.3, lower panel). The number of bacteria ranged from 8 to 9×10^6 cells/ml, which was as stable as but slightly lower than the metalimnion. ANF were not detected at all at the depth of 9 m. Heterotrophic nanoflagellates decreased from 8.4 to 4.8×10^3 HNF/ml. A mean clearance rate of 0.14 nl HNF^{-1} h^{-1} indicated comparably lower activity and grazing (average 0.75×10^4 cells ml^{-1} h^{-1}) could hardly equal the low bacterial production (average 1.4×10^4 cells ml^{-1} h^{-1}). Consequently, any evident effect of HNF grazing on the bacteria could not be observed.

The Microbial Loop in Lake Plußsee

The relationship between bacteria and nanoflagellates in the pelagic zone of the Plußsee revealed the importance of nanoflagellates to turnover of organic matter processes in the lake. Nanoflagellates consumed a large part of or even the entire bacterial production. The NF may keep the bacterial density at a threshold for grazing efficiency that leads to a mutual or feedback control of bacteria and flagellates. During periods with high numbers of NF, not only the entire production but additionally a considerable part of the bacterial standing stock was also removed by grazing. At certain times the course of the curves of NF and bacteria suggested a prey–predator relationship with a phase difference of roughly 2 weeks. When phytoplankton biomass markedly increased during bloom, large amounts of DOM favored high bacterial secondary production. In these cases the bacteria were able to grow and were apparently unaffected by nanoflagellate consumption.

Interestingly, HNF ingesting bacteria can also be found in the anoxic hypolimnion. They may be rather microaerophilic, using minor and temporary fluxes of oxygen from the upper layers. However, control of bacteria by HNF grazing does not seem to exist during anoxia. During the period of the spring phytoplankton bloom, when the hypolimnion still contained oxygen, the number and activity of HNF were similar to the values found in the epi- and metalimnion during other seasons. Moreover, this is the only time ANF were observed in the hypolimnion. Mass developments of HNF and ANF that occurred in the upper water layers of the lake did not usually occur in the hypolimnion.

The study shows that autofluorescent (mixotrophic) nanoflagellates (ANF) may reach clearance rates comparable to those of the HNF. High ANF numbers during the spring phytoplankton bloom led to significant grazing rates, which sometimes were greater than the HNF. ANF tend to have higher clearance rates in the metalimnion relative to the epilimnion, which emphasizes that phagotrophy of ANF becomes more important during light-limitation conditions.

Carbon Transfer and Nutrient Regeneration

An overall estimation of the carbon transfer from DOC $\rightarrow$ bacteria $\rightarrow$ HNF $\rightarrow$ ciliates $\rightarrow$ macrozooplankton will be given. Such figures must be regarded with caution because of the many conversion factors applied. They are, however, useful for demonstrating tendencies.

In the epilimnion of the Plußsee during the autumn–winter period only 10% of the bacterial carbon was finally available to ciliates or cladocerans, whereas 33% of the bacterial carbon was available during spring phytoplankton bloom. In the metalimnion these values were 11% and 44%, respectively. The relatively high percentages in spring can be explained by the high density and grazing rates of HNF, which acted as a temporary stock of bacterial particulate organic carbon. The HNF were preyed on quantitatively at the beginning of the clear-water phase. This was the only stage of significant carbon transfer throughout the microbial loop.

Whether this carbon finally becomes a significant source for macrozooplankton metabolism depends on whether it has to pass ciliates, an additional trophic level that increases respiration losses. This seems to be at least in part the case. During the clear-water phase a short circuit exists from bacteria to cladocera, but no conclusions on its efficiency are available from this study.

The portion of carbon transferred in the epilimnion and metalimnion was only 7% during summer stagnation. Concentrations of dissolved P-PO_4^{3-} were 3–6 μg P/L in the 0- to 1-m layer and 4–57 μg P/L at 5-m depth. In spite of this measurably depleted nutrient store, the phytoplankton was able to increase to a summer bloom with the highest chlorophyll content recorded throughout the year. This indicates indirectly that during this stage nutrients are effectively regenerated by the grazing of nanoflagellates on bacteria.

From the study it is evident that the microbial loop in the Plußsee is neither a link nor a sink exclusively. As a component within the pelagic food web, its function depends on the number, activity, and trophic state of the phytoplankton installed before, of the macrozooplankton connected after, and the nutrient and physical conditions in the respective stratum.

Pelagic Ciliates

The qualitative and quantitative distribution of pelagic ciliates throughout the entire water column of the Plußsee was studied in weekly intervals from March to December 1981. A total of 103 ciliate species were determined (Reck 1987). The greatest part of the pelagic ciliates belonged to saprobic or sapropelic species consuming free-living bacteria (Kahl 1930–1935). Only 30% of the pelagic species of the Plußsee were found as obligate or facultative pelagic ciliates feeding frequently on phytoplankton (Bick 1972).

Figure 12.4 gives a general view of the distribution of predominating pelagic ciliates in the Plußsee throughout 1981. The greatest abundance of ciliates (more than 10,000 ind/L) was found in the epilimnion during the spring phytoplankton bloom (April–May). Turnover of the lake in October–November was also characterized by a high density of ciliates moving to greater depths after the anaerobic hypolimnion (nearly free of ciliates) was reaerated in the autumn. High numbers of ciliates were found temporarily in the metalimnion.

During the spring bloom of diatoms and phytoflagellates (Chapter 1, this volume), *Tintinnidium pusillum* reached the absolute highest number of the entire investigation period (34,750 ind/L) with high density up to 25 m depth (Fig. 12.4A). Grazing experiments with ciliates sampled from the depths of 5, 6, and 9 m showed a strong grazing impact of ciliates on *Stephanodiscus hantzschii* and *Rhodomonas* (Table 12.2, Fig. 12.5). At greater depths, the species probably grazed on bacteria. Experiments indicated a selective grazing, which may be different in the single compartments of the lake. The bacterivorous *Astylozoon fallax* also reached its highest point in development in spring, with maximal densities of approximately 5000 ind/L.

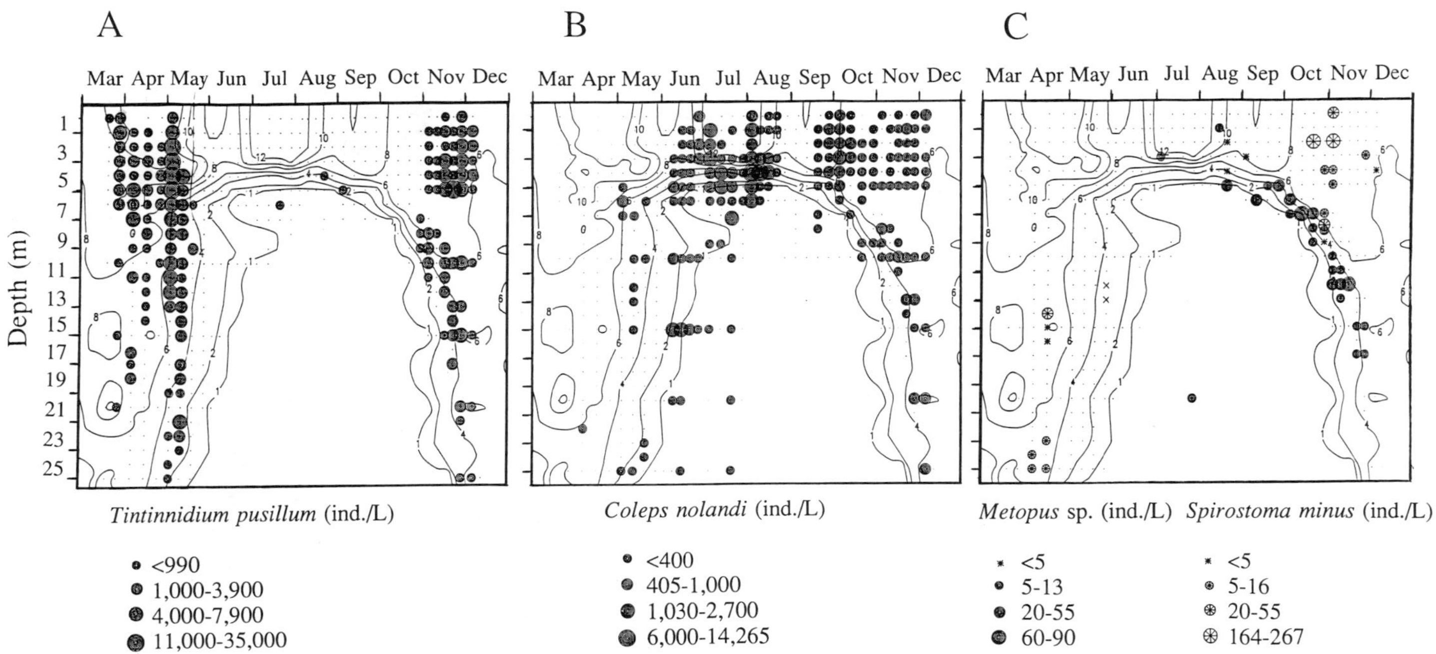

Figure 12.4. Spatial and temporal distribution of (A) *Tintinnidium pusillum*, (B) *Coleps nolandi*, and (C) *Metopus* sp. and *Spirostomum minus* in the Plußsee, 1981. Curves show isolines (interval 2 mg O_2/L) of dissolved oxygen.

Table 12.2. Ciliate Grazing Activity during Spring 1983

Assay	Depth (m)		
	5	6	9
Ciliates (ind/L)	720	290	220
Rhodomonas sp. (cells/ml)	994	640	190
Ingestion rate (cells/h)	11	10	4
Filtration rate (cells/h)	11	15	24
% of grazing[a]	19	11	12
Stephanodiscus sp. (cells/ml)	310	220	210
Ingestion rate (cells/h)	4	4	8
Filtration rate (cells/h)	12	19	40
% of grazing[a]	22	13	21

[a]Percentage of the initial density of cells.

The community structure of ciliates differed highly during summer stratification. Mesosaprobic and polysaprobic species dominated in the aerobic part of the epilimnion and the upper part of the metalimnion. The largest densities of *Coleps nolandi* were found during summer (Fig. 12.4B). In this omnivorous species

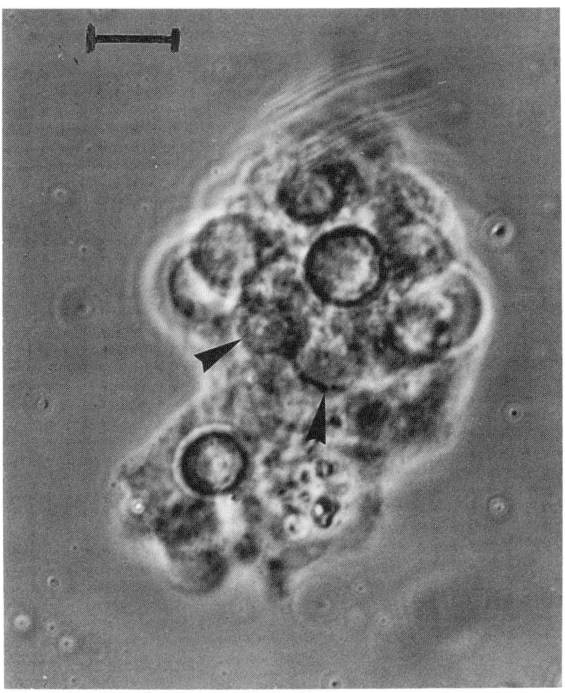

Figure 12.5. *Tintinnidium pusillum* ingesting cells of *Stephanodiscus* sp. Bar, 10 μm.

zoochlorellae were always present, which was probably a result of the oxygen production of these photosynthetic algae. Therefore, *Coleps nolandi* may also live in the anaerobic hypolimnion.

The ciliate community at the beginning of fall mixing of the lake water was composed of different dominating species. On the boundary between O_2 and H_2S, ciliates were represented by many sapropelic genera (genus *Metopus* and *Caenomorpha*), which specialized in grazing on phototrophic bacteria (Fig. 12.4C). The decline of the thermocline together with the degradation of DOM by bacteria was accompanied by an increase in density of ciliates. A maximum reached in this area was 11,120 ind/L of *Hancoma setosas*. Parallel to the density maximum of bacterivorous ciliates in the hypolimnion, in the epilimnion a remarkable development of phytoplankton-consuming ciliates was observed, especially *Tintinnidium pusillum* (Fig. 12.5). When the total circulation began at the end of November, ciliates were uniformly distributed throughout the entire water column.

The highest diversity of ciliates was found during the spring (phytoplankton-consuming species in the epilimnion and sapropelic bacterivorous species in the hypolimnion). When the oxygen was depleted in the hypolimnion, the diversity of ciliates decreased. The sapropelic ciliates reached the highest diversity of taxa and species during the active microbial degradation of DOM during autumn.

Table 12.2 shows results of grazing experiments with natural assemblages of ciliates that were enriched 10-fold. Fifty liters of lake water were prefiltered through 200-μm gauze to separate the macrozooplankton, followed by an enrichment of microzooplankton with 30-μm gauze. The experiments were continued for 6 h (for further details, see Reck 1987). *Tintinnidium pusillum* was the main grazer of *Rhodomonas* sp. and *Stephanodiscus* sp. during the spring phytoplankton bloom. The decreasing cell numbers of *Rhodomonas* and *Stephanodiscus* species in deeper parts of the lake were compensated by an increased filtration rate of the ciliates. At all depths, ciliates that consumed phytoplankton removed approximately 20–25% of the biomass of unicellular algae at all depths.

References

Anagnostidis K, Overbeck J (1966) Methanoxidierer und hypolimnische Schwefelbakterien. Studien zur Ökologischen Biocönotik der Gewässermikroorganismen. Ber Deutsch Bot Ges 79/3:163–174

Andersen P, Fenchel T (1985) Bacterivory by microheterotrophic flagellates in seawater samples. Limnol Oceanogr 30:198–202

Andersson A, Larsson U, Hagström Å (1986) Size-selective grazing by a microflagellate on pelagic bacteria. Mar Ecol Prog Ser 33:51–57

Azam F, Fenchel T, Field JG, Gray JS, Meyer-Reil LA, Thingstad F (1983) The ecological role of water-column microbes in the sea. Mar Ecol Prog Ser 10:257–263

Bell RT (1990) An explanation for the variability in the conversion factor deriving bacterial cell production from incorporation of [^{3}H]-thymidine. Limnol Oceanogr 35:910–915

Bell RT, Ahlgren GM, Ahlgren I (1983) Estimating bacterioplankton production by measuring [^{3}H]-thymidine incorporation in an eutrophic Swedish lake. Appl Environ Microbiol 45:1709–1721

Bennett SJ, Sanders RW, Porter KG (1990) Heterotrophic, autotrophic, and mixotrophic nanoflagellates: Seasonal abundances and bacterivory in a eutrophic lake. Limnol Oceanogr 35:1821–1832

Bick H (1972) Ciliata. In *Binnengewässer Mitteleuropas XXVI. Das Zooplankton der Binnengewässer,* 1. Teil. Stuttgart, pp 31–83

Bloem J, Albert C, Bär-Gilissen MJB (1989) Bacterial activity and grazing potential in a stratified lake. Limnol Oceanogr 34:297–309

Bloem J, Starink M, Bär-Gilissen MJB, Cappenberg TE (1988) Protozoan grazing, bacterial activity, and mineralization in two stage continuous cultures. Appl Environ Microbiol 54:3113–3121

Bloem J, Bär-Gilissen MJB, Berman T, Cappenberg TE (1989a) Nutrient cycling through phytoplankton, bacteria and protozoa, in selectively filtered Lake Vechten water. J Plankton Res 11:119–131

Bloem J, Ellenbroek FM, Bär-Gilissen MJB, Cappenberg TE (1989b) Protozoan grazing and bacterial production in stratified lake Vechten estimated with fluorescently labeled bacteria and by thymidine incorporation. Appl Environ Microbiol 55:1787–1795

Bratbak G (1985) Bacterial biovolume and biomass estimations. Appl Environ Microbiol 49:1488–1493

Chranowski TH, Šimek K (1990) Prey-size selection by freshwater flagellated protozoa. Limnol Oceanogr 35:1429–1436

Fenchel T (1982a) Ecology of heterotrophic microflagellates. II. Bioenergetics and growth. Mar Ecol Prog Ser 8:225–231

Fenchel T (1982b) Ecology of heterotrophic microflagellates. III. Adaptations to heterogeneous environments. Mar Ecol Prog Ser 9:25–33

Fenchel T (1982c) Ecology of heterotrophic microflagellates. IV. Quantitative occurrence and importance as bacterial consumers. Mar Ecol Prog Ser 9:35–42

Fenchel T, Finlay BJ (1990) Anaerobic free-living protozoa: Growth efficiencies and the structure of anaerobic communities. FEMS Microb Ecol 74:269–276

Fenchel T, Kristensen LD, Rasmussen L (1990) Water column anoxia: Vertical zonation of planktonic protozoa. Mar Ecol Prog Ser 62:1–10

Fuhrman JA, Azam F (1980) Bacterioplankton secondary production estimates for coastal waters of British Columbia, Antarctica, and California. Appl Environ Microbiol 39: 1085–1095

Fuhrman JA, Azam F (1982) Thymidine incorporation as a measure of heterotrophic bacterioplankton production in marine surface waters: Evaluation and field results. Mar Biol 66:109–120

Gonzalez JM, Sherr EB, Sherr BF (1990) Size-selective grazing on bacteria by natural assemblages of estuarine flagellates and ciliates. Appl Environ Microbiol 56:583–589

Güde H (1986) Loss processes influencing growth of planktonic bacterial populations in Lake Constance. J Plankton Res 8:795–810

Güde H (1988) Direct and indirect influences of crustacean zooplankton on bacterioplankton of Lake Constance. Hydrobiologia 159:63–73

Jürgens K, Güde H (1991) Seasonal changes in the grazing impact of phagotrophic flagellates on bacteria in Lake Constance. Mar Microb Food Webs 5:27–37

Kahl A (1930–1935) Wimpertiere oder Ciliata (Infusoria). In Dahl F (ed) *Die Tierwelt Deutschlands,* Teile 18 (1930), *21* (1931), *25* (1932), *30* (1935).

Kjelleberg S, Hermansson M (1984) Starvation-induced effects on bacterial surface characteristics. Appl Environ Microbiol 48:497–503

Krambeck C, Krambeck H-J, Overbeck J (1981) Microcomputer-assisted biomass determination of plankton bacteria on scanning electron micrographs. Appl Environ Microbiol 42:142–149

Kuuppo-Leinikki P (1990) Protozoan grazing on planktonic bacteria and its impact on bacterial population. Mar Ecol Prog Ser 63:227–238

Lampert W (1978) Release of dissolved organic carbon by grazing zooplankton. Limnol Oceanogr 23:831–834

Lampert W, Fleckner W, Rai H, Taylor B (1986) Phytoplankton control by grazing zooplankton: A study on the spring clear-water phase. Limnol Oceanogr 3:478–490

McManus GB, Fuhrman JA (1988) Clearance of bacteria-sized particles by natural populations of nanoplankton in the Chesapeake Bay outflow plume. Mar Ecol Prog Ser 42:199–206

Meffert ME, Overbeck J (1985) Dynamics of chlorophyll and photosynthesis in natural phytoplankton associations. II. Primary productivity, quantum yields and photosynthetic rates in small Northgerman lakes. Arch Hydrobiol 104:363–385

Münster U, Chróst RJ (1990) Origin, composition, and microbial utilization of dissolved organic matter. In Overbeck J, Chróst RJ (eds) *Aquatic Microbial Ecology: Biochemical and Molecular Approaches*. Springer-Verlag, New York, pp 8–46

Nagata T (1988) The microflagellate–picoplankton food linkage in the water column of Lake Biwa. Limnol Oceanogr 33:504–517

Ohle W (1952) Die hypolimnische Kohlendioxyd-Akkumulation als produktionsbiologischer Indikator. Arch Hydrobiol 46:153–285

Ohle W (1962) Der Stoffhaushalt der Seen als Grundlage einer allgemeinen Stoffwechseldynamik der Gewässer. Kieler Meeresforschungen 18:107–120

Porter KG, Feig YS (1980) The use of DAPI for identifying and counting aquatic microflora. Limnol Oceanogr 24:943–948

Reck E (1987) Zur Ökologie der pelagischen Ciliaten des Plußsees. PhD thesis, University of Kiel, Germany

Sanders RW, Porter KG (1988) Phagotrophic phytoflagellates. Adv Microb Ecol 10:167–192

Sanders RW, Porter KG, Bennett SJ, DeBiase AE (1989) Seasonal patterns of bacterivory by flagellates, ciliates, rotifers, and cladocerans in a freshwater planktonic community. Limnol Oceanogr 34:673–687

Schröder D, Krambeck H-J (1991) Advances in digital image analysis of bacterioplankton with epifluorescence microscopy. Verh Int Ver Limnol 24:2601–2604

Sell A (1990) Ökophysiologische Untersuchungen am Phytoplankton des Plußsees. Diploma thesis, University of Kiel, Germany.

Sell A, Overbeck J (1992) Exudates: Phytoplankton–bacterioplankton interactions in Plußsee. J Plankton Res 14:1199–1215

Sherr BF, Sherr EB, Fallon RD (1987) Use of monodispersed, fluorescently labeled bacteria to estimate in situ protozoan bacterivory. Appl Environ Microbiol 53:958–965

Sherr BF, Sherr EB, Pedros-Alio C (1989) Simultaneous measurement of bacterioplankton production and protozoan bacterivory in estuarine water. Mar Ecol Prog Ser 54:209–219

Sieburth JMcN (1979) *Sea Microbes*. Oxford University Press, New York

Šimek K, Chrzanowski TH (1992) Direct and indirect evidence of size-selective grazing on pelagic bacteria by freshwater flagellates. Appl Environ Microbiol 58:3715–3720

Šimek K, Straskrabova V (1992) Bacterioplankton production and protozoan bacterivory in a mesotrophic reservoir. J Plankton Res 14:773–787

Šimek K, Macek M, Seda J, Vyhnalek V (1990) Possible food chain relationships between bacterioplankton, protozoans, and cladocerans in a reservoir. Int Rev Ges Hydrobiol 75:583–596

Tranvik L, Porter KG, Sieburth JMcN (1989) Occurrence of bacterivory in *Cryptomonas*, a common freshwater phytoplankter. Oecologia 78:473–476

Weisse T (1990) Trophic interactions among heterotrophic microplankton, nanoplankton, and bacteria in Lake Constance. Hydobiologia 191:111–122

Weisse T (1991) The annual cycle of heterotrophic freshwater nanoflagellates: Role of bottom-up versus top-down control. J Plankton Res 13:167–185

Wikner J, Rassoulzadegan F, Hagström Å (1990) Periodic bacterivore activity balances bacterial growth in the marine environment. Limnol Oceanogr 35:313–324

13. Viruses

Karl-Paul Witzel, Jutta Demuth, and Christian Schütt

Lytic Bacteriophages and Other Viruses

Bacteriophages are viruses that infect bacteria and cyanobacteria, their host cells. Since their discovery at the beginning of this century (Twort 1915, d'Hérelle 1914), studies on the interaction between phage and host cells have significantly contributed to our understanding of modern molecular biology. Phages have been isolated for virtually all of the known bacterial genera and from all habitats where bacteria can exist (Ackermann and DuBow 1987). Some of them have been studied in great detail. However, little is known about those phages that infect native bacterial populations, especially in freshwater habitats.

Until recently, bacteriophages were thought to be ecologically unimportant because of their low numbers, as determined with specific host bacteria. Moreover, the cell density and growth rates of indigenous host bacteria in aquatic environments seemed to be below a presumptive threshold necessary for multiplication of the phages and significant effects within aquatic ecosystems (Wiggins and Alexander 1985).

The application of electron microscopy to count aquatic viruses revealed high numbers of viral particles in a number of aquatic environments, including marine

---

The section on Lytic Bacteriophages and Other Viruses was written by Karl-Paul Witzel and Jutta Demuth. The section on Temperate Bacteriophages was written by Christian Schütt.

and freshwater habitats. Torrella and Morita (1979) were the first to demonstrate that the number of phage particles in seawater was at least 10^4 phages/ml and was often significantly higher than the bacterial counts. These authors stated: "the lytic and genetic role of phages on the natural bacterial population and microbial taxonomy in near-shore water cannot be disregarded in the future." It was one decade before these results were confirmed by Bergh et al. (1989), who studied several marine and freshwater habitats and reported viral counts exceeding the bacterial numbers by a factor between 3 and 50. In a Plußsee sample from May 1988, they found 2.5×10^8 per ml of virus-like particles and 6.2×10^6 cells/ml of bacteria.

In this chapter, we will try to summarize published results on the ecological role of bacteriophages and other viruses in aquatic habitats and compare them with our results. We have isolated and characterized several phage/host systems from the Plußsee and other Holstein lakes. The distribution of these phages and their host bacteria was followed by plaque assay and epifluorescence microscopy in the Plußsee, and the morphological variability of natural phage populations was studied using direct electron microscopy.

Methods

The host bacteria were isolated from the Plußsee in January 1990 on a casein–peptone–starch (CPS) medium (Collins and Willoughby 1962) at 28°C. Bacteriophages were counted, isolated, and cultivated with the conventional agar overlay technique (Adams 1959). When necessary, the phages were concentrated by adsorption onto their host cells and centrifugation (modified from Purdy et al. 1984). Bacteria were counted on CPS medium. Colonies of the host bacteria were identified by their characteristic color and morphology. DNA probes (see below) and polyclonal antibodies were used to confirm the identification.

For epifluorescence microscopic counts, water samples were prefiltered with a 0.2-μm pore-size polycarbonate membrane (Nuclepore) and stained with DAPI (4′,6-diamidino-2-phenylindole). The small DNA-containing particles were collected on a 0.02-μm pore-size Anopore filter and counted in a Zeiss epifluorescence microscope at a magnification of 1,500×.

For electron microscopy, phages from water samples were directly adsorbed onto a carbon film without enrichment, negatively stained with uranyl acetate, and viewed with a Philips EM 300 transmission electron microscope at a primary magnification of 114,000×. Details of all the above methods will be published elsewhere (Demuth et al. 1993).

Results and Discussion

We studied several phage/host pairs from the Plußsee and the Belauersee (Witzel, unpublished). Some of the phages and their hosts were strongly bound to a specific habitat; others were found in a great variety of different aquatic habitats. Phages lysing the bacterial strain R116 which originated from the Belauersee were found in that lake only during the spring, which is the time when the bacteria were isolated. By contrast, phages for the host strain B18 which originated from the Plußsee were widespread throughout different aquatic environments. B18 phages were isolated from ponds, rivers, and lakes in different areas of Germany as well as from marine habitats of the Baltic Sea and Lake Kinneret in Israel. The phage isolates were different in the aspects of plaque morphology, fine structure, and host range.

Some of the isolated phages had a surprisingly wide host range. At times, B18 phages could lyse up to 90% of the bacterial colonies from the Plußsee water when grown on CPS agar plates (Witzel, unpublished). For other phages, these values were lower: C23 lysed up to 43%, and D8 lysed up to 18%. In general, it turned out that phages were found only in those habitats where the host bacteria existed. So the widespread occurrence of certain phages might be influenced not only by their broad host range but mainly by the high adaptability of their host bacteria to different environmental conditions. Nevertheless, the highest phage densities were observed in samples where the number of sensitive host bacteria was low, and vice versa.

Using different host strains that had been isolated from the same habitat to determine the number of plaque-forming units (PFU) in the Plußsee, we observed between fewer than 33 and 10,000 PFU/L from March to July 1990 (Table 13.1). Even higher phage densities (results not shown) were observed with strain B18 as host bacterium. These numbers ranged from 1×10^4 PFU/L (September 29, 1989) to 5.5×10^6 PFU/L (June 3, 1991). Plaque numbers were very dynamic. Shortly after the time the host strain was isolated, they were at their maximum and later disappeared. Ideally host strains should be isolated from the same habitat at the same time to ensure that they are sensitive to the prevailing phages. Because the phages are eliminated from the water at a rapid rate (Ahrens 1971, Heldal and Bratbak 1991, Witzel, unpublished), they need a high production rate to maintain their population. Therefore, suitable host strains must be present at the same time.

Table 13.1. Number of Bacteriophages (Plaque-Forming Units; PFU/L) in the Plußsee That Lysed Various Host Bacteria Isolated from This Lake

Sampling Date, 1990	Host Bacterial Strains				
	A16	B13	C8	C23	D8
March 7	167	267	nd[a]	+[b]	<33[c]
March 14	267	467	0	150	2,183
March 26	250	167	267	nd[a]	nd[a]
April 9	<33[c]	<33[c]	33	33	<33[c]
April 10	<33[c]	117	533	10,000	33
July 2	<33[c]	<33[c]	<33[c]	<33[c]	<33[c]
July 7	<33[c]	<33[c]	<33[c]	<33[c]	<33[c]

[a]Not determined.
[b]Phages present but not counted.
[c]Below detection limit of 33 PFU/L.

Table 13.2. Phage Numbers Determined as Plaque-Forming Units (PFU/ml) with Specific Host Strains in Various Aquatic Habitats

Habitat	Host Strain	PFU/ml	Reference
Seawater			
Kiel Bight	*Agrobacterium*[a]	<8,100 36,500[b]	Ahrens (1971)
North Sea	isolates[a]	0–30 715[b]	Moebus (1991)
		0–10 1,500[b]	Moebus (1992)
Freshwater			
River Avon, England	*Aeromonas hydrophila*[c]	97	Logan et al. (1980)
	Hyphomicrobium[c]	447	
	Aeromonas hydrophila	16–12,896	Seeley et al. (1979)
River in England	*Pseudomonas aeruginosa*[c]	760–2,040	Primrose and Day (1977)
	Aeromonas hydrophila[c]	240–2,080	
Plußsee	Isolates[a]	10–5,500	Witzel (unpublished)

[a]Isolates from the same habitat.
[b]Exceptionally high numbers at areas of high pollution.
[c]Strains of unknown origin.

Large numbers of plaque-forming units have been reported from other aquatic habitats as well (Table 13.2). In all studies, the phage numbers were highly variable. The highest values were reported for freshwater habitats and the inner part of the Kiel Bight. At offshore stations in the North Sea, phage numbers were significantly lower. In those studies where host strains and their phages were counted simultaneously (Ahrens 1971, Witzel, unpublished), the number of phages could temporarily exceed the number of host bacteria. However, for an evaluation of such numbers, more studies are needed to reveal short-term fluctuations and the factors regulating phage–host interactions.

Plaque assays to count phages with host bacteria will always reveal lower numbers than direct counts with transmission electron microscopy (TEM) or epifluorescence light microscopy (ELM) because they are counting only that part of the total assemblage that is active against one specific host. It is generally accepted that the small (<0.2 μm) fluorescent particles that can be seen with the light microscope in preparations from DAPI-stained water samples are viruses (Hara et al. 1991, Proctor and Fuhrman 1992). However, it is not possible with this technique to distinguish phages from other viruses, which may be abundant in water samples too (Padan and Shilo 1973, Proctor and Fuhrman 1990, Suttle et al. 1990, 1991, Cottrell and Suttle 1991, Suttle and Chan 1993). Similarly, published virus counts with TEM do not distinguish phages from viruses in general. It is understood that direct counts of viruses by either of these two methods reveal numbers that are at least two or three orders of magnitude higher than the highest plaque counts.

Meanwhile, after the first report of Bergh et al. (1989), it is well accepted that the total number of viruses in diverse aquatic habitats is generally above 10^6 particles/ml (Table 13.3). Usually, freshwater habitats have higher numbers than near-shore marine habitats, and these have higher numbers than open-ocean sites. Values from the highly oligotrophic Sargasso Sea are significantly lower than all others in comparison. Counts with ELM give results similar to TEM counts (Hara et al. 1991, Proctor and Fuhrman 1992).

With the ELM technique, we determined virus numbers from 1.6×10^7 to 3.7×10^8 particles/ml in the Plußsee. This agrees well with the values determined by Bergh et al. (1989) for the same lake and Heldal and Bratbak (1991) for Lake Kalandsvannet (see Table 13.3) by using TEM after ultracentrifugation onto microscopic grids. Also, from our studies with TEM, despite the fact that they were not calibrated for counting, we can conclude that the phage numbers should be of the order of 10^8 phages/ml.

Our electron microscopic observations showed a broad spectrum of morphologically distinct phages in the Plußsee (Demuth et al. 1993). Nearly all of them had tails of different lengths and were easily distinguishable from viruses of eukaryotic organisms such as green algae, fish, or insects. Some tails were flexible, some rigid, some with or without a contractile sheath. The diameter of the heads varied between 41 and 117 nm, the length of the tails from 11 to 636 nm. We could not find indications of typical morphological differences for depth or time (Fig. 13.1).

The fact that phages and other viruses are permanently present in numbers that exceed the host density and that they are highly dynamic and have high morphological variability supports the view that they may have a significant ecological function in aquatic habitats (Bratbak et al. 1990, Proctor and Fuhrman 1992, Suttle and Chan 1993; Fig. 13.2).

Unlike grazers, viruses are more or less host specific and thus alter particular populations out of the total assemblage. Special defense mechanisms, such as modification of cell surface receptors, the expression of restriction enzymes, or infection by a temperate phage (see below), can lead to immunity against reinfection or at least prevent lysis (Hill et al. 1990, Romashko and Bylinskii 1992). Thus, a permanent infection pressure by viruses should increase the diversity and dynamics of certain host populations, which might lead to a stabilization of the system. Vice versa, we can assume that virus populations themselves may be diverse and composed of different clones. This was shown for viruses infecting the eukaryotic alga *Micromonas pusilla* (Cottrell and Suttle 1991), for phages infecting the marine cyanobacterium *Synechococcus* sp. (Suttle and Chan 1993), and it is well known for phages of lactic streptococci. So, we can assume that phage populations infecting heterotrophic aquatic bacteria should behave in a similar manner.

After lysis of the host cell, dissolved organic material may leak into the pool of dissolved carbon, phosphorus, and nitrogen (Fig. 13.2). Also, as a part of the host

Table 13.3. Number of Viruses (or Phages) and the Ratio of Viral Number to Bacterial Number (*V*/*B*) in Various Aquatic Habitats Determined by Direct Counting Techniques with the Use of Transmission Electron Microscopy (TEM) or Epifluorescence Light Microscopy (ELM)

Habitat	Viruses/ml	*V*/*B*	Preparation	Reference
Near-shore seawater				
Yaquina Bay	ca. 10^4	nd[a]	TEM after 0.2-μm pore-size filtration	Torrella and Morita (1979)
Raunefjorden	$<0.01 \times 10^6$ to 9.9×10^6	10–50	TEM after ultracentrifugation	Bergh et al. (1989)
Long Island Sound	7.3×10^7 to 7.1×10^8	nd[a]	ELM after ultrafiltration	Proctor and Fuhrman (1992)
Chesapeake Bay	2.6×10^6 to 1.4×10^8	3–25	TEM after ultracentrifugation	Wommack et al. (1992)
Southern California Bight	1.0×10^6 to 2.8×10^7	5–13	TEM after ultracentrifugation	Cochlan et al. (1993)
Laguna Madre (hypersaline)	3.3×10^7	nd[a]	ELM after ultrafiltration	Suttle et al. (1990)
Osaka Bay	3.5×10^7	18	TEM	Hara et al. (1991)
Otsuchi Bay	1.9×10^6 to 5.0×10^6	7–15	TEM	Hara et al. (1991)
Open ocean				
North Atlantic	1.5×10^7	50	TEM after ultracentrifugation	Bergh et al. (1989
Gulf of Mexico	2.0×10^6	nd[a]	ELM after ultrafiltration	Suttle et al. (1990)
Caribbean Sea	3.8×10^6 to 6.0×10^6	nd[a]	ELM after ultrafiltration	Proctor and Fuhrman (1992)
Eastern Pacific	2.6×10^6	nd[a]	ELM after ultrafiltration	Proctor and Fuhrman (1992)
Gulf of Bothnia	1.9×10^6 to 52.2×10^6	14	TEM after ultracentrifugation	Cochlan et al. (1993)
Kuroshio area	1.2×10^6 to 2.0×10^6	2.8–5.3	TEM	Hara et al. (1991)
Sargasso Sea	3×10^3	nd[a]	TEM after ultrafiltration	Proctor and Fuhrmann (1990)
Freshwater				
Plußsee	2.5×10^8	40	TEM after ultracentrifugation	Bergh et al. (1989)
Plußsee	$>10^8$	nd[a]	TEM without enrichment	Demuth et al. (1993)
Plußsee	1.6×10^7 to 3.7×10^8	2–50	ELM after 0.02 μm pore-size filtration	Witzel et al. (unpublished)
Lake Kalandsvannet	1.9×10^7 to 2.0×10^8	24–67	TEM after ultracentrifugation	Heldal and Bratbak (1991)

[a]Not determined.

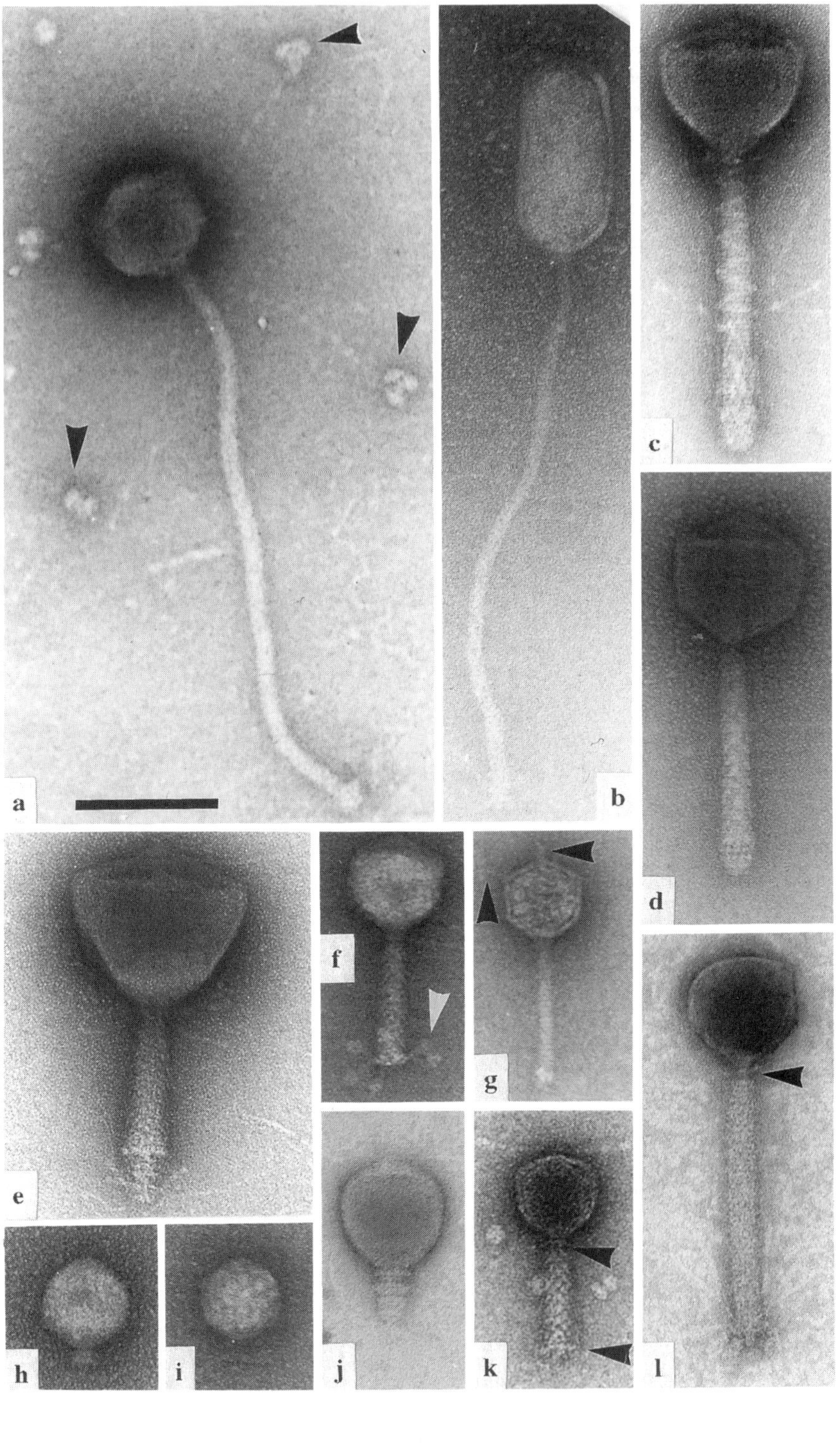
a
b
c
d
e
f
g
h
i
j
k
l

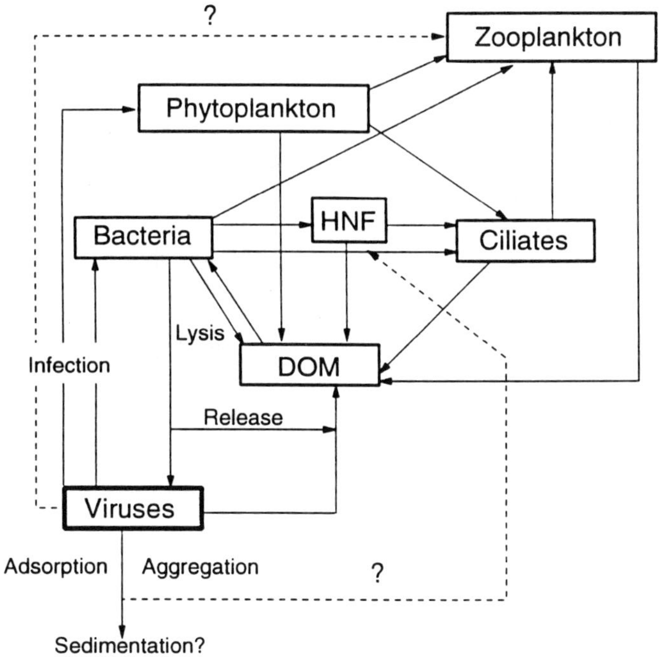

Figure 13.2 Potential role of bacteriophages and other viruses within major components of the microbial loop. The scheme was modified and completed from Bratbak et al. (1990) and Weisse et al. (1990). HNF, heterotrophic nanoflagellates; DOM, dissolved organic matter.

cell's production, the viral biomass will be channeled to fine particulate material, which is not distinguished by the currently applied routine techniques for DOM determination from the real dissolved pool of organic matter. If we assume that total virus densities are of the order of 10^8 particles/ml, this could be a significant part of the DOM that is not directly available for heterotrophic bacteria but may be grazed by heterotrophic nanoflagellates (González and Suttle 1993).

At present, no routine technique is available to distinguish between viruses infecting different components of the lake ecosystem (e.g., heterotrophic bacteria, cyanobacteria, phytoplankton, macrophytes, fish), especially the microbial loop.

Figure 13.1. Bacteriophages from the Plußsee: (a) with long flexible tail and four antennae (arrows) with trilobate ends; (b) with elongated head and long, flexible tail; (c) with contractile tail and a large spiral filament; (d) with contractile tail, showing scale-like substructures; (e) with large head, contractile tail with scale-like substructures; (f) with thick noncontractile tail, tail fibers with knob-like structures (arrow); (g) with striated tail and two antennae at the head (arrows); (h) with short tail; (i) without tail, head with capsomeres; (j) swollen head with collar and disks; (k) with collar (arrow), contractile tail with scale-like substructures and a base plate (arrow); (l) with collar (arrow), contractile tail with striated substructures. Bar, 100 nm.

Usually it is assumed that most of the viruses are bacteriophages infecting heterotrophic bacteria (Bratbak et al. 1990, Heldal and Bratbak 1991, Wommack et al. 1992, Cochlan et al. 1993). Nevertheless, it is well established that other viruses that influence cyanobacteria (Proctor and Fuhrman 1990, Suttle and Chan 1993), primary production (Suttle et al. 1990), or other organisms can be readily detected in aquatic environments. At present, these facts are setting limits to the approaches to quantify the influence of viruses on the microbial loop organisms. In addition to the lytic effects of phages, they could be important vehicles for the transfer of genetic material within and between bacterial populations. The remainder of this chapter will deal with these aspects.

Temperate Bacteriophages

The mechanisms of genetic adaptation to environmental changes (temperature, pH, ionic strength, toxic heavy metals, and recalcitrant organic compounds) are essential to the survival of bacteria in nature. Yet, their asexual propagation does not allow an exchange of nucleic acids between organisms. However, it is well established that bacteria are not genetically stable. Several cell-borne and DNA transfer mechanisms create genetic plasticity. Stal et al. (1989) listed the major mechanisms, including gene expression and its control, mutations that may trigger silent genes (Beachem 1987), gene amplification that can function to overcome inhibition caused by toxic compounds, transposons with the ability to regulate specific genes, extrachromosomal plasmid DNA that codes for additional physiological traits, and gene transfer by conjugation, transformation, or transduction followed by segregation. However, the frequencies of these mechanisms in nature are still unknown. An outline of the adaptational role of plasmids has become visible (Schütt 1990, 1991).

Temperate bacteriophages are maintained in lysogenic bacteria. It can be assumed that lysogenic bacteria may provide other potential hosts with foreign DNA. Hence, natural aquatic microbial communities are likely to be subjected to dynamic genetic adaptation processes caused by gene transfer via temperate phages. Besides natural plasmid transfer among bacteria (Schütt 1991), transduction with temperate phages seems to be a major mechanism of gene exchange (Sherr 1989). There is evidence that, at least among lactic acid bacteria, temperate phages transduce not only their own viral genes but also mobilized chromosomal and plasmid DNA of their hosts (Davidson et al. 1990, Hill et al. 1991). However, the ecological role of viruses in aquatic habitats is still enigmatic. Molecular techniques are required to detect and trace nucleic acids of phage in bacterial host genomes and among natural bacterioplankton. Recent advances in molecular biology developed excellent tools for the elucidation of the ecological role of phages.

This work provides evidence for the abundance of one typical temperate phage/host system among the natural bacterial community of Lake Plußsee. Three phage/host systems were isolated from the lakes Plußsee and Belauersee; both lakes are located in Schleswig-Holstein. Temperate phages and their bacterial hosts were traced by nonradioactive DNA–DNA hybridization. Data on the occurrence and seasonal distribution of temperate phages and their lysogenic hosts present the first hints of their importance in natural gene transfer. This chapter gives no details; they will be published elsewhere.

The elucidation of the role of temperate phages in an aquatic environment requires the detection of genetic differences between uninfected host bacteria and lysogenic bacteria harboring temperate phages. This is particularly important since temperate phages are invisibly integrated into their bacterial host genomes. Usually, morphological differences between uninfected and infected cells cannot be observed. Unquestionably, classical microbiological techniques do not allow rapid tracing and identification of host bacteria and their virus particles in natural bacterial communities. Molecular approaches are efficient, rapid, specific, and sensitive in tracing nucleic acids of bacteria and phages unambiguously. A key technique is DNA probing, which includes the isolation of phage/host systems, phage purification, DNA extraction and electrophoretic separation, nonradioactive probe labeling, and detection of hybrid DNA. A major assumption for this approach was that there is no genetic relatedness between bacterial hosts and their phages. This was examined first. Specific DNA probes of phages and their hosts are required to detect phage/host systems in nature. Many ecologists may not be acquainted with molecular approaches; hence, it may be helpful to give some general information on major techniques we applied here.

Materials and Methods

The investigation focused on the abundance of three typical bacterial hosts and their bacteriophages among the natural bacterial population during fall 1989. Phage/host systems were recovered from the Plußsee and the Belauersee (see above). Figure 13.3 illustrates the major experimental steps for the detection of lysogens and their temperate phages.

Four water samples (taken at a depth of 1 m) were collected during fall 1989 from the Plußsee. Three hundred sixty-six representative bacterial colonies were recovered from the water samples. Isolates were examined for lysogeny and the occurrence of temperate phages by DNA–DNA hybridization.

Three phage/host systems were used for the experiments; they were collected from the Plußsee (Ps) and the Belauersee (Bs). Host bacteria and phages were bacterium B2 (Ps) and its phage P2 (Bs), B18 (Ps) and its phage P18 (Ps), and BR116 (Bs) and its phage PR116 (Bs) (laboratory designations). Isolates B2 and B18 were yellow, and BR116 pinkish pigmented. None of the bacterial hosts carried plasmid DNA.

Genomic DNA was extracted from bacterial hosts B2, B18, and BR116. For preparation of genomic probes, crude DNA extracts were separated by gel electrophoresis. Genomic DNA, purified by two-step cesium chloride gradient centrifugation, was subjected to restriction analysis. Purified bacterial genomes of isolates B2, B18, and BR116 were digested for genome comparison with HindIII restriction endonuclease.

Phage-host system

↓

Development of DNA probes
of bacterial hosts and its phages

↓

Genetic relatedness
between bacteria and phages
by cross-hybridization

↓

Temperate phages and lysogens
among phage-host systems
detection of phage DNA in host genomes
(restriction, electrophoretical separation,
Southern DNA-transfer, hybridization)

↓

Induction of lysis by UV-light

↓

Isolation of native bacteria
(preparation of master plates)

↓

Distribution of phage-host systems
by colony hybridization
(colony transfer, to nylon membranes, alkaline cell lysis,
DNA-denaturation to single strand, DNA-fixation, hybridization)

Figure 13.3. Diagram showing the major experimental steps to examine the distribution of lysogenic host bacteria and its temperate phages (phage/host systems were provided by K. P. Witzel).

Phages P2, P18, and PR116 were propagated in bacterial hosts B2, B18, and BR116 on CPS agar according to Adams (1959). Phages were purified by cesium chloride gradient centrifugation (Sambrook et al. 1989) and digested by HindIII restriction endonuclease.

Target DNA from hosts and phages was separated by pulsed-field gel electrophoresis with a clamped homogeneous electric field system. Genomic and phage target DNA were transferred from gels to nylon blots by vacuum transfer. Purified genomic DNA and phage DNA were selected for preparation of probes. Host DNA was isolated according to Sambrook et al. (1989) and labeled by random primed techniques with nonradioactive digoxigenin-11-dUTP (Haas and Fleming 1986). Genetic relatedness and probe specificity between bacterial hosts and their phages were determined by cross-hybridization on nylon filters. Each DNA species of the three hosts and their phages was used as target and as probe.

Three hundred sixty-six unidentified isolates of the Plußsee water samples were examined for the presence of host strains B2, B18, and BR116 and their phages P2, P18, and PR116 by DNA–DNA colony hybridization (Schütt 1990). Water samples were plate spread and incubated. Colonies were transferred to agar masterplates and incubated. The colonies on masterplates were transferred to nylon sheets. Bacteria adhering to membranes were lysed. Bacterial DNA was denatured by alkaline treatment, neutralized, air dried, and fixed to membranes by UV cross-linking. Hybridizations were performed in plastic bags under stringent conditions. Hybrid DNA was detected on filters after binding of antidigoxigenin alkaline phosphatase conjugate. Washing steps removed nonhybridized DNA probes. Subsequent enzyme-catalyzed color reaction with 5-bromo-4-chloro-3-indolyl phosphate (x-phosphate) and nitroblue tetrazolium salt (NBT) produced blue precipitate, indicating hybrid DNA (Boehringer provides complete protocols). Specificity of genomic probes can be an obstacle. Homologous genes are common among related bacteria, where probes do not discriminate between related isolates. To avoid misinterpretation of ambiguous detection signals, the pigmented strains B2, B18, and BR116 were also examined for colony appearance and cell morphology. The DNA species from different bacteria and phages were subjected to restriction analysis. The DNA probes were specific, and signals were generated.

Strains B2, B18, and BR116, infected by phages P2, P18, and PR116, were examined for lysogeny. Surviving cells were cultivated in CPS broth, harvested, and washed. A dilution series was plate spread on agar and incubated. Colonies were transferred to masterplates and incubated. Resulting colonies from masterplates were transferred to nylon membranes. DNA of uninfected host bacteria was used as negative and phage DNA as positive controls. Digoxigenin-11-UTP-labeled DNA from all three phages served as probes.

Results

The results comprised three aspects: (1) probe specificity as evidence that phage DNA was not integrated in uninfected genomes of host bacteria; (2) presence of lysogenic bacteria and their temperate phages among the phage/host systems used; (3) abundance and distribution of hosts and temperate phages in bacterioplankton of the Plußsee.

Cross-Hybridization of Phage and Host DNA

A prerequisite for the experiments was the absence of DNA sequences of phages P2, P18, and PR116 in the uninfected host bacteria B2, B18, and BR116. This examination was essential to discriminate between host bacteria and their phages. All three phages contained double-stranded DNA, which was DNase I sensitive and RNase A stable. The yellow-pigmented bacterial hosts B2 and B18 were closely related genetically (positive signals for both strains). These isolates even displayed identical restriction patterns. However, isolate BR116 did not show relatedness to the other B2/B18 stains.

Lysogenic Bacteria among Phage Host Systems

The phage probes P2, P18, and PR116 were used for the detection of temperate phages in the genomes of their infected host bacteria. None of them could be digested with restriction enzymes. One successful attempt unveiled DNA of temperate phage P18 in the genome of its infected host bacterium B18. Lysogens of B2 and BR116 contained phages P2 and PR116, respectively, and could not be detected. Temperate phage P18 was integrated in the genome of bacterial host B18. The membrane was loaded with restricted DNA of infected host bacterium B18 and hybridized with phage P18 probe, resulting in one band of approximately 23 kb, representing phage P18. No hybrid DNA was detected in the uninfected strain B18. Lysis of the infected host B18 was induced by UV light treatment. Plaque hybridization confirmed that all plaques were generated by bacteriophage P18.

Abundance of Phage/Host Systems among the Natural Bacterial Community of the Plußsee

Genomic and phage probes were applied to identify B2/B18 and BR116 strains and their phages P2, P18, and PR116 among bacterioplankton of the Plußsee.

Three hundred sixty-six unidentified bacteria from the Plußsee were collected on CPS masterplates. Each colony was transferred from the masterplates to different sets of nylon membranes and hybridized with genomic DNA of B18 and BR116. The other sets of filters were hybridized with probes of phages P2, P18, and PR116. Results of all hybridizations are given in Table 13.4. Host B18 represents the dominant bacterium among culturable bacterioplankton. Most of the B18 isolates were lysogens and contained phage P18. The data imply seasonal shifts in total numbers of B18 strains and in numbers of the B18 lysogen fractions. During the investigation during fall 1989, the other bacterial host, BR116, and the phages P2 and PR116 seem to play a minor ecological role. Host strain BR116 was not detected among the isolates, and phage P2 was only detected in a few isolates of samples from October 17 (8% of the isolates) and October 27, 1989 (2%). Their

Table 13.4. Detection of Host Bacteria B2/B18 and Their Phages, Which Contain Lysogens, from among 366 Isolates of Bacteria of the Plußsee during Fall 1989[a]

Sampling Date (1989)	Number of Isolates	Host Bacterium B18 (%)	Infected B18 (%)
October 17	114	56	91
October 27	111	83	97
November 17	98	73	70
December 1	43	54	52

[a]Host strain B2 shows genetic homology with bacterium B18. Only a few isolates harbored phage P2 (8% in samples from October 17 and 2% from October 27, 1989). Their host strains were not B2/B18. Host, BR116, and phage PR116 were not detected among the isolates.

bacterial hosts were not the B2/B18 type. Phage PR116 was detected in none of the 366 isolates.

Discussion

Colony hybridization experiments unveiled the occurrence and distribution of temperate phage P18 among freshly isolated bacteria collected during fall 1989 from the Plußsee. The examination displayed extremely high numbers, between 54% and 85%, for dominant strain B2/B18 type of 366 representative bacterial isolates. A striking majority, between 52% and 97%, were lysogens containing detectable DNA of bacteriophage P18. From the experiments it is not clear whether parts or the whole phage DNA was integrated in bacterial host genomes. Lysogeny of the P18/B18 phage/host system seems to be host specific, since hybridization data showed that only yellow-pigmented B18 strains were lysogenized. No other Plußsee isolates contained P18 phage DNA. A narrow host range of temperate phages seems to be frequent. Colony hybridization results with DNA probes of bacteriophages P2 and PR116 confirmed this indirectly. Bacteriophage P2 was detected in very few bacterial colonies, which differed considerably from the strain B2/B18 type. None of the colonies tested harbored DNA of PR116 bacteriophage. Literature data also reported a narrow host range for lactic acid bacteria (Davidson et al. 1990). There is a first hint of the existence of habitat-specific phage/host systems. Phage/host system BR116/PR116 was absent among the bacterial population of the Plußsee. This system originated from the Belauersee, a different limnological environment.

The present state of interpretation suggests seasonal changes in numbers of host and phages. Samples from January 1990 indicated considerably decreased numbers of yellow-pigmented bacteria (data not shown). This is consistent with the findings of Witzel et al. (1982), who showed seasonal variation of bacterial communities in the Plußsee. However, structural changes in bacterial communities also may alter the spectrum of bacteriophages. Bergh et al. (1989) reported seasonal shifts in the concentration of viruses in natural waters.

Detection of high numbers of lysogenic strain B18, which harbored temperate phage P18, suggests dynamic genetic transfer within this particular phage/host system. Ogunseitan et al. (1990) also found dynamic genetic interactions among phages and host bacteria in their microcosm experiments. Phages may play a similar role in genetic adaptation processes as mobile plasmid DNA. It can be assumed that transfer of phage or plasmid DNA plays a significant role in genetic adaptation processes in natural bacterioplankton. The gain of additional, useful physiological and regulatory activities provides a selective advantage over other organisms that lack these useful phage- or plasmid-encoded traits (Reanney 1977, Schütt 1990). The genetic adaptation to extreme habitats and to natural or man-made environmental changes can be interpreted as survival strategy. Gene transfer is essential for horizontal evolution (Anderson 1966). Lack of gene transfer within natural bacterial communities would decrease genetic diversity (Stewart and Carlsson 1986). It is very likely that genetic adaptation processes strongly influence structures and functions of bacterioplankton. Unfortunately, physiological and/or regulatory functions of naturally transferred nucleic acids are still enigmatic. Further research on other phage/host systems and decoding the presently unknown genetic functions will be among the major future tasks in microbial ecology.

Acknowledgments. We thank Prof. Dr. J. Overbeck, Max-Planck Institute of Limnology, Plön, for continuous support. Help and supervision with electron microscopy by Horst Neve and the expert technical assistance of Liselotte Schöler and Antje Hay are gratefully acknowledged. Part of the work was supported by funds from the Ministry of Research and Technology (F.R.G.).

References

Ackermann H-W, DuBow MS (1987) *Viruses of Prokaryotes, vol I.* CRC Press, Boca Raton, Florida

Adams MH (1959) *Bacteriophages.* Interscience, New York

Ahrens R (1971) Untersuchungen zur Verbreitung von Phagen der Gattung *Agrobacterium* in der Ostsee. Kiel Meeresforsch 27:102–112

Anderson ES (1966) Possible importance of transfer factors in bacterial evolution. Nature 209:637–638

Beachem IR (1987) Silent genes in procaryotes. FEMS Microbiol Rev 46:409–417

Bergh O, Børsheim KY, Bratbak G, Heldal M (1989) High abundance of viruses found in aquatic environments. Nature 340:467–468

Bratbak G, Heldal M, Norland S, Thingstad TF (1990) Viruses as partners in spring bloom microbial trophodynamics. Appl Environ Microbiol 56:1400–1405

Cochlan WP, Wikner J, Steward GF, Smith DC, Azam F (1993) Spatial distribution of viruses, bacteria and chlorophyll *a* in neritic, oceanic and estuarine environments. Mar Ecol Prog Ser 92:77–87

Collins VG, Willoughby LG (1962) The distribution of bacteria and fungal spores in Blelham Tarn with particular reference to an experimental overturn. Arch Mikrobiol 43:294–307

Cottrell MT, Suttle CA (1991) Wide-spread occurrence and clonal variation in viruses which cause lysis of a cosmopolitan, eukaryotic marine phytoplankter, *Micromonas pusilla.* Mar Ecol Prog Ser 78:1–9

Davidson BE, Powell IB, Hillier AJ (1990) Temperate bacteriophages and lysogeny in lactic acid bacteria. FEMS Microbiol Rev 87:79–90

Demuth J, Neve H, Witzel K-P (1993) Morphological diversity of bacteriophage populations in lake Plußsee, studied by direct electron microscopy. Appl Environ Microbiol 59: 3378–3384

González JM, Suttle CA (1993) Grazing by marine nanoflagellates on viruses and virus-sized particles: Ingestion and digestion. Mar Ecol Prog Ser 94:1–10

Haas MJ, Fleming D (1986) Use of biotinylated DNA probes in colony hybridization. Nucleic Acid Res 14:3976

Hara S, Terauchi K, Koike I (1991) Abundance of viruses in marine waters: Assessment by epifluorescence and transmission electron microscopy. Appl Environ Microbiol 57:2731–2734

Heldal M, Bratbak G (1991) Production and decay of viruses in aquatic environments. Mar Ecol Prog Ser 72:205–212

d'Hérelle F (1917) Sur un microbe invisible antagoniste des bacilles dysentériques. CR Acad Sci Paris 165:373–375

Hill C, Miller LA, Klaenhammer TR (1990) Cloning, expression, and sequence determination of a bacteriophage fragment encoding bacteriophage resistance in *Lactococcus lactis.* J Bacteriol 172:6419–6426

Hill C, Miller LA, Klaenhammer TR (1991) In vivo genetic exchange of a functional domain from a type IIA methylase between lactococcal plasmid pTR2030 and a virulent bacteriophage. J Bacteriol 173:4363–4370

Logan KB, Rees GE, Seeley ND, Primrose SB (1980) Rapid concentration of bacteriophages from large volumes of freshwater: Evaluation of positively charged, microporous filters. J Virol Methods 1:87–97

Moebus K (1991) Preliminary observations on the concentration of marine bacteriophages in the water around Helgoland. Helgoländer Meeresunters 45:411–422

Moebus K (1992) Further investigations on the concentration of marine bacteriophages in the water around Helgoland, with reference to the phage-host system encountered. Helgoländer Meeresunters 46:275–292

Ogunseitan OA, Sayler GS, Miller RV (1990) Dynamic interactions of *Pseudomonas aeruginosa* and bacteriophages in lake water. Microb Ecol 19:171–185

Padan E, Shilo M (1973) Cyanophages—Viruses attacking blue-green algae. Bacteriol Rev 37:343–370

Primrose SB, Day M (1977) Rapid concentration of bacteriophages from aquatic habitats. J Appl Bacteriol 42:417–421

Proctor LM, Fuhrman JA (1990) Viral mortality of marine bacteria and cyanobacteria. Nature 343:60–62

Proctor LM, Fuhrman JA (1992) Mortality of marine bacteria in response to enrichments of the virus size fraction from seawater. Mar Ecol Prog Ser 87:283–293

Purdy RN, Dancer BN, Day MJ, Stickler DJ (1984) A novel technique for the enumeration of bacteriophage from water. FEMS Microbiol Lett 21:89–92

Romashko AM, Bylinskii AF (1992) Analysis of phage resistance of pseudomonads. Microbiology 61:337–342

Reanney DC (1977) Gene transfer as a mechanism of microbial evolution. Bioscience 27:340–344

Sambrook J, Fritsch EF, Maniatis T (1989) *Molecular Cloning: A Laboratory Manual.* Cold Spring Harbor Laboratory, Cold Spring Harbor, New York

Schütt C (1990) Plasmids and their role in natural aquatic bacterial communities. In Overbeck J, Chróst RJ (eds) *Aquatic Microbial Ecology: Biochemical and Molecular Approaches.* Springer-Verlag, New York, pp 160–183

Schütt C (1991) Ecogenetics: A new concept of aquatic microbial ecology at genetical level. Verh Int Ver Limnol 24:2593–2597

Seeley ND, Hallard G, Primrose SB (1979) A portable device for concentrating bacteriophages from large volumes of freshwater. J Appl Bacteriol 47:145–152

Sherr EB (1989) And now, small is plentiful. Nature 340:429

Stal LJ, Bock E, Bower EJ, et al. (1989) Cellular physiology and interactions of biofilms. In Characklis WG, Wilderer PA (eds) *Structure and Function of Biofilms.* John Wiley & Sons, Chichester, pp 269–289

Stewart GJ, Carlsson CA (1986) The biology of natural transformation. Annu Rev Microbiol 40:211–235

Suttle CA, Chan AM (1993) Marine cyanophages infecting oceanic and coastal strains of *Synechococcus:* Abundance, morphology, cross-infectivity and growth characteristics. Mar Ecol Prog Ser 92:99–109

Suttle CA, Chan AM, Cottrell MT (1990) Infection of phytoplankton by viruses and reduction of primary productivity. Nature 347:467–469

Suttle CA, Chan AM, Cottrell MT (1991) Use of ultrafiltration to isolate viruses from seawater which are pathogens of marine phytoplankton. Appl Environ Microbiol 57:721–726

Torrella F, Morita RY (1979) Evidence by electron micrographs for a high incidence of bacteriophage particles in the waters of Yaquina Bay, Oregon: Ecological and taxonomical implications. Appl Environ Microbiol 37:774–778

Twort FW (1915) An investigation on the nature of ultra-microscopic viruses. Lancet 2:1241–1243

Weisse T, Müller H, Pinto-Coelho R, Schweizer A, Springmann D, Baldringer G (1990) Response of the microbial loop to the phytoplankton spring bloom in a large prealpine lake. Limnol Oceanogr 35:781–794

Wiggins BA, Alexander M (1985) Minimum bacterial density for bacteriophage replication: Implications for significance of bacteriophages in natural ecosystems. Appl Environ Microbiol 49:19–23

Witzel K-P, Overbeck J, Moaledj K (1982) Microbial communities in lake Plußsee—an analysis with numerical taxonomy of isolates. Arch Hydrobiol 94:38–52

Wommack KE, Hill RT, Kessel M, Russek-Cohen E, Colwell RR (1992) Distribution of viruses in the Chesapeake Bay. Appl Environ Microbiol 58:2965–2970

14. Phototrophic Bacteria in the Plußsee: Ecology of the Sulfuretum

Rita Oberhäuser-Nehls, Konstantinos Anagnostidis, and Jürgen Overbeck

Introduction

A eutrophic lake like the Plußsee is divided into the aerobic epilimnion, the metalimnion as defined by the thermocline, and the anaerobic hypolimnion. In the upper zone of the hypolimnion phototrophic sulfur bacteria are temporarily abundant if there is a supply of light and H_2S. They are represented by the families Chromatiaceae (purple sulfur bacteria) and Chlorobiaceae (green sulfur bacteria). Their ability to oxidize reduced sulfur compounds without oxygen consumption is ecologically very important for the lake metabolism. In contrast to the phototrophic sulfur bacteria, the Rhodospirillaceae (purple nonsulfur bacteria) rarely occur in spectacular mass accumulations. In the dark part of the hypolimnion, moreover, a great variety of different colorless anaerobic bacteria are present; among them are sulfate reducers, which depend on a supply of substrate from phototrophic bacteria.

The "sulfur bacteria" do not represent a homogeneous taxonomic group but rather an ecologically defined one (Bavendamm 1924, Schwartz 1958). Their common feature is the ability to settle in the biotopes that are more or less rich in hydrogen sulfide. In his report "Die sapropelische Lebewelt" Lauterborn (1915) described several new genera and species derived from smaller ponds, and Koppe (1924) was the first to study the organisms living in sediments of East Holstein lakes. Hence, numerous notes on the occurrence of planktonic sulfur bacteria in lakes with an anaerobic hypolimnion temporarily containing hydrogen sulfide are found in Utermöhl's (1925) "Limnologische Phytoplanktonstudien." This kind of biotope, low in H_2S content and inhabited by a diversity of bacterial assemblages, was designated as Sulfuretum (Baas-Becking 1925).

In East Holstein lakes, these conditions appear during summer stagnation and induce development of a significant thiobiocoenosis (Anagnostidis and Overbeck 1966). The development of both thiobiotope and thiobiocoenosis is initiated by the formation of anaerobic zones arising from the sediment. Bacterial sulfate reduction in the anaerobic hypolimnion leads to the formation of H_2S in the deeper parts of the lakes. The sulfide, on the other hand, gives rise to the development of blooms of phototrophic bacteria by means of light energy. A number of review articles on the ecology of phototrophic bacteria have appeared recently (Pfennig 1978, Pfennig and Trüper 1981, Walsby 1981, Van Gemerden and Beeftink 1981, 1983, Gorlenko et al. 1983, Tindall and Grant 1986, Lindholm 1987).

The phototrophic and colorless bacteria connected with the sulfur cycle of the Plußsee as well as their accompanying microorganisms are the subjects of the first part of this chapter. The ecology of purple nonsulfur bacteria is discussed in the second part of this chapter.

Materials and Methods

Sampling

In the Plußsee, weekly studies were carried out between March 1977 (melting of the ice cover) and December 1978. Samples were collected at a site 25 m deep; sampling depths were adapted to the lake's stratification conditions. Water samples for all investigations were taken by means of a 2-L Ruttner sampler. Glass or polyethylene bottles were filled to the top to exclude oxygen from air bubbling and kept cool and dark until further processing in the laboratory.

Physical and Chemical Analysis

Temperature profiles were recorded from a continuously operating station in the Plußsee or directly read from a mercury thermometer installed in the Ruttner sampler. Water transparency was estimated with a Secchi disk, diameter 25 cm. Light attenuation rates were recorded by means of a selenium underwater photometer (Dr. Lange, Berlin) for "white light" (disk of opal glass) as well as for the red, green, and blue spectral ranges with the corresponding glass filters (RG 2: 660 nm; VG 9: 525 nm; BG 12: 435 nm; Schott, Mainz). Conductivity, pH, and oxygen of water samples were determined routinely. Redox potential was determined under continuous N_2 flow by means of a redox electrode (Pt 405–85; Ingold, Frankfurt), which was connected with a potentiograph E 436 (Metrohm, Herisau, Switzerland). All data represent relative values without correction for temperature or pH. Reproducible data on the electrode potentials were controlled by means of redox buffers (Ingold, Frankfurt). Sulfide (in samples fixed immediately after collection with Zn-acetate) was determined by conversion to methylene blue after addition of dimethyl-*p*-phenylendiamine. Samples were measured photometrically against a blank that contained 35% H_2SO_4 instead of the reagent (*Deutsche Einheitsverfahren,* D 7, 1975). Inorganic nutrients (nitrite, nitrate, ammonia, phosphate, and sulfate) were determined with the use of Autoanalyzer (Technicon, Frankfurt) according to Albrecht and Overbeck (1969) and Albrecht (1973).

Uptake of Organic Substrates

The uptake of organic compounds by heterotrophic bacteria was studied by means of labeled substrates (Wright and Hobbie 1965, 1966). In order to estimate the kinetic parameters, 25, 50, 100, or 200 µl of substrates were added to 50 ml of lake water. Each isotope was diluted to 1 µCi/ml (Amersham Buchler, Braunschweig): D-[U-^{14}C]glucose, sp. act. 16.7 µCi/mg (30, 60, 120, and 240 µg/L); D-[U-^{14}C]lactate (Na-salt), sp. act. 186 µCi/mg (2.5, 5, 10, and 20 µg/L); and [U-^{14}C]acetate (Na-salt), sp. act. 696 µCi/mg (0.72, 1.44, 2.88, and 5.76 µg/L).

All experiments were done with three parallels plus one blank, which was fixed immediately with 0.5 ml Lugol's solution; incubation time was 1 h (15°C). The samples were filtered through 0.2-µm pore-size membrane filters, and their radioactivity was measured after the addition of 10 ml Liquid Scintillator KL 454 (Zinsser, Frankfurt/M.), on a scintillation counter (Berthold BF 5000).

For the various substrate concentrations, the uptake velocity *v* was determined using the formula of Parsons and Strickland (1962). Additionally, the "direct uptake" (Wright 1978) was determined by using a single but higher substrate concentration (500 µl of substrate solution, 0.5 µCi, 600 µg glucose/L, 50 µg lactate/L, and 14.4 µg acetate/L). In this case the kinetic approach, after an incubation time of 20 min at 15°C, was calculated according to the formula

$$v\ (\mu g\ L^{-1}\ h^{-1}) = (f \times A)/t$$

in which *f* means that fraction of the added substrate *A* taken up during the incubation time *t*.

As standard substrate for the determination of the "heterotrophic potential" D-[U-^{14}C]glucose is used routinely. As is well known, the actual heterotrophic microbial metabolic activity cannot be determined by applying only one substrate, and often glucose is not an optimal substrate. But there is apparently a clear correlation between primary production of phytoplankton or concentration of dissolved organic material and the metabolic activity of bacteria (Gocke 1977; Overbeck 1979).

Monthly measurements to characterize the heterotrophic activity in the sulfuretum of the Plußsee were carried out between August and December 1978 at depths of 5 m (upper metalimnion), 12 m (upper border of the phototrophic bacteria), 17 m (maximum of phototrophic bacteria), and 25 m (above the sediment). The correlations between uptake velocities and substrate concentrations followed sigmoid curves in numerous cases; however, part of the results could not be interpreted at all. Only 60% of the experiments yielded saturation curves that corresponded approximately to Michaelis–Menten kinetics. The investigation of glucose uptake by Overbeck (1973, 1975) in Plußsee showed similar results.

Results

Physical and Chemical Properties of Lake Water and Phytoplankton

Temperature

Figure 14.1 shows the temperature conditions in the Plußsee during 1977 and 1978. In March 1977, the temperature (4°C) was nearly homogeneous throughout the vertical profile. The increase of the surface temperature in spring led to the development of thermal stratification. During the summer months (June–August) the thermocline was found between 2 and 4 m. The temperature of the epilimnion persisted above 18°C and reached maximum values of 21–22°C in the middle of June. The hypolimnion below 6–7 m had a temperature of about 4–6°C. In September–October, a progressive cooling down of the epilimnion took place, followed by a wind-induced drop of the thermocline, which led to complete circulation at the end of December (temperature at the surface 4.2°C, above sediment 3.9°C). In January 1978, under a temporary ice cover, a winter stratifica-

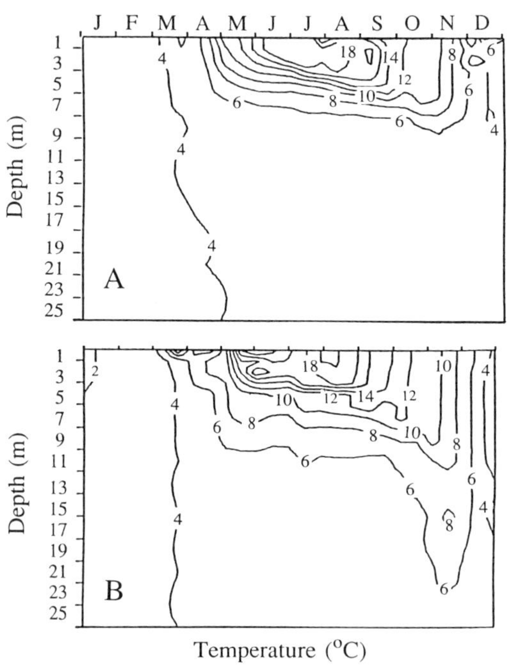

Figure 14.1. Spatial and temporal variations of Plußsee water temperature in (A) 1977 and (B) 1978. Isolines of temperature are at 2°C intervals.

tion developed showing an inverse temperature profile (2.3°C at the surface, 3.4°C above the sediment), which was discharged by a spring turnover in February.

The development of the summer stratification in 1978 occurred in a similar manner, although the metalimnion exceeded by far its size of the year before (ca. 3 m in 1977 and 6 m in 1978). The 6°C isoline stayed 9–10 m deep throughout the summer and sank down into deeper layers by the first days of October. In 1978 the mixing of the water column increased gradually, finally leading to isothermic conditions (4°C) at the end of December, which was in contrast with the sudden fall turnover in 1977.

Transmission of Light and Phytoplankton

The attenuation of light intensity with water depth was measured as a percentage of surface values during the summer stagnation of 1978. As usual, the red and especially the green fractions exceeded the transmission of the blue light. In general, the transmission of light in lake water did not depend only on light-absorbing dissolved substances but was mostly influenced by the seasonal variations in composition and density of phytoplankton.

In the Plußsee the following successions of phytoplankton could be distinguished. The first algal bloom in April–May consisted of diatoms (*Asterionella* sp., *Stephanodiscus* sp.) and Cryptomonads (*Cryptomonas* sp., *Rhodomonas* sp.); the Secchi disk transparency at that time was very low (<1.0–1.5 m). The blue light, mostly absorbed by the algal chlorophyll$_a$ (absorption maximum ca. 450 nm), reached the 1% level by a depth of 3 m. The breakdown of the spring phytoplankton in May–June was accompanied by an increase in water transparency (3.5–4.1 m). In this clear-water phase the 1% value of blue light was measured at a depth of 7 m. During early summer (June–July) a highly diverse phytoplankton community developed (*Aphanizomenon* sp., *Gomphosphaeria* sp., *Cryptomonas* sp., as well as several Chlorophytes). Again the Secchi disk visibility decreased to 1.0–1.4 m, and the blue light (1% value) penetrated to a depth of only 3 m. This community of high diversity was followed by *Aphanocapsa* sp. in September. Water transparency amounted to 2.4–3.0 m, and the blue light reached its 1% value at a depth of 5 m. During winter, after the disappearance of phytoplankton, a Secchi disk transparency of 4.6 m was measured.

pH and Conductivity

The maximum pH values in the epilimnion increased to above 9.0 (Fig. 14.2) simultaneously with the phytoplankton blooms in spring and summer. In the hypolimnion pH values fluctuated between 7.0 and 7.6 throughout the whole depth profile.

Conductivity in the Plußsee reached about 260 μS_{20}/cm during circulation periods. During summer stratification, values increased up to 290 μS_{20}/cm above the sediment and decreased to 220–240 μS_{20}/cm in the trophogenic layer (Fig. 14.3).

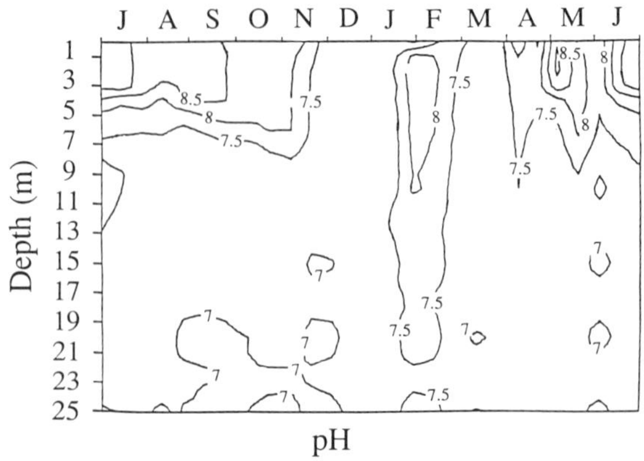

Figure 14.2. Spatial and temporal variations of pH in the Plußsee in 1977–1978. pH isolines are at pH 7.0, 7.5, 8.0, 8.5, and 9.0.

Oxygen

After the melting of the ice cover in March 1977, the oxygen amounted to only 58% of saturation at the surface and to about 5% near the sediment (Fig. 14.4A). The oxygen deficiency in the lower parts of the lake was not equilibrated by a spring circulation. In April 1977, the surface waters gained a supersaturation of up to 150% (i.e., 14 mg O_2/ml, Fig. 14.4A) from high phytoplankton photosynthetic activity.

At the same time, a complete oxygen depletion at the sediment–water interface began; it extended to the upper strata of the hypolimnion and was to a great extent caused by methane-oxidizing bacteria (Anagnostidis and Overbeck 1966). The oxygen curve ran parallel to the temperature curve with a clinograde profile typical for eutrophic lakes (Åberg and Rodhe 1942). The oxygen content decreased rapidly in the metalimnion, and the hypolimnion below 8 m in depth remained anaerobic throughout the summer months. After the fall turnover in the

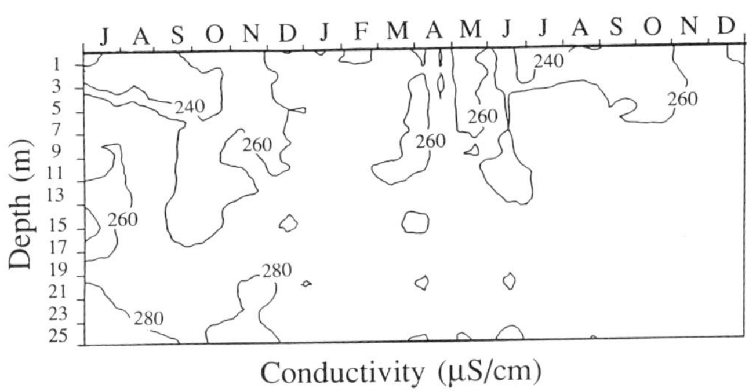

Figure 14.3. Variations of conductivity of water in the Plußsee in 1977–1978.

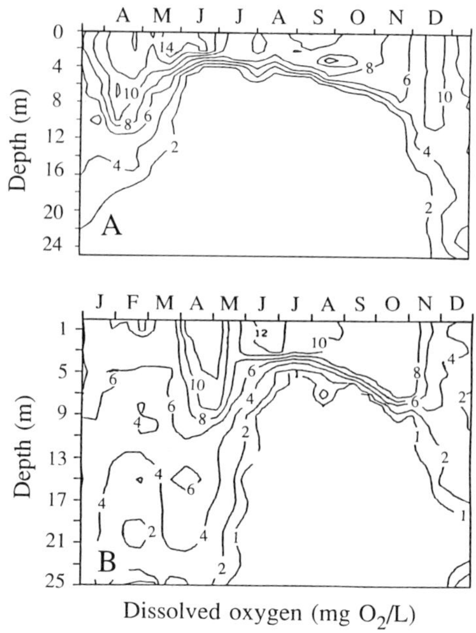

Figure 14.4. Concentration of dissolved oxygen in the entire water column of the Plußsee in (A) April–December 1977 and (B) January–December 1978.

middle of December, the whole water column was oxygenated (19% saturation). Oxygen saturation increased to 41% in January and again decreased to 3–4% above the sediment under the ice cover of the lake in winter 1977–1978. However, under the ice in the light gradient, because of the activity of winter phytoplankton, the saturation reached 78%. After spring turnover saturation values between 46% and 50% were observed.

A comparison of the two oxygen profiles of 1977 and 1978 reveals a much more extended anaerobic hypolimnion in the first year in the temporal as well as in the spatial dimensions. In 1978, the depletion of oxygen was much slower as a result of different climatic conditions (Fig. 14.4B); e.g., in 1977 the 2-mg isoline reached a depth of 5 m in early June. In 1978, because of stormy weather, it reached a depth of 5 m 1 month later. The different meteorological conditions are also reflected in the temperature profiles.

Hydrogen Sulfide

The isolines of hydrogen sulfide (Fig. 14.5) are essentially complementary to those of oxygen. Sulfide first appeared in lower concentrations, increasing with time and expanding to higher strata, which coincided with the consumption of oxygen above the sediment surface. At the time of complete thermal stratification the whole hypolimnion was filled with H_2S. The highest concentrations (ca. 3 mg S^{2-}/L) were measured above the sediment in August and September 1977. The sulfide was oxidized with the successive aeration of the hypolimnion in autumn. During winter stagnation and under ice cover, small amounts of sulfide (0.02–

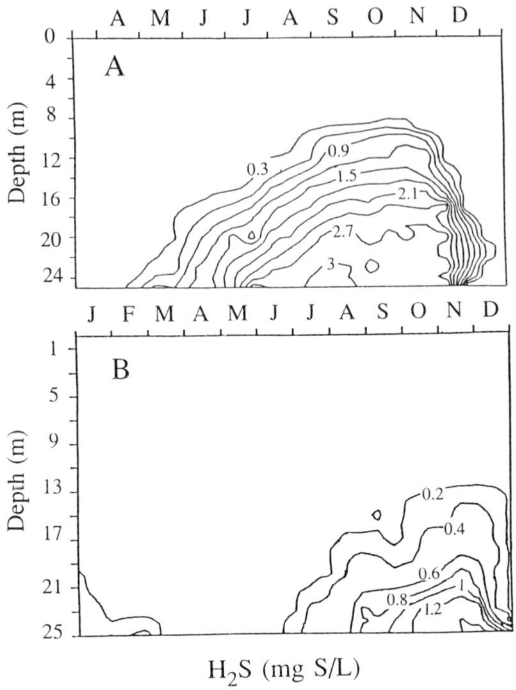

Figure 14.5. Spatial and temporal distribution of H_2S in the entire water column of the Plußsee in (A) April–December 1977 and (B) January–December 1978.

0.15 mg S^{2-}/L) were measured above the sediment, although some oxygen was present at these depths.

In spring 1978 the sulfide production was retarded in comparison to 1977; a maximum value of only 1.7 mg S^{2-}/L was not reached until the end of November. During the 1978 summer stagnation, the sulfide rose to an upper border of only 13 m depth under a 3- to 4-m-wide region that lacked both H_2S and O_2.

Biological oxidation of sulfide is carried out by phototrophic bacteria under anaerobic conditions or by chemoautotrophic thiobacilli under aerobic conditions. Investigations of Van Gemerden (1967) led to the conclusion that in the Plußsee microbial oxidation of sulfide was done mainly by bacterial photosynthesis. There were diurnal variations of the sulfide concentrations in the upper hypolimnion, where sulfide was present at night but not detectable during the day.

The abiotic sulfide oxidation by dissolved oxygen does not occur until the end of an induction phase, depending on such reaction conditions as initial concentrations or pH (Chen and Morris 1972). Hence, it is possible that H_2S and O_2 may coexist in one system. This coexistence mainly favors the aerobic *Thiobacillus* and *Beggiatoa* species, whose metabolism requires both dissolved gases. The simultaneous occurrence of small amounts of O_2 and H_2S is well known, e.g., in the Black Sea (Sorokin 1972) and in Solar Lake (Jørgensen et al. 1979). In stromatolites high O_2 production by cyanobacterial photosynthesis as well as an intensive sulfate reduction by *Desulfovibrio* sp. led to a simultaneous occurrence of 0.4 mg O_2/L and up to 25 mg H_2S/L (Krumbein et al. 1979).

Redox Potential

The redox potential describes the intensity of electron-transferring processes in a system. It depends on temperature as well as on pH values and represents an index for the ratio of oxidizing to reducing matter in that system without giving evidence of their absolute concentrations (Allgeier et al. 1941, ZoBell 1946). The most important inorganic factors influencing the redox potential in waters are oxygen, hydrogen sulfide, iron, and manganese. Additionally, it is affected by dissolved organic material that can be oxidized by heterotrophic microorganisms, thus leading to a decrease of redox values.

A quantitative interpretation of redox potentials in a natural habitat is not possible, because the measured values at best represent a diversity of potentials. A detailed discussion of methodological problems and limits of interpretation is given by Stumm (1967), who in spite of all limitations concludes, "such an 'apparent electrode potential' is a valuable environmental parameter even if the readings have no value an analytical chemist would desire."

The data of the Plußsee (Fig. 14.6) should be interpreted in this sense. The isolines show a substantial relationship to seasonal changes in oxygen and hydrogen sulfide. When O_2 concentrations in water increased during intense photosynthetic activity of spring or summer phytoplankton, highly positive redox values were measured in the epilimnion. In contrast, during summer stagnation with successive O_2 depletion and increasing H_2S production, the hypolimnion revealed negative electrode potentials.

The boundary between aerobic and anaerobic layers, clearly defined only in 1977, was approximately parallel with the 0-mV isoline (Fig. 14.6). Redox potentials in the fully aerated lake varied between +20 and +50 mV during the winter months.

Eventually, the redox values also reflected the diverging O_2 and H_2S conditions during summer stagnation in 1977 and 1978 (Figs. 14.4 and 14.5). The low sulfide concentrations above the sediment as well as an anaerobic layer lacking H_2S in the upper hypolimnion resulted in a less pronounced redox gradient in 1978.

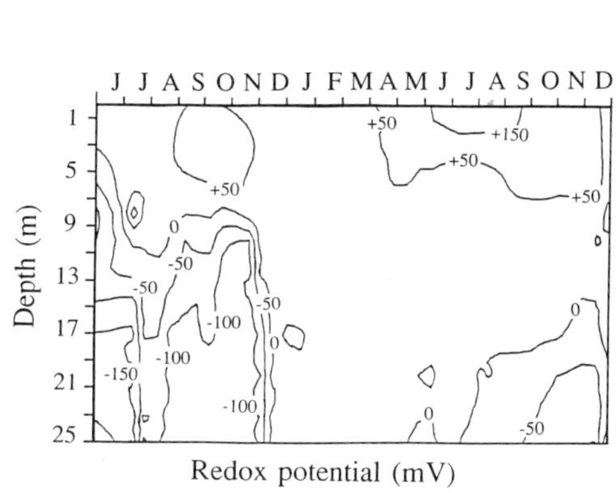

Figure 14.6. Variations of water redox potential in the Plußsee in period of June 1977 to December 1978.

Inorganic Nutrients

The distribution of the most important inorganic nutrients, i.e., phosphorus and nitrogen compounds, was directly related to the diversity of biological processes in lake water and therefore was the basis of considerable seasonal and spatial changes. For detailed information on nutrients in the Plußsee, see Chapter 1 (this volume). Comprehensive studies have been done by Ohle (1962, 1964), Overbeck (1972), Albrecht et al. (1978), Golachowska (1979), and others. Therefore, only a brief summary of the situation at the end of summer stagnation is given here (Fig. 14.7).

Within the epilimnion, phosphate as well as ammonia are depleted during the phytoplankton successions in spring and summer. Phosphate was, at least temporarily, a limiting factor with concentrations of only a few micrograms P/L (Chróst and Overbeck 1987, Chróst 1988). An exchange of nutrients between the nutrient-rich hypolimnion and the nutrient-depleted epilimnion was rather restricted because of the stability of the stratification.

The accumulation of ammonia above the sediment reflected the intensive anaerobic decomposition of organic substances, mainly of proteins. At the same time nitrification was inhibited by the low redox potential in this area, which was also responsible for the high hypolimnetic phosphate concentrations because of the well-known Fe–P relations. Phosphate was linked to $Fe(OH)_3$ in the aerobic part of the lake and released in the anaerobic hypolimnion after reduction of Fe^{3+} to Fe^{2+} together with the precipitation of iron sulfide.

At the end of the vegetation period only 30–50 μg N/L nitrate was measured, or nitrate was even depleted in lake water. The decline of nitrate at a depth of 15–25 m in the Plußsee was a result of denitrification. The remarkably small

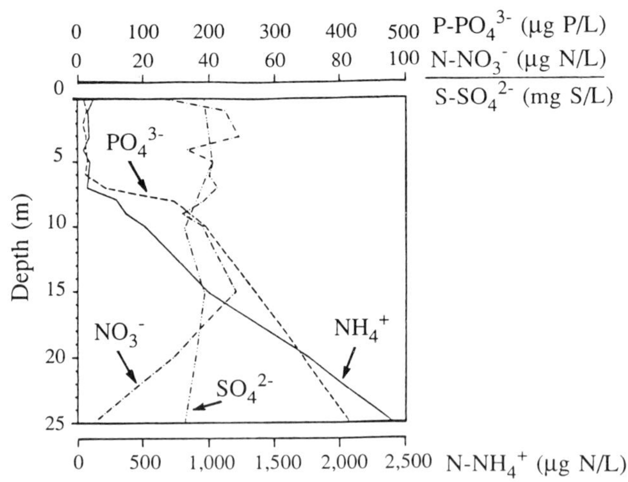

Figure 14.7. Inorganic phosphate, nitrate, ammonia, and sulfate concentration in the water column of the Plußsee (September 25, 1978).

concentrations of nitrate at 8–10 m depth were probably caused by metabolic activity of *Thiobacillus denitrificans* or by facultative anaerobic heterotrophic bacteria (Overbeck 1967, 1968).

In contrast to the nutrients described above, sulfate showed very little variation throughout the depth profile (the same is true for the seasonal changes). The average concentrations in the Plußsee were 30–32 mg S/L. The lowest value (26 mg S/L) was measured in autumn directly above the sediment, coinciding with increased sulfide concentrations. The highest concentration of sulfate (41 mg S/L in autumn) occurred in the metalimnion from sulfide oxidation (Ohle 1954).

Microorganisms of Sulfuretum

The organisms discussed here are classified into three major groups: (1) phototrophic sulfur bacteria, (2) colorless organisms of the sulfuretum (which often are only morphologically described and not well characterized concerning their taxonomy and physiology), and (3) sulfate-reducing bacteria. Besides several original descriptions, essentially the publications of Van Niel (1932), Skuja (1956), Anagnostidis (1968), Buchanan and Gibbons (1974), Fjerdingstad (1979), Häusler (1982), Pfennig and Trüper (1989), and Hirsch (1989) were used for the analyses of bacterial specimens.

Phototrophic Sulfur Bacteria: Chromatiaceae and Chlorobiaceae

The species of Chromatiaceae (purple sulfur bacteria) and Chlorobiaceae (green sulfur bacteria) that have been observed in the Plußsee are summarized in Table 14.1. Some representatives are shown in Figures 14.8–14.10.

The last two forms in Table 14.1 are "consortia," barrel-shaped aggregates consisting of numerous rod- to vibrio-shaped colored organisms surrounding a colorless flagellated central bacterium and probably living in symbiosis. In *Chlorochromatium aggregatum* Lauterborn, ca. 7–12 × 2.5–5.0 μm in size, the covering bacteria (1.0–2.5 × 0.5–1.0 μm) are green (Utermöhl 1925, "*Schmidlea luteola*"; Van Niel 1957, "*Pelodictyon aggregatum*") and are more or less regularly arranged in four to eight longitudinal and two to five transverse rows on the surface of the endobacterium. The smaller *Pelochromatium roseum* Lauterborn (4–8 × 2.5–4.0 μm) reveals a peripheral covering of 20–30 red- to brown-colored bacteria (Utermöhl 1924, "*Lauterborniola minima*") irregularly surrounding the central bacterium (Fig. 14.11).

Both consortia are actively motile, and in lake water they may occur alone or in groups of four to eight units. They could be enriched (Pfennig 1967) but have not been in pure culture yet. Mechsner (1957) succeeded in isolating the green components of *Chlorochromatium aggregatum* and identified them as members of the genus *Chlorobium*. As is well known, there are also brownish variants of *Chlorobium* (Pfennig 1978). The central bacterium (Utermöhl 1924, "*Endosoma palleum*"), 3.5–5.0 μm long and 1–2 μm in diameter, could not be classified up to now. In a transmission electron microscope (TEM) study on *Chlorochromatium aggregatum,* Caldwell and Tiedje (1975a) found hexagonally arranged cup-shaped

Table 14.1. Major Characteristics of Chromatiaceae and Chlorobiaceae of Lake Plußsee

Species	Cell Size (μm)	Color	Cell Inclusions	Other Characteristics
Chromatium okenii (Ehrenberg) Perty	4–5 × 6–8	Purple-red	Evenly distributed sulfur granules, no gas vacuoles	Cells always single, ovoid to bean-shaped, polarly flagellated
Thiocystis violacea Winogradsky	Diameter 2.5–4.0	Pink to violet	Evenly distributed sulfur granules, no gas vacuoles	Single cells spherical to ovoid, polarly flagellated; often forming colonies of 10–20 nonmotile cells commonly in slime sheath
Lamprocystis roseo-persicina (Kützing) Schroeter	Diameter 2.5–3.5	Pink to violet	Spherical central gas vacuole, peripheral sulfur globules	Single cells spherical to ovoid, mostly building irregularly formed colonies, always nonmotile
Thiopedia rosea Winogradsky	2.0–2.5 × 2.5–3.0	Pale pink	Spherical central gas vacuole, peripheral sulfur globules	Cells spherical to ovoid, nonmotile, regularly arranged in plates of 4, 8, 16, 32 or 64 cells
Amoebobacter roseus Winogradsky (*Rhodothece conspicua* Skuja)	Diameter 2.5–3.5	Pink-red	Lobed central gas vacuole, peripheral sulfur globules	Cells spherical to ovoid, nonmotile, occurring singly or as aggregates of 2, 4, or more individuals
Chlorobium limicola Nadson	0.5–0.8 × 1.0–1.5	Light green	None	Ovoid to rod-sphaped cells forming helical cell chains
Chlorobium phaeobacteroides Pfennig	0.5–0.8 × 1.0–1.5	Pink to brownish	None	See above: brownish variant of *C. limicola*
Pelodictyon clathratiforme (Szafer) Lauterborn	0.7–1.2 × 1.5–3.0	Yellow-green	Central gas vacuole	Individual cell edged rods; binary as well as ternary fissions give rise to the formation of large three-dimensional nets
Pelodictyon phaeum Gorlenko	0.7–1.0 × 2.0–3.5	Brown	Central gas vacuole	See above: probably brownish variant to *P. clathratiforme*
Clathrochloris sulfurica (Szafer) Geitler	Diameter 0.5–0.9	Yellow-green	Central gas vacuole	Spherical to ovoid, slightly edged cells, formation of large trellis-like aggregates
Ancalochloris perfilievii Gorlenko	Diameter 0.5–1.0	Yellow-green	Central gas vacuole	Star-shaped cells with four to six appendices of varying length, formation of trellis-like aggregates like "*Clathrochloris*"[a]
Chlorochromatium aggregatum Lauterborn	See text	Green	None	See text
Pelochromatium roseum Lauterborn	See text	Pink-brown	None	See text

[a]Cells with appendices (so-called "*prosthecae*"; Staley 1968) were described for two genera of Chlorobiaceae: *Prosthecochloris* sp. and *Ancalochloris perfilievii*. *Prosthecochloris* sp. was isolated from brackish water, had up to 20 appendices per cell, and lacked gas vacuoles (Gorlenko and Zhilina 1968, Gorlenko 1970). The star-shaped cells of *Ancalochloris perfilievii* (Gorlenko and Lebedeva 1971) had a central gas vacuole and four to six prosthecae. The cells formed characteristic aggregates, corresponding to the form found in the Plußsee. *Ancalochloris perfilievii* differentiation from *Clathrochloris* sp. is difficult because the length of the appendices may vary strongly depending on light conditions and other environmental factors (Gorlenko and Lebedeva 1971). In Germany, bacteria of this type were first observed by Anagnostidis and Overbeck (1966) in Plußsee samples.

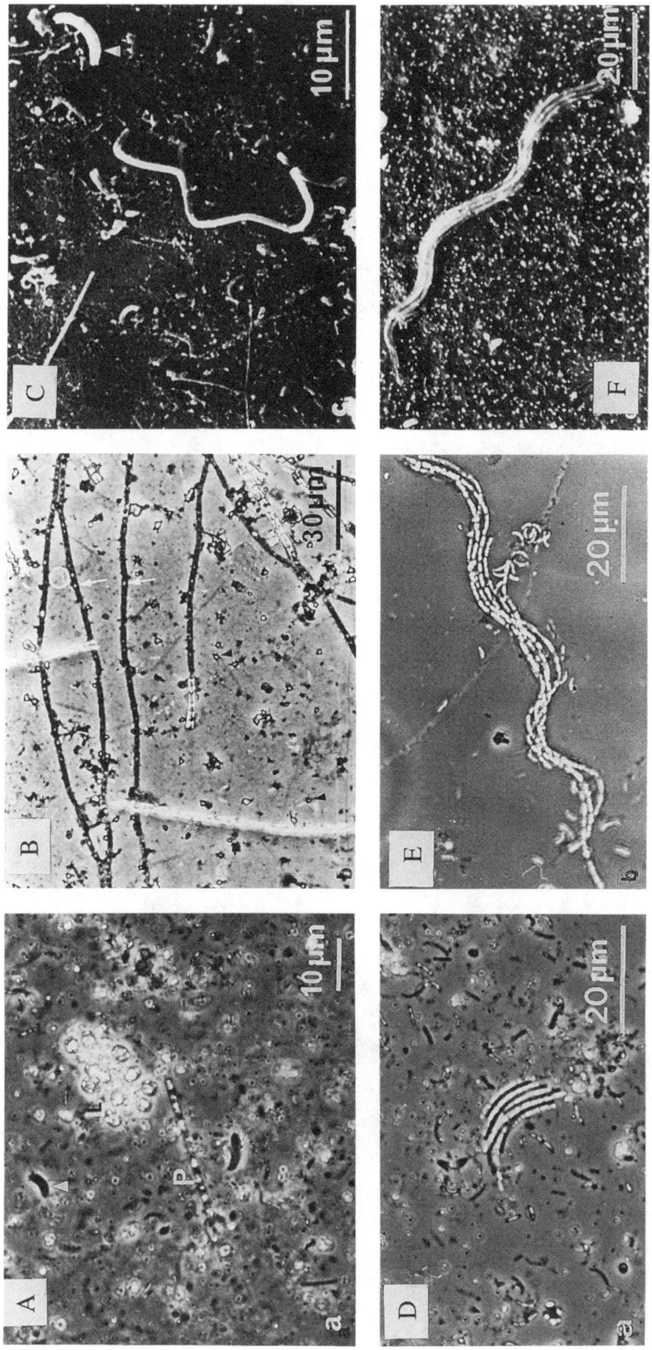

Figure 14.8. Representative sulfuretum species of the Plußsee. (A) *Pelonema tenue* (p) with *Lamprocystis roseopersicina*, photomicrograph. (B) *Pelonema pseudovacuolatum*, photomicrograph. (C) *Pelonema spirale* and Caldwell's T_2, scanning electron micrograph. (D) *Peloploca undulata*, photomicrograph. (E) *Peloploca pulchra*, photomicrograph. (F) *Peloploca pulchra*, scanning electron micrograph.

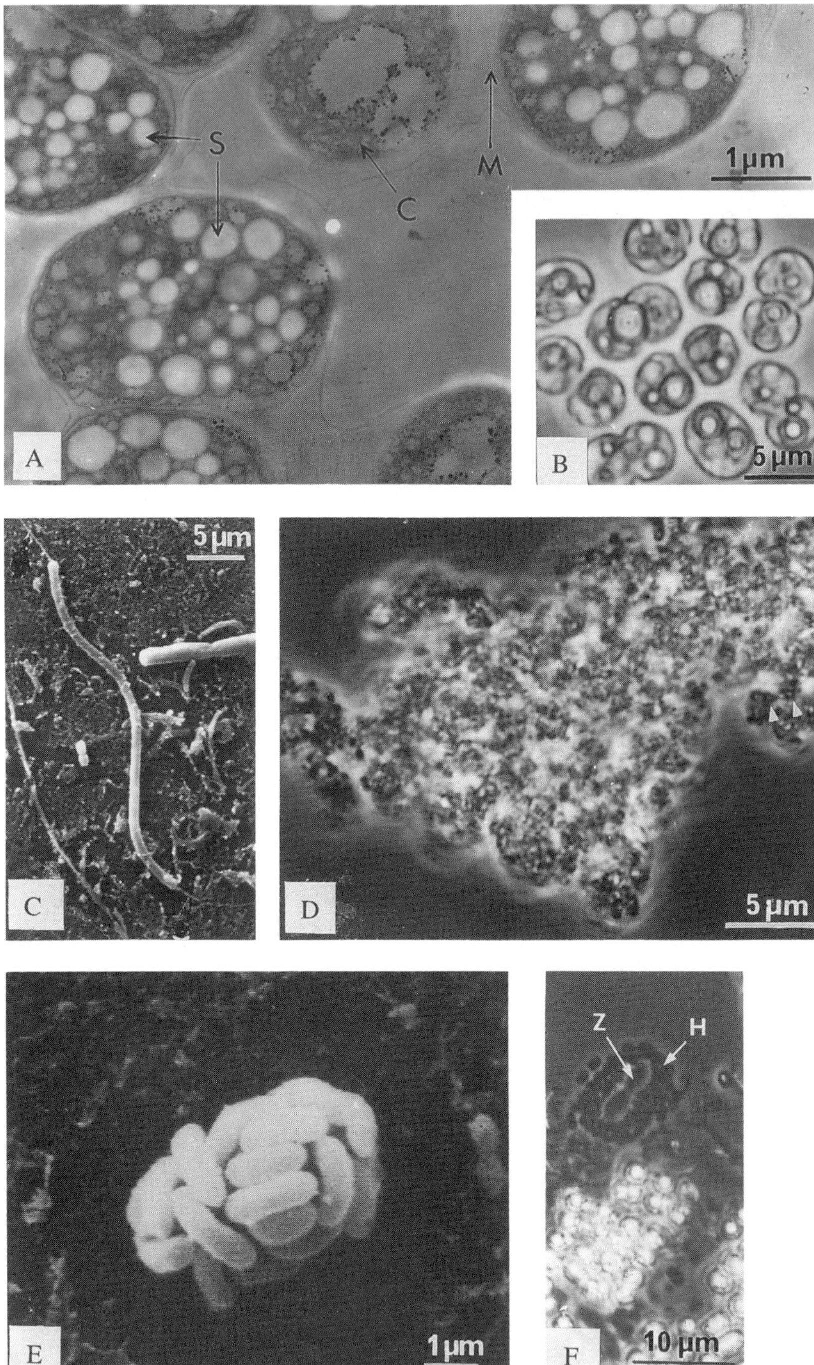
S
S
C
M
1 μm
A
B
5 μm
5 μm
C
D
5 μm
E
1 μm
Z
H
F
10 μm

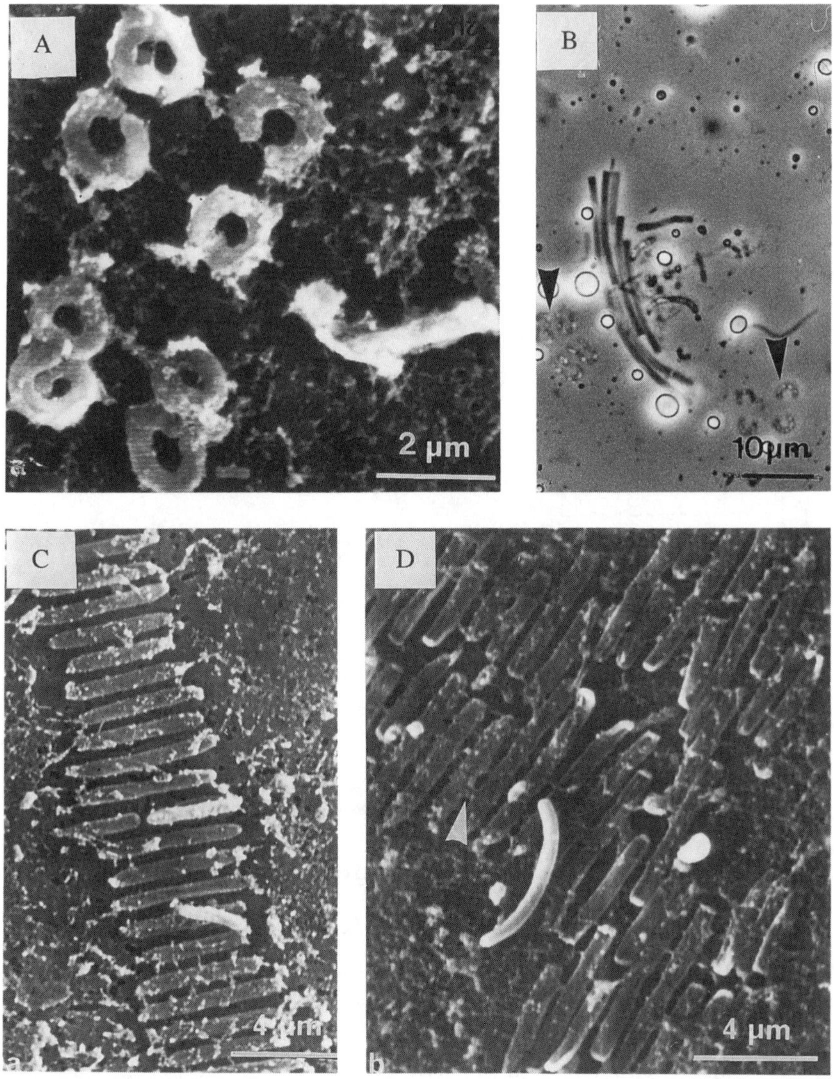

Figure 14.10. Representative sulfuretum species of the Plußsee. (A) Caldwell's T_3 cells in slime sheath, scanning electron micrograph. (B) T_3 cells with gas vesicles (arrows), in the center of the photomicrograph *Peloploca* sp. (C) and (D) palisade-like aggregates of T_4, scanning electron micrographs.

Figure 14.9. Representative sulfuretum species of the Plußsee. (A) *Thiocystis violacea,* transmission electron micrograph: S, sulfur granules; C, vesicular chromatophores; M, mucus sheath. (B) *Thiocystis violacea,* photomicrograph. (C) *Thiospirillum jenense,* scanning electron micrograph. (D) *Ancalochloris perfilievii,* photomicrograph. (E) *Pelochromatium roseum,* scanning electron micrograph of consortium with vibrio-like bacteria surrounding the central bacterium. (F) *Pelochromatium roseum,* photomicrograph: Z, central bacterium in the stage of cell division; H, the outer bacteria.

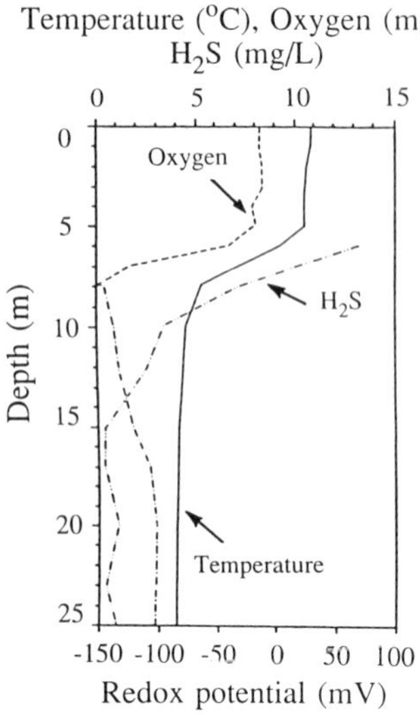

Figure 14.11. Vertical distribution of dissolved oxygen, H_2S, and temperature in the Plußsee (October 17, 1978).

structures on the surface of the endobacterium that probably function in the stability of the conglomerate.

As long as the taxonomic questions remain open, the names of species may be regarded only as preliminary (Trüper and Pfennig 1971). Yet without any doubt these consortia have to be interpreted as functional units capable of simultaneous fission of all the constituent parts (see also Anagnostidis 1968). However, both components, at least the phototrophic partners, are viable separately, as observed by Utermöhl (1925) as well as by Mechsner (1957).

Colorless Bacteria of Sulfuretum

The microorganisms classified in this group resisted cultivation until now and are mainly characterized by morphological criteria. In the Plußsee the Pelonemataceae (Skuja 1956, 1964, 1974) are represented by several species of the two genera *Pelonema* and *Peloploca* established by Lauterborn (1913, 1915, Anagnostidis and Overbeck 1966), (Table 14.2). Their cells mostly contain gas vacuoles, and they build unbranched cell chains, which are often of considerable length. The chains may occur singly (genus *Pelonema*) or arranged as bands or bundles (genus *Peloploca*) and may be surrounded by a thin slime sheath. The trichomes lack polar differentiation; they are propagated by fragmentation. Skuja (1956) and Maiden and Jones (1984) observed gliding motility of single filaments or individual bundles with rotation about their long axis. In the original description by Lauterborn (1913, 1915) these organisms were designated as bacteria. However, Skuja (1956) considered them to represent a colorless, heterotrophic collateral line of hormogonal blue-green algae ("*Cyanophytorum apochromaticorum*") because of their resemblance to members of Oscillatoriales.

The features of the several Pelonemataceae "species" are gathered in Table 14.2 and demonstrated in Figure 14.8. The quotation of species designations

Table 14.2. Characteristics of *Pelonema* and *Peloploca* Species

Species	Cell Size (μm)	Characteristics
Pelonema tenue Lauterborn	8–12 × 2	Single, straight filaments, up to 300 μm long occasionally gliding, not constricted at cross-walls; each cell contains one irregularly shaped gas vacuole
Pelonema pseudovacuolatum Lauterborn	4 × 1.5–2	Single, straight filaments, 250–300 (–500) μm long, immobile, weakly constricted at cross-walls, small square to roundish distinct gas vacuoles
Pelonema spirale Lauterborn	0.8–1.5	Trichomes spirally wound in 10- to 14-μm intervals, hardly detectable cross-walls without constrictions, filament length 40–150 μm, numerous small, irregularly arranged gas vacuoles
Peloploca taeniata Lauterborn	3–7 × 0.5–0.9	Several filaments laterally grown together to flat or nodulated bands whose ends appear fibrous, common slime sheath; cross-walls hardly constricted. Bands appear lattice-like because of the arrangement of gas vacuoles. Length of the bands 60–150 μm.
Peloploca undulata Lauterborn	3–7 × 0.5–0.9	*P. undulata* is more regularly undulated than *P. taeniata*
Peloploca fibrata Skuja	2–2.5 × 0.4–0.6	Fragile filaments building flat bands, often with fibrous ends, total width 2.5–3.0 μm; because of fragmentation the bands are often short and consist of only few cells; gelatinous sheath hardly detectable
Peloploca pulchra Skuja	0.4–0.8	Four to 15 parallel filaments in regularly undulating bundles with fibrous ends, gelatinous sheath; no gas vacuoles, cross-walls hardly detectable; length of bundles up to 400 μm; mobile

is not correct until the organisms have been cultivated, but the traditional names are preserved here in order to distinguish the morphological forms. It should also be pointed out that the inability to cultivate organisms isolated from a natural environment strongly indicates the complexity of metabolic interactions within the ecosystem, which is hitherto largely unknown.

Some other hypolimnetic bacteria occurring in the Plußsee were also described in several Michigan lakes. Since they, too, could not be clearly determined until now, the different morphological types preliminarily were supplied with numbers (Caldwell and Tiedje 1975a). The T_2 prokaryotes are relatively large fusiform rods or vibrios (4–12 × 1.2–1.5 μm) without any cellular differentiation under the light microscope (Fig. 14.8C). The T_3 bacteria (2–6 μm × 0.5–0.6 μm) form characteristic plates composed of 2, 4, 8, 16, or 32 single cells in a common slime sheath (Fig. 14.10A). The cells are strongly curved so that the opposite cell poles may touch or even overlap each other, thus presumably inducing binary fission. In the curvature of the cells as well as in the possession of gas vacuoles (Fig. 14.10A and B), T_3 resembles the singularly occurring cells of certain *Microcyclus aquaticus* strains (Van Ert and Staley 1971) but differs from this in its anaerobic mode of life and in its assemblage in cell plates.

The T_4 cells (3–8 × 0.6–0.8 μm) are of very low contrast and hence difficult to recognize under the light microscope. They also assemble in more or less large aggregates with palisade-like arrangement of the single cells and are surrounded by a slime layer (Fig. 14.10C and D). The scanning electron microscope micrographs reveal cell structures indicating a cross-connection between adjacent cells (Fig. 14.10D). Caldwell and Tiedje (1975a) identified gas vesicles in TEM preparations of T_4; furthermore, they described a special mode of cell division proceeding with a thickening in the middle of the cell.

The numerically dominating colorless microorganisms in the hypolimnion of the Plußsee are more or less strongly curved vibrios ("*Microspira desulfuricans*," Anagnostidis and Overbeck 1966). The fusiform cells measure 3–8(10) × 0.5–0.8 μm and are characterized by a large central gas vacuole, which occasionally occupies most of the cell volume. Because of their size, morphology, habitat, and cell density in the hypolimnetic bacterial community, they largely correspond to the T_5 prokaryotes described by Caldwell and Tiedje (1975a) and to the R_3 bacteria observed in British lakes by Clark and Walsby (1978a). Apparently these vibrios are not sulfate reducers; their classification remains uncertain, and the neutral term T_5 should provisionally be preserved.

Schmidt (1979) studied the distribution of a presumptive iron bacterium called *Metallogenium* sp., which probably oxidizes iron and especially manganese. During turnover processes in the lake, *Metallogenium* propagated throughout the whole vertical profile. With the development of a summer stratification, *Metallogenium* migrated with the chemocline to the upper strata, where it usually was associated with the phototrophic bacterial community (see Chapter 15, this volume).

Sulfate-Reducing Bacteria

Sulfate-reducing bacteria were enriched using the medium B after Baars (1930) and medium C after Starkey (1938) and Butlin et al. (1949); all were modified after Postgate (1967). For isolation of pure cultures a standard medium S was developed from data of Postgate (1963), Pfennig (1965), Widdel (1980), and others. Lactate was used as an organic substrate, 4.5 g Na_2SO_4/L (electron acceptor), and 15 ml 0.1 mol/L $Na_2S \cdot 9H_2O$ (reducing agent); for details see Oberhäuser (1981). In enriched cultures with samples from the hypolimnion of the Plußsee (25 m depth), a black precipitate of FeS was evident occasionally after 3 to 4 days. In media without free Fe^{2+}, growth of sulfate reducers was indicated by turbidity; the H_2S content in the culture medium at that time amounted to 15–18 mg S^{2-}/L. The purification of the sulfate reducers was performed in Hungate tubes, which were inoculated after Postgate (1965). As indicator Fe^{2+} was added to the medium S.

Table 14.3. Characteristics of *Desulfovibrio desulfuricans* (Postgate and Campbell 1966) and Strain 78 Isolated from Plußsee

Trait	*Desulfovibrio desulfuricans*	Strain 78
Morphology	Vibrio, sigmoid	Rod, vibrio
Flagellation	Polar, monotrichal	Subpolar, monotrichal
Gram staining	Negative	Negative
Dominant cytochrome	C_3	C_3
Desulfoviridin	+	+
G + C (mol %)	55	54.6
Hibitane resistance (mg/L)	10–25	10
Meso/thermophilic	Mesophilic	Mesophilic
Utilization of:		
Lactate + sulfate	+	+
Glucose + sulfate	±	–
Acetate + sulfate	–	–
Formate + sulfate	±	–
Pyruvate + sulfate	+	+
Electron acceptors:		
SO_4^{2-}	+	+
$S_2O_3^{2-}$	+	+
SO_3^{2-}	+	+
S^0	±	–

A comparison of the pure culture isolated from the Plußsee (strain 78) and *Desulfovibrio desulfuricans* (Postgate and Campbell 1966) is shown in Table 14.3. Hence, in all major features the isolated strain from the Plußsee was identical to *Desulfovibrio desulfuricans*. Since extensive studies using different lactate/sulfate media always gave the same result, i.e., enrichment of the same species, it can be concluded that the sulfate reduction in the Plußsee was performed essentially by *Desulfovibrio desulfuricans*, although the participation of other species cannot be excluded.

Development of the Sulfuretum and Its Colonization

As was already mentioned, there had not been a complete turnover of the Plußsee basin in early 1977. Hence, at the beginning of the investigation period in March, there was an oxygen gradient between 58% saturation at the surface and 5% above the sediment. In these only weakly aerated layers Cyanobacteria species as well as "*Metallogenium*" (Schmidt 1979, Schmidt and Overbeck 1984; Chapter 13, this volume) were found. Diatoms and green algae were found sporadically. After the rapid depletion of oxygen in the hypolimnion from the middle of April 1977 (Figs. 14.4 and 14.5), the anaerobic strata were filled up with hydrogen sulfide, which was attributed mostly to the metabolic activity of sulfate-reducing bacteria in the sediment surface.

In early May, H_2S had already reached the 20 m stratum. Among the short trichomes of Cyanobacteria, the first typical organisms of the sulfuretum like T_5, some *Lamprocystis roseopersicina, Amoebobacter roseus,* and Caldwell's T_4 appeared. Parallel to the increase of H_2S in space and concentration, the following successions of species were observed during the summer months: *Peloploca* and *Pelonema* species, T_2, T_3, the consortia of *Pelochromatium roseum* and *Chlorochromatium aggregatum,* were found in June–July; in August *Pelodictyon phaeum* and *Clathrochloris/Ancalochloris* were found also (both genera are combined here because of their similarity of morphology under the light microscope; see footnote of Table 14.1). From the middle of September *Chlorobium limicola* and *Chlorobium phaeobacteroides* were found as well (Figs. 14.8–14.10). The H_2S concentration at this time reached ca. 1 mg/L at a depth of 10–12 m.

The main development of the thiobiocoenosis in the Plußsee coincided with the climax of thermic and chemical stratification in October (Fig. 14.11). The epilimnion extended down to a depth of 6 m with a uniform temperature of 11°C; the oxygen concentration, after the breakdown of a *Limnothrix* (*Oscillatoria*) *redekei* bloom decreased to 8 mg/L (i.e., 70% saturation). The metalimnion, between 6 and 8 m, was characterized by a steep gradient of temperature, oxygen, and redox potential. A completely anaerobic hypolimnion extended below 8 m with temperatures of 4.0–4.5°C, negative redox values, and increasing H_2S concentrations.

Table 14.4 summarizes the distribution of the dominating sulfuretum organisms in the hypolimnion, which was distinctly divided into two compartments. The upper layer with H_2S concentrations of 0.5–1.5 mg/L was mostly the domain of phototrophic bacteria, which require both light energy and H_2S. The composition of the community of phototrophic bacteria in the Plußsee was characterized by a predominance of Chlorobiaceae species. Here they were obviously represented by their pink or brownish colored variants. Some species were also found in the dark part of the hypolimnion near the sediment surface, where photosynthesis was not possible. The dark sulfuretum was predominated by the colorless heterotrophic organisms designated as "community C" by Caldwell and Tiedje (1975b). The occurrence of members of the Pelonemataceae was clearly restricted to the deepest water layers. The *Peloploca* species, i.e., *P. taeniata, P. undulata,* and *P. pulchra,* were found exclusively at depths of highest H_2S concentrations and strongly negative redox potentials, whereas the *Pelonema* sp. were occasionally scattered throughout the upper hypolimnion. The T_3 and the other colorless procaryotes reached their maximum in these depths, but they were present in lower numbers throughout the hypolimnion (Tables 14.4 and 14.5). For example, the T_5 cells ascended to the upper border of H_2S but were most abundant directly above the sediment (Fig. 14.5). The palisade-like aggregates were found just below the chemocline by Clark and Walsby (1978a: R_5), but in the Plußsee they developed mainly in the lower part of the hypolimnion (T_4: Caldwell and Tiedje 1975b). The cell plates of T_3 showed a maximum slightly above T_4, and the T_2 vibrios, sharing only about 1% of the whole community, were evenly distributed throughout the hypolimnion. Some further species such as *Thiocystis violacea, Chromatium okenii, Thiopedia rosea,* and *Pelodictyon clathratiforme* only sporadically oc-

Table 14.4. Distribution of Dominant Sulfuretum Organisms in the Hypolimnion of Plußsee in October 1977

Depth (m)	Dominant Sulfuretum Organisms[a]													
	Lr	Ar	Cl	Cp	Pp	C/A	Ca	Pr	P	Pl	T3	T4	T5	M
6														✦✦✦
8	✦✦✦	✦✦	✦✦✦	✦✦✦	✦✦	✦✦	✦✦✦	✦✦✦					✦	✦✦
10	✦✦✦	✦✦	✦✦✦	✦✦✦	✦✦✦	✦✦	✦✦✦	✦✦✦	✦				✦✦	✦
12	✦✦✦	✦✦	✦✦	✦✦	✦✦	✦✦✦	✦✦✦	✦✦✦	✦		✦	✦✦	✦✦	
15	✦✦	✦✦	✦✦	✦✦	✦✦	✦✦	✦✦	✦✦	✦	✦	✦✦	✦✦	✦✦✦	
17	✦✦	✦✦	✦✦	✦✦	✦	✦	✦	✦✦	✦	✦	✦✦	✦✦✦	✦✦✦	
20	✦✦	✦	✦	✦	✦	✦		✦	✦✦	✦	✦✦	✦✦✦	✦✦✦	
23	✦	✦	✦	✦				✦	✦✦	✦✦	✦	✦✦	✦✦✦	
25	✦	✦	✦					✦	✦✦✦	✦✦	✦	✦✦	✦✦✦	

[a](Lr) *Lamprocystis roseopersicina,* (Ar) *Amoebobacter roseus,* (Cl) *Chlorobium limicola,* (Cp) *Chlorobium phaeobacteroides,* (Pp) *Pelodictyon phaeum,* (C/A) *Clathrochloris/Ancalochloris,* (Ca) *Chlorochromatium aggregatum,* (Pr) *Pelochromatium roseum,* (P) *Pelonoma* sp., (Pe) *Peloploca* sp., (T3) Caldwell's T_3, (T4) Caldwell's T_4, (T5) Caldwell's T_5, (M) "*Metallogenium.*"

Table 14.5. Distribution of Dominant Sulfuretum Organisms in the Hypolimnion of Plußsee in October 1978

Depth (m)	Dominant Sulfuretum Organisms[a]													
	Lr	Ar	Cl	Cp	Pp	C/A	Ca	Pr	P	Pl	T3	T4	T5	M
8														✦✦✦
10		✦												✦✦✦
12	✦	✦✦	✦		✦	✦	✦	✦						✦✦✦
15	✦	✦✦	✦✦	✦	✦	✦✦	✦✦	✦✦				✦	✦	✦✦
17	✦✦	✦✦	✦✦	✦	✦✦	✦	✦	✦✦	✦		✦	✦✦	✦✦	✦✦
20	✦✦	✦	✦	✦	✦	✦	✦	✦✦	✦	✦	✦✦	✦✦	✦✦✦	✦
23	✦✦	✦	✦	✦				✦	✦✦	✦✦	✦✦	✦✦	✦✦✦	
25	✦	✦		✦					✦	✦✦	✦✦	✦✦✦	✦✦✦	

[a](Lr) *Lamprocystis roseopersicina,* (Ar) *Amoebobacter roseus,* (Cl) *Chlorobium limicola,* (Cp) *Chlorobium phaeobacteroides,* (Pp) *Pelodictyon phaeum,,* (C/A) *Clathrochloris/Ancalochloris,* (Ca) *Chlorochromatium aggregatum,* (Pr) *Pelochromatium roseum,* (P) *Pelonema* sp., (Pe) *Peloploca* sp., (T3) Caldwell's T_3, (T4) Caldwell's T_4, (T5) Caldwell's T_5, (M) "*Metallogenium.*"

curred in the Plußsee; therefore, they are not listed in Table 14.4. In the thermochemocline a pronounced bloom of "*Metallogenium*" developed (see Chapter 15, this volume).

This structure of the thiobiocoenosis only existed for a period of a few weeks; with the beginning of the fall turnover, combined with increasing aeration of the Plußsee from above, the community of phototrophic bacteria shifted to deeper strata. After total aeration in the middle of December, resulting in a complete oxidation of hydrogen sulfide, the sulfuretum disappeared. Its place was occupied essentially by short trichomes of Cyanobacteria and "*Metallogenium*," but a few *Amoebobacter roseus,* which occurred in the Plußsee throughout the year, also were found (Anagnostidis and Overbeck 1966). These organisms were also present in January–February 1978 in water layers between 20 and 25 m; these layers were characterized by the coexistence of O_2 and H_2S. *Lamprocystis roseopersicina,* T_4 and a remarkable development of T_5, *Pelonema* sp. and *Peloploca* sp. were other microorganisms found in the hypolimnion during winter 1977–1978. *Peloploca fibrata* (Fig. 14.8) was found here for the first time. These sulfuretum organisms were maintained near the sediment until the spring circulation in March 1978. Thereafter, "*Metallogenium*" structures, Cyanobacteria, and diatoms (*Stephanodiscus* sp., *Asterionella* sp.) in addition to *Amoebobacter* sp. dominated in a similar way as in 1977.

In 1978, the Plußsee was completely aerated in spring, thus differing from 1977 (50% saturation throughout the water column). The H_2S production was very poor, so the development of a sulfuretum was delayed. The thermic stratification was the same as in 1977 (Fig. 14.1), but the upper part of the hypolimnion, down to a depth of 12 m, was still aerated, whereas in 1977 the whole hypolimnion was filled with hydrogen sulfide. Even below 20 m only half of the 1977 concentration levels of H_2S were found. Consequently, the conditions for the development of phototrophic bacteria were extremely unfavorable. The numbers of Rhodospirillales in 1978 (Table 14.5) were smaller than those of 1977 with a main development at a depth of 15–20 m without light, where they probably existed by means of a fermentative metabolism using elemental sulfur as electron acceptor. Only *Amoebobacter roseus* propagated in the upper layers of the hypolimnion (10–12 m depth), indicating its ability to tolerate microaerophilic conditions and to utilize organic compounds, thiosulfate, or H_2 as H^+ donor instead of sulfide (Kämpf and Pfennig 1980).

After decay of an epilimnetic algal bloom at the end of October 1978, followed by an increase of H_2S to 0.2–0.4 mg/L at a depth of 15–17 m, a further increase of phototrophic bacteria, especially of the two consortia *Pelochromatium roseum* and *Chlorochromatium aggregatum* was observed. It is possible that the input of dissolved organic matter enhanced the activity of the sulfate reducers. The complete turnover in December 1978 destroyed the thiobiocoenosis.

For both groups of organisms, i.e., phototrophic and colorless bacteria, the consortia of *Pelochromatium roseum* and the cells of T_5 were counted at weekly intervals between August 1977 and December 1978 (Figs. 14.12 and 14.13), reflecting the seasonal and spatial variations of major sulfuretum microorganisms. The numbers are mean values of the indicated months.

Figure 14.12. Distribution and abundance of *Pelochromatium roseum* Ltb. in the Plußsee in 1977–1978. Black vertical bars show water redox potential less than 0 mV.

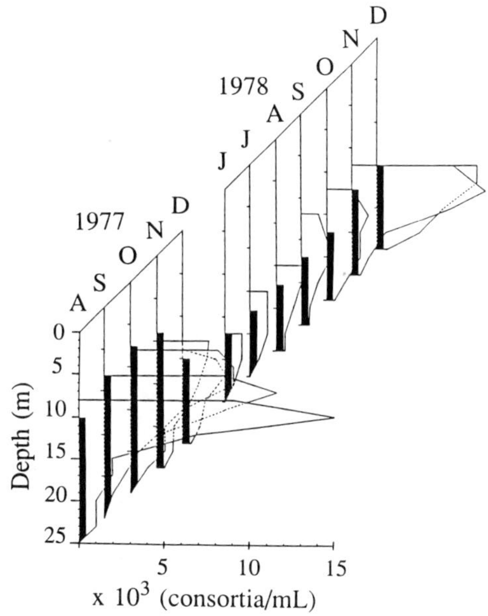

Uptake Kinetics

Uptake of Glucose

In late summer (August–September) maximum values of V_{max}, 0.91 and 0.95 µg L^{-1} h^{-1}, and shortest turnover times were measured in the upper metalimnion (5 m depth) and above the sediment (25 m depth), respectively (Fig. 14.14). In October, the values were highest at the sediment surface (1.45 µg L^{-1} h^{-1}) but very low in

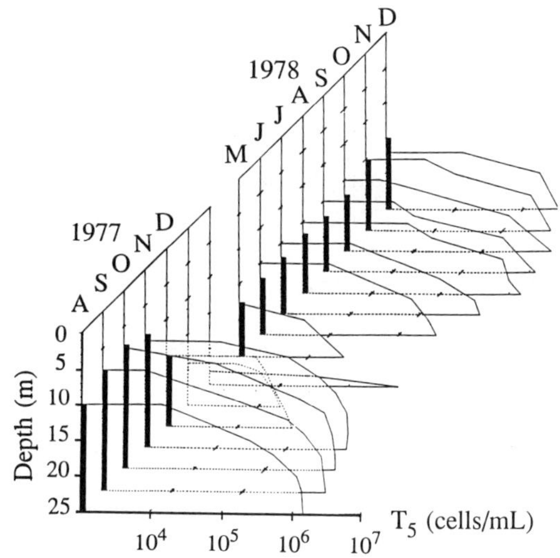

Figure 14.13. Distribution and abundance of T_5 in the Plußsee in 1977–1978. Black vertical bars show water redox potential less than 0 mV.

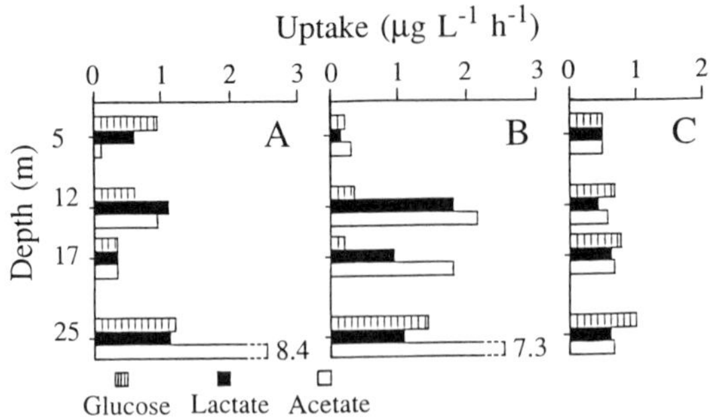

Figure 14.14. Uptake of glucose, lactate, and acetate in the water column of the Plußsee during (A) phytoplankton bloom (September), (B) bloom breakdown (October), and (C) after lake mixing (December).

the water column. However, here they increased again in the water column during December. The different modes of uptake apparently reflect the physiological state of the bacterial assemblages ("physiological compartments," Overbeck 1973, 1975, 1979). During phytoplankton blooms, i.e., high photosynthetic activities (August–September), high V_{max} values were found in the metalimnion. In October, the breakdown of the summer plankton was reflected by low V_{max} values in the metalimnion but high values in the deeper parts of the lake. After the turnover of the lake in December, accompanied by liberation of DOM from bacterial breakdown, the V_{max} again increased. Blauw (1988), in his study of microbial processes in the sediment of the Plußsee, also found a rapid and strong increase of glucose uptake at the sediment surface as a result of sedimentation of organic material after the breakdown of phytoplankton blooms (Chapter 8, this volume).

Uptake of Lactate

Lactate is one of the predominant electron donors for the group of anaerobic bacteria capable of dissimilatory sulfate reduction (Postgate 1979). The changes in the vertical profile (Fig. 14.14) as well as the seasonal changes of lactate utilization agree with the assumption that at least some of the results of lactate uptake in situ may be correlated with sulfate-reducing bacteria.

Usually two maxima of sulfate reducers were found in the thermal and chemically stratified lakes. One is found at the sediment surface, and the second one is observed just below the upper border of hydrogen sulfide because of the metabolic coupling between sulfate reducers and photosynthetic bacteria (Van Gemerden 1967, Cappenberg 1972, Gorlenko et al. 1973, and others). This distribution pattern explains the high rate of lactate uptake at a depth of 25 m in the Plußsee (>1 $\mu g\ L^{-1}\ h^{-1}$). The relatively high values in August and September at 12 m must be attributed to other organisms. The considerable increase in the Plußsee during October at 12–17 m may again indicate the activity of sulfate-reducing bacteria.

Uptake of Acetate

The uptake of acetate in the hypolimnion exceeded by far that of the two other substrates, thus emphasizing the significance of this prevalent product of anaerobic decomposition. Highest uptake rates were measured directly above the sediment (V_{max} ca. 7 µg L^{-1} h^{-1}) but also at depths of 12 m and 17 m (in October). Following the turnover process in December, the uptake rates showed little variation within the vertical profile (Fig. 14.14). Acetate is a convenient substrate for photoorganotrophic growth of red and green sulfur bacteria (Trüper and Genovese 1968, Sirevåg 1975, Biebl and Pfennig 1978, Bergstein et al. 1979). The distribution pattern of acetate uptake and phototrophic bacteria reflect these conditions. In the Plußsee in August and September the uptake of acetate was low at depths of 12 and 17 m. But in October at these depths, the V_{max} values increased considerably simultaneously with the development of phototrophic bacteria. In studies of the sulfuretum of the Kolksee, maximum uptake rates of acetate were measured at a depth of 5–6 m, which was attributed to high numbers of Chromatiaceae (Olah et al. 1973, Oberhäuser 1981).

Sulfate reducers are known to utilize acetate as a convenient carbon source (Sorokin 1966a,b,c). Therefore, the high uptake of acetate near the surface of the sediment may be correlated with the distribution pattern of sulfate reducers. In surface sediments of a hypereutrophic shallow lake, sulfate reducers were responsible for about 14% of the total acetate mineralization (Smith and Klug 1981).

Discussion

General Aspects

Our results confirmed that the hypolimnion of the Plußsee creates a typical freshwater sulfuretum according to the original definition of Baas-Becking (1925) and usually exists for only a few months during summer stagnation. A low content of H_2S (few milligrams per liter), a high species diversity, and a division into an upper, light-exposed and a lower, dark compartment are characteristic of a freshwater sulfuretum. During circulation periods the occurrence of sulfur bacteria and their accompanying organisms is mostly restricted to the sediments, which remain anaerobic throughout the year in temperate eutrophic lakes and exhibit permanent active sulfate reduction (Ohle 1953). After the depletion of oxygen in the deeper parts of the lake, the organisms are able to leave the sediment to propagate in the free water, a way of life designated as "meroplanktonic" (Utermöhl 1925).

Obviously the sulfuretum organisms occurring regularly in the Plußsee are cosmopolitans. The species composition as well as the structure of the thiobiocoenoses are comparable to the anaerobic microflora of numerous lakes in Sweden (Skuja 1956, 1964), Greece (Anagnostidis 1968, Anagnostidis and Economou-Amilli 1980), Great Britain (Clark and Walsby 1978a,b, Maiden and Jones 1984), Russia (Gorlenko and Kusnetzow 1971, 1972, Dubinina et al. 1973, Gorlenko et al. 1973), Estonia (Gorlenko and Lokk 1979), and the United States (Walsby 1974, Caldwell and Tiedje 1975a,b).

The particular species of the thiobiocoenosis gather in a characteristic vertical position according to their individual physicochemical requirements. The capability of active chemo- and/or phototactic locomotion by means of flagella is restricted to the Chromatiaceae (Table 14.1). Some species may occur as flagellated single cells under low sulfide concentrations and low light intensities. On the other hand, gas vacuolation, permitting a regulation of density and thus regulation of buoyancy, is a common feature of the majority of colorless Pelonemataceae (Table 14.2) as well as phototrophic hypolimnetic organisms (Van Ert and Staley 1972, Walsby 1974, Caldwell and Tiedje 1975a,b, Clark and Walsby 1978a,b, Maiden and Jones 1984, Pfennig 1989). Even if the regulation of density does not allow for the ability to cover long distances, gas-vacuole-containing cells may, because of their lower specific gravity, passively rise within the cold (4–8°C) turbulence-free water of the hypolimnion (Walsby 1972) and significantly reduce their sinking rate. This last aspect may also be a reason for the widespread tendency among hypolimnetic bacteria to gather in large cell aggregates. Another mechanism of locomotion is realized among the *Chlorobium* species, which lack flagella and gas vacuoles. In association with a colorless flagellated partner, i.e., in the *Chlorochromatium* and *Pelochromatium* consortia, the phototrophic components are able to move toward more favorite positions within the H_2S and light gradients.

Photosynthetic Pigments and Distribution of Phototrophic Sulfur Bacteria

Photoautotrophic sulfur bacteria in the light part of the sulfuretum are characterized by anoxigenic photosynthesis. The most common electron donors for the reduction of NAD are reduced sulfur compounds, molecular hydrogen, or simple organic compounds. Several bacteriochlorophylls are photosynthetically active pigments. Certain carotenoids serve as "antenna pigments," collecting light energy that is transferred to bacteriochlorophyll within the real photochemical reaction centers. This is generally true for both families of phototrophic sulfur bacteria; however, there are some differences of fundamental importance for the ecology of Chromatiaceae and Chlorobiaceae.

The red as well as the green sulfur bacteria may oxidize the primary electron donor H_2S directly to sulfate. In higher sulfide concentrations, elemental sulfur is formed and is deposited extracellularly by Chlorobiaceae but can be stored intracellularly (Fig. 14.9A) by Chromatiaceae. Besides reducing sulfur compounds and molecular hydrogen, the Chromatiaceae are potentially photoorganotrophic using organic acids, alcohols, and carbohydrates as electron donors. Several species are able to grow facultatively chemoautotrophically in the dark and can even tolerate low oxygen concentrations (Kämpf and Pfennig 1980, Pfennig 1989). The Chlorobiaceae may photoassimilate simple organic compounds for anabolic processes, but only in the presence of sulfide and bicarbonate; all species are obligate phototrophs and strict anaerobes (Pfennig 1989).

The photosynthetic pigments of the sulfur bacteria are characterized by in vivo absorption maxima in the near infrared (bacteriochlorophylls) and in the blue-green (carotenoids) spectral ranges (Pfennig 1967, 1978, 1989). The Chromatiaceae contain mostly bacteriochlorophyll$_a$ (800–890 nm) as well as the carotenoids spirilloxanthin, rhodopinal, and okenone, which are responsible for the yellow, red, or violet colors, respectively. The Chlorobiaceae also possess small amounts of bacteriochlorophyll$_a$ within the reaction centers, but mainly bacteriochlorophyll$_c$ and bacteriochlorophyll$_d$ (725–760 nm) as well as the carotenoid chlorobactene. As previously indicated (Table 14.1), in numerous Chlorobiaceae genera green as well as brownish pigmented species have been described. These last mentioned forms are characterized by containing not only bacteriochlorophyll$_a$ but also bacteriochlorophyll$_e$ (715–725 nm; Gloe et al. 1975) and the carotenoid isorenieratene, which causes the pink-brownish colors.

The pigments of the Chromatiaceae are located in chromatophores (Fig. 14.9), which correspond to invaginations of the cytoplasmic membrane and may occupy large parts of the cell volume (Drews 1991, Drews and Imhoff 1991). In the Chlorobiaceae only the bacteriochlorophyll$_a$-containing reaction centers are associated with the cytoplasmic membrane. The "antenna pigments" are situated in so-called chlorobium vesicles (chlorosomes) centripetally accumulated on the cytoplasmic membrane. In either group low light intensities may cause an increase of both the photosynthetic cell structures and the specific pigment content (Pfennig 1978), an ability more pronounced in Chromatiaceae than in the Chlorobiaceae species (Broch-Due et al. 1978).

Collectively, the Chromatiaceae thus achieve much greater ecological flexibility than the Chlorobiaceae (Pfennig 1989). The accumulation of the O_2-sensitive chromatophores within the cell core concedes, to a certain extent, protection against oxygen, which permits the organisms (supported by the ability of locomotion) to search for microaerobic regions with more advantageous light conditions. Additionally, one of the potential electron donors, i.e., S^0, may be stored intracellularly and reactivated under H_2S deficiency. Furthermore, several alternative electron donors may be utilized, and in the dark chemoautotrophic growth is possible.

Compared with this the Chlorobiaceae seem to be at a disadvantage. They are dependent not only on strictly anaerobic conditions (because of the peripheral arrangement of their chlorosomes) but also on the existence of reduced sulfur compounds and light within their habitat. Nevertheless, the Chlorobiaceae may successfully compete with the Chromatiaceae by means of higher growth rates (Pfennig 1965), a greater affinity for H_2S (Van Gemerden and Beeftink 1981, Van Gemerden 1984, 1987, Veldhuis and Van Gemerden 1986, Pfennig 1989), and more effective exploitation of lower light intensities by using their chlorosomes as light-harvesting units. The last mentioned item is especially true for those Chlorobiaceae species attaining a considerable increase in their cell surface by the formation of prosthecae (Table 14.1) densely filled with chlorosomes. A further ecological advantage is the ability of syntrophism with anaerobic heterotrophic organisms, which was shown not only by the existence of the consortia but also in experiments with mixed cultures of sulfur- as well as sulfate-reducing bacteria (Biebl and Pfennig 1978).

Provided the conditions in a stratified lake are very favorable, photosynthetic bacteria may propagate to such a degree that they cause a visible coloring of the water. Such bacterial plates, which are mostly limited to a few decimeters or meters below the chemocline, were often found in the meromictic (Kusnetzow 1958, Trüper and Genovese 1968, Culver and Brunskill 1969, Dubinina et al. 1973, Sorokin and Donato 1975, Guerrero et al. 1985) but were also occasionally found in holomictic lakes Plußsee (Biebl 1973), Kolksee (Olah et al. 1973), Kinneret (Bergstein et al. 1979), and Cisó (Guerrero et al. 1985). Usually, such mass developments were dominated by one single species.

In stratified lakes, two more or less separate layers of phototrophic bacteria may be distinguished within the vertical gradient of light intensity and H_2S concentration. To satisfy their individual physiological requirements the Chlorobiaceae regularly settle below the Chromatiaceae. This is true for the free water (Gorlenko and Kusnetzow 1971, 1972) as well as for sufficiently illuminated H_2S-containing sediments (Fenchel and Straarup 1971, Jørgensen and Fenchel 1974, Blackburn et al. 1975). In the course of these studies, however, neither a distinct bacterial plate nor a pronounced spatial separation between Chromatiaceae and Chlorobiaceae was obvious in the Plußsee. The light sulfuretum was inhabited by mixed assemblages of red and green sulfur bacteria. As was pointed out above, both families may successfully compete with each other; their ability to coexist under relatively low sulfide concentrations was also confirmed in laboratory experiments (Van Gemerden and Beeftink 1981). Occasionally a microstratification of nonsulfur and purple sulfur bacteria could be observed throughout the oxygen–H_2S gradient (Table 14.6).

The chlorophyll$_a$ of algae and Cyanobacteria shows two absorption maxima at 680 and 450 nm, respectively. Consequently, the orange and especially the blue spectral parts of the light, largely filtered out by the epilimnetic phytoplankton, are attenuated within a few meters of depth to about 1% of the surface value. Higher transmission values were measured for red (>700 nm) and green light (ca. 540–590 nm). Light at these wavelengths is mainly absorbed by the bacteriochlorophylls. The absorption spectra of algal chlorophylls and of the

Table 14.6. Phototrophic Sulfur Bacteria (PSB) and Rhodospirillaceae (RS), and Oxygen and Inorganic Phosphorus in the Metalimnion of Plußsee[a]

	25 September 1968				24 October 1968			
Depth (m)	PSB[b]	RS (cells/L)	O_2 (mg/L)	PO_4^{3-} (µg P/L)	PSB[b]	RS (cells/L)	O_2 (mg/L)	PO_4^{3-} (µg P/L)
5		1,720	7.6	5		1,810	4.4	37
6		1,090	4.3	5		1,930	4.5	42
7	✦	180	0.7	6	✦	2,070	3.8	51
8	✦✦	410	0.3	88	✦✦	440	0.6	191
9		260	0.2	190	✦✦	340	0.0	233
10		140	0.0	205		200	0.0	270

[a]Data from Biebl (1973).
[b]Relative abundance.

chlorophylls of phototrophic bacteria are complementary to each other (Schlegel and Pfennig 1961).

In the Plußsee even under the most favorable conditions (i.e., when the entire hypolimnion contained sulfide, as in 1977) the upper border of the sulfuretum did not exceed a depth of 8 m, a region that received at most 1% of red and green light. Obviously the light intensity at these depths was sufficient to permit a considerable development of phototrophic bacteria in 1977. The community included the Chromatiaceae genera *Lamprocystis* and *Amoebobacter*. If the distribution of sulfide restricts the living space of the thiobiocoenosis to even deeper strata (as in 1978), only some Chlorobiaceae may carry out active photosynthesis under these low light intensities. In both years, mainly the brownish-colored species (and some green ones) propagated here; because of their carotenoid content, they could absorb the blue-green light and thus inhabit an ecological niche nearly free from competitors.

Such selective effects of light quantity as well as light quality on the composition of a phototrophic bacterial community have been described for numerous lakes (Parkin and Brock 1980a,b) in which, as was the case in the Plußsee, both thermocline and chemocline were located far from the lake's surface. The hypolimnia of these lakes were characterized by extremely low light intensities (<1% of surface) and light of wavelengths of 450–470 nm. Mostly the brown pigmented *Chlorobium* (*C. phaeobacteroides* and *C. phaeovibrioides;* Trüper and Genovese 1968, Dubinina et al. 1973, Gorlenko et al. 1973, Sorokin and Donato 1975, Bergstein et al. 1979, Montesinos et al. 1983) and *Pelodictyon* (*P. luteum* and *P. phaeum;* Gorlenko et al. 1973) species as well as the consortia (Gorlenko and Kusnetzow 1971, 1972) were the constituting organisms.

Secondary Primary Production

The development of blooms of phototrophic bacteria within the light–sulfide gradient of the stratified lake signifies that the predominant end products of anaerobic decomposition, i.e., CO_2, H_2S, and H_2, which ultimately stem from primary production, are used for biomass production. This so-called "secondary primary production" may in certain cases reach the same order of magnitude as the primary production of the phytoplankton (Sorokin and Donato 1975, Lawrence et al. 1978) or even exceed it. The highest values (74–91% of the total production) were recorded from Solar Lake (Israel; Cohen et al. 1977), Fayetteville Green Lake (USA; Culver and Brunskill 1969), and Lake Mogilnoye (Russia; Gorlenko et al. 1978). Smaller values were found in Lake Vechten (3.9–17.5%; Steenbergen 1982, Steenbergen and Korthals 1982) as well as in several Japanese lakes (3–25%; Takahashi and Ichimura 1968). A relatively high portion of 25% was also measured in the Plußsee (Biebl 1973) within a Chlorobiaceae "plate" during the breakdown of the epilimnetic phytoplankton at the end of summer stagnation. The secondary primary production is thus subject to strong seasonal variations (Lawrence et al. 1978, Steenbergen 1982).

In most lakes the contribution of bacterial photosynthesis to total annual production is rather low (ca. 3–6%; Parkin and Brock 1981, Biebl and Pfennig 1979).

Very little correlation was found concerning lakes' circulation type (high values often derive from meromictic lakes), trophic state (Biebl and Pfennig 1979), or H_2S concentration (Takahashi and Ichimura 1970). However, a relationship was established for the light conditions within the zone of thermocline and chemocline (Parkin and Brock 1980a). On that account the comparison of chlorophyll and bacteriochlorophyll concentrations was not instructive, since the specific activity decreased severely within the cold and mostly light-limited hypolimnion (Parkin and Brock 1981).

When nutrient supply was sufficient (inorganic and organic), which it usually was in the hypolimnion, the concentration of hydrogen sulfide and, above all, the light conditions were the most decisive ecological factors determining the sulfuretum. This is true for the structure (i.e., species composition and species distribution) as well as for the function (especially bacterial photosynthesis) of thiobiocoenosis. Further physical or chemical parameters such as temperature, pH value, conductivity, salinity, etc. at most produced modifying effects.

Conclusions

The organic carbon compounds entering the anaerobic hypolimnion generally are degraded when three types of bacterial metabolism are present:

1. The fermentative bacteria can decompose complex organic matter into compounds of lower molecular weight, i.e., mainly organic acids.
2. The sulfate-reducing bacteria usually oxidize these substrates incompletely to acetate plus CO_2 (and H_2O, depending on the substrate), combined with the stoichiometric reduction of several sulfur compounds to H_2S (for results of experiments and a detailed discussion of this topic, see Oberhäuser 1981).
3. The methanogenic bacteria convert fermentation products of other anaerobes (notably CO_2, H_2, acetate) to CH_4. The latter may not be converted under anaerobic conditions but is reconducted into the carbon cycle only by the O_2-consuming activity of methane-oxidizing bacteria.

However, certain strains of sulfate-reducing bacteria were isolated (Widdel and Pfennig 1977, 1981, Widdel 1980) that are capable of the complete oxidation of acetate and higher fatty acids to CO_2, thus replacing the methane producers in the final step of anaerobic mineralization. Consequently, the sole end products of this process (CO_2 and H_2S) may, by means of light energy, be converted into biomass plus sulfate even under reducing conditions.

In this respect the cycle of matter within the anaerobic light-exposed sulfuretum runs parallel to that in the aerobic compartment. The mineralization of biomass either under O_2 reduction by aerobic heterotrophic organisms or under SO_4 reduction by sulfate reducers resulted in the formation of $CO_2 + H_2O$ or $CO_2 + H_2S$, respectively. The countercurrent autotrophic processes, oxygenic photosynthesis and anoxygenic photosynthesis, permitted, by exploitation of light energy, the regeneration of cell substance as well as oxygen or sulfate. This means that an anaerobic ecosystem, like a sulfuretum, may be considered a self-supporting system driven by passing the different oxidation stages of sulfur. The oxidation of reduced sulfur compounds without oxygen consumption by purple and green sulfur bacteria is undoubtedly one of the most important features of aquatic ecosystems.

Numerous "aerobic" organisms such as Cyanobacteria, green algae (Biebl 1973), and zooplankton were encountered within the population of "typical" sulfuretum prokaryotes. These organisms therefore must exhibit a considerable resistance to hydrogen sulfide (Durner et al. 1965). As documented, *Oscillatoria limnetica* (Cohen et al. 1975a,b) = *Phormidium hypolimneticum* (Campell 1985) = *Pseudanabaena hypolimnetica* (sensu Anagnostidis and Komárek 1988) and *Scenedesmus quadricauda* (Vincent 1980a,b) are capable of certain metabolic adaptations to anaerobic environments.

Zooplankton (e.g., ciliates, rotifers, copepods, cladocerans) frequently are observed below the chemocline (Takahashi and Ichimura 1968, Culver and Brunskill 1969, Sorokin 1970, Pfennig 1975, Sorokin and Donato 1975, Bergstein et al. 1979, Biebl and Pfennig 1979, Reck 1987). In the course of their diurnal migrations they appear within or directly above the zone of phototrophic bacteria and are temporarily active as grazers. After return to the upper aerated layer they are exposed to predation themselves. In this way the secondary primary production is transferred via microbial loops back to the classical food chains.

Purple Nonsulfur Bacteria: Rhodospirillaceae

In contrast to the purple sulfur bacteria, the Rhodospirillaceae rarely showed spectacular mass developments. This is one of the main reasons for our meager information about their distribution patterns in aquatic habitats. Van Niel (1944) writes in his monograph: "The material which seems to harbor the richest flora of Athiorhodaceae are, generally speaking, muds in which considerable microbial decomposition is going on, or the overlaying water." Biebl (1973) studied the ecology of this group with regard to their occurrence and the importance in the Plußsee and other East Holstein lakes. For enrichment of Rhodospirillaceae, water samples were filtered through membrane filters, which were spread on mineral agar with D,L-malate as the carbon source. This organic substrate can be used rather universally by Rhodospirillaceae even if they are isolated with other organic media. After filtration of 10 and 100 ml, respectively, the filters were incubated anaerobically in a light thermostat. The filters could be evaluated quantitatively and qualitatively, because the CFUs (colony-forming units) of the species were different in color and growth. The colors of the colonies of the different species ranged from olive and light brown to dark red and brown.

The Rhodospirillaceae are a small group of bacteria comprising three genera (*Rhodopseudomonas, Rhodospirillum, Rhodomicrobium*) with a total of 13 species. The dominant species in the Plußsee were *Rhodopseudomonas palustris, Rhodopseudomonas gelatinosa,* and *Rhodopseudomonas viridis. Rhodopseudomonas gelatinosa* occurred predominantly in the semiaerobic layers, *Rps. viridis* in the anaerobic ones. The numbers of *Rhodopseudomonas gelatinosa* increased in the deeper parts of the Plußsee after the fall turnover. *Rhodomicrobium vannielii, Rhodopseudomonas capsulata, Rhodospirillum rubrum, Rhodomicrobium fulvum,* and *Rhodomicrobium tenue* were found rather infrequently.

Distribution of Rhodospirillaceae in the Plußsee

When grown using CO_2 as the carbon source, Rhodospirillaceae (as well as Cyanobacteria, Chromatiaceae, and Chloroflexaceae) employ the Calvin cycle for CO_2 fixation. Rhodospirillaceae grow anaerobically on acetate in the light but also aerobically. Several species *(Rhodopseudomonas palustris, Rhodospirillum tenue, Rhodomicrobium vannielii*) use the glyoxalate cycle as an anaplerotic sequence. Experiments using enriched cultures as well as pure cultures of isolates from the Plußsee revealed the conditions under which Rhodospirillaceae can successfully compete with other bacteria. Rhodospirillaceae require anaerobic compartments, light, low-molecular organic substances, and low concentrations of sulfate to grow. When concentrations of sulfate are high, sulfate reducers will develop H_2S in the presence of a suitable organic substrate, thus inhibiting growth of purple nonsulfur bacteria.

The distribution pattern of purple sulfur and purple nonsulfur bacteria in the Plußsee (and more generally in aquatic environments) is nearly the opposite. A typical seasonal cycle in the Plußsee is shown in Figure 14.15. Initially, the Rhodospirillaceae are probably introduced continuously into the pelagic region of the lake from the littoral zone. Temporarily there was a high population density (13 March 1969; 2360 bacteria/ml, with 50% *Rhodopseudomonas viridis* and 40%

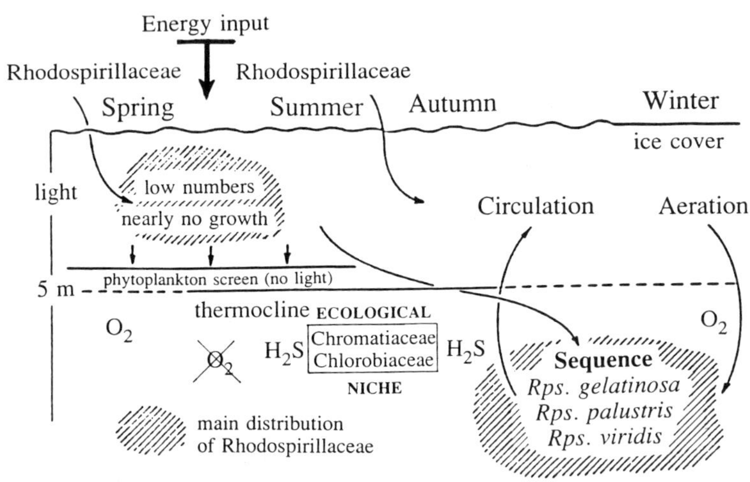

Figure 14.15. Seasonal distribution of purple sulfur and purple nonsulfur bacteria in the Plußsee. Sequence: *Rps. gelatinosa, Rhodospirillum gelatinosa* (aerobe); *Rps. palustris, Rhodospirillum palustris* (facultative anaerobe); *Rps. viridis, Rhodospirillum viridis* (anaerobe).

Rhodopseudomonas palustris). The number of Rhodospirillaceae in the littoral zone of the Plußsee was generally 10-fold higher than in the pelagic zone, where low numbers were usually found (1 to 2 cells/ml). This and the small variation in numbers, which were mostly not related to the successions of phytoplankton, indicate that these bacteria probably are not autochthonous bacteria of the pelagic zone. They seem to originate from the mud surface of the bright littoral zone. In contrast to the purple sulfur bacteria, they have no specific significance for the productivity of the lake.

The unexpected absence of Rhodospirillaceae in the upper part of the anaerobic hypolimnion with no or only low amounts of H_2S (Fig. 14.15) was apparently a result of the strong attenuation of light by the dense screen of spring and summer phytoplankton. Light was the limiting factor for growth of Rhodospirillaceae in spite of the availability of organic substrates from the metabolism of phytoplankton.

Because of their metabolic flexibility, the purple nonsulfur bacteria do not occupy a special ecological niche, as is the case with purple sulfur and green sulfur bacteria. After the fall turnover in 1968 they developed a dense population in the deeper aerobic parts of the Plußsee. It is interesting to note that the growth of the bacteria begins only after the end of fall turnover under the ice cover. The number of *Rhodopseudomonas gelatinosa* cells at the end of October 1968, on average, was 38 cells/ml. At the end of January their number rose to 8000 cells/ml. This bacterium disappeared during the winter, whereas *Rhodopseudomonas palustris* and *Rhodopseudomonas viridis* were present in the anaerobic hypolimnion throughout the following summer, but apparently with diminished physiological activity.

In the thermocline of the Plußsee we occasionally observed a microstratification of phototrophic bacteria depending on the presence or absence of H_2S (Table 14.6). The purple nonsulfur bacteria were apparently not adapted to the low concentration of dissolved organic matter in the pelagic zone. They developed predominantly in biotopes in which larger amounts of organic substances were available, e.g., during and after fall turnover. It should be pointed out that all species of phototrophic bacteria (sulfur and nonsulfur) were able to utilize organic substrates, or at least they were able to take up acetate. When interpreting their distribution pattern, photoheterotrophic competition for dissolved organic substrates should be kept in mind.

References

Åberg B, Rodhe W (1942) Über die Milieufaktoren in einigen südschwedischen Seen. Symp Bot Ups V 3:256

Albrecht D (1973) Verbesserte Methoden zur automatischen Wasseranalyse in der Limnologie. Vom Wasser 41:129–135

Albrecht D, Overbeck J (1969) Wasseranalysen in der Limnologie mit Hilfe des Technicon Auto-Analyzers. Technicon Wasserkolloquium, Bad Nauheim 1969:1–15

Albrecht D, Münster U, Stabel HH (1978) Untersuchungen zum Stoffhaushalt des Plußsees. Verh Ges Ökol Kiel 1977:109–118

Allgeier RJ, Hafford BC, Juday C (1941) Oxidation–reduction potentials and pH of lake waters and of lake sediments. Trans Wisc Acad Sci Arts Lett 33:115–133

Anagnostidis K (1968) Untersuchungen über die Salz- und Süßwasser-Thiobiocönosen (Sulphuretum) Griechenlands. Wiss Jb Phys-Math Fak Univ Thessaloniki 10:406–860

Anagnostidis K, Economou-Amilli A (1980) Limnological studies on Lake Pamvotis (Ioannina), Greece. I. Hydroclimatology, phytoplankton-periphyton with special reference to the valency of some microorganisms from sulfurate as bioindicators. Arch Hydrobiol 89:313–342

Anagnostidis K, Komárek J (1988) Modern approach of the classification system of Cyanophytes. 3. Oscillatoriales. In Anagnostidis K, Golubic S, Komórek J, Lhotsky O (eds) *Cyanophyta (Cyanobacteria) Morphology, Taxonomy, Ecology.* Arch Hydrobiol Suppl 80, Algolog Stud 50–53:327–472

Anagnostidis K, Overbeck J (1966) Methanoxidierer und hypolimnische Schwefelbakterien. Studien zur ökologischen Biocönotik der Gewässermikroorganismen. Ber Deutsch Bot Ges 79:163–174

Baars JK (1930) *Over Sulfaatreductie Door Bacterien.* W. D. Meineman, Delft, Holland

Baas-Becking LGM (1925) Studies on sulphur bacteria. Ann Bot 39:613–561

Bavendamm W (1924) *Die Farblosen und Roten Schwefelbakterien des Süß- und Salzwassers.* Fischer, Jena

Bergstein T, Henis Y, Cavari, BZ (1979) Investigations on the photosynthetic bacterium *Chlorobium phaeobacteroides* causing blooms in Lake Kinneret. Can J Microbiol 25:999–1007

Biebl H (1973) Die Verbreitung der schwefelfreien Purpurbakterien im Plußsee und anderen Seen Ostholsteins. PhD Thesis, Freiburg, Germany

Biebl H, Pfennig N (1978) Growth yields of green sulfur bacteria in mixed cultures with sulfur and sulfate reducing bacteria. Arch Mikrobiol 117:9–16

Biebl H, Pfennig N (1979) Anaerobic CO_2 uptake by phototrophic bacteria. A review. Arch Hydrobiol Beih Ergeb Limnol 12:48–58

Blackburn TH, Kleiber P, Fenchel T (1975) Photosynthetic sulfide oxidation in marine sediments. Oikos 26:103–108

Blauw TS (1988) Nitrogen fixation (acetylene reduction) in the sediments of the Pluss-See with special attention to the role of sedimentation. Arch Hydrobiol Suppl 81:359–489

Broch-Due M, Ormerod JG, Strand Fjerdingen B (1978) Effect of light intensity on vesicle formation in *Chlorobium.* Arch Microbiol 116:269–274

Buchanan RE, Gibbons NE (1974) (eds) *Bergey's Manual of Determinative Bacteriology,* 8th ed. Williams & Wilkins, Baltimore

Butlin KR, Adams ME, Thomas M (1949) The isolation and cultivation of sulfate-reducing bacteria. J Gen Microbiol 3:46–59

Caldwell DE, Tiedje JM (1975a) A morphological study of anaerobic bacteria from the hypolimnion of two Michigan lakes. Can J Microbiol 21:362–376

Caldwell DE, Tiedje, JM (1975b) The structure of anaerobic bacterial communities in the hypolimnia of several Michigan lakes. Can J Microbiol 21:377–385

Campbell S (1985) "*Oscillatoria limnetica*" from Solar Lake, Sinai is a *Phormidium* (Cyanophyta or Cyanobacteria). Arch Hydrobiol Suppl 71, Algolog Stud 38/39:175–190

Cappenberg TE (1972) Ecological observations on heterotrophic, methane oxidizing and sulfate reducing bacteria in a pond. Hydrobiologia 40:471–485

Chen KY, Morris JC (1972) Kinetics of oxidation of aqueous sulfide by O_2. Environ Sci Technol 6:529–537

Chróst RJ (1988) Phosphorus and microplankton development in a eutrophic lake. Acta Microbiol Pol 37:205–225

Chróst RJ, Overbeck J (1987) Kinetics of alkaline phosphatase activity and phosphorus availability for phytoplankton and bacterioplankton in Lake Plußsee (north German eutrophic lake). Microb Ecol 13:229–249

Clark AE, Walsby AE (1978a) The occurrence of gas-vacuolate bacteria in lakes. Arch Microbiol 118:223–228

Clark AE, Walsby AE (1978b) The development and vertical distribution of populations of gas-vacuolate bacteria in a eutrophic, monomictic lake. Arch Microbiol 118:229–233

Cohen Y, Jørgensen BB, Padan E, Shilo M (1975a) Sulphide-dependent anoxigenic photosynthesis in the cyanobacterium *Oscillatoria limnetica*. Nature 257:489–492

Cohen Y, Padan E, Shilo M (1975b) Facultative anoxygenic photosynthesis in the cyanobacterium *Oscillatoria limnetica*. J Bacteriol 123:855–861

Cohen Y, Krumbein WE, Shilo M (1977) Solar Lake (Sinai). 2. Distribution of photosynthetic microorganisms and primary production. Limnol Oceanogr 22:609–620

Culver DA, Brunskill GJ (1969) Fayetteville Green Lake, New York. V. Studies of primary production and zooplankton in a meromictic lake. Limnol Oceanogr 14:862–873

Deutsche Einheitsverfahren zur Wasser-, Abwasser- und Schlammuntersuchung (1975) Verlag Chemie, Weinheim

Drews G (1991) Cytology and morphogenesis in the prokaryotic cell. Prog Bot 52:1–9

Drews G, Imhoff JF (1991) Phototrophic purple bacteria. In *Variations in Autotrophic Life*. Academic Press, London, pp 51–97

Dubinina GA, Gorlenko VM, Suleimanov YI (1973) A study of microorganisms involved in the circulation of manganese, iron, and sulfur in meromictic Lake Gek-Gel. Microbiology 42:918–924

Durner G, Römer R, Schwartz W (1965) Untersuchungen über die Lebensgemeinschaften des Sulphuretums. Z Allg Mikrobiol 5:106–221

Fenchel T, Straarup BJ (1971) Vertical distribution of photosynthetic pigments and the penetration of light in marine sediments. Oikos 22:172–182

Fjerdingstad E (1979) *Sulfur bacteria. ASTM Special Technical Publication 650*. ASTM, Washington

Gloe A, Pfennig N, Brockmann H, Trowitzsch W (1975) A new bacterial chlorophyll from brown-colored Chlorobiaceae. Arch Microbiol 102:103–109

Gocke K (1977) Heterotrophic activity. In Rheinheimer G (ed) *Microbial Activity of a Brackish Water Environment*. Springer-Verlag, Heidelberg, pp 198–222

Golachowska JB (1979) Phosphorus forms and their seasonal changes in water and sediments of Lake Plußsee. Arch Hydrobiol 86:217–241

Gorlenko VM (1970) A new phototrophic green sulfur bacterium—*Prosthecochloris aestuarii* nov. gen. nov. spec. Z Allg Mikrobiol 10:147–149

Gorlenko VM, Kusnetzow SI (1971) Vertical distribution of photosynthetic bacteria in the Konon'er Lake of the Mari ASSR. Microbiology 40:651–652

Gorlenko VM, Kusnetzow SI (1972) Über die photosynthetisierenden Bakterien des Kononjer-Sees. Arch Hydrobiol 70:1–13

Gorlenko VM, Lebedeva EV (1971) New green sulfur bacteria with appendages. Microbiology 40:1035–1039

Gorlenko VM, Lokk SI (1979) Vertical distribution and characteristics of species composition of microorganisms in several stratified lakes in Estonia, USSR. Microbiology 48:283–290

Gorlenko VM, Zhilina TW (1968) Study on the ultrastructure of green sulfur bacteria, strain SK-413. Microbiology 37:892–897

Gorlenko VM, Chebotarev EN, Kachalkin VI (1973) Microbiological processes of oxidation of hydrogen sulfide in the Repnoe Lake (Slavonic Lakes). Microbiology 42:643–646

Gorlenko VM, Vainstein MB, Kachalkin VI (1978) Microbiological characteristic of Lake Mogilnoye. Arch Hydrobiol 81:475–492

Gorlenko VM, Dubinina GA, Kuznetsov SI (eds) (1983) *The Ecology of Aquatic Microorganisms*. Schweizerbart, Stuttgart

Guerrero E, Montesinos E, Pedrós-Alió C, et al. (1985) Phototrophic sulfur bacteria in two Spanish lakes: Vertical distribution and limiting factors. Limnol Oceanogr 30:919–931

Häusler J (1982) Schizomycetes. In Ettl H, Gerloff J, Heynig H (eds) *Süßwasserflora von Mitteleuropa, 20,* Fischer, Stuttgart, p 558

Hirsch P (1989) "Pelonemataceae" Skuja 1956. In Staley JT, Bryant MP, Pfennig N, Holt JG (eds) *Bergey's Manual of Systematic Bacteriology 3.* Williams & Wilkins, Baltimore, pp 2112–2118

Jørgensen BB, Fenchel T (1974) The sulfur cycle of a marine sediment model system. Mar Biol 24:189–201

Jørgensen BB, Kuenen JG, Cohen Y (1979) Microbial transformations of sulfur compounds in a stratified lake (Solar Lake, Sinai). Limnol Oceanogr 24:799–822

Kämpf C, Pfennig N (1980) Capacity of Chromatiaceae for chemotrophic growth. Specific respiration rates of *Thiocystis violacea* and *Chromatium vinosum.* Arch Microbiol 127:125–135

Koppe F (1924) Die Schlammflora der ostholsteinischen Seen und des Bodensees. Arch Hydrobiol 14:619–672

Krumbein WE, Buchholz H, Franke P, Giani D, Giele C, Wonneberger K (1979) O_2 and H_2S coexistence in stromatolites. A model for the origin of mineralogical lamination in stromatolites and banded iron formations. Naturwissenschaften 66:381–389

Kusnetzow SI (1958) A study of the size of bacterial populations and of organic matter formation due to photo- and chemosynthesis in water bodies of different types. Verh Int Ver Limnol 13:156–169

Lauterborn R (1913) Zur Kenntnis einiger sapropelischer Schizomyceten. Allg Bot Z 19:97–100

Lauterborn R (1915) Die sapropelische Lebewelt. Ein Beitrag zur Biologie des Faulschlammes natürlicher Gewässer. Verh Naturhist Med Ver Heidelberg 13:395–481

Lawrence JR, Haynes RC, Hammer UT (1978) Contribution of photosynthetic green sulphur bacteria to total primary production in a meromictic saline lake. Verh Int Ver Limnol 20:201–207

Lindholm T (1987) Ecology of photosynthetic prokaryotes with special reference to meromictic lakes and coastal lagoons. Acta Acad Aboensis 47:13–27

Maiden, MFI, Jones JG (1984) Gleding motility of *Peloploca* spp., and their distribution in the sediment and water column of a eutrophic lake. Arch Microbiol 140:44–49

Mechsner K (1957) Physiologische und morphologische Untersuchungen an Chlorobakterien. Arch Mikrobiol 26:32–51

Montesinos E, Guerrero R, Abella C, Esteve I (1983) Ecology and physiology of the competition for light between *Chlorobium limicola* and *Chlorobium phaeobacteroides* in natural habitats. Appl Environ Microbiol 46:1007–1016

Oberhäuser R (1981) Untersuchungen über die hypolimnische Schwefelbakterien-Vegetation in zwei ostholsteinischen Seen. PhD Thesis, Kiel

Ohle W (1953) Der Vorgang rasanter Seenalterung in Holstein. Naturwissenschaften 40:153–162

Ohle W (1954) Sulfat als "Katalysator" des limnischen Stoffkreislaufes. Vom Wasser 21: 13–32

Ohle W (1962) Der Stoffhaushalt der Seen als Grundlage einer allgemeinen Stoffwechseldynamik der Gewässer. Kiel Meeresforsch 18:107–120

Ohle W (1964) Interstitiallösungen der Sedimente, Nährstoffgehalt des Wassers und Primärproduktion des Phytoplanktons in Seen. Helg Wiss Meeresunters 10:411–429

Olah J, Biebl H, Overbeck J (1973) Photoorganotrophic utilization of acetate in a stratified eutrophic lake. Hidrol Közl 1973:21–27

Overbeck J (1967) Zur Bakteriologie des Süßwassersees—Ergebnisse und Probleme. Gas Wasserfach 108:1258–1260

Overbeck J (1968) Bakterien im Gewässer—Ein Beispiel für die gegenwärtige Entwicklung der Limnologie. Mitt Max-Planck-Ges 3:165–182

Overbeck J (1972) Zur Struktur und Funktion des aquatischen Ökosystems. Ber Dtsch Bot Ges 85:553–577

Overbeck J (1973) Über die Kompartimentierung der stehenden Gewässer—Ein Beitrag zur Struktur und Funktion des limnischen Ökosystems. Verh Ges Ökol, Saarbrücken 1973:211–223

Overbeck J (1975) Distribution pattern of uptake response in a stratified eutrophic lake (Plußsee ecosystem study IV). Verh Int Ver Limnol 19:2600–2615

Overbeck J (1979) Studies on heterotrophic functions and glucose metabolism of microplankton in Plußsee. Arch Hydrobiol Beih Ergeb Limnol 13:56–76

Parkin TB, Brock TD (1980a) Photosynthetic bacterial production in lakes: The effects of light intensity. Limnol Oceanogr 25:711–718

Parkin TB, Brock TD (1980b) The effects of light quality on the growth of phototrophic bacteria in lakes. Arch Microbiol 125:19–27

Parkin TB, Brock TD (1981) Photosynthetic bacterial production and carbon mineralization in a meromictic lake. Arch Hydrobiol 91:366–382

Parsons TR, Strickland JDH (1962) On the production of particulate organic carbon by heterotrophic processes in sea water. Deep Sea Res 8:211–222

Pfennig N (1965) Anreicherungskulturen für rote und grüne Schwefelbakterien. Zentralbl Bakteriol Suppl 1:179–189,503–505

Pfennig N (1967) Photosynthetic bacteria. Annu Rev Microbiol 21:285–324

Pfennig N (1975) The phototrophic bacteria and their role in the sulfur cycle. Plant Soil 43:1–16

Pfennig N (1978) General physiology and ecology of photosynthetic bacteria. In Clayton RK, Sistrom R (eds) *The Photosynthetic Bacteria*. Plenum Press, New York, pp 3–18

Pfennig N (1989) Ecology of phototrophic purple and green sulfur bacteria. In Schlegel HG Bowien B (eds) *Autotrophic Bacteria*. Science Tech, Madison, Wisconsin, pp 97–116

Pfennig N, Trüper HG (1981) Isolation of members of the families Chromatiaceae and Chlorobiaceae. In Starr MP, Stolp H, Trüper HG, Balows A, Schlegel HG (eds) *The Procaryotes*. Springer-Verlag, Berlin, pp 279–289

Pfennig N, Trüper HG (1989) Anoxygenic phototrophic bacteria. In Staley JT, Bryant MP, Pfennig N, Holt JG (eds) *Bergey's Manual of Systematic Bacteriology 3*. Williams & Wilkins, Baltimore, pp 1635–1761

Postgate JR (1963) Versatile medium for the enumeration of sulphate-reducing bacteria. Appl Microbiol 11:265–267

Postgate JR (1965) Enrichment and isolation of sulphate-reducing bacteria. Zentralbl Bakteriol Abt I Suppl 1:190–197

Postgate JR (1967) Media for sulfur bacteria. Lab Pract 15:1239–1244

Postgate JR (1979) *The Sulfate-Reducing Bacteria*. Cambridge University Press, Cambridge

Postgate JR, Campbell LL (1966) Classification of *Desulfovibrio* species, the nonsporulating sulfate-reducing bacteria. Bacteriol Rev 30:732–738

Reck E (1987) Zur Ökologie der pelagischen Ciliaten des Plußsees. PhD Thesis, University of Kiel, Germany

Schlegel HG, Pfennig N (1961) Die Anreicherungskultur einiger Schwefelpurpurbakterien. Arch Mikrobiol 38:1–39

Schmidt WD (1979) Morphologie und Physiologie manganoxidierender Mikroorganismen—Kultur- und in situ-Untersuchungen zur ökologisch-mikrobiologischen Charakterisierung von *Metallogenium sp.* und *Siderocapsa geminata* im Plußsee. PhD Thesis, University of Kiel

Schmidt WD, Overbeck J (1984) Studies of "iron bacteria" from Lake Pluss. 1. Morphology, finestructure and distribution of *Metallogenium* sp. and *Siderocapsa geminata*. Z Allg Mikrobiol 24:329–339

Schwartz W (1958) Die Schwefelspezialisten unter den Mikroorganismen. In Ruhland W (ed) *Handbuch der Pflanzenphysiologie IX*. Springer-Verlag, Berlin, pp 89–102

Sirevåg R (1975) Photoassimilation of acetate and metabolism of carbohydrate in *Chlorobium thiosulphatophilum*. Arch Microbiol 104:105–111

Skuja H (1956) Taxonomische und biologische Studien über das Phytoplankton schwedischer Binnengewässer. Nova Acta Reg Soc Sci Ups Ser IV 16(3):1–404

Skuja H (1964) Grundzüge der Algenflora und Algenvegetation der Fjeldgegenden um Abisko in Schwedisch-Lappland. Nova Acta Reg Soc Sci Ups Ser 4 18:1–465

Skuja H (1974) Family Pelonemataceae. In Buchanan RE, Gibbons NE (eds) *Bergey's Manual of Determinative Bacteriology,* 8th ed. Williams & Winkins, Baltimore. pp 122–127

Smith RL, Klug MJ (1981) Electron donors utilized by sulfate-reducing bacteria in eutrophic lake sediments. Appl Environ Microbiol 42:116–121

Sorokin YI (1966a) Sources of energy and carbon for biosynthesis in sulfate-reducing bacteria. Microbiology 35:643–647

Sorokin YI (1966b) Investigation of the structural metabolism of sulfate-reducing bacteria with ^{14}C. Microbiology 35:806–814

Sorokin YI (1966c) Role of carbon dioxide and acetate in biosynthesis by sulfate-reducing bacteria. Nature 210:551–552

Sorokin YI (1970) Interrelations between sulphur and carbon turnover in meromictic lakes. Arch Hydrobiol 66:391–446

Sorokin YI (1972) The bacterial population and the process of hydrogen sulphide oxidation in the Black Sea. J Cons Int Explor Mer 34:423–454

Sorokin YI, Donato N (1975) On the carbon and sulphur metabolism in the meromictic Lake Faro (Sicily). Hydrobiologia 47:241–252

Staley J (1968) *Prosthecomicrobium* and *Ancalomicrobium:* New prosthecate fresh water bacteria. J Bacteriol 95:1921–1942

Starkey RL (1938) A study of spore formation and other morphological characteristics of *Vibrio desulfuricans.* Arch Mikrobiol 9:268–304

Steenbergen CLM (1982) Contribution of photosynthetic sulphur bacteria to primary production in Lake Vechten. Hydrobiologia 95:59–64

Steenbergen CLM, Korthals HJ (1982) Distribution of phototrophic microorganisms in the anaerobic and microaerophilic strata of Lake Vechten (The Netherlands). Pigment analysis and role in primary production. Limnol Oceanogr 27:883–895

Stumm W (1967) Redox potential as an environmental parameter; conceptual significance and operational limitation. In Jaag O (ed) *Advances in Water Pollution Research, Vol 1.* Pergamon Press, Oxford, pp 283–308

Takahashi M, Ichimura S (1968) Vertical distribution and organic matter production of photosynthetic sulfur bacteria in Japanese lakes. Limnol Oceanogr 13:644–655

Takahashi M, Ichimura S (1970) Photosynthetic properties and growth of photosynthetic sulfur bacteria in lakes. Limnol Oceanogr 15:929–944

Tindall BJ, Grant WD (1986) The anoxygenic phototrophic bacteria. In Barnes EM, Mead GC (eds) *Anaerobic Bacteria in Habitats Other Than Man.* Blackwell Science Publishers, Oxford, p 444

Trüper HG, Genovese S (1968) Characterization of photosynthetic sulfur bacteria causing red water in Lake Faro (Messina, Sicily). Limnol Oceanogr 13:225–232

Trüper HG, Pfennig N (1971) Family of phototrophic green sulfur bacteria: Chlorobiaceae Copeland, the correct family name; rejection of *Chlorobacterium* Lauterborn; and the taxonomic situation of the consortium-forming species. Int J Syst Bacteriol 21:8–10

Utermöhl H (1924) Phaeobakterien (Bakterien mit braunen Farbstoffen). Biol Zentralbl 43:605–610

Utermöhl H (1925) Limnologische Phytoplanktonstudien. Arch Hydrobiol Suppl 5:1–527

Van Ert M, Staley JT (1971) Gas vacuolated strains of *Microcyclus aquaticus.* J Bacteriol 108:236–240

Van Ert M, Staley JT (1972) A new gas vacuolated heterotrophic rod from freshwaters. Arch Mikrobiol 80:70–77

Van Gemerden H (1967) On the bacterial sulfur cycle of inland waters. PhD Thesis, University of Leiden, The Netherlands

Van Gemerden H (1984) The sulfide affinity of phototrophic bacteria in relation to the location of elemental sulfur. Arch Microbiol 139:289–294

Van Gemerden H (1987) Competition between purple sulfur bacteria and green sulfur bacteria: Role of sulfide, sulfur and polysulfides. In Lindholm T (ed) *Ecology of Photosynthetic Procaryotes with Special Reference to Meromictic Lakes and Coastal Lagoons*. Abo Academy Press, Abo, pp 13–27

Van Gemerden H, Beeftink HH (1981) Coexistence of *Chlorobium* and *Chromatium* in a sulfide-limited continuous culture. Arch Microbiol 129:32–34

Van Gemerden H, Beeftink HH (1983) Ecology of phototrophic bacteria. In Ormerod JG (ed) *The Phototrophic Bacteria: Anaerobic Life in the Light*. Blackwell Science Publishers, Oxford, pp 146–185

Van Niel CB (1932) On the morphology and physiology of the purple and green sulphur bacteria. Arch Mikrobiol 3:1–112

Van Niel CB (1944) The culture, general physiology, morphology and classification of the non-sulphur purple and brown bacteria. Bacteriol Rev 8:1–118

Van Niel CB (1957) Suborder I. Rhodobacteriinae. In Breed RS, Murray EGD, Smith NR (eds) *Bergey's Manual of Determinative Bacteriology*. Williams & Wilkins, Baltimore, pp 35–66

Veldhuis MJW, van Gemerden H (1986) Competition between purple and brown phototrophic bacteria in stratified lakes: Sulfide, acetate, and light as limiting factors. FEMS Microbiol Ecol 38:31–38

Vincent WF (1980a) The physiological ecology of a *Scenedesmus* population in the hypolimnion of a hypertrophic pond. I. Photoautotrophy. Br Phycol J 15:27–34

Vincent WF (1980b) The physiological ecology of a *Scenedesmus* population in the hypolimnion of a hypertrophic pond. II. Heterotrophy. Br Phycol J 15:35–41

Walsby AE (1972) Structure and function of gas vacuoles. Bacteriol Rev 36:1–32

Walsby AE (1974) The identification of gas vacuoles and their abundance in the hypolimnetic bacteria of Arco Lake, Minnesota. Microb Ecol 1:51–61

Walsby AE (1981) Gas-vacuolate bacteria (apart from cyanobacteria). In Starr MP, Stolp H, Trüper HG, Balows A, Schlegel HG (eds) *The Prokaryotes*. 1. Springer-Verlag, Berlin, pp 441–447

Widdel F (1980) Anaerober Abbau von Fettsäuren und Benzoesäure durch neu isolierte Arten Sulfat-reduzierender Bakterien. PhD Thesis, University of Göttingen

Widdel F, Pfennig N (1977) A new anaerobic, sporing acetate-oxidizing, sulfate-reducing bacterium, *Desulfotomaculum* (emend.) *acetoxidans*. Arch Microbiol 112:119–122

Widdel F, Pfennig N (1981) Studies on dissimilatory sulfate-reducing bacteria that decompose fatty acids. I. Isolation of new sulfate-reducing bacteria enriched with acetate from saline environments. Description of *Desulfobacter postgatei* gen. nov., sp. nov. Arch Microbiol 129:395–400

Wright RT (1978) Measurement and significance of specific activity in the heterotrophic bacteria of natural waters. Appl Environ Microbiol 36:297–305

Wright RT, Hobbie JE (1965) The uptake of organic solutes in lake water. Limnol Oceanogr 10:22–28

Wright RT, Hobbie JE (1966) Use of glucose and acetate by bacteria and algae in aquatic ecosystems. Ecology 47:447–464

ZoBell CE (1946) Studies on redox potential of marine sediments. Bull Am Assoc Pet Geol 30:477–513

15. Iron Bacteria

Wolf-Dieter Schmidt and Jürgen Overbeck

Introduction

Biological precipitation of Fe oxide and/or Mn oxide is associated with bacteria of different taxonomic groups. Since the cultivation of these so-called "iron bacteria" still is problematic, and pure culture studies have not been verified, their taxonomy is based on morphological properties. Therefore, the taxonomy of many "iron bacteria" is still uncertain and needs further study (Aristovskaya 1961, Mulder and van Veen 1963, Hirsch 1967, 1974, Tyler and Marshall 1967, Dubinina 1970, 1984, Dubinina and Zhdanov 1975, Zavarzin and Hirsch 1974, Zavarzin 1989).

Siderocapsa geminata (Skuja 1954/57) is one of the organisms that were regularly found in the Plußsee. The species of genus *Siderocapsa* are classified with use by morphological criteria (Molisch 1910). Since ecological observations (Hardman and Henrici 1939, Drake 1965, 1966) and culture studies (Dubinina and Zhdanov 1975, Zhdanov and Dubinina 1975, Gorlenko et al. 1977, Schmidt and Overbeck 1984, Schmidt 1984) indicate a strong cell pleomorphism, the whole taxonomy of this group appears to be rather questionable (Tuovinen et al. 1989).

Another "iron bacterium" that was first observed in the Plußsee by Utermöhl and Gabe (mentioned by Beger and Bringmann 1952/54) was later named *Leptothrix echinata* (Beger 1935). He postulated that these microbes are related to the sheathed bacteria and especially to *Leptothrix lopholea* (Dorff 1934). Dubinina suggested that they are closely related to or even synonymous with *Metallogenium*

(Perfilev and Gabe 1961), which may be a mycoplasma (Dubinina 1969, 1970, Walsh and Mitchell 1972, Gregory et al. 1980) or even nonliving matter representing morphologically distinct Mn oxide precipitations (Bayliss 1924, Ruttner 1938, Crecelius 1974, Klaveness 1977).

Materials and Methods

Weekly samples were taken from 1975 to 1979 in the middle of the Plußsee using a Ruttner or Sorokin sampler. The stratification of the lake was recorded by continuous measurements of temperature, pH, conductivity, and oxygen content (Schmidt 1979). Water samples for microscopic studies were incubated in the laboratory under aerobic conditions (4°C) to allow for a biogenic Mn/Fe precipitation. Microscopic examinations were done twice a day. The inverted microscope was used for continuous observation of individual cells. Iron and manganese oxides were determined by the "Berliner-Blau" reaction and bencidine hydrochloride or leucoberbelin blue I, respectively (Hardman and Henrici 1939, Altmann 1972). The "iron bacteria" were identified on the basis of their morphology according to *Bergey's Manual* using light, transmission electron, and scanning electron microscopy (Schmidt and Overbeck 1984).

Results and Discussion

During the summer stratification, Mn- and/or Fe-precipitating bacteria were found in the Plußsee in the metalimnion and/or hypolimnion, where oxygen was found in very low concentrations or was absent (Table 15.1). A short aerobic incubation of hypolimnetic samples induced a strong oxide deposition resulting in tiny, reddish-brown to black pieces of flock precipitates, as described by Beger (1935), Minder (1939), Pankin (1940/41), Rehackova (1963), Kalbe et al. (1965), and Clasen (1969). Samples from the depths of 4–14 m contained oxides after an incubation period of 8–14 h. The flocks consisted of clumped phototrophic bacteria and reddish-brown to yellow, encrusted "stars" that resembled morphologically both *Leptothrix echinata* and *Metallogenium* sp. In accordance with Gorlenko and Kusnetzov (1972) and Overbeck (unpubl.), *Metallogenium* sp. was also attached to algal trichomes and to detritus. At metalimnetic oxygen concentrations below 1 mg O_2/L *Metallogenium* reached highest densities of about 7000 "coenobia"/ml at 5- to 6-m depth (Schmidt 1976).

Samples from depths of 15–19 m showed precipitation after a lag phase of 6 to 7 days. Simultaneously *Siderocapsa geminata* (Skuja 1954/57) appeared and increased in number. Maximal density of cells (21,300 cells/ml) was found in anaerobic layers of water at depths of 14–17 m (Table 15.1). Both *Metallogenium* sp. and *Siderocapsa geminata* were abundant throughout the year. From autumn circulation until the spring period, the "iron bacteria" constituted the most conspicuous and frequent populations of microorganisms in the entire water column of the Plußsee.

Table 15.1. Distribution of *Metallogenium* sp. and *Siderocapsa geminata* During Summer Stratification of the Plußsee (August 13–September 9, 1975)[a]

Depth (m)	Temperature (°C)	*Metallogenium* sp.	*Siderocapsa geminata*
0	24.0		
1	24.3		
2	23.4	Absent	
3	18.1	Absent	
4	11.4	Present	
5	8.1	Abundant	
6	6.5	Abundant	
7	5.2	Very abundant	Absent
8	5.1	Abundant	Absent
9	4.9	Rare	Very rare
10	4.8	Present	Absent
11	4.8	Rare	Absent
12	4.8	Rare	Rare
13	4.8	Present	Very rare
14	4.8	Present	Abundant
15	4.8	Absent	Very abundant
16	4.8	Absent	Abundant
17	4.8		Present
18	4.8		Rare
19	4.8		Rare
20	4.8		Absent
21	4.8		Very rare
22	4.8		Very rare
23	4.8		Very rare
24	4.6		Absent

[a]Data from Schmidt and Overbeck (1984).

Morphology and Fine Structure of *Metallogenium* sp.

Under anaerobic conditions, *Metallogenium* sp. shows a trichospheric, tiny structure without strong Mn and Fe precipitation (Fig. 15.1). These aggregations, so-called "microcolonies" or "coenobia" (Dubinina 1970), vary in size from 7 to 20 μm. They are composed of innumerable fine fibrils radiating from an almost round common center. The fibrils could only be observed in the electron microscope and possess a width of 0.03–0.07 μm and a length of 3–9 μm. With the light microscope we could only detect the condensed round centers (Figs. 15.2 and 15.3).

The ribbon-like fibrils appeared to be flexible and showed no fine structure. Nonmineralized parts were hard to identify and had no distinct borders (Fig. 15.1). The morphology of these fibrils was comparable to "arais" described by Zavarzin

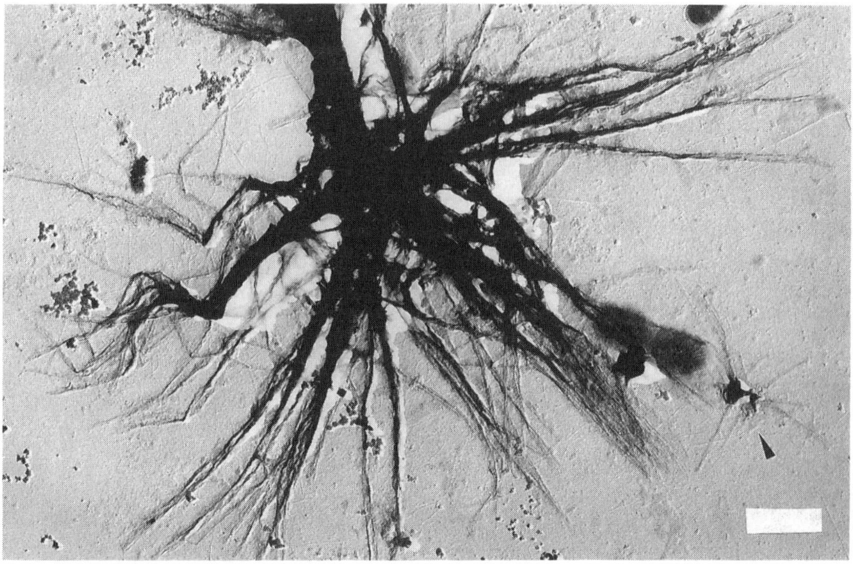

Figure 15.1. Transmission electron microscope slide of nonencrusted *Metallogenium* sp. with single, tiny fibrils and a "young colony" (arrow). Bar, 1 μm. (Modified from Schmidt 1976.)

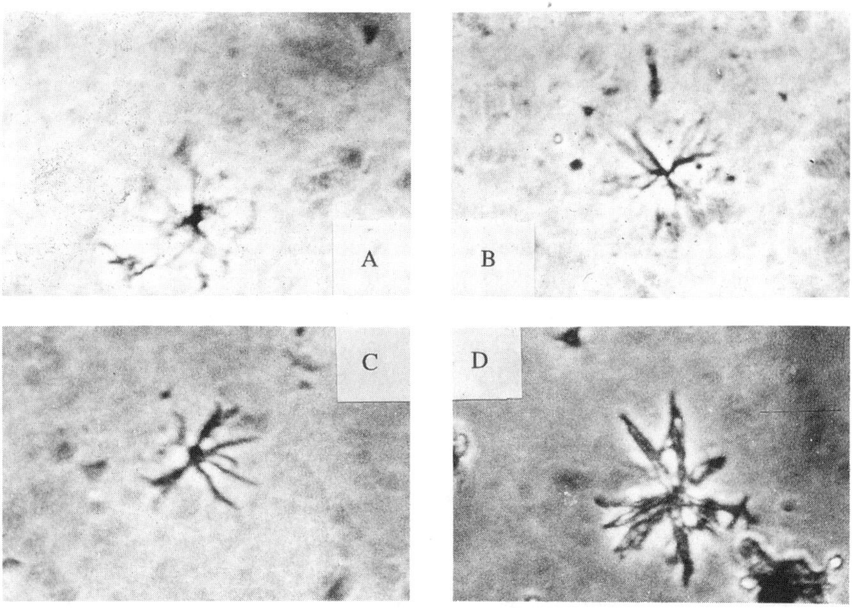

Figure 15.2. Progressive encrustation of *Metallogenium* sp.: (A) immediately after sampling, (B) 2 h after sampling, (C) 8 h after sampling, and (D) 16–20 h after sampling (light microscope, 800×; from Schmidt 1976).

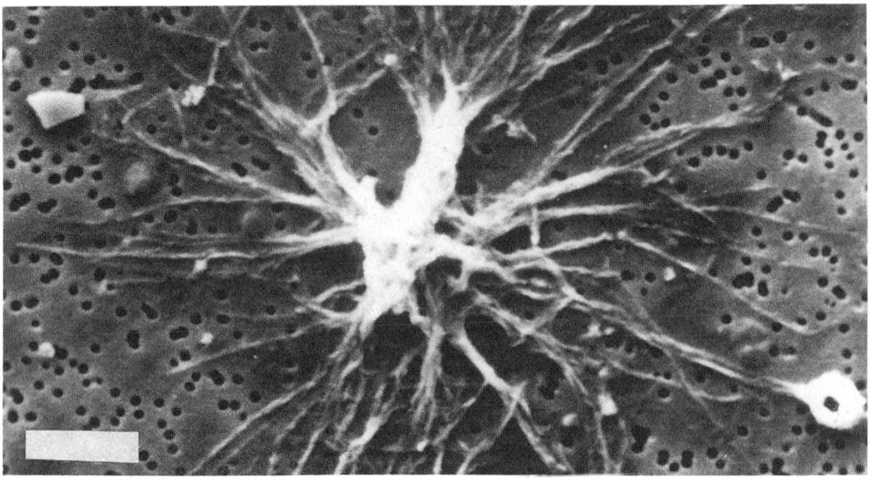

Figure 15.3. Scanning electron microscope slide of encrusted *Metallogenium* sp. with round bodies connected with the tips of the fibrils. Bar, 1 μm. (From Schmidt 1976.)

(1963) and Dubinina (1970). Occasionally the fibrils carried small round bodies (diameter 0.3–0.4 μm) on their twisted tips (Fig. 15.1, arrow). These "bodies" were the most encrusted parts of *Metallogenium* sp. Size and shape of microbes, especially a central "hole" (Fig. 15.3), were typical for the genus *Metallogenium* (Dubinina and Derjugina 1972, Kutuzova 1972, 1974, Kutuzova et al. 1972). They are supposed to be the reproduction units of *Metallogenium* (Dubinina 1970, Zavarzin and Hirsch 1974) from which small "colonies" may grow as shown in Figure 15.1.

Metallogenium of the Plußsee belongs to the preponderant manganese-precipitating "iron bacteria." The mineralization always starts on their round bodies. Proceeding centripetally, the hydroxide encrustation joins together the fine fibrils to form thick, mineralized bundles with structures sometimes identical to "empty sheaths" as described from *Leptothrix echinata* (Beger and Bringmann 1952/54). Thus, encrustation converts the irregular shape of *Metallogenium* into an almost regular "mace" with 3–30 stiff, pointed fibril bundles (Fig. 15.2), showing its development from an undetectable structure to *Metallogenium,* which is readily detectable under the light microscope.

Ultrathin sections of *Metallogenium* sp. showed no sheath but structures similar to those described by Dubinina (1969, 1970), Kutuzova et al. (1972), Klaveness (1977), and Gregory et al. (1980). None of the sections revealed structures resembling cellular constituents or organelles. An electron-poor central area detected by Klaveness (1977) was not present. The whole structure seemed to consist of an electron-poor matrix with ribbon-like substructures. The arais showed one or more uninterrupted axial filaments. We assume that the electron-dense oxides were deposited in or on this preformed filamentous matrix. Cutting it gives the impression that the minerals were precipitated as needles or "sheaths" with an empty interspace, which is actually filled by an electron-poor matrix. For more details see Schmidt (1976, 1979).

The viability of naturally occurring *Metallogenium*-like structures (Margulis et al. 1983, Maki et al. 1987, Emerson et al. 1989) is still under discussion. A stable culture has not been reported. Schmidt (1979) isolated a strain of Agaricales (Basidiomycetes) from enriched cultures of *Metallogenium* that correlated strongly to Mn oxidation and to increase of mycelium biomass. Apparently Mn^{2+} is taken up by the growing tips of the hyphae of this fungus. Dubinina (1984) showed that some Mn-oxidizing bacteria, e.g., *Arthrobacter pascens,* could be infected with *Metallogenium symbioticum,* which was parasitizing a fungus.

Our results indicate that the *Metallogenium symbioticum* structures are connected with extracellular slime from a fungus that possesses an extracellular enzymatic activity for Fe/Mn oxidation for more than two months after separation from the hyphae. *Metallogenium*-like structures also apparently develop in association with heterotrophic bacteria and fungi. Autoradiographic studies with [^{3}H]organic substrates showed no biochemical activities. Our results could not prove *Metallogenium* was an organism with a corresponding metabolism. Perhaps *Metallogenium* is a nonliving organic matrix catalyzing the star-like deposits of Mn oxide in the Plußsee (Schmidt 1976, 1979). The results are supported, e.g., by Emerson et al. (1989), who have argued "that *Metallogenium*-like structures represent morphologically distinctive Mn-oxide deposits that result from the growth and activity of viable organisms." The matrix represents a fine structure that is quite similar to bands of *Gallionella* (Balashova 1967, Hanert 1970, 1971, Blauw 1973).

The ecological importance of *Metallogenium* lies in its involvement in the cycling of manganese and iron in the lake, especially by precipitating out encrusted structures from the aerobic part of the stratified lake. Future studies have to elucidate the role of heterotrophic microorganisms for the development and manganese oxidation of *Metallogenium* (Miyajima 1992).

Morphology of *Siderocapsa geminata*

Siderocapsa sp. is characterized by a mineralized capsule with central cells (Skuja 1956). A single cell is 0.4–0.5 × 0.4–0.7 μm and of elongated to oval shape. The well-defined capsule is typical for *Siderocapsa geminata.* This planktonic "iron bacterium" occurs in the hypolimnion of the Plußsee (Table 15.1). One or two cells are situated in the middle of an almost round capsule generally encrusted by Mn and Fe oxide. A nonmineralized "halo" surrounds the cells (Fig. 15.4). Encrusted filaments, which connect the cells with the encrusted slime area, were sometimes observed and are typical for *Siderocapsa* sp. and other slime-producing microorganisms. The coat of Mn oxide is approximately 0.1 μm thick, giving the capsule a reddish-brown color. For more details see Schmidt (1976, 1979, 1984, Schmidt and Overbeck 1984).

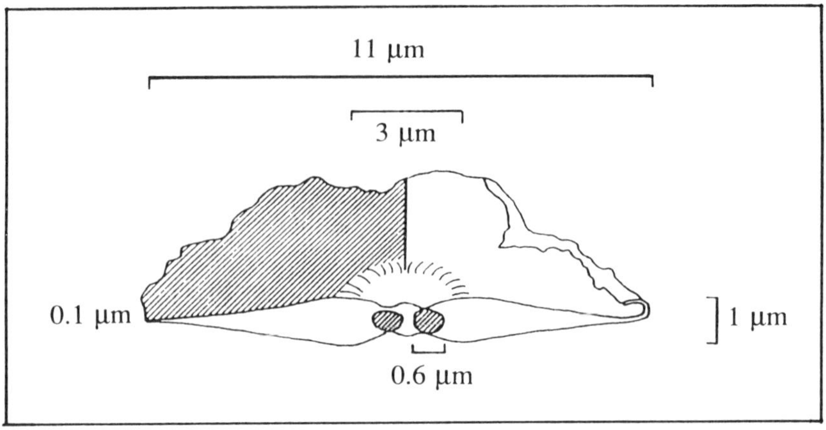

Figure 15.4. A schematic cross section of *Siderocapsa geminata;* right side not encrusted, left side encrusted. (From Schmidt and Overbeck 1984.)

Samples of *Siderocapsa geminata* exhibited a typical cell evolution after 6 days (Fig. 15.5). The cell division stopped within the next 3 days. The enlargement of the capsule began (Fig. 15.5, stage b–h), and the division area was premineralized before the whole capsule was divided. Cells without this premineralization process were horseshoe shaped (Fig. 15.5, stage g′–h′) and morphologically resembled *Oochrobium* sp. (Perfilev 1927). After approximately 10 days, the solitary, resting cells catalyzed a very strong Mn and Fe precipitation. They were the most

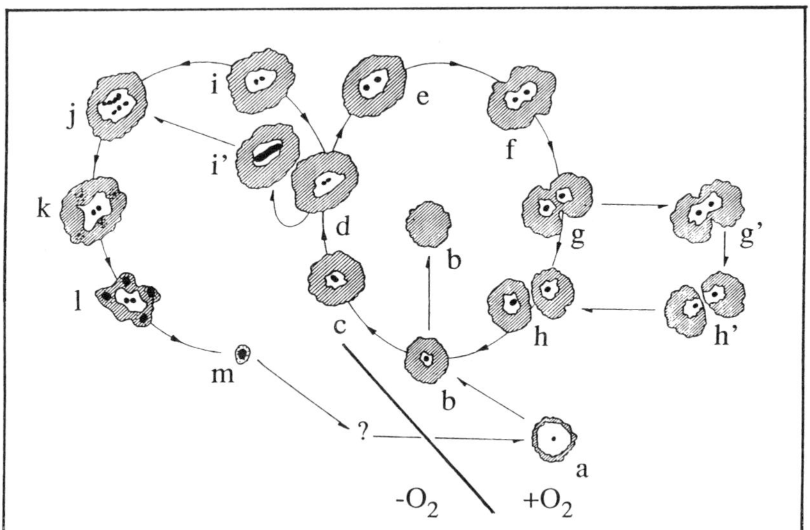

Figure 15.5. Evolution cycle of *Siderocapsa geminata* under aerobic conditions: stages a–h, 6–22 days; stages i–m, 22–26 days. (From Schmidt and Overbeck 1984.)

active "iron bacterium" state of *Siderocapsa geminata* (Schmidt 1976). After the mineralization phase and a stagnation period, a sudden cell division began (Fig. 15.5, phase i–l). The coccoid cells grew into long motile rods (0.5–0.7 × 2–3 μm) or swollen, oval stages (0.8 μm) and separated themselves once again into small coccoid cells. The cocci showed a distinct manganese encrustation of the cell surfaces (Fig. 15.5, phase i′). Further cell divisions destroyed the whole "*Siderocapsa geminata*" aggregate, and the "iron bacterium" vanished. The hydroxides were removed.

When this "culture" was incubated under anaerobic conditions for more than 1 month and then returned to aerobic conditions, *Siderocapsa geminata* again developed along nearly the same life cycle as described above. The precipitation of Mn and Fe hydroxides was associated with special growth stages and slime excretion by the pleomorphic cells. Mineralization could be catalyzed only in the stationary phase, when slime production was quite common. Similar results were obtained in Mn-oxidizing cultures (Bromfield 1974, Dubinina and Zhdanov 1975, Jung 1975, Schulze and Schweisfurth 1975, Bromfield and David 1976, Jung and Schweisfurth 1976).

All these facts indicate that *Siderocapsa geminata* is only a cell stage of a morphologically insignificant but pleomorphic bacterium that normally does not accumulate Mn and Fe oxide. Oxide reduction might be utilized by growing cells and would explain the "halo" within the capsule. The resting cells are the manganese-precipitating cell stages. Both facts could be proved by a pure culture of "*Siderocapsa geminata*" (Schmidt 1979). This points out that "iron bacteria" can switch their enzymatic activity from intracellular iron/manganese reduction to extracellular iron/manganese oxidation. Iron hydroxide can be precipitated afterward in the capsule in a redox reaction. The pleomorphic cells of *Siderocapsa geminata* did not show cell stages, which would conform to the opinion that all *Siderocapsa* spp. belong to the genus *Arthrobacter* (Clasen 1969, Dubinina and Zhdanov 1975). It is more likely that this bacterium is a manganese-precipitating microorganism generally found among many genera (Schweisfurth 1973, 1976, Schmidt 1979).

References

Altmann HJ (1972) Bestimmung von im Wasser gelöstem Sauerstoff mit Leucoberbelinblau. I. Z Anal Chem 26:97–99

Aristovskaya TV (1961) Accumulation of iron in breakdown of organomineral humus complexes by microorganism. Dokl Akad Nauk SSSR 23:954–957

Balashova VV (1967) Structure of "stem" fibres of laboratory *Gallionella filamenta* culture. Mikrobiologija 123:1050–1053

Bayliss IR (1924) Manganese in Baltimore water works supply. J Am Water Works Assoc 12:211–233

Beger H (1935) *Leptothrix echinata,* ein neues vorwiegend Mangan fällendes Eisenbakterium. Zbl Bakteriol Abt II 92:401–406

Beger H, Bringmann G (1952/54) Die Scheidenstruktur des Abwasserpilzes *Sphaerotilus* und des Eisenbakteriums *Leptothrix* im elektronischen Bilde und ihre Bedeutung für die Systematik dieser Gattungen. Zbl Bakteriol Abt II 107:318–334

Blauw TS (1973) *Gallionella ferruginea*, Ehrenberg and *Metallogenium symbiotikum*, Zavarzin. Doktor verslaag, Wageningen University, The Netherlands

Bromfield SM (1974) Bacterial oxidation of manganous ions as affected by organic substrate concentration and composition. Soil Biol Biochem 6:383–398

Bromfield SM, David DJ (1976) Sorption and oxidation of manganous ions and reduction of manganese oxide by cell suspensions of a manganese oxidizing bacterium. Soil Biol Biochem 8:37–43

Clasen J (1969) *Leptothrix echinata* und die Mangankonzentration in der Wahnbachtalsperre. Städtehyg 20:171–174

Crecelius EA (1974) The geochemical cycle of arsenic in Lake Washington and its relation to other elements. Limnol Oceanogr 20:441–451

Dorff P (1934) Die Eisenbakterien, Systematik und Morphologie. Pflanzenforschungen 16:1–62

Drake CH (1965) Occurrence of *Siderocapsa treubii* in certain waters of the Niederrhein. Gewäss Abwäss 39/40:41–63

Drake CH (1966) A note on the distribution of *Siderocapsa treubii* in certain waters of the Eifel region. Gewäss Abwäss 43:92–93

Dubinina G (1969) Inclusion of *Metallogenium* among the *Mycoplasmatales*. Dokl Akad Nauk SSSR 184:87–90

Dubinina G (1970) Untersuchungen über die Morphologie von *Metallogenium* und die Beziehungen zu *Mycoplasma*. Z Allg Mikrobiol 10:309–320

Dubinina G (1984) Infection of prokaryotic and eurykaryotic microorganisms with *Metallogenium*. Curr Microbiol 11:349–356

Dubinina G, Derjugina Z (1972) Vergleichende Untersuchungen der Mikroflora von Moorerz und Seewasser. Arch Hydrobiol 71:90–102

Dubinina G, Zhdanov AV (1975) Recognition of the iron bacteria "*Siderocapsa*" as *Arthrobacter* and description of *Arthrobacter siderocapsulatus* sp. nov. Int J Syst Bacteriol 21:340–350

Emerson D, Garen RE, Ghiorse WC (1989) Formation of *Metallogenium*-like structures by a manganese-oxidizing fungus. Arch Microbiol 151:223–231

Gorlenko WM, Kusnetzov SI (1972) Über die photosynthetisierenden Bakterien des Kononjer-Sees. Arch Hydrobiol 70:1–13

Gorlenko WA, Dubinina G, Kusnetzov SI (1977) *Ecology of Water Microorganisms*. Nauka, Moscow

Gregory E, Perry RS, Staley JT (1980) Characterisation, distribution and significance of *Metallogenium* in Lake Washington. Microb Ecol 6:125–140

Hanert H (1970) Struktur und Wachstum von *Gallionella ferrugines* Ehrenberg am natürlichen Standort in den ersten 6 Stunden der Entwicklung. Arch Mikrobiol 75:10–24

Hanert H (1971) Beginn und Verlauf der Eisen(III)hydroxid-Ablagerungen an *Gallionella*-Bändern. Z Allg Mikrobiol 11:679–682

Hardman Y, Henrici ST (1939) Study of freshwater bacteria. V. The distribution of *Siderocapsa treubii* in some lakes and streams. J Bacteriol 37:97–104

Hirsch P (1967) Biology of budding bacteria. IV. Epicellular deposition of iron by budding bacteria. Arch Mikrobiol 60:201–216

Hirsch P (1974) Budding bacteria. Annu Rev Microbiol 28:391–444

Jung WK (1975) Manganoxidierende Bakterien—mikrobiologische, biochemische und geomikrobiologische Untersuchungen. Dissertation, University of Saarbrücken.

Jung WK, Schweisfurth R (1976) Manganoxidierende Bakterien III. Z Allg Mikrobiol 16:587–597

Kalbe L, Keil R, Thiele M (1965) Licht- und elektronenoptische Studien an Arten von *Leptothrix* und *Planctomyces*. Arch Protistenkd 108:29–40

Klaveness E (1977) Morphology, distribution and significance of the manganese-accumulating microorganism *Metallogenium* in lakes. Hydrobiology 56:25–33

Kutuzova RS (1972) Electron microscopic studies of ooze-dwelling microorganisms. Mikrobiologija 41:859–861

Kutuzova RS (1974) Electron microscopic study of ooze overgrowths of an iron-oxidizing coccus related to *Siderococcus limnoliticus,* Dorff. Mikrobiologija 44:156–159

Kutuzova RS, Gabe DR, Kravikna IM (1972) Electron microscope study of overgrowths from ooze. Mikrobiologija 41:1099–1102

Maki JS, Tebo BM, Palmer FE, Nealson KH, Staley JT (1987) The abundance and biological activity of manganese-oxidizing bacteria and *Metallogenium*-like morphotypes in Lake Washington, USA. FEMS Microbiol Ecol 45:21–29

Margulis L, Grosovsky BDD, Stolz JF, et al. (1983) Distinctive microbial structures and the pre-phanerozoic fossil record. Precamb Res 20:443–477

Minder L (1939) Ein manganspeichernder Organismus im Hypolimnion des Züricher Sees. Verh Int Ver Limnol 9:301–307

Miyajima T (1992) Biological manganese oxidation in a lake II. A thermodynamic consideration of the habitat utilization of *Metallogenium* sp. Arch Hydrobiol 124:411–426

Molisch H (1910) *Die Eisenbakterien.* Gustav Fischer Verlag, Jena

Mulder EG, van Veen WL (1963) The *Sphaerotilus–Leptothrix* group. Antonie van Leeuwenhoek 29:121–153

Pankin W (1940/41) Über ein Vorkommen von *Leptothrix echinata,* Berger, im Tiefenplankton des Scharmützelsees bei Breskow. Zbl Bakteriol Parasitenkd Abt II 92:401–406

Perfilev BV (1927) Die Rolle der Mikroben in der Erzbildung. Verh Int Ver Limnol 3: 350–359

Perfilev BV, Gabe DR (1961) Capillary method of investigating microorganisms. Dokl Akad Nauk SSSR, 1–534

Rehackova V (1963) *Leptothrix echinata,* Berger in dem Stausee Klieava. Zbl Bakteriol Abt II 117:189–196

Ruttner F (1938) Limnologische Studien an einigen Seen der Ostalpen. Arch Hydrobiol 32:167–319

Schmidt WD (1976) Zur Biologie der Eisenbakterien *Siderocapsa geminata* und *Leptothrix echinata.* MSc Thesis, University of Kiel

Schmidt WD (1979) Morphologie und Physiologie manganoxidierender Mikroorganismen. Kultur- und in situ Untersuchungen zur ökologisch-mikrobiologischen Charakterisierung von *Metallogenium* sp. und *Siderocapsa geminata* im Plußsee. PhD Thesis, University of Kiel, Germany

Schmidt WD (1984) Die Eisenbakterien des Plußsees II. Morphologie und Feinstruktur von *Siderocapsa geminata* (Skuja 1954/57). Z Allg Mikrobiol 24:391–396

Schmidt WD, Overbeck J (1984) Studies of "iron bacteria" from Lake Pluss. I. Morphology, finestructure and distribution of *Metallogenium* sp. and *Siderocapsa geminata.* Z Allg Mikrobiol 24:329–339

Schulze H, Schweisfurth R (1975) Eisenoxidierende stäbchenförmige Bakterien. I. Untersuchungen unter Verwendung von Fe(II)oxalat im Medium. Z Allg Mikrobiol 15:605–614

Schweisfurth R (1971) Manganoxidierende Pilze. I. Z Allg Mikrobiol 11:415–430

Schweisfurth R (1973) Manganoxidierende Bakterien. I. Z Allg Mikrobiol 13:341–347

Schweisfurth R (1976) Manganoxidierende Bakterien. II. Z Allg Mikrobiol 16:133–147

Skuja H (1954/57) Taxonomische und biologische Studien über das Phytoplankton schwedischer Binnengewässer. Nova Acta Reg Soc Sci Ups Ser IV 16:3

Tuovinen OH, Hirsch P, Zavarzin GA (1989) Family "Siderocapsaceae" Pribam 1929. In Staley JT, Bryant MP, Pfennig N, Holt JG (eds) *Bergey's Manual of Systematic Bacteriology,* 9th ed. Williams & Wilkins, Baltimore, pp 1874–1878

Tyler PA, Marshall KC (1967) Pleomorphy in stalked, budding bacteria. J Bacteriol 93: 1132–1136

Walsh F, Mitchell R (1972) An acid-tolerant iron-oxidizing *Metallogenium*. J Gen Microbiol 72:369–376
Zavarzin GA (1963) Der Bau von *Metallogenium*. Mikrobiologija 32:1020–1023
Zavarzin GA (1989) Genus "*Metallogenium*" Perfilev and Gabe, 1961, 50. In Staley JT, Bryant MP, Pfennig N, Holt JG (eds) *Bergeys's Manual of Systematic Bacteriology*, 9th ed. Williams & Wilkins, Baltimore, pp 1986–1989
Zavarzin GA, Hirsch P (1974) Genus *Metallogenium* Perfilev and Gabe. In Staley JT, Bryant MP, Pfennig N, Holt JG (eds) *Bergey's Manual of Determinative Bacteriology*. Williams & Wilkins, Baltimore, pp 163–165
Zhdanov VA, Dubinina G (1975) *Arthrobacter siderocapsulatus* isolated from lake water. Mikrobiologiya 44:714–719

16. Metabolism of Methane and the Carbon Cycle in the Plußsee

Monir Naguib

Introduction

The carbon flow in an aquatic ecosystem can be described, from the ecophysiological point of view, in terms of the carbon cycling. The dynamic state of carbon during its cycling is reflected in its continuous conversion from the oxidized inorganic state as CO_2 to different states of the reduced forms ending with methane. The carbon cycle is not a simple one but is intersected by many loops and subloops. Therefore, for its quantification, it is necessary to have proper estimates of the turnover rates and potentials of the major reacting pathways. The decisive operating loops are aerobic heterotrophy and anaerobic heterotrophy, which includes the methane cycle. Special emphasis is placed on the methane cycle as an important integral of the whole carbon cycle. The main pathways framing the methane cycle are the methanogenesis and methane oxidation (methanotrophy). The methane cycle, which was underestimated in earlier C-balance studies, was also recognized in studies on a eutrophic shield lake and discussed by Heyer (1990).

Experimental Design and Methods

Gas Chromatographic Determination of Methane

A TC Becker-Unigraph gas chromatograph fitted with a molecular sieve and porapack Q columns was used throughout these investigations. O_2, N_2, and CH_4

were separated with the molecular sieve column, and O_2/N_2, CH_4, and CO_2 were separated with the porapack Q column (100/120 mesh). Concentrations of gases were integrated with a joined Shimadzu Chromatopack C-R2A integrator. Gas samples were injected into the columns using gas-tight microsyringes from the gas headspaces, which were already prepared for analysis.

The Gas Phase

In all estimations the gases were collected directly from the lake, produced in incubation experiments during methanogenesis, or produced from gas mixtures used in methane oxidation experiments. The pressure of the gas-tight vials used was first equalized with atmospheric pressure by connecting them to a U-shaped water manometer.

The Dissolved Phase

A quick method for estimating the amounts of dissolved methane is described by Naguib (1978). The extraction of dissolved gases in a gas head space was accomplished on the spot in previously prepared gas-tight vials. This enabled reliable and quick estimations of horizontal and vertical profiles of dissolved methane. Gas chromatographic analyses were performed in the laboratory.

Estimation of dissolved oxygen in collected water samples from the vertical profiles of the lake was done both with an oxygen probe routinely used for this purpose and with a parallel iodometric Winkler titration method.

Biochemical Oxygen Demand and Methane Oxidation Demand

The methods used to estimate biochemical oxygen demand (BOD) and methane oxidation demand (MOD) were described in detail by Naguib (1981a). The collected water samples (about 2 L) from different water levels and surface sediments were at first saturated with air by bubbling air through the samples. Standard BOD bottles were filled with the prepared lake water in a series of at least 10 bottles for each water level. In half of each series, 1–2 ml of methane gas was injected into the bottles using a special injection method. The prepared bottles with and without a methane bubble were submerged in the lake at the corresponding depth of the treated water or sediment samples. This procedure ensured more or less natural incubation conditions at prevailing temperature conditions. Analysis of dissolved O_2 was carried out at different intervals of some days with a modified Winkler method (Naguib 1981a). The difference between the BOD with and without a methane bubble is the Methane Oxidation Demand (MOD) for vertical profiles in the different sampling areas.

Isolation of Methane-Oxidizing Bacteria: Strain M 102

Difficulties encountered in isolating methane oxidizers by conventional bacteriological methods were discussed by Naguib and Overbeck (1970). The most reliable method for the isolation of pure cultures was single-cell isolation by micromanipulation (Naguib 1970a). The strain M 102 used in this study was isolated from the Plußsee by the micromanipulation method.

Stoichiometry of Methane Oxidation

This was estimated from data obtained by two different measurements. First, turnover of metabolic gases was measured in incubation experiments with pure and naturally enriched cultures. The second measurement consisted of O_2 uptake tracings, which corresponded to the oxidation of different concentrations of injected methane "solution" as revealed with a polarized Clark micro-oxygen probe in a Gibson Oxygraph (Naguib 1975, 1976, 1980).

The temperature-stabilized (27°C) reaction cell of 2.5 ml was filled with air-saturated phosphate buffer continuously stirred with a micromagnetic stirrer, and the baseline was adjusted. A defined concentrate of cell suspension was injected with a microsyringe into the reaction cell. The endogenous O_2 uptake was registered and was soon leveled to a new baseline. A measured volume of a defined "methane solution" was injected into the reaction cell. A positive reaction registered on the movable recording paper as a declining curve that ended in a parallel baseline. The O_2 uptake (V) for each methane concentration (S) was calculated from the slope of the curves obtained in a defined time. The kinetic parameters K_m and V_{max} were calculated by the $1/V$ versus $1/S$ reciprocal plot method. An example of such estimations is shown in Figure 16.3 along with estimations of cell suspensions of the strain M 102.

Estimation of Methane Ebullition

Methane ebullition was estimated by hanging inverted funnels equipped with suitable gas-collection devices at the surface and just over the sediments at different sites in the lake. The volume of collected gases was measured after equilibration with atmospheric pressure. Gases were analyzed in the laboratory.

Influence of Substrates on Methanogenesis in Sediments

The techniques include incubation of collected sediments with different substrate additives and gas chromatographic analysis of gases; they have been described in detail by Naguib (1988).

Chemical Analyses

Chemical oxygen demand (COD), dry weight, loss after ignition, protein content, and pH were routinely measured by conventional laboratory methods.

Methanotrophy

A detailed complete survey about methanotrophs is not provided here. References are made to many publications and books available that deal with the morphological, physiological, biochemical, and biotechnological aspects of methanotrophs (Ribbons et al. 1970, Quayle 1972, Anthony 1982, King 1992).

Methane-Oxidizing Bacteria (Methanotrophs)

The first investigations into the biology of methane-oxidizing bacteria in German lakes were described by Overbeck and Ohle (1964) and Naguib and Overbeck (1970). Methane oxidation by methanotrophs is an aerobic process requiring the presence of molecular oxygen. Indications of anaerobic methane oxidation utilizing the bound O_2 in nitrate or sulfate have been claimed by some investigators (Reeburgh 1979, 1980, Zehnder and Brock 1980). However, there has not been a single report until now regarding the cultivation or growth of methane oxidizers in mixed or pure cultures in the absence of molecular oxygen. Our experience with anaerobic methane oxidation, when using nitrate as an electron acceptor in already grown aerobic cultures, showed that there is simultaneous reduction of nitrate to nitrite and no sign of growth or cell multiplication. Similar findings were reported later by Smith et al. (1991). The phenomenon of net consumption of methane, which presumably does occur in anoxic marine waters and sediments, needs further clarification. The metabolic methane oxidation route goes from methanol → formaldehyde → formate → CO_2 (Fig. 16.1). The metabolism of CO_2 in methylotrophic bacteria was discussed by Overbeck (1977) and Naguib (1978, 1979).

Enrichment Cultures

Methane-oxidizing bacterial enrichment cultures were obtained from all water samples from the pelagic zone and from the sediments. The ingredients in the nutrient solutions used were a mixture of selected inorganic salts and trace elements (Overbeck 1964). Methane gas was the only organic ingredient added and functioned as the sole carbon and energy source. In closed vials, the inoculated medium was gassed with a mixture of CH_4 and O_2 in a volumetric ratio approximating 1:3 and continuously shaken to avoid the formation of a surface biofilm. Colonies on agar or silica plates were of different shapes and colors. The presence

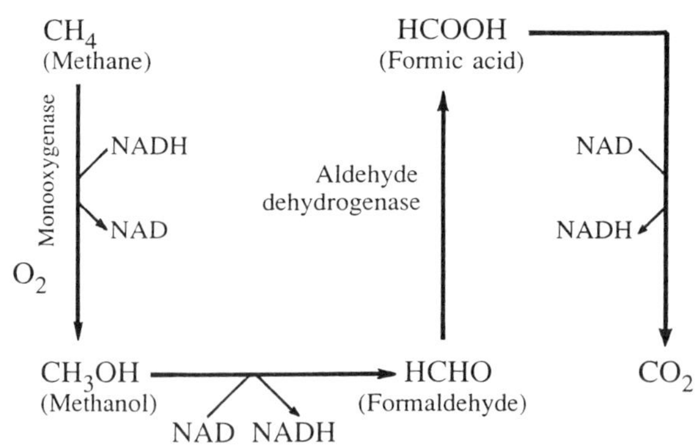

Figure 16.1. Enzymatically mediated oxidation steps of methane → methanol → formaldehyde → formate → carbon dioxide performed by methane oxidizers.

of the "pink" colonies seems to be typical in enrichments from freshwater biotopes (Overbeck 1964, Naguib and Overbeck 1970; Chapter 7, this volume).

Isolation of Pure Methanotrophic Bacteria

Isolation of pure strains by conventional bacteriological methods from agar or silica gel plates was time consuming. Colonies transferred to liquid media were either contaminated or unable to grow under methane/oxygen atmosphere. The best method was the isolation of single cells or optically controlled colonies from inoculated plates by micromanipulation (Naguib 1970a). Many pure cultures have since then been isolated, characterized, and classified by Whittenbury et al. (1970, 1977). Phylogenetic relationships based on 16 S rRNA sequences (Tsuji et al. 1990) reflected the classical taxonomic classification based on phenotypic characteristics. A different phylogeny that resulted from 5 S rRNA analyses was presented by Bulygina et al. (1990). These authors concluded that the phylogenetic relations within groups of obligate methano- and methylotrophic bacteria revealed by comparative 5 S rRNA sequence analysis do not support the current classification. However, the classification by Whittenbury et al. (1970, 1977) still remains the fundamental basis for the classification of methano- and methylotrophs. In experienced laboratories, the isolation of pure cultures of methylotrophs or specifically methanotrophs is no longer difficult because of modified standard bacteriological methods. Purified agar or silica gel solid media are inoculated with an inoculum of enriched methane-oxidizing culture. The plates are then incubated in gas-tight jars in a gas mixture of CH_4 and air (1:3 v/v) at 27°C. In these investigations, CH_4 must be secured as the only organic ingredient added during incubation. Grown colonies are then successively and frequently picked up and reinoculated on the solid media. Our best pure culture isolation technique used a Leitz micromanipulator to pick up optically controlled colonies or for single cell isolation (Naguib 1970a).

The M 102 Strain

Of the numerous pure strains isolated, our studies concentrated on one of the most active strains, M 102. The M 102 strain is an obligate methane oxidizer growing prototrophically on CH_4 as the only carbon and energy source. As will be shown later, the M 102 strain shows outstanding growth conditions and has been internationally patented for use in the industrial conversion of natural gas into protein-rich biomass (Naguib 1973, 1980, 1981b).

The cells of the M 102 strain are coccoid rods that have dimensions of 0.5–0.6 × 0.6–1.0 μm; the variation in size is dependent on the growth phase. The cells are slightly gram-negative, catalase-positive, peroxidase-positive, and acid-fast. On agar plates, the cells do not build proper colonies but rather undefined cell aggregates, forming a thin film on the surface of the agar. That is perhaps the reason why these cells have been overlooked by conventional isolation techniques. Electron microscopically, thin sections have revealed a peripheral arrangement of intracytoplasmic membranes, indicating a type II methanotroph (Fig. 16.2).

Some enzymes have been found to be localized in the membrane system (Naguib 1979). The guanine and cytosine content of 63% would confirm this membership.

The cells of the M 102 strain tend to build poly-β-hydroxybutyrate (PHB), especially in the stationary growth phase and unfavorable growth conditions. In freeze-etch preparations, the PHB polymer is stretched out of the cells and appears, depending on the degree of polymerization, as projected spikes. The exact classification of the M 102 strain is not certain. A morphological similarity with the genus *Methylocystis* should not be overlooked. However, other characteristics of the genus *Methylocystis,* e.g., the colony form and the formation of extracellular pigments, do not agree with properties of thc M 102 strain. The classification of Whittenbury et al. (1970) has been criticized by some Russian authors (Trotsenko 1976, Shishkina et al. 1976) as being tentative, since they found that the serine metabolic pathway characteristic of type II methanotrophs would function in some type I methylotrophs under certain growth conditions.

Interesting culture characteristics of thc M 102 strain are shown in the growth rates, substrate affinities, and protein content (Naguib 1975). The maximum growth rate was found to be on the order of 0.17 h^{-1}, corresponding to a maximal doubling time of 4 h. The doubling time has been determined recently with computerized controlled fermentation systems. The substrate affinity constant K_m and the maximal reaction velocity V_{max}, measured by polarographic O_2 tracings, showed values in the range of 10 μmol/L and 110–150 nmol O_2 min^{-1} mg dry $weight^{-1}$, respectively (Fig. 16.3), indicating a very high affinity and conversion rate of the specific substrate methane for the M 102 strain. These favorable data suggested the possibility of converting methane into protein-rich biomass. This idea has been adopted by some other groups for the conversion of natural gas into single-cell protein (SCP). This will be dealt with below.

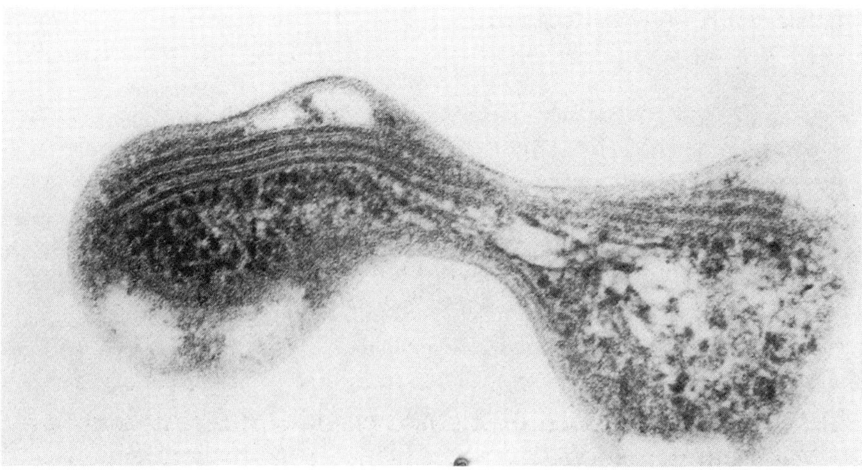

Figure 16.2. An electron microscopically thin section of a dividing cell of the M 102 strain showing the peripheral arrangements of the intracytoplasmic membrane system.

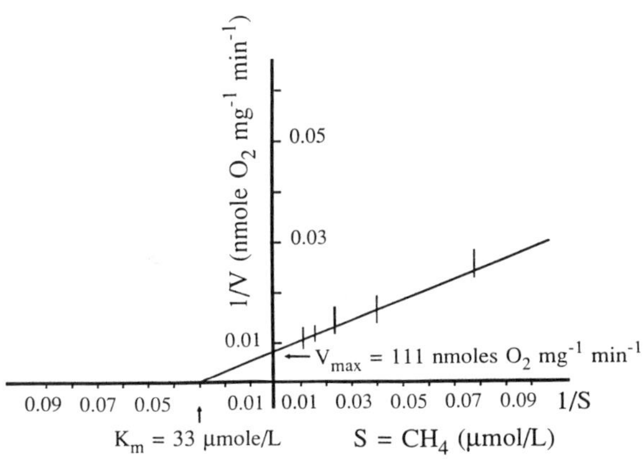

Figure 16.3. Lineweaver–Burk reciprocal plot of data obtained during oxygraph measurements of specific O_2 uptake rates (V) corresponding to the oxidation of defined methane concentrations (S). Oxygen uptake rates are per milligram of dry weight.

The kinetic parameters estimated for whole-cell suspensions of the M 102 strain, or generally enriched cultures, were not at all constant and sometimes showed no reaction to the injected methane. Several O_2 uptake measurements were gathered together from individual experiments and plotted by the Lineweaver–Burk reciprocal plot method. The results gave linear plots crossing each other as shown in Figure 16.4. The interpretation of these findings, in the sense of enzymatic inhibition reactions, gives two distinguishable inhibition patterns: competitive inhibition (the same K_m and different V_{max} values) and noncompetitive inhibition (the same V_{max} and different K_m values). One has to suppose that in the course of cell growth the enzymes responsible for the methane oxidation processes underlie one or more inhibition reactions that lead to the deviations noticed in the growth and kinetic parameters. In fact, it has been proven that the extracellular products produced during cell growth exert an inhibitory effect on the growth potential itself (Naguib 1975, 1980). The exact nature of the extracellular product that causes this inhibition phenomenon is not known. Also, the

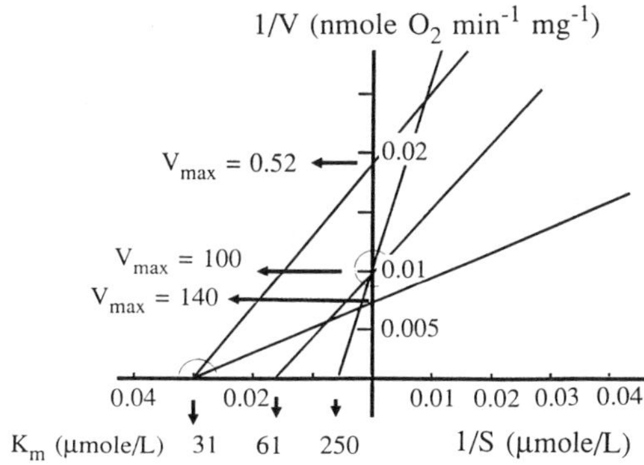

Figure 16.4. Lineweaver–Burk reciprocal plots of four experiments carried out under the same conditions. The plots reveal two different inhibition mechanisms: noncompetitive inhibition (similar K_m values and different V_{max} values) and competitive inhibition (similar V_{max} values and different K_m values). Specific velocity (V) is expressed per milligram of dry weight.

regulation mechanisms that operate to cause such inhibition patterns—feedback inhibition, for example—must be studied. However, the removal of these products by coupling a dialysis apparatus has resulted in the elimination of considerable inhibition potential, and the cell densities of the growing M 102 strain, which were grown using continuous fermenter techniques, increased rapidly from 6 g to almost 35 g dry weight per liter of culture (Naguib 1974). The inhibition occasionally was seen during growth of the methane-oxidizing bacterial cultures and raised some questions about the reproducibility of continuous cultivation. In general, such phenomena originate in nonoptimal cultivation conditions, i.e., the composition of the gas mixture, the nature and concentration of the N_2 source, absence of some trace elements, deviation of the pH from the optimal value during growth. Computerized continuous fermentation processes controlling the optimal growth conditions have led to significant improvements in the reproducibility of attained productivities.

Stoichiometry of the Biological Methane Oxidation

The molar metabolic relation of the involved gases, viz. O_2, CH_4, and CO_2, estimated for enrichment and pure cultures (see Experimental Design and Methods) is given by the following formula (see also Naguib 1970b, 1971, 1976):

$$1CH_4 + 1.2O_2 = 0.3CO_2 + \text{Water}$$

Single-Cell Protein Production

Chemical analyses showed that the raw protein content in lyophilized cells of the M 102 strain amounted to 60–70% of the dry weight and reached 80% under optimal growth conditions. The proteins revealed a broad amino acid spectrum containing all the essential amino acids in concentrations fulfilling the requirements of the FAO for biological SCP food and fodder materials (Naguib 1980, 1981b). Problems encountered in this process, including the explosive nature of the gas mixture, the irregular growth intensities, and the possibility of contaminants (bacteria, fungi, or protozoa), have been solved. The conversion of natural gas into SCP edible protein is running successfully in pilot projects, and the production of SCP may reach industrial maturity and produce enough to cover the protein deficiency in food and fodder in some parts of the world.

The Aerobic Heterotrophy

The aerobic heterotrophy is the sum of the oxygen consumed in biochemical processes taking place in the pelagic and surface sediments (1 cm thickness) and is measured in terms of biochemical oxygen demand (BOD). This is standard practice in every laboratory. An additional oxygen demand is specifically considered here resulting from the methane oxidation and is termed the methane oxidation demand (MOD). The ultimate end product in these processes is CO_2. In this chapter, the oxygen demand is differentiated into three categories: the specific, absolute, and effective BOD/MODs. The specific BOD/MOD (directly measured in mg O_2/L) was extrapolated to give the absolute BOD/MOD (in tons O_2 lake^{-1} y^{-1}). The effective BOD/MOD (meaning the actual oxygen demand taking place in situ) was calculated according to the measured O_2 and CH_4 vertical profiles and is given in tons O_2 lake^{-1} y^{-1}.

The MOD estimations can be converted to values representing the methane-oxidizing potential (MOP) using the estimated stoichiometry of biological methane oxidation. The MOP can then be defined as the biological methane uptake in the investigated biotope. This is an important item in the determination of the total budget of methane output. It should be emphasized again that in this chapter the generally known BOD measurements are extended to cover the methane-oxidizing demand (MOD). We have shown that the MOD estimations reveal a considerable potential that has been neglected and was not previously considered in oxygen balance calculations.

The BOD values were measured after incubation of pretreated water or sediment samples in the presence and absence of an injected methane gas bubble. The airtight sealed bottles were then submerged in the lake at the same water level from which the samples had been collected. The difference between the BOD+M (with methane bubble) and the BOD–M (without a methane bubble) gives the MOD for the given period of incubation time. An example of such measurements, indicating the specific potential, is given in Table 16.1.

This profile showed a pronounced high specific MOD beginning in the metalimnion, at about 10 m, at the O_2 and CH_4 interface. This trend was propagated into the hypolimnion, a pattern characteristic of eutrophic lakes during

Table 16.1. Vertical Profile of the Biochemical Oxygen Demand (BOD) of Samples Containing Methane (BOD+M) and without Methane (BOD–M), and the Methane Oxidation Demand of Oxygen (MOD) in the Plußsee (September 1979)[a]

Depth (m)	10 Days			19 Days		
	BOD+M	BOD–M	MOD	BOD+M	BOD–M	MOD
0	2.0	1.5	0.5	3.9	1.9	2.0
5	1.7	1.3	0.4	4.0	2.0	2.0
7	2.8	0.7	2.1	3.8	1.8	2.0
10	2.0	0.3	1.6	9.6	2.0	7.6
15	3.4	1.0	2.4	9.2	2.0	7.2
20	3.1	0.9	2.2		1.8	
25	3.2	1.9	1.3	9.2	2.0	7.2
28.8	4.0	1.2	2.8	9.5	3.3	6.2
29	850	480	370	2100	600	1500

[a]MOD is the difference between (BOD+M) and (BOD–M). Samples were incubated for 10 days and 19 days. BOD and MODO are expressed in mg O_2 L^{-1} d^{-1}. Initial oxygen concentration: 10 mg O_2 L^{-1}.

Table 16.2. Annual Average Estimates of the Specific, Absolute, and Effective (Actual) Potentials in the Pelagic Zone and Surface Sediments of the Plußsee[a]

Potential	Water		Sediments				
	BOD_w	MOD_w	BOD_s	MOD_s	MOD_s/ MOD_w	BODs/ BOD_w	Year
Specific ($mg\ O_2\ L^{-1}\ d^{-1}$)	0.13	0.33	36	50	150	275	
Absolute ($tons\ O_2\ lake^{-1}\ y^{-1}$)	67	148	9	14	0.1	0.13	
Effective (actual) ($tons\ O_2\ lake^{-1}\ y^{-1}$)	58	25	5.2	6.5	0.26	0.09	1979
	53	15	4.8	4.9	0.32	0.09	1987–1988
Effective/absolute (%)	86.6	16.9	57.8	46.4			
	79	10	53.3	35			

Converting the O_2 values in C equivalents, using the ratio 12:32 as for the CO_2 molecule, we arrive at the following estimates of C for the above O_2 values of the effective potential. These values are needed for balancing the C cycle.

Tons ($C\ lake^{-1}\ y^{-1}$)	21.8	9.37	1.95	2.44	(Total = 35.56)	1979
	19.9	5.6	1.8	1.8	(Total = 29.1)	1987–1988

[a]Specific potential is represented by the direct laboratory estimates in mg $O_2\ L^{-1}\ d^{-1}$. Absolute potentials are the extrapolation of the specific potentials to cover the whole lake and are given in tons $O_2\ lake^{-1}\ year^{-1}$. Effective (actual) potentials are derived from calculations based on both laboratory measurements and the distribution of O_2 and CH_4 in vertical and horizontal profiles. Ratios are approximate. w, Water; s, sediment.

periods of thermal stratification. The highest specific potentials of the BOD and MOD, however, are definitely localized in the surface sediments and could reach capacities of more than 200 times that in the free water.

The display of these specific potentials, to indicate the effective potential, depends on the distribution of the limiting factors O_2 and CH_4 and the relative volumetric and area dimensions. It is evident that a wide range of analytical data, measured in vertical and horizontal profiles, is needed. Nevertheless, there remains a certain approximation in the proper estimation of the dynamics of the oxygen uptake. Periodic monthly measurements, as well as the proper interpretation and extrapolation of the accumulated data, lead to an objective estimation. Results of measurements in the Plußsee (samples were taken in 1979 and 1987–1988) are given in Table 16.2. In 1979 meromictic conditions (incomplete turnover of the lake) were prevailing as compared to 1988, thus reflecting a relatively higher O_2 demand in the BOD and MOD values for 1979. The relationship between the effective and the absolute potentials for the whole lake were different. BOD and MOD showed variable values in different localities of the Plußsee according to the prevailing environmental conditions, which differed from year to year.

Now, in order to establish a carbon balance, the dimensions of the total effective heterotrophic potentials of the lake, aerobic and anaerobic, must be evaluated. The O_2 values have to be converted to C values using a conversion factor given in the relationship of the oxidation product of C, namely CO_2. It is of interest to note

Table 16.3. Possible Dimensions of the Methane Oxidation Potentials (Specific, Absolute, and Effective) in the Plußsee

Methane Oxidation Potential	Water	Sediment	Total
Specific ($mg\ CH_4\ L^{-1}\ d^{-1}$)	0.14	21	21.14
Absolute ($tons\ CH_4\ lake^{-1}\ y^{-1}$)	62	5.8	67.8
Effective ($tons\ CH_4\ lake^{-1}\ y^{-1}$)	10.4	2.7	13.1
Ratio (%): effective to absolute	16.8	46.6	19

that other methods used for determining the aerobic heterotrophy did not deviate much from the simple BOD method employed here. The pelagic aerobic heterotrophy using the ^{14}C method (Overbeck 1979) revealed an almost constant potential of about 20 tons C/year in the Plußsee. In the present investigations, the pelagic aerobic heterotrophic potential was calculated to be within the range of 24 tons C/year for 1977 and 22 tons C/year for 1987–1988.

The estimated MOD values can be converted into "methane" units to represent the MOP (Table 16.3). The factors for this conversion were derived from the stoichiometry of the biological methane oxidation estimated in laboratory experiments and previously described. The ratios of effective to absolute potential of the surface sediments and water from the Plußsee were 46% and 17%, respectively. This is interesting because the periods of O_2 "invasion" into the surface sediments are limited to periods of full circulation. Under meromictic conditions, the values for the sediments are much lower. However, the effective MOD or MOP in the free water column was more than three times that of the surface sediments.

It is beyond the scope of this survey to go into detail for all of the calculations. One should remember that the absolute potentials are estimations based on extrapolating the specific potentials directly measured, whereas the effective potentials are calculated on the basis of both laboratory measurements and the distribution of O_2 and CH_4 in vertical and horizontal profiles.

Methanogenesis

Methane formation is an anaerobic microbiological process performed by the methanogens that belong to Archaebacteria. Methanogenic activities are confined to anaerobic habitats, found mainly in the sediments of aquatic biotopes. The upper few centimeters, 1–5 cm, are the most active because of the sporadic supply of autochthonous and allochthonous organic substances. The intensity of the gas production is dependent on the physicochemical conditions prevailing in the sites

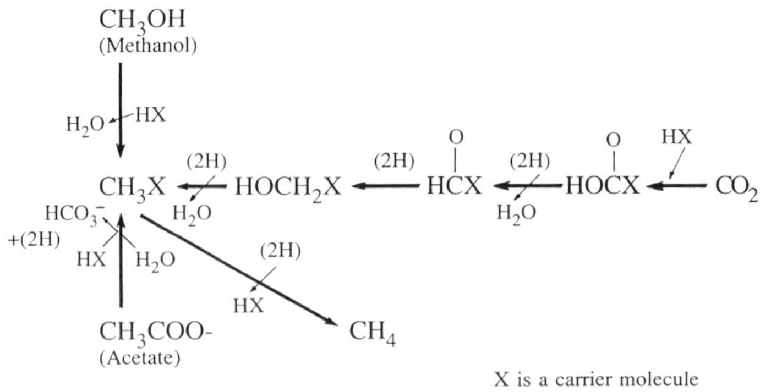

Figure 16.5. The metabolic reductive pathways of the methane precursors: CO_2, methanol, and acetate to methane.

of the methanogenic activities (Naguib 1991). Our measurements do not allow any conclusion regarding possible methanogenic activities in the water column, though theoretically it is possible that some exist in anaerobic microniches. Anaerobic sediments are the site of intensive metabolic couplings subjected variously to competition and inhibition, with CH_4 and CO_2 as the final end products if the electron flow is not directed to some noncarbon acceptors.

The methanogenic process proceeds and remains active under defined physicochemical and biological conditions. The methanogens are specialized mainly in

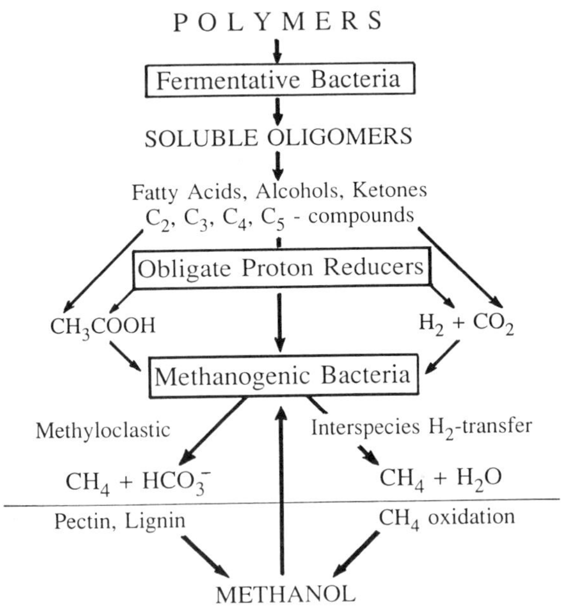

Figure 16.6. A simplified scheme of the biodegradative conversion of organic polymers to methane.

converting a certain group of compounds with low molecular structure, the methane precursors, to methane. These methane precursors are mostly monomers (e.g., CO_2, formate, methanol, acetate, methylamines) undergoing variable reduction steps until the methane stage (Fig. 16.5). They are formed at the end of previous successive bacterial degradation reactions. A simple scheme of this biodegradation chain starting with organic polymer, with some intermediate reactions, is represented in Figure 16.6. Restrictions in the reaction flow of this chain will be reflected in the intensity of the end product, methane.

Methane Precursor Preference

Extensive investigations were performed to follow and verify the degradation chain in collected sediments. Of special interest was the question of the methane precursor preference. Methane production in incubation experiments on sediments augmented with the methane precursors acetate, methanol, and CO_2/H_2 was most pronounced with methanol. Methylated amines, being also substrates for methane production (King et al. 1983), were not considered in these investigations. The clear preference for methanol, compared to acetate and CO_2/H_2, was consistent in methanogenic studies with sediments of other regional eutrophic lakes. The addition of peptone to all precursor-augmented sediments increased methane production even more. The role of peptone as a stimulant is speculative and will be dealt with later. In Table 16.4 some literature data demonstrate that the methane precursor preference in diverse anaerobic habitats is variable and selective.

The methane production from methanol and acetate, with and without peptone, was found to follow Michaelis–Menten kinetics (Naguib 1988). In analogy to enzyme kinetics, the Lineweaver–Burk reciprocal plots reveal clearly defined kinetic patterns during the methane production from acetate and methanol, as shown in Figure 16.7. Acetate showed a competitive behavior with different K_m values of 7 and 2 mmol/L in the absence and presence of peptone, respectively, with a common V_{max} of 0.3 µmol CH_4 g wet weight^{-1} d^{-1}. Methanol showed a noncompetitive pattern with a common K_m 16.5 mmol/L and different V_{max} values of 0.2 and 0.5 µmol CH_4 g wet weight^{-1} d^{-1} in aged and fresh sediments, respectively, which both increased to a common V_{max} of 0.8 µmol CH_4 g wet weight^{-1} d^{-1} in the presence of peptone. The acetate pool in eutrophic sediments lies far below the determined K_m values, which means that acetate is limiting to methanogens. The acceleration of sediment methanogenesis by acetate augmentation and stimulants such as peptone lowered the K_m values and rendered methanogens potentially more favorably competitive. Stimulation of methanol methanogenesis was reflected in the increase in V_{max} with no influence on K_m values. The stimulating effect of protein hydrolysates, whether they function as H_2 donors or as sources of important growth and metabolic ingredients or both, caused a decrease in the substrate coefficient K_m in the presence of acetate and an increase in the V_{max} in the presence of methanol. In this context, it is worthwhile mentioning that proper application of altered kinetic parameters in anaerobic systems, e.g., in sewage

Table 16.4. Overview of the Methane Precursor Preference

Biotope	Acetate	CO_2/H_2	MeOH[a]	Formate	References
Anaerobic digestion	+		–		Jerris and McCarty (1965)
	+				Smith and Mah (1966)
	+				Huser et al. (1982)
	+	+			Jain (1980)
	+				Bhadra et al. (1984)
Rumen		+			Hungate (1975)
		+			Jain (1980)
Aquatic environments					
Lake Mendota	–	+			Zeikus and Winfrey (1976)
			–		Winfrey et al. (1977)
Wintergreen	+	+		+	Strayer and Tiedje (1978)
	+	+			Lovley and Klug (1982)
	+				Lovley and Klug (1983)
Russian Lakes		+			Belayev et al. (1975)
River Saar		+			Zaiss (1981)
Plußsee	+	+	+		Naguib (1982)
Ukleisee	+	–	+		Naguib (1984)

[a]Methanol as methane precursor was found to be negative or was not investigated.

S (mmol/L)	V (μmole CH_4 g^{-1} d^{-1}) Control	+ Peptone
2.5	0.08	0.15
3.5	0.09	0.16
5	0.13	0.20
10	0.20	0.24
25	0.23	0.28
50	0.26	0.29

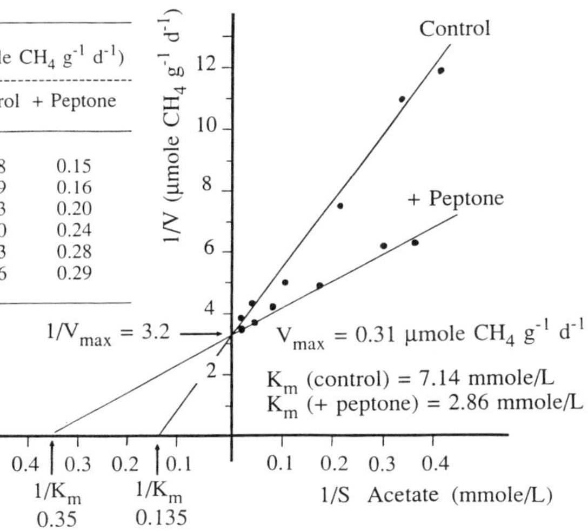

Figure 16.7. Effect of peptone on kinetics of acetate conversion to methane by surface sediments of the Plußsee. Methane production velocity (V) is expressed per gram of wet weight of sediments. Water content of the sediments was 84%.

treatment anaerobic towers or in landfills, proved to be useful in effecting increased biodegradability and biogas production (Naguib 1988).

Anaerobic Heterotrophy

The end products of carbon anaerobic heterotrophy are CH_4 and CO_2, resulting from anaerobic oxidation and fermentation processes that take place in the sediments and anaerobic niches. Some methanogenic aspects have been already described. No measurements of produced CO_2 are available, so that a direct evaluation of the CO_2 produced cannot be provided. However, an approximate estimation can be undertaken. From methanogenic experiments, the CO_2 produced during methanogenesis is about 30–40% of the volume of CH_4 produced. The CO_2 was then evaluated by calculation.

The Methane Budget

Precise information about the morphometry of the lake is needed and is a prerequisite for carrying out these estimations. Complete topographic data for the Plußsee were provided by Krambeck (1979; Chapter 1, this volume).

The fate of the methane produced in anaerobic sediments can be described and localized as follows:

1. Some of the methane gas is released from the sediments as bubbles into the atmosphere. This ebullition can be measured by collecting evolved gases in inverted funnels (Ohle 1958) stationed at different levels in the lake. The eutrophic Plußsee, with a surface area of 142,905 m^2, releases about 1 ton C/year as methane to the atmosphere. Analysis of collected bubbles at the surface of the 29-m-deep lake showed that methane comprised more than 75% of the gases reaching the surface (Fig. 16.8). The CO_2 part was substantially higher in gases released directly at the production sites and would almost instantaneously be involved in solubility and conversion reactions with the prevailing phases of CO_2 from the CO_2 pool. The intensity of the gases released underlies seasonal fluctuations. The trend of the seasonal rhythm and magnitude for collected gases, as shown in Figure 16.9 for two successive years, may be incidental and was not reproducible. This is correlated to the fact that methanogenesis is influenced by a number of variable factors.

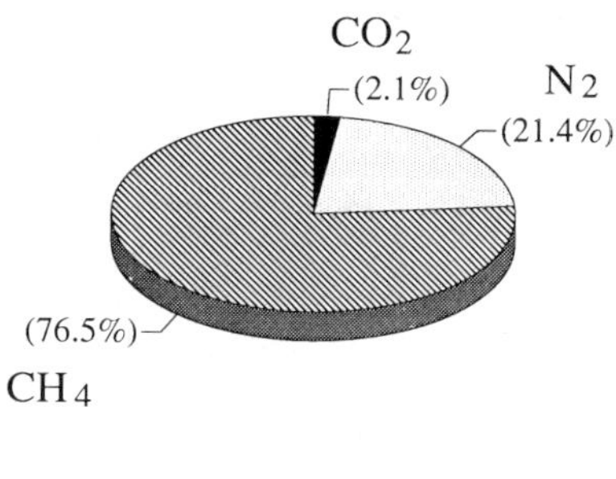

Figure 16.8. Gas chromatographic analysis of gas bubbles collected from the surface of the Plußsee.

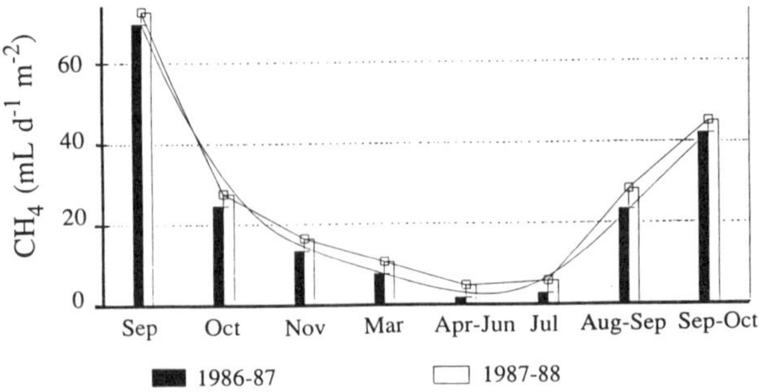

Figure 16.9. Seasonal variation in gas ebullition collected from the surface of the Plußsee during 1986–1987 and 1987–1988. The curves show that the pattern of gas ebullition during the study period was not significantly different.

2. Some of the gas bubbles produced are trapped in sediment pockets. This is designated as the "sediment stagnant pool." The exact estimation of this item is difficult and requires skilled SCUBA divers. Representative areas of the sediments were localized and marked. SCUBA divers whirled the selected areas to get them degassed. At 2-month intervals, the divers would set an inverted funnel over these marked areas and gently poke the surface sediments to collect evolving gases, which were then measured and chromatographically analyzed in the laboratory. The sediment stagnant pool of "freshly" produced methane is extrapolated from these results. The annual stagnant pool in the Plußsee amounted to about 2 tons C/year with no considerable divergences over several years.

3. Some gas was dissolved in the water and, by diffusion, built a concentration gradient in which its upper border coincided, in most cases, with the lower border of the overlying oxygen concentration gradient. In O_2-free zones, methane is trapped in a pelagic stagnant pool. This can be directly estimated by measuring the vertical distribution of methane and oxygen. In winter, methane is trapped in the whole water column under the ice cover, which forms a barrier to the atmospheric oxygen. In summer, the barrier is represented by the metalimnion. The vertical profile will begin with a condition of supersaturation in the sediments, declining upward to the thermocline. In this mixed zone the dissolved methane undergoes a highly dynamic process of continuous oxidation. This has already been quantified in the MOD values and is included in "aerobic heterotrophy." With regular measurements, trapped methane can be estimated fairly precisely even in cases when the lake becomes meromictic. With the complexity of the calculations involved, we computed a methane concentration between 2 and 2.6 tons C per year.

Accordingly, the anaerobic heterotrophic potential would amount to 3.7 tons CH_4, corresponding to 2.8 tons C. The accompanying amount of CO_2 would amount to 1.6 tons C/year, so that the total anaerobic heterotrophic potential for the year 1987–1988 would be in the order of 4.8 tons C/year.

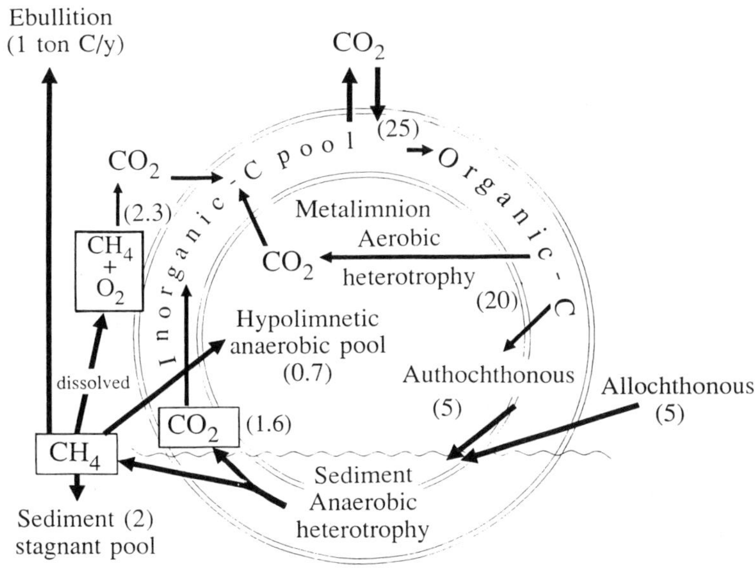

Figure 16.10. Estimates of the carbon flow in the Plußsee.

Balance of the Carbon Cycle

Balancing the C input, indicated in the autotrophic potential of 25 tons C/year, with the output, indicated in the total heterotrophic potential of 30 tons C/year, would produce a deficit of about 5 tons C/year. This would mean that during the year 1987/88, an amount of 5 tons C has invaded the lake as allochthonous material. The figures given in the C-cycle diagram (Fig. 16.10) show an attempt to visualize these relationships.

References

Anthony C (1982) *The Biochemistry of Methylotrophs*. Academic Press, London

Belayev SS, Finkelstein ZI, Ivanov MV (1975) Intensity of bacterial methane formation in ooze deposits of certain lakes. Microbiology 44:272–275

Bhadra A, Mukhopadhyay SN, Ghose TK (1984) A kinetic model for methanogenesis of acetic acid in a multireactor system. Biotechnol Bioeng 26:257–264

Bulygina ES, Galchenko VF, Govorukhina NI, et al. (1990) Taxonomic studies on methylotrophic bacteria by 5 S ribosomal RNA sequencing. J Gen Microbiol 136:441–446

Heyer J (1990) *Der Kreislauf des Methans. Mikrobiologie, Ökologie, Nutzung*. Akademie-Verlag, Berlin

Hungate RE (1975) The rumen microbial ecosystem. Annu Rev Ecol Syst 6:39–66

Huser BA, Wuhrmann K, Zehnder AJB (1982) *Methanothrix soengenii* gen. nov. sp. nov. A new acetotrophic non-hydrogen-oxidizing methane-bacterium. Arch Microbiol 132:1–9

Jain MK (1980) Microbial methanogenesis. Int J Ecol Environ Sci 6:5–17

Jerris JS, McCarty PL (1965) The biochemistry of the methane fermentation using ^{14}C tracers. J Water Pollut Control Fed 37:178–192

King GM (1992) Ecological aspects of methane oxidation, a key determinant of global methane dynamics. Adv Microb Ecol 12:431–468

King GM, Klug MJ, Lovley DR (1983) Metabolism of acetate, methanol and methylated amines in intertidal sediments of Lowes Cove, Maine. Appl Environ Microbiol 45:1848–1853

Krambeck HJ (1979) *Plussee Daten Sammlung und Computer Auswertung*. MPI für Limnologie, Plön.

Kusnetzow SI (1959) *Die Rolle der Mikroorganismen im Stoffkreislauf der Seen*. Deutscher Verlag der Wissenschaft, Berlin.

Lovley DR, Klug MJ (1982) Intermediary metabolism of organic matter in the sediments of a eutrophic lake. Appl Environ Microbiol 43:552–560

Lovley DR, Klug MJ (1983) Methanogenesis from methanol and methylamines and acetogenesis from hydrogen and carbon dioxide in the sediments of a eutrophic lake. Appl Environ Microbiol 45:1310–1315

Naguib M (1970a) Eine Technik zur Isolierung von Bakterienreinkulturen mit Hilfe des Mikromanipulators. Leitz-Mitt Wiss Techn 2:60–62

Naguib M (1970b) On methane-oxidizing bacteria in fresh waters. II. A method for the estimation and statistical evaluation of the metabolic turnover of gases by methane-oxidizing bacteria. Z Allg Mikrobiol 10:627–636

Naguib M (1971) On methane-oxidizing bacteria in fresh waters. III. The capacity of methane utilization by methane-oxidizing enrichment cultures as revealed by gas chromatographic analyses. Z Allg Mikrobiol 11:39–47

Naguib M (1973) Über Methan-oxidierende Bakterien in Binnengewässern und deren moegliche oekonomische Bedeutung. Verh Int Ver Limnol 18:1287–1299

Naguib M (1975) Overall metabolic regulations in cultures of the obligate methane-oxidizing strain M 102. In *Proceedings International Symposium on Microbial Growth on C1-Compounds*. Tokyo, pp 203–212

Naguib M (1976) On the stoichiometry of methane-oxidation in the methane-oxidizing strain M 102 under the influence of various CH_4/O_2 mixtures. Z Allg Mikrobiol 16:437–444

Naguib M (1978) A rapid method for the quantitative estimation of dissolved methane and its application in ecological research. Arch Hydrobiol 82:66–73

Naguib M (1979) Alternative carboxylation reactions in type II methylotrophs and the localization of carboxylase activities in the membrane system. Z Allg Mikrobiol 19:333–342

Naguib M (1980) The methanotrophic strain M 102 and its application for SCP production. In *Proceedings OPEC Symposium on Petroprotein*, OPEC, Kuwait. Society of Fermentation Technology, Japan, pp 253–277

Naguib M (1981a) Die Dimension des Methan-Oxidations-Potentials in einem eutrophen See. Verh GFO 11:211–222

Naguib M (1981b) Eiweiß aus Erdgas. Umschau Wiss Tech 81(11):344–345.

Naguib M (1982) Methanogenesis in sediments of inland waters. I. Methanol as the dominant precursor in the sediments of an eutrophic lake. Arch Hydrobiol 95:317–329

Naguib M (1984) Methanogenesis in sediments of eutrophic lakes. II. Comparative investigations on the methane-precursor-preference. Verh Int Ver Limnol 22:951–955

Naguib M (1988) Kinetics of acetate and methanol conversion into methane in eutrophic sediments and its application in anaerobic systems. Water Sci Tech 20:61–67

Naguib M (1991) The potentials of gas metabolism in the sediments of eutrophic and acidified lakes. Application in the treatment of anaerobic systems. In Fliermans CB, Hazen TC (eds) *Proceedings 1st International Symposium on Microbiology of the Deep Subsurface*. WSRC Information Services, Aiken, South Carolina, pp 4113–4117

Naguib M, Overbeck J (1970) On methane-oxidizing bacteria in fresh waters. 1. Introduction to the problem and investigations on the presence of obligate methane oxidizers. Z Allg Mikrobiol 10:17–36

Ohle W (1958) Die Stoffwechseldynamik der Seen in Abhängigkeit von der Gasausscheidung ihres Schlammes. Vom Wasser 25:127–149

Overbeck J. (1964) Über Anreicherung und Isolierung Methan-oxydierender Bakterien aus dem Süßwasser. Zbl Bakteriol Suppl 1:139–147

Overbeck J (1977) Some remarks on the ecology of the CO_2-metabolism of heterotrophic and methylotrophic bacteria. In Schlegel HG, Gottschalk G, Pfennig N (eds) *Microbial Production and Utilization of Gases.* Akad. Wissenschaft, Göttingen, pp 263–266

Overbeck J (1979) Studies on heterotrophic functions and glucose metabolism of microplankton in Plussee. Arch Hydrobiol Beih Ergeb Limnol 13:56–76

Overbeck J, Ohle W (1964) Contributions to the biology of methane oxidizing bacteria. Verh Int Ver Limnol 15:535–543

Quayle JR (1972) The metabolism of one-carbon compounds by microorganisms. Adv Microbiol Physiol 7:119–203

Reeburgh WS (1979) A major sink and flux control for methane in marine sediments: Anaerobic consumption. In Fanning K, Manheim ET (eds) *The Dynamic Environment of the Ocean Floor.* DC Heath, Lexington, Mass.

Reeburgh WS (1980) Anaerobic methane oxidation: Rate depth distribution in Skan Bay sediments. Earth Planet Sci Lett 47:345–352

Ribbons DW, Harrison JE, Wadzinski AM (1970) Metabolism of single carbon compounds. Annu Rev Microbiol 24:135–158

Shishkina VN, Yurchenko VV, Romanovskaya VA, Malashenko YR, Trotshenko YA (1976) On the alternativity of methane assimilation pathways in obligate methylotrophs. Mikrobiologija 3:417–419

Smith MH, Mah RA (1966) Kinetics of acetate metabolism during sludge digestion. Appl Microbiol 14:368–371

Smith RL, Howes BL, Garabedian SP (1991) In situ measurement of methane oxidation in ground water by using natural-gradient tracer tests. Appl Environ Microbiol 57:1997–2004

Strayer RF, Tiedje JM (1978) Kinetic parameters of conversion of methane precursors to methane in a hypereutrophic lake sediment. Appl Environ Microbiol 36:330–340

Trotsenko YA (1976) Isolation and characterization of obligate methanotrophic bacteria. In Schlegel HG, Gottschalk G, Pfennig N (eds) *Symposium.* Göttingen, pp 329–336.

Tsuji K, Tsien HC, Hanson RS, De Palma SR, Scholtz R, LaRoche S (1990). 16 S ribosomal RNA sequence analysis for determination of phylogenetic relationship among methylotrophs. J Gen Microbiol 136:1–10

Whittenbury R, Phillips KC, Wilkinson JF (1970) Enrichment, isolation and some properties of methane-utilizing bacteria. J Gen Microbiol 61:205–218

Whittenbury R, Colby J, Dalton H, Reed HL (1977) Biology and ecology of methane oxidizers. In Schlegel HG, Gottschalk G, Pfennig N (eds) *Microbial Production and Utilization of Gases,* Akad. Wissenschaft, Göttingen, pp 281–292

Winfrey MR, Nelson DR, Klevickis SC, Zeikus JG (1977) Association of hydrogen metabolism with methanogenesis in Lake Mendota sediments. Appl Environ Microbiol 33:312–318

Zaiss U (1981) Seasonal studies of methanogenesis and desulfurication in sediments of the river Saar. Zbl Bakteriol Hyg 2:76–89

Zehnder AJB, Brock TD (1980) Anaerobic methane oxidation: Occurrence and ecology. Appl Environ Microbiol 39:194–204

Zeikus JG, Winfrey MR (1976) Temperature limitation of methanogenesis in aquatic sediments. Appl Environ Microbiol 31:99–107

17. Bacterial Life in the Plußsee: General Remarks on Aquatic Microbial Ecology

Ryszard J. Chróst and Jürgen Overbeck

Until recently it was thought that heterotrophic bacteria in aquatic ecosystems played only a trivial role in organic fluxes and in the functioning of the ecosystem. It was commonly accepted that phytoplankton primary production was utilized predominantly by herbivores (mainly zooplankton) and passed on to the organisms that comprise the grazing food chain. Even in recently published, widely distributed books on limnology, bacteria are still treated as only remineralizers and/or decomposers of organic matter in the energy and organic matter fluxes throughout a lake ecosystem (Wetzel 1983, Cole 1988, Schwoerbel 1987, Sommer 1989, Lampert and Sommer 1993). Very little attention has been directed to further implications of bacterial dynamics for the trophic relationships between these microorganisms and higher trophic levels and the functioning of the lake ecosystem. This notion has changed significantly during the last decade with results from new methods of measuring bacterial production, biomass and activity of bacteria, and their in situ growth rates. In an ecosystem context, these new findings portray bacteria as a very dynamic metabolic and trophic component representing a major pathway for organic matter and energy flux in the aquatic food web (Sherr and Sherr 1988, Ducklow and Carlson 1992, Simon et al. 1992, Chróst 1992). We believe that the study of microorganisms in the Plußsee, described in this volume, contributes new findings to a contemporary concept of aquatic microbial ecology of lake ecosystems.

Microbial Ecology of the Plußsee

Aquatic microbial ecology has developed rapidly since the middle of the 1970s, when this science was recognized as a separate discipline and when a variety of methods and approaches were first applied in a consistent way (Overbeck et al. 1991). Having gone through a tedious lag phase, microbial ecology has now emerged as a bristling area of field and laboratory studies. The ultimate aim of microbial ecology is to answer fundamental questions: "How do microorganisms cope with an aquatic environment, and how do the chemical and physical environmental conditions combined with other biota influence the activity and function of microbial communities in a variety of aquatic ecosystems?" Aquatic microbial ecology has successfully solved many general aspects of the above queries during the last decade. Many detailed questions regarding unique features of microbial life in aquatic habitats, however, were born during the course of progress in approaches and methods now available in microbial ecology. Many questions remain unanswered.

This book reflects the development of studies and the progress in techniques and approaches of aquatic microbial ecology. The first three chapters inform the reader of the general limnological state of the Plußsee, i.e., physical factors, inorganic nutrients, plankton composition and succession, and fish fauna. Composition and concentration of dissolved organic matter and the process of photosynthetic production of organic matter by phytoplankton (i.e., factors that are prerequisites for growth and survival of aquatic heterotrophic bacteria) complete the introductory and descriptive chapters of the book. The successive chapters of the book present the dynamics of structure, function, and metabolic capabilities of the microbial populations and communities in the Plußsee. Different ecophysiological, biochemical, and molecular biological approaches (Overbeck and Chróst 1990, Chróst 1991a,b) were used for these studies.

The presentation of a heterogeneous microbial community in the Plußsee concerns temporal and spatial distribution patterns of heterotrophic bacteria in the lake, metabolic coupling between bacteria and phytoplankton or cyanobacteria, bacterial secondary production, and the fate of the produced biomass in the lake (i.e., grazing by protozoans and zooplankton, lysis by bacteriophages). Enzymatic transformation and degradation of organic matter and its utilization, biochemical and molecular aspects of enzyme regulation, and potential mechanisms of the gene transfer within microbial populations (viruses and bacteriophages) are also addressed. The major goal of these studies was to examine the different aspects of in situ metabolism of the microbial community (mainly bacteria) that are involved in the ecological processes (adaptation, succession, nutrient cycling, methanogenesis, etc.) in the lake. We did not aspire to do a complex study of the ecosystem of the Plußsee. However, many control mechanisms of the individual metabolism of microbial populations and communities that were examined and described in the book are also extremely important for the overall functioning of the lake ecosystem.

Two conceptually distinct approaches have been used to study microbial activity and processes in aquatic environments. The first approach employs pure cultures of microorganisms and defined incubation conditions in homogeneous growth medium. Earlier studies used batch cultures; nowadays a chemostat is generally used as the model system. The implicit rationale for the "culture approach" is that the natural system is too complex to be tractable, but there are many fundamental difficulties with the "culture approach" in aquatic microbial ecology (Jannasch 1974). It is therefore essential to use the second approach, which is the "in situ approach" or the "ecosystem approach" (Azam and Cho 1987), to study how microorganisms cope with an aquatic habitat. Often the "mesocosm approach," which is transitional to the "culture" and "ecosystem approach" is used. The mesocosm approach is very useful in aquatic microbial ecology. It is especially practical for model studies (Bratbak et al. 1992, Höfle 1992, Chróst and Rai 1993). A better understanding of metabolic capabilities of aquatic microbial communities very often necessitates combined studies of microbial metabolism both in the natural environment and under laboratory conditions. The in situ and culture approaches were both used in studies of microorganisms in the Plußsee.

Bacterial Microenvironment

Let us draw a simplified picture of a "bacterial aquarium" model, miniaturized to the volume of 1 µl and placed in the euphotic zone in the Plußsee during summer stratification. If all bacteria and other planktonic organisms, particulate detritus and soluble detritus, are randomly distributed, then each microliter of the volume of a "bacterial aquarium" will contain approximately 6200 bacteria (weighting roughly 132 pg), 5–10 algae, 3–10 cyanobacterial cells, 3–4 heterotrophic nanoflagellates, and more than 100,000 virus particles. Ciliates and macrozooplankton will be present in this 1 µl for only a small fraction of time. Bacterial biomass will constitute one-third of the total biomass of microplankton.

The "bacterial aquarium" will contain a large variety of organic compounds. In terms of dissolved organic carbon there will be approximately 10 ng DOC per microliter, of which no more than 250 pg will be readily utilizable by bacteria. Different heterotrophic bacteria will compete for the utilizable DOC because only 40 fg of DOC is available to a single bacterium. More than 90% of the bacteria are very small cells (cell volume smaller than 0.1 μm^3), and each average bacterium is composed of roughly 20 fg C. In the absence of new DOC inputs the amount of utilizable DOC that is available per bacterium would maintain no more than one cell division (assuming that more than 50% of available substrate would be respired). Bacteria divide frequently, usually every 6–13 h, because accompanying algae and cyanobacteria support them with readily utilizable photosynthetically produced and extracellularly released organic compounds (EOC). Approximately 60–70% of the organic carbon demand for bacterial growth and metabolism will be supplied by photosynthetic microorganisms that mutually coexist with heterotrophic bacteria. This also means that approximately 40–60% of phytoplankton primary production will be channeled to heterotrophic bacteria.

The exudation of organic substrates by photosynthetic microorganisms will be most profitable to bacteria that are in proximity to excreting algae or cyanobacteria and create the "phycosphere" surrounding a given photoautotroph. Those bacteria outside the phycosphere will use chemotaxis and actively swim toward the exuded substrates. On the way to EOC, swimming bacteria will absorb additional substrates and growth stimulators (e.g., vitamins) from the water. Their absorption capacity is very well developed because they are small, often round cells, but they have a large cell surface area that is nonproportional to their cell size. From time to time it will happen that photoautotroph will be lysed by viruses, and the high-molecular-weight constituents (HMWC) of the cell (protein, polysaccharides, nucleic acids, lipids, etc.) will be liberated into the "bacterial aquarium." After the death of an algal or cyanobacterial cell, neighboring heterotrophic bacteria will be organic substrate limited by the lack of exudation of readily utilizable EOC. A similar situation happens during the night, when photosynthetic production of bacterial substrates does not continue. Bacteria will colonize fragments of an algal or cyanobacterial cell or other particulate detritus. To be able to utilize HMWC, both in particulate and dissolved fractions, heterotrophic bacteria begin to synthesize ectoenzymes that will hydrolyze and transform HMWC to smaller subunits that are transportable to the bacterium. Thus, in the absence of active exudation of organic substrates by photoautotrophs, bacteria, after enzymatic transformation, will efficiently assimilate nonutilizable HMWC.

To supply the bacterial metabolism with EOC and/or enzymatically transformed organic substrates will result in active growth and division of bacterial cells. Bacteria will produce their biomass at the approximate rate of 1 pg C μL^{-1} h^{-1}, which is equal to about 40–60 of the "young" bacterial cells "born" during 1 h in every microliter of water of the "bacterial aquarium." Divisions and growth of bacteria, however, will not markedly influence their population density or the heaviness of the biomass. About 32–48 bacteria that are large in size will be recognized simultaneously by bacterivorous micrograzers (HNF) and eliminated from 1 μl of water per hour because of grazing. Additionally, few bacteria (2–5 cells) can be eaten when ciliate or macrozooplankton are temporarily present for a short time period in the "bacterial aquarium." Moreover, except for the grazers, which will eliminate about 80% of bacterial biomass produced per unit of time, several bacteria may be infected by phages, lysed, and liberate new phages into the "bacterial aquarium." Thus, the growth and death of aquatic bacteria are tightly coupled processes.

The crude picture of the microbial world in the Plußsee as described above is intended to give an idea of the composition and function of the unstructured microenvironment of a bacterium. In reality, it is reasonable to assume that this microenvironment will be structured, thus greatly increasing the complexity of processes and relationships between bacteria and other aquatic biota.

Strategies for Adaptation of Aquatic Bacteria

Because of the small volume of cells and the great capacity of their metabolism, bacteria are adapted to live in microenvironments that very often are characterized by distinct gradients of environmental conditions. The stable stratification of the Plußsee creates a variety of compartments of high diversity, e.g., an aerobic epilimnion with depletion of inorganic and organic nutrients and an anaerobic hypolimnion rich in high-molecular-weight dissolved organic substrates (Chapter 2, this volume) and H_2S. The supply by photosynthetically produced organic substrates (Chapters 4 and 10, this volume) and microbial enzymatic degradation and utilization of organic matter (Chapter 6) regulate the distribution of oligotrophic and saprophytic bacteria (Chapter 7). The distribution pattern of uptake kinetics in the water column gives an indication of the complexity of metabolic stratification of the Plußsee (Chapter 8). Thus, a lake's microflora is not uniform but is divided into very specialized communities and populations; as examples, there are phototrophic bacteria (Chapter 14), iron bacteria (Chapter 15), and methanotrophic bacteria (Chapter 16).

The complexity of microbial metabolism consisting of manifold catabolic, amphibolic, and anabolic pathways can be interpreted as a response of the cell to the great diversity of information from the environment. Reception of extracellular signals from the surrounding environment, exchange of the signals within the cells, and signal processing at the cellular level are steps in the continuous flow of information to which microorganisms must respond. For example, a change in conformation of an allosteric enzyme is triggered by cellular signals. Substrates of the glycolytic pathway may be the signals that act as positive effectors in the case of the allosteric enzyme phosphoenolpyruvate carboxylase and regulate changes in the anabolic and catabolic metabolism (Overbeck and Sako 1989). Such signals may also control the expression of genes and the translation of membrane proteins or enzymes. Moreover, it is also known that synthesis of many catabolic enzymes, such as ectoenzymes involved in degradation and transformation of high-molecular-weight substrates (Chróst 1990), is explained by the "signal hypothesis" (Blobel et al. 1979). The "signal peptides" control the maturation of the secreted precursors of the ectoenzymes (Verner and Schatz 1988).

The response of catabolic enzymes to alterations in growth rate in continuous cultures has been widely studied, as reviewed by Harder and Dijkhuizen (1983). It was frequently observed that decreasing substrate concentration led to an increase in the synthesis of the enzymes involved in the catabolism of that substrate. As a consequence, the utilization of the substrate was more effective. Similarly, a number of bacteria synthesized different substrate-utilizing enzymes when the substrate was limited, resulting in a net increase in the organism's affinity for the substrate (Neijssel et al. 1975). In the lake environment, the range of potential substrates available to bacteria is large (Chapter 2, this volume). It is also recognized that substrates do not enter the aquatic ecosystem at a constant rate (e.g., day and night studies of the concentration of glucose and glycolic acid in the Plußsee; Chapter 2, this volume); thus, they are not constantly available to bacteria in the same amounts and composition. In the natural environment, aquatic bacteria show rapid adaptation to changes in substrate quantity and quality, and they can induce enzyme synthesis for a substrate metabolism within a period of a few hours of nutrient pulsing (Koch 1979, Chróst 1991a,b). It is unfortunate that little is known regarding enzyme affinities and metabolic interactions involved in mixed substrate utilization by aquatic bacteria. In addition to specific metabolic pathways, it has long been known that competitiveness under conditions of nutrient limitation is related to a maintenance energy requirement (Morita 1982).

Generally, we have to differentiate between metabolic and molecular levels of mechanisms of adaptation. In physiologically based studies of microbial ecology, the metabolic adaptation has to be considered. However, both mechanisms of adaptation are coupled. At the metabolic level, adaptations are connected with the uptake of substrate and the subsequent metabolism of the carbon or inorganic nutrient sources. Allosteric enzymes play a dominant role in metabolic adaptations. Environmental changes, e.g., transition from aerobic to anaerobic conditions, from steady state to exponential growth, changes in temperature, produce effector molecules within the microbial cell. Acceptors of these effector molecules are allosteric enzymes, mostly key enzymes of metabolism. The effectors can adjust the metabolic rates in terms of fine-tuning to the actual needs of a microbial cell living under ambient conditions (Chapter 9, this volume). The understanding of the regulation of a metabolic sequence in connection with environmental changes requires knowledge of the properties of these "bottleneck enzymes" (Hecker and Babel 1988). Apart from considerations regarding the metabolic background of adaptations, one of the principal mechanisms for survival and resource competition is the high metabolic diversity of microorganisms. The bacterial communities in the Plußsee react to manifold environmental changes and are quite different throughout the distinct seasons (Chapter 7, this volume). One reason may be the tightness of the metabolic coupling between phytoplankton and accompanying bacteria, but other control mechanisms such as grazing on bacteria, unknown mutations, effect of phages, and quality and quantity of organic and inorganic substrates must also be taken into consideration.

The adaptation of bacteria to frequent and dynamic changes in conditions in the aquatic environment very often involves adjustment to low nutrient concentrations for their growth and metabolism. One of the developed strategies for adaptation is attachment of bacteria to surfaces. Large numbers of aquatic bacteria produce variable amounts of adhesive polysaccharide for attachment. The physical nature of a solid–liquid interface results in the concentration of nutrients and ions at this site, thus creating a high nutrient gradient, which attracts an attachment of bacterial cells. Attached bacteria utilize the nutrients, which are adsorbed at the particle's surface, or, in the case of detrital organic particles, bacteria produce a variety of ectoenzymes with high specific activity that solubilize colonized organic particles (Chapter 6, this volume).

Aquatic bacteria have also developed survival strategies under conditions of poor nutrient availability. Many microorganisms can survive extended periods of starvation. Many gram-positive bacteria respond to starvation by forming dormant spores, which are generally resistant to extreme environmental conditions. Nonsporulating gram-negative bacteria, which are typical aquatic autochthonous microflora (Chapter 7, this volume), remain metabolically active but also develop increased resistance to a variety of environmental stresses after exponential growth has stopped and cells enter a stationary phase. In gram-negative bacteria, the overall rate of protein synthesis decreases, but distinct sets of proteins are induced on entry into a stationary phase. Some of these proteins protect the cell against environmental challenges such as oxidative damage; others are necessary for maintaining viability. The molecular mechanism of this response involves the induction of at least one regulon, defined by the genes whose expression depends on the putative stationary-phase-specific sigma factor σ^s, the product of the *rpoS* gene also known as *katF.* Recently Zambrano et al. (1993) reported that mutations in *rpoS* can have profound effects on the ability of bacteria to compete and survive under certain conditions in stationary-phase cultures of *E. coli*. They demonstrated that mutants from the aged cultures of *E. coli* that could grow eventually took over the population, which resulted in the death of the cells from the young cultures. This phenotype was conferred by mutations in *rpoS,* which encodes a putative stationary-phase-specific sigma factor. These rapid population shifts may have important implications for the studies of microbial evolution and ecology.

The existence in gram-negative bacteria of a wide range of periplasmically located substrate-binding proteins has been clearly demonstrated (Pardee 1968). These proteins possess no demonstrated enzymatic activity. Their function is to bind a substrate, such as an amino acid (Morita 1982). Substrate-binding proteins of bacteria appear to have three different physiological functions in bacteria, which seem to have extremely important regulatory and metabolic roles for bacterial adaptation to aquatic environments (Morgan and Dow 1986). First, together with the transmembrane electrochemical proton gradient, the binding proteins function in driving substrate uptake systems. Second, they are involved in chemotactic responses of the bacterium. Third, binding proteins are also the sensors of the environment with a potential role in metabolic regulation and cell type expression.

Because acquisition of substrates is the critical process for growth and metabolism, the aquatic bacteria have developed multiphasic transport systems for substrate uptake (Azam and Hodson 1981) that are distinctly different in high and low nutrient concentrations. It appears that high-affinity–low-capacity substrate transport systems may be responsible for scavenging of substrates from the dilute bulk phase of the DOM pool. There is extensive evidence for high-affinity substrate transport systems in a range of bacteria that utilize a range of substrates (e.g., glucose, certain amino acids, phosphate) and also evidence for the involvement of binding proteins in high-affinity uptake systems (Rosen 1971). The low-affinity–high-capacity transport systems, however, may enable the bacteria to optimize substrate uptake from high microenvironmental concentrations, e.g., in the vicinity of enzymatically solubilized POM. Our studies clearly indicated that EOC, enzymatically solubilized POM, and transformed DOM are the major sources of substrates for bacteria in the Plußsee (Chapters 4, 5, and 6, this volume). They are taken up as they become available to bacterial transport systems. Thus, in light of the studies mentioned above, the bacterial hydrolysis–uptake coupling system may be an important strategy for bacteria in aquatic environments.

A wide range of bacteria, when living in temporarily nutrient-rich environments, are known to accumulate excess nutrients and to store them intracellularly as polymers (e.g., polyphosphates, lipids, polyhydroxybutyrate), which are very slowly utilized during starvation conditions (Dawes 1976). Konnings and Veldkamp (1983) speculated that certain bacteria have the ability to block their transport systems below critical proton-motive force levels and hence to prevent efflux of accumulated nutrients. Nutrient gradients are widespread in aquatic environments, and bacteria use chemotaxis to acquire substrates. Sensory transduction in chemotactic bacteria, mediated by specific binding proteins for a number of chemoattractants, is documented (Morita 1982), but there are few existing studies of chemotactic responses in bacterial competitiveness in low-nutrient microenvironments in the aquatic environment. Chemotaxis has been shown to be important in *Caulobacter* and has been partially characterized at the molecular level (Gomes and Shapiro 1984), but ecological data are lacking.

Outlined above are the main metabolic adaptations for bacterial growth and survival in aquatic environments, many of which we described or postulated throughout this volume of our studies in the Plußsee. The preceding short summary describes the main strategies bacteria use to achieve the "best fit" to the aquatic environment and clearly indicates the necessity of different methods and approaches that have to be used in the study of microbial ecology. The microbial ecologist must always use various approaches in his or her efforts to answer questions on the structure and function of microbial communities and their responses to environmental stress. In view of the inherent difficulties of almost all microbiological methods, one must realize that very few if any of these methods will produce the "absolute truth." Using different approaches to solve any one problem requires consistency in interpreting the uncertain results obtained from these methods. Frequently, the need for consistency will enable us to draw more far-reaching conclusions than those we might otherwise have dared to form by examining the results of the different methods separately. There is still a strong need for a multidisciplinary approach to aquatic microbial ecology. Integration is a necessary part of future studies in microbial ecology. Not only it is important in space and time, but integration is equally important among various disciplines involved in environmental research.

References

Azam F, Cho BC (1987) Bacterial utilization of organic matter in the sea. In Fletcher M, Gray TRG, Jones JG (eds) *Ecology of Microbial Communities*. Cambridge University Press, Cambridge, pp 261–281

Azam F, Hodson RR (1981) Multiphasic kinetics for D-glucose uptake by assemblages of natural marine bacteria. Mar Ecol Prog Ser 6:213–222

Blobel G, Walter P, Chang CN, Goldman BM, Erickson AH, Lingappa VR (1979) Translation of proteins across membranes: The signal hypothesis and beyond. Symp Soc Exp Biol 33:9–36

Bratbak G, Heldal M, Thingstad TF, Riemann B, Haslund OH (1992) Incorporation of viruses into the budget of microbial C-transfer. A first approach. Mar Ecol Prog Ser 83:273–280

Chróst RJ (1990) Microbial ectoenzymes in aquatic environments. In Overbeck J, Chróst RJ (eds) *Aquatic Microbial Ecology. Biochemical and Molecular Approaches*. Springer-Verlag, New York, pp 47–78

Chróst RJ (1991a) Environmental control of the synthesis and activity of aquatic microbial ectoenzymes. In Chróst RJ (ed) *Microbial Enzymes in Aquatic Environments*. Springer-Verlag, New York, pp 29–59

Chróst RJ (ed) (1991b) *Microbial Enzymes in Aquatic Environments*. Springer-Verlag, New York

Chróst RJ (1992) Significance of bacterial ectoenzymes in aquatic environments. Hydrobiologia 243/244:61–70

Chróst RJ, Rai H (1993) Ectoenzyme activity and bacterial secondary production in nutrient-impoverished and nutrient-enriched freshwater mesocosms. Microb Ecol 25:131–150

Cole GA (1988) *Textbook of Limnology*. Waveland Press, Prospect Heights, Illinois

Dawes EA (1976) Endogenous metabolism and the survival of starved prokaryotes. In Gray TRG, Postgate JR (eds) *The Survival of Vegetative Microbes*. Cambridge University Press, Cambridge, pp 19–53

Ducklow HW, Carlson CA (1992) Oceanic bacterial production. Adv Microb Ecol 12:113–181

Gomes SL, Shapiro L (1984) Differential expression and positioning of chemotaxis methylation proteins in *Caulobacter*. J Mol Biol 178:551–568

Harder W, Dijkhuizen L (1983) Physiological responses to nutrient limitation. Annu Rev Microbiol 37:1–23

Hecker M, Babel W (eds) (1988) *Physiologie der Mikroorganismen*. Gustav Fischer Verlag, Jena

Höfle MG (1992) Bacterioplankton community structure and dynamics after large-scale release of nonindigenous bacteria as revealed by low-molecular-weight-RNA analysis. Appl Environ Microbiol 58:3387–3394

Jannasch HW (1974) Steady-state and the chemostats in ecology. Limnol Oceanogr 19:716–720

Koch AL (1979) Microbial growth in low concentrations of nutrients. In Shilo M (ed) *Strategies of Microbial Life in Extreme Environments*. Verlag Chemie, Weinheim, pp 341–356

Konnings WN, Veldkamp H (1983) Energy transduction and solute transport mechanisms in relation to environment occupied by micro-organisms. In Slater JH, Whittenbury R, Wimpenny JWT (eds) *Microbes in Their Natural Environments*. Cambridge University Press, Cambridge, pp 153–186

Lampert W, Sommer U (1993) *Limnoökologie*. Georg Thieme Verlag, Stuttgart

Morgan P, Dow CS (1986) Bacterial adaptations for growth in low nutrient environments. In Herbert RA, Codd GA (eds) *Microbes in Extreme Environments*. Academic Press, London, pp 187–214

Morita RY (1982) Starvation-survival of heterotrophs in the marine environment. Adv Microb Ecol 6:171–198

Neijssel OM, Hueting S, Crabbendam KJ, Tempest DW (1975) Dual pathways of glycerol assimilation in *Klebsiella aerogenes* NCIB 418. Their role and possible functional significance. Arch Microbiol 104:83–87

Overbeck J, Chróst RJ (eds) (1990) *Aquatic Microbial Ecology. Biochemical and Molecular Approaches*. Springer-Verlag, New York

Overbeck J, Sako Y (1989) Ecological aspects of enzyme regulation in aquatic bacteria. In *Proceedings ISME-5*, Japan Scientific Societies Press, Kyoto, pp 378–382

Overbeck J, Rheimheimer G, Gunkel W, Krumbein WE, Weyland H (1991) Stand und Perspektiven der Gewässermikrobiologie in der Bundesrepublik Deutschland. Naturwissenschaften 78:543–556

Pardee AB (1968) Membrane transport proteins. Science 162:632–637
Rosen BP (1971) Basic amino-acid transport in *Escherichia coli*. J Biol Chem 246:3653–3662
Schwoerbel J (1987) *Handbook of Limnology*. Ellis Horwood, New York
Sherr BF, Sherr EB (1988) Role of microbes in pelagic food webs: A revised concept. Limnol Oceanogr 33:1225–1227
Simon M, Cho BC, Azam F (1992) Significance of bacterial biomass in lakes and the ocean: Comparison to phytoplankton biomass and biogeochemical implications. Mar Ecol Prog Ser 86:103–110
Sommer U (ed) (1989) *Plankton Ecology: Succession in Plankton Communities*. Springer-Verlag, New York
Verner K, Schatz G (1988) Protein translocation across membranes. Science 241:1307–1313
Wetzel RG (1983) *Limnology*, 3rd ed. WB Saunders, Philadelphia
Zambrano MM, Siegele DA, Almirón M, Tormo A, Kolter R (1993) Microbial competition: *Escherichia coli* mutants that take over stationary phase cultures. Science 259:1757–1760

Index

Entries occurring in figures are followed by an f; those occurring in tables, by a t.

A

Abramis brama, 18
Acetate, 160, 349
  heterotrophic potential and, 192, 196, 198
  in sulfuretum, 310–311, 316, 319
Acid phosphatase hydrolysis, 32
Actinomycetes sp., 186, 187
Adenosine, total dissolved, 142, 143
Adenosine diphosphate (ADP), 137, 142, 207–208
Adenosine monophosphate (AMP), 122, 123
  5PNase and, 137, 138, 139, 140, 142, 143
  secondary production and, 111
Adenosine triphosphate (ATP), 122, 123, 125
  APase and, 136
  PEP carboxylase and, 206
  5PNase and, 137, 138–141, 142
  secondary production and, 106
Adenosine triphosphate (ATP) phosphohydrolase, 137
ADP, *see* Adenosine diphosphate
Aerobic conditions
  bacterial adaptation strategies in, 361
  iron bacteria and, 327
  in sulfuretum, 294, 296
Aerobic heterotrophy, 344–347, 352
Agaricales, 331
Alanine, 220
Alanyl-serine, 160
Albumin, 160
Alburnus alburnus, 19
Algae, 19, 63–79, 87, 95, 251, 359
  AMPase and, 154, 159, 160, 161
  APase and, 128–129, 130, 131, 132–134, 135, 136, 165
  autotrophic, 19
  blue-green, 302
  chlorococcalean, 70
  chlorophyll$_a$ and, 63–69, 67, 70, 71, 72–75, 76–78
  dephosphorylation and, 141
  DIP and, 127

Algae (*cont.*)
  DOP and, 126
  EOC of, *see* Extracellularly released organic carbon
  eukaryotic, 82
  filamentous, 85–86
  GLCase and, 150, 151, 152
  grazing on, 85, 259, 260–261
  green, 305, 317
  oligotrophic bacteria and, 180
  PEP carboxylase and, 203
  pH and, 5
  photosynthesis and, 5, 63–65, 69–76, 78–79, 82, 105, 136, 160
  planktonic, 9
  5PNase and, 138, 143, 144
  saprophytic bacteria and, 180
  secondary production and, 95, 105, 110–111
  SRP and, 5–6
  in sulfuretum, 291, 314–315
  vertical stratification of, 10
Algae assemblages, 70, 71, 73–76, 78
Algae biomass, 123, 167, 267
  AMPase and, 157
  APase and, 134
  chlorophyll$_a$ and, 76
  DFCHO and, 40
  GLCase and, 146, 147
  5PNase and, 142
  secondary production and, 94, 104
Algae maxima, 67, 217
Algae minima, 217
Alkaline hydrolysis, 32
Alkaline phosphatase (APase), 87, 121, 122, 127–136, 144, 165
  activity in lake water, 138–139
  in euphotic zone, 128–132, 134, 135, 136, 139
  kinetic data analysis of, 123
  microplankton and, 131, 132, 134, 135–136
  5PNase and, 137–138
  in water column, 132–135
Allochthonous dissolved organic matter (DOM), 26, 201
Allosteric enzyme regulation, 201–214, 360, 361; *see also* Phosphoenolpyruvate carboxylase
Amberlite XAD-2, 36–37, 38
AMC, *see* 7-Amino-4-methylcoumarin
Amicon hollow fiber concentrator, 231
Amino acids, 119, 362
  AMPase and, 160, 161
  APase and, 136
  dissolved, 33, 34
  dissolved combined, *see* Dissolved combined amino acids
  dissolved free, *see* Dissolved free amino acids
  dissolved total, 34, 36, 39
  free, 159–160
  Limnothrix redekei and, 220, 221, 226
  secondary production and, 104
7-Amino-4-methylcoumarin (AMC), 122
Ammonia, 4, 8, 289, 296
Ammonium nitrogen, 219, 222, 225, 226
Amoebobacter roseus, 306, 308
Amoebobacter sp., 308, 315
AMP, *see* Adenosine monophosphate
AMPase, *see* Leucine-aminopeptidase
Anabaena flos-aquae, 9, 196
Anabaena mendotae, 9
Anabaena sp., 10
  grazing resistance of, 85
  photosynthesis and, 70–71, 72, 75
Anabolic pathways, 201
Anaerobic conditions
  bacterial adaptation strategies in, 361
  DON and, 30
  grazing and, 257
  iron bacteria and, 328
  methanogenesis in, 347, 349–351
  in sulfuretum, 294, 295
  see also Anoxic conditions
Anaerobic heterotrophy, 351
Anaerobic methane oxidation, 340
Ancalochloris sp., 306
ANF, *see* Autofluorescent nanoflagellates
Anguilla anguilla, 17
Anisogamic copulation, 9
Ankistrodesmus sp., 240
Annual cycle, *see* Seasonal variation
ANOVA, 97
Anoxic conditions, 5
  APase in, 135
  grazing and, 262

SRP and, 6
see also Anaerobic conditions
Antenna pigments, 312, 313
APase, *see* Alkaline phosphatase
Aphanizomenon flos aquae, 71
Aphanizomenon gracile, 9, 12, 28, 217
Aphanizomenon sp., 10, 70, 291
Aphanocapsa sp., 291
Aphanothece clathrata, 85
Apyrase, 137, 139
Archaebacteria, 347
Arsenate, 137
Arthopod exoskeletons, 145
Arthrobacter pascens, 331
Ascarides, 145
Ascinomycetes, 198
Aspartate, 206, 208, 211, 214
Asplanchna girodi, 14
Asplanchna priodonta, 13, 14
Asplanchna sp., 244
Asterionella formosa, 9, 10
Asterionella sp., 10, 235
photosynthesis and, 70–71, 72, 73, 75
in sulfuretum, 291, 308
Astylozoon fallax, 261, 264
ATP, *see* Adenosine triphosphate
Aulacoseira, 9
Autochthonous bacteria, 175
Autochthonous dissolved organic carbon (DOC), 28, 50
Autochthonous dissolved organic matter (DOM), 26
Autofluorescent nanoflagellate (ANF) grazing, 252, 253, 257–260, 262, 263
Autotrophic algae, 19
Autotrophic bacteria, 179
Autotrophic microorganisms, 48, 56, 153
Autotrophic picoplankton, 84
Autotrophic processes, 84
Autumn
algae in, 19
APase in, 128, 131, 135, 136
C:N ratio in, 32
DCCHO in, 49
dephosphorylation in, 141
DFAA in, 40
DFCHO in, 40
dissolved oxygen in, 5
dissolved polyphenols in, 36
DTCHO in, 34
EOC in, 85
grazing in, 254–257, 258, 263, 264
iron bacteria in, 327
oligotrophic bacteria in, 177
phytoplankton in, 10
phytoplankton primary production in, 66
saprophytic bacteria in, 177
SRP in, 6
sulfuretum in, 292–293, 308, 319
zooplankton in, 14
Auxosporulation, 9
Axenic cultures
grazing in, 230
Limnothrix redekei in, 217–218, 219, 220, 221t, 223, 225–226

B

Bacillariophyceae, 257
Bacillus subtilis, 230, 235, 240, 245
Bacteria, 1, 356–363
adaptation strategies of, 360–363
AMPase and, 159–162
APase and, 128–129, 131–132, 134–136
autochthonous, 175
autotrophic, 179
colorless, 287, 297, 302–304, 308
copiotrophic, 175
cyano-, *see* Cyanobacteria
dephosphorylation and, 141
DFCHO and, 54
DIP and, 127
DOC and, 39
DOM and, 24, 26, 51, 53, 55, 118–120, 163–165
DON and, 30
DOP and, 126
ectoenzymate production by, 162–163
EOC and, 83, 84–87
glycolic acid and, 43
gram-negative, 120, 164, 175, 186, 362
gram-positive, 175, 361–362
grazing on, *see* Zooplankton grazing
green sulfur, 287, 297, 311, 312, 314, 317
heterotrophic, *see* Heterotrophic bacteria
heterotrophic potential of, *see* Heterotrophic potential
iron, *see* Iron bacteria

Bacteria (*cont.*)
 lactic acid, 278, 283
 LDOCP and, 40, 41
 Limnothrix redekei and, 181, 217–227; *see also Limnothrix redekei*
 lysogenic, 278, 282, 283
 manganese-oxidizing, 331
 methane-oxidizing, *see* Methane-oxidizing bacteria
 methanogenic, 316
 microenvironment of, 358–359
 oligocarbophilic, 198
 oligotrophic, *see* Oligotrophic bacteria
 pelagic, 252
 PEP carboxylase and, 201, 203, 207–208
 photosynthesis and, 315–316
 phototrophic, 196, 327
 phototrophic sulfur, *see* Phototrophic sulfur bacteria
 5PNase and, 142–143
 polyphenols and, 45
 purple nonsulfur, 287, 288, 314, 317–319
 purple sulfur, 287, 297, 314, 317, 318, 319
 R3, 304
 red sulfur, 311, 312, 314
 spore-forming, 175
 SRP and, 6
 sulfate-reducing, 287, 297–302, 304–305, 310, 311, 313, 316
 sulfur-reducing, 313
 zymogenous, 175
Bacteria assemblages
 GLCase and, 150
 heterotrophic potential of, *see* Heterotrophic potential
 5PNase and, 137
 secondary production and, 94, 101
 in sulfuretum, 310
Bacteria biomass, 92, 111–112, 118, 166, 167, 356, 357, 358
 AMPase and, 154–155, 157
 APase and, 135
 biopolymers and, 45
 DOM and, 107, 111–112, 123
 GLCase and, 146–147, 150–151
 grazing on, 109, 242, 245, 359
 heterotrophic potential and, 198
 oligotrophic bacteria in, 179
 PEP carboxylase and, 203–204, 209
 rate of production of, 359
 regulators of, 108–111
 saprophytic bacteria in, 179
 secondary primary production and, 315
 secondary production and, 92, 93, 96, 104, 105, 106, 107
 UDOM in, 93, 119
Bacterial abundance
 bacteriophages and, 282
 GLCase and, 147
 of oligotrophic bacteria, 176–178
 of saprophytic bacteria, 176–178
 secondary production and, 96, 98, 105
Bacterial biovolume, 96, 123
Bacterial maxima, 217
Bacterial minima, 217
Bacterial organic carbon demand (BOCD), 94, 108, 110–111
Bacterial paradox, 119
Bacterial secondary production, *see* Secondary production
Bacteriochlorophyll, 312, 316
Bacteriochlorophyll$_a$, 313
Bacteriochlorophyll$_c$, 313
Bacteriochlorophyll$_d$, 313
Bacteriophages, 10, 56, 357, 361
 lytic, 120, 143, 270–278, 282, 283, 357
 temperate, 278–284
Bacterioplankton
 bacteriophages and, 278, 281, 282
 biopolymers and, 45
 DFAA and, 41
 DFCHO and, 41, 53
 DOC and, 39
 grazing on, 231, 244, 245, 246
 Limnothrix redekei and, 222
 PEP carboxylase and, 208, 211, 213, 214
Baltic Sea, 272
Basidiomycetes, 331
Beggiatoa sp., 294
Belauersee, 278–279, 283
Benthos fauna, 17, 18
Berliner-Blau reaction, 327
BGP, *see* β-Glycerophosphate
Bicarbonate, 97, 222, 312
Binnensee, 77
Biochemical oxygen demand (BOD), 338, 344–347
Biological oxygen demand, 5

Biomass, 19–21
algae, *see* Algae biomass
bacteria, *see* Bacteria biomass
cyanophyte, 220
fish, 81
fish fauna in, 18
heterotrophic, 19
methane conversion to, 342
microplankton, *see* Microplankton biomass
phytoplankton, *see* Phytoplankton biomass
Biomass maximum, 42
Biopolymers, 45–51, 56
distribution and dynamics of, 49–51
fractionation experiments with, 46–49
MHE and, 54–55
Black Sea, 294
Bleak, 19
Blue-green algae, 302
BOCD, *see* Bacterial organic carbon demand
BOD, *see* Biochemical oxygen demand
Borate, 137
Bosmina coregoni, 13, 235, 237–238, 240
Bosmina longirostris, 13, 14
Bosmina sp., 242, 246
Bothkamper See, 64, 77
Bottom-up regulation, 108, 111, 112
Brachionus angularis, 13
Brachionus calyciflorus, 13
Brachionus leydigi, 232
Brachionus sp., 13
Bream, 18
5-Bromo-4-chloro-3-indolyl phosphate (x-phosphate), 281

C

Cadmium, 154
Caenomorpha, 267
Calcium, 4, 8
Calvin cycle, 82, 318
CAM, *see* Crassulacean acid metabolism
cAMP, *see* Cyclic adenosine monophosphate
Cannibalism, 19
Capillary gas chromatography (GC), 44
Carassius carassius, 19
Carbohydrates, 119, 226
dissolved, 33, 34, 144
dissolved combined, *see* Dissolved combined carbohydrates
dissolved total, 34, 36, 39
polymeric, 144, 150
Carbon
AMPase and, 161
bacteriophages and, 274
ectoenzymes and, 120, 164
grazing and, 233, 235, 245
inorganic, 222, 225
Limnothrix redekei and, 227
low-molecular-weight, 87
in microbial loop, 263–264
organic, *see* Organic carbon
proteins as source of, 153
in sulfuretum, 311, 318
Carbonate-hydrogen carbonate buffer system, 5
Carbon balance calculations, 29
Carbon 14-bicarbonate method, 97
Carbon cycle
methane and, 337–353; *see also* Methane
oligotrophic bacteria in, 186
in sulfuretum, 316
Carbon dioxide
glycolic acid and, 42, 43
Limnothrix redekei and, 218
methane and, 337, 340, 344, 348, 349, 351
microbial enzymatic degradation and, 119
pH and, 5
photosynthetic assimilation of, 67
in sulfuretum, 318
Carbon dioxide dark uptake
PEP carboxylase and, 203–204, 206–207, 208
secondary production and, 104
Carbon dioxide fixation
dark, 48, 49, 66
PEP carboxylase and, 207
photoautotrophic, 251
in sulfuretum, 318
Carbon fixation
photosynthetic, 24, 43, 50, 81
phytoplankton primary production and, 85
see also Carbon dioxide fixation
Carbon flow, 45, 104
Carbon flux
DOC and, 29

Carbon flux (*cont.*)
  DOM and, 24, 53, 56
  secondary production and, 107–108
Carbon 14-labeled bacteria, 229, 230–232, 233
Carbon:nitrogen ratio, 32, 33
Carbon:phosphorus ratio, 32, 33
Carboxylase reaction, 82
Carboxylation, 87
Carboxymethylcellulose, 145
Carboxypeptidases, 153
Carex, 18
Carotenoids, 312–313
Carp, 19
Carteria, 10
Catabolic enzymes, 164, 360–361
Catabolic pathways, 201
Catabolic repression, 120, 152, 160–161
Caulobacter sp., 186, 363
Cellobiose, 136, 150
Celluhexose, 145
Cellulose, 26, 120, 144, 145, 163
Ceratium furcoides, 9
Ceratium hirundinella, 9, 12, 85, 86
Ceratium rhomvoides, 9
Ceratium sp., 10, 12
  chlorophyll$_a$ and, 78
  grazing on, 85, 235
  photosynthesis and, 70–71, 72, 75, 76f
Ceriodaphnia quadrangula, 13, 16, 237, 240
Ceriodaphnia sp., 246
CFUs, *see* Colony-forming units
Chelators, 82
Chemical characterization
  of DOM, 33–36
  of lake water, 3–8
  of sulfuretum, 288–289, 290–297
Chemical oxygen demand (COD), 339
Chemical stratification, 306
Chemoautotrophs, 294, 312
Chemocline, 314, 315, 316
Chemostats, 358
Chemotaxis, 359, 362, 363
Chitin, 120, 144–145, 163
Chlamydomonas sp., 10
Chlorella minutissima, 10
Chlorella pyrenoidosa, 232, 245
Chlorobactene, 313
Chlorobiaceae, 287, 297–302
  photosynthetic pigments and, 312, 313, 314
Chlorobium limicola, 306
Chlorobium phaeobacteroides, 306, 315
Chlorobium phaeovibrioides, 315
Chlorobium sp., 297, 312
Chlorochromatium aggregatum, 297, 306, 308
Chlorochromatium sp., 312
Chlorococcalean algae, 70
Chloroflexaceae, 318
Chlorophyll
  of ANF, 252
  grazing and, 264
  secondary production and, 97
  in sulfuretum, 314–315, 316
  see also Bacteriochlorophyll
Chlorophyll$_a$, 48, 63–66, 71, 73–75, 76–78, 79, 124
  APase and, 135
  EOC and, 84, 86
  GLCase and, 146, 147, 149
  grazing and, 254, 257, 258, 260
  Limnothrix redekei and, 222, 223f, 225
  5PNase and, 142
  quantum yields of, 69, 70, 72–73
  seasonal variation in, 66–69
  in sulfuretum, 314–315
Chlorophyll maxima, 40, 66
Chlorophyll minima, 70, 75
Chlorophyll$_a$ maxima, 72–73, 75, 77–78
Chlorophyll$_a$ minima, 77–78
Chlorophytes, 9, 10, 291
Chlorosomes, 313
Chromatiaceae, 287, 297–302, 311, 318
  photosynthetic pigments and, 312–313, 314, 315
Chromatium okenii, 306
Chromatophores, 313
Chrysochromulina parva, 10
Chrysochromulina sp., 258, 259
Chrysophytes, 9, 10
Chydorus sp., 242
Chydorus sphaericus, 13, 16, 232, 235, 245, 246
Ciliate grazing, 229, 242, 252, 259, 260, 261, 262, 263, 264–267, 359
Ciliates, 112, 317, 358

pelagic, 264–267
Citric acid cycle, 84
Cladocerans, 12, 13, 16
grazing by, 229, 235, 240, 242, 245, 246, 257, 259, 262, 263, 264
in sulfuretum, 317
vertical stratification, 14
Clathrochloris sp., 306
Clear-water phase
AMPase in, 157
EOC in, 85
GLCase in, 147
grazing in, 257, 259, 260, 263, 264
phytoplankton in, 10
5PNase in, 138
secondary production in, 111
sulfuretum in, 291
zooplankton in, 13
Closterium acutum var. *variabile,* 10
Closterium sp., 10
5′-CMP, 137
Cobalt, 137
Cobitis taenia, 19
COD, *see* Chemical oxygen demand
Coenzyme A, 125, 205, 206
Coleps nolandi, 265f, 267
Coleps sp., 262
Colony-forming units (CFUs), 175, 177, 317
Colorless bacteria, 287, 297, 302–304, 308
Cometabolism, 188
Common carp, 19
Concanavalin A, 137
Conductivity
iron bacteria, 327
of lake water, 4, 8
in sulfuretum, 288, 291
Conochilus natans, 13
Conochilus unicornis, 13
Cooxidation, 188
Copepods, 12
grazing by, 229, 237, 240, 242
in sulfuretum, 317
Copiotrophic bacteria, 175
Copper, 154
Cosmarium, 10
Crassulacean acid metabolism (CAM), 202, 204
Cross-hybridization, 281
Crustaceans, 14, 16
Cryptomonads, 191, 196, 257, 291
Cryptomonas sp., 193, 262, 291
Cryptophyceae, 75, 257
Cryptophytes, 10
Culture approach, 358
Cyanobacteria, 9, 357, 358
bacteriophages and, 10, 270, 277, 278
DOC and, 28
EOC and, 86
grazing resistance of, 85
iron-limited, 82
Limnothrix redekei and, 217–227; *see also Limnothrix redekei*
microenvironment of, 359
photosynthetic pigments and, 314
in sulfuretum, 294, 305, 306, 308, 317, 318
Cyanophyte biomass, 220
Cyanophytes, 9, 12
filamentous, 9, 10
Limnothrix redekei and, 217–227
see also Limnothrix redekei
Cyanophytorum apochromaticorum, 302
Cyclic adenosine monophosphate (cAMP), 136
Cyclotella sp., 9, 10
Cyprinid, 18
Cyprinus carpio, 19
Cysteine, 30
Cytosine, 342

D

DAA, *see* Dissolved amino acids
Dalny Lake, 180
Daphnia cucullata, 13, 235, 237, 238, 244, 245
Daphnia galeata, 13, 245
Daphnia hyalina, 14
Daphnia longispina, 232, 235, 245, 246
Daphnia magna, 232, 245, 246
Daphnia pulex, 242, 245, 246
Daphnia schoedleri, 240
Daphnia sp., 13, 242, 246
DAPI, *see* 4,6-Diamidino-2-phenylindol
Dark respiration, 65, 72
DCAA, *see* Dissolved combined amino acids
DCCHO, *see* Dissolved combined carbohydrates

DCHO, *see* Dissolved carbohydrates
DCMU, *see* 3(3,4-Dichlorophenyl)-1,1-dimethylurea
DCPPhOH, *see* Dissolved combined polyphenols
Deep waters, 39, 42, 157
Denitrification, 179, 296
Deoxysugars, 145
Dephosphorylation, 136, 139–141
Depolymerization, 120
Desmids, 10
Desulfovibrio desulfuricans, 305
Desulfovibrio sp., 294
DFAA, *see* Dissolved free amino acids
DFCHO, *see* Dissolved free carbohydrates
DFPhOH, *see* Dissolved free monomeric polyphenolic compounds
DFPPhOH, *see* Dissolved free polyphenols
DHM, *see* Dissolved humic matter
Diacyclops bicuspidatus, 14
4,6-Diamidino-2-phenylindol (DAPI), 252, 254, 271, 273
Diaphanosoma brachyurum, 13, 235, 237, 238, 240
Diaphanosoma sp., 246
Diatoms, 9–10, 145
  DFCHO and, 42
  grazing and, 257, 264
  heterotrophic potential and, 193, 195
  photosynthesis and, 75
  silicon in, 8
  in sulfuretum, 291, 305, 308
Dicarboxylic acids, 202
3(3,4-Dichlorophenyl)-1,1-dimethylurea (DCMU), 226
5-(4,6-Dichlorotriazinyl)aminofluorescein (DTAF), 252, 253
Dicotyledonae, 202
Digoxigenin-11-dUTP, 280, 281
Dimethyl-*p*-phenylendiamine, 289
Dinoflagellates, 9, 12, 85, 87
Dinophytes, 85–86
DIP, *see* Dissolved inorganic phosphorus
Dipeptides, 160
Dissolved amino acids (DAA), 33, 34
Dissolved carbohydrates (DCHO), 33, 34, 144
Dissolved combined amino acids (DCAA), 29–30, 49, 50–51, 124
  AMPase and, 154, 157, 159, 161, 162
Dissolved combined carbohydrates (DCCHO), 49–51, 124, 163
Dissolved combined (DC) galactose, 55
Dissolved combined (DC) glucose, 46, 55, 150
Dissolved combined (DC) leucine, 46
Dissolved combined polyphenols (DCPPhOH), 44–45
Dissolved free amino acids (DFAA), 25, 29, 31, 40–42, 43, 51, 55
  heterotrophic activities and, 52–53
Dissolved free carbohydrates (DFCHO), 25, 40–42, 43, 50, 51, 55
  heterotrophic activity and, 53–54
Dissolved free extracellular released organic carbon (EOC_d), 84, 86
Dissolved free (DF) galactose, 44
Dissolved free (DF) glucose, 44
  biopolymers and, 49
  heterotrophic activity and, 53–54
Dissolved free (DF) glycine, 44
Dissolved free (DF) mannose, 44
Dissolved free monomeric polyphenolic compounds (DFPhOH), 45
Dissolved free polyphenols (DFPPhOH), 38–39, 44–45
Dissolved humic matter (DHM), 25, 35, 49, 188
Dissolved inorganic phosphorus (DIP)
  APase and, 128, 129, 130, 131, 134, 135, 136, 138
  microbial enzymatic degradation of, 124–125, 126, 128, 129, 130, 131, 134, 135, 138, 143, 144
  5PNase and, 143, 144
Dissolved methane, 337
Dissolved organic acids (DOAc), 33, 39, 42
Dissolved organic carbon (DOC), 25–26, 27–29, 30, 31, 32, 33, 34, 35t, 36, 39–45, 55–56, 124, 358–359
  AMPase and, 159
  autochthonous, 28, 50
  enrichment experiments and, 37
  GLCase and, 150, 151
  grazing and, 254, 256, 260, 263
  heterotrophic potential and, 196
  heterotrophic activities and, 51–52

high-molecular-weight, *see* High-molecular-weight dissolved organic carbon
HTCO analysis of, 27, 28–29
low-molecular-weight, 25, 55
oligotrophic bacteria and, 187–188
5PNase and, 142
secondary production and, 97, 102, 110
WCO analysis of, 27–28, 29
Dissolved organic carbon (DOC) maxima, 28
Dissolved organic matter (DOM), 24–56
allochthonous, 26
allosteric enzyme regulation and, 201–214; *see also* Allosteric enzyme regulation
autochthonous, 26
bacteria biomass and, 107, 111–112, 123
bacteriophages and, 277
bulk-phase characterization of, 26–29, 362
chemical characterization of, 33–36
elemental composition of, 32–33
enrichment experiments and, 26, 36–39
grazing and, 259, 260, 262, 263, 267
heterotrophic activities and, 51–55
heterotrophic potential and, 198
microbial enzymatic degradation of, 118–167; *see also* Microbial enzymatic degradation
in microbial loop, 26, 56, 251–252
oligotrophic bacteria and, 175, 179, 181, 185, 186, 188
proteins in, 153
secondary production and, 92–94, 111–112
in sulfuretum, 289, 310, 319
utilizable, *see* Utilizable dissolved organic matter
Dissolved organic nitrogen (DON), 4, 29–31, 32, 225
Dissolved organic phosphorus (DOP), 31–32, 124–127
APase and, 128, 129
high-molecular-weight, 127
5PNase and, 127, 142
Dissolved organic phosphorus (DOP) maxima, 31
Dissolved oxygen, 4–5, 254, 257
Dissolved polyphenols, 34–36
Dissolved reactive phosphate, 4
Dissolved silicon dioxide, 4
Dissolved total amino acids (DTAA), 34, 36, 39
Dissolved total carbohydrates (DTCHO), 34, 36, 39
Dissolved total phosphorus, 4
Dissolved total polyphenols (DTPPhOH), 36, 39, 44–45
Diurnal variation
DFAA and, 40, 41–42
DFCHO and, 40, 41–42, 53, 54
glycolic acid and, 44
grazing and, 254
HMWDOC and, 50–51
DNA, 49, 185
bacteriophages and, 278–284
double-stranded, 281
[^{3}H]TdR incorporation of, *see* [^{3}H-Methyl]thymidine analysis
5PNase and, 143
secondary production and, 103
DNA hybridization, 279, 281, 282–283
DNA probing, 271, 279–284
DNase, 143
DNase I, 281
DOAc, *see* Dissolved organic acids
DOC, *see* Dissolved organic carbon
DOM, *see* Dissolved organic matter
DON, *see* Dissolved organic nitrogen
DOP, *see* Dissolved organic phosphorus
Double-stranded DNA, 281
DTAA, *see* Dissolved total amino acids
DTAF, *see* 5-(4,6-Dichlorotriazinyl)aminofluorescein
DTCHO, *see* Dissolved total carbohydrates
DTPPhOH, *see* Dissolved total polyphenols
Dy-Gyttja, 17

E

Ecosystem approach, 358
Ecosystems
microbial enzymatic degradation significance in, 165–167
PEP carboxylase control in, 208–214
Ectoenzymes, 164–165
AMPase and, 161
DOM and, 120–121

Ectoenzymes (*cont.*)
POM solubilization by, 162–163, 164, 165, 166–167
UDOM and, 93
Edebergsee, 63
chlorophyll$_a$ in, 77, 78
Limnothrix redekei in, 222
phytoplankton primary production in, 68, 69, 71, 73, 77, 78
Eels, 17, 18–19
EHP, *see* Enzymatically hydrolyzable phosphate
Electron microscopy, 270–271, 341
ELM, *see* Epifluorescence light microscopy
Elodea sp., 17
Endopeptidases, 153
Endosoma palleum, 297
Enrichment
DOM and, 26, 36–39
methane-oxidizing bacteria and, 340–341
oligotrophic bacteria and, 182, 185
saprophytic bacteria and, 182
in sulfuretum, 304, 317, 318
Enterobacter aerogenes, 204, 205, 206, 209
Enzfitter program, 123
Enzymatically hydrolyzable phosphate (EHP), 127, 131, 136
Enzyme assays, 121–123
Enzyme diversion, 138
Enzyme effectors, 202, 205–207, 361
Enzymes
bottleneck, 361
catabolic, 164, 360–361
ecto, *see* Ectoenzymes
extracellular, 120
hydrolytic, 120
kinetic data analysis of, 123
microbial hydrolytic, 25, 51, 54–55
proteolytic, 153
EOC, *see* Extracellularly released organic carbon
Epifluorescence light microscopy (ELM), 271, 273–274
Epilimnion, 251
AMPase in, 157
APase in, 132–134, 135
bacterial adaptation strategies in, 360
biopolymers in, 47
chlorophyll$_a$ in, 64
DFAA in, 40
DFCHO in, 40
dissolved oxygen in, 4
DOC in, 27, 28, 39
DOP in, 31
DTAA in, 34
DTCHO in, 34
EOC in, 83
fish fauna in, 17
GLCase in, 149
grazing in, 233, 234t, 242, 254–255, 257–259, 260, 261–262, 263, 264, 267
heterotrophic potential and, 193
inorganic nitrogen in, 8
oligotrophic bacteria in, 179
pH and, 5
phosphorus in, 19
phytoplankton in, 10–12
phytoplankton primary production in, 69, 83
saprophytic bacteria in, 179
secondary production in, 105
SRP in, 6
sulfuretum organisms in, 287, 290, 291, 295, 296, 314, 315
zooplankton in, 14, 15, 16–17
Escherichia coli, 128, 203, 206, 362
Esox lucius, 17
Eudiaptomus gracilis, 16, 235, 237
Eudiaptomus graciloides, 245
Eukaryotic algae, 82
Euphotic zone, 121, 358
AMPase in, 154–157, 159, 160, 161
APase in, 128–132, 134, 135, 136, 139
chlorophyll$_a$ in, 65
DCAA in, 50
dephosphorylation in, 141
DFAA in, 41
DOP in, 125, 126–127
ectoenzymes in, 165
GLCase in, 146–147, 150
heterotrophic potential and, 196
inorganic nitrogen in, 6
LDOCP in, 40
phosphate in, 21
phytoplankton primary production in, 66, 73
5PNase in, 138, 143, 144

secondary production in, 95, 96, 100, 101, 102, 105, 106, 108, 109, 110, 111, 112
SRP in, 5, 6
Eutrophic lakes, 175
EOC in, 84
grazing in, 244–245, 258
heterotrophic potential in, 193
MOD in, 345–346
PEP carboxylase in, 208
phytoplankton primary production in, 68
sulfuretum organisms in, 292, 311
zooplankton in, 13
Exoectoenzymes, 154
Exopeptidases, 153
Extracellular enzymes, 120
Extracellularly released organic carbon (EOC), 40, 41, 42, 44, 56, 81–87, 123–124, 358–359, 362
AMPase and, 159, 160, 161
APase and, 136
dissolved free, 84, 86
GLCase and, 150, 152
grazing and, 254, 261
heterotrophic potential and, 196
oligotrophic bacteria and, 180, 182, 188
release and utilization of, 83
saprophytic bacteria and, 180–181, 182
secondary production and, 95, 97, 110–111, 112
total, 84, 86
Fall, *see* Autumn
Fayetteville Green Lake, 315
Fecal pellets, 163
Fibrous proteins, 153
Filamentous algae, 85–86
Filamentous cyanophytes, 9, 10
Filinia hofmanni, 14, 15f
Filinia longiseta, 13
Filinia sp., 14
Filinia terminalis, 13, 14, 15f
Fish, 245, 277
Fish biomass, 81
Fish fauna, 17–19
Flagellates, 10, 28, 193, 242, 245; *see also* Nanoflagellates
FLB, *see* Fluorescently labeled bacteria
Fluorescently labeled bacteria (FLB), 252–262
Fluoride, 137
Food chains
DOM in, 24, 25, 51, 56, 118, 119, 121
grazing and, 92, 166, 229, 252, 356
phytoplankton in, 81
RDOCP in, 44
secondary production in, 104, 108, 111, 112
Formate, 349
Fractionation experiments, 46–49
Fragillaria, 235
Free amino acids, 159–160
Freshwater
bacteriophages in, 271, 273
biopolymers in, 49
DOC in, 39
DOM in, 25
LDOCP in, 40
secondary production in, 92, 107
Fructose, 205–206
Fructose-1,6-bisphosphate, 205
Fructose-6-phosphate, 205
Fulvic acids, 187
Fungi, 119, 120, 145, 162, 166, 331

G

Galactose, 44, 55
β-Galactosidase (GALase), 54–55
GALase, *see* β-Galactosidase
Gallionella sp., 331
Gas chromatography (GC), 44, 337–338
Gas-liquid chromatography (GLC), 34, 35, 37, 38, 39, 53
Gas-phase hydrolysis, 30
GC, *see* Gas chromatography
Gel permeation chromatography (GPC), 38–39, 46–48, 181
Gene amplification, 278
Gene exchange, 278–279
Gene expression, 202, 278
Gene transfer, 283–284
Geopolymers, 45
Glacial periods, 1
GLC, *see* Gas-liquid chromatography
GLCase, *see* β-Glucosidase
Global radiation, 1–3, 64, 73, 75, 76, 77, 78
Globular proteins, 153
Gluconeogenesis, 211

Glucose, 50, 362
AMPase and, 160–161
APase and, 128, 136
biopolymers and, 48–49
dissolved combined, 46, 55, 150
dissolved free, *see* Dissolved free glucose
ectoenzymatic solubilization and, 162–163
GLCase and, 145, 150, 151, 152
heterotrophic activity and, 53–54
heterotrophic potential and, 192–199
Limnothrix redekei and, 226
MHE and, 55
oligotrophic bacteria and, 187, 188
PEP carboxylase and, 203, 205–206, 208, 209–211, 212t, 213, 214
phytoplankton primary production and, 66
in sulfuretum, 289, 290, 309–310
uptake of in sediments, 198–199
uptake of in water column, 193–198, 199
Glucose-6-phosphatase, 137, 139
Glucose-6-phosphate, 128, 136, 205, 206, 212t, 213t, 214
β-Glucosidase (GLCase), 46, 50, 54–55, 145–152, 161, 162, 163
DOM and, 121, 123
environmental control of, 150–152
in euphotic zone, 146–147
kinetic data analysis of, 123
in water column, 149
Glutamate, 208
Glutamine, 30, 220
Glyceria, 18
β-Glycerophosphate (BGP), 122, 138, 139
Glycine, 42, 44, 220
Glycogen, 144
Glycolate, 82
Glycolate oxidase, 43
Glycolic acids, 25, 42–44, 56
Glycolipids, 145
Glycoproteins, 145
Glyoxalate, 42
G model, 24
Gomphosphaeria, 291
GPC, *see* Gel permeation chromatography
Gram-negative bacteria, 120, 164, 175, 186, 362
Gram-positive bacteria, 175, 361–362
Grazing, *see* Zooplankton grazing
Grazing pressure, 179, 242, 244–245, 259
Green algae, 305, 317
Green sulfur bacteria, 287, 297, 311, 312, 314, 317
Großer Binnensee, 64
Großer Plöner See, 63, 77, 78
Guanine, 342
Gulf of Calvi, 101
Gymnocephalus cernua, 18

H

Hancoma setosas, 267
Haptophytes, 10
Hemicellulose, 144
Herbivorous consumers, 118
Herbivorous zooplankton, 13, 82, 85, 126, 356
Heterotrophic activities
DOM as target of, 51–55
EOC and, 84
terms used to describe, 84
Heterotrophic bacteria, 19, 356, 357
AMPase and, 154
bacteriophages and, 277, 278
DOM and, 118, 119, 120, 121, 163, 166–167
iron bacteria and, 331
Limnothrix redekei and, 217–227; *see also Limnothrix redekei*
microenvironment of, 359
PEP carboxylase and, 202, 203–204
polysaccharides and, 144–152
secondary production by, 92, 93, 94, 104
in sulfuretum, 289–290, 297
vertical stratification of, 179–181
Heterotrophic biomass, 19
Heterotrophic fungi, 119, 120, 166, 331
Heterotrophic microorganisms, 4, 48, 56, 153
Heterotrophic nanoflagellate (HNF) grazing, 229, 251–264, 259
bacteria biomass and, 109, 359
bacteriophages in, 277
FLB technique and, 252–262
on oligotrophic bacteria, 179
on saprophytic bacteria and, 179
Heterotrophic nanoflagellates (HNF), 111, 112, 358
Heterotrophic potential, 192–199
methane and, 346–347

PEP carboxylase and, 208
in sulfuretum, 289
Heterotrophic processes, *see* Heterotrophic activities
Heterotrophic zooplankton, 19
Heterotrophy
aerobic, 344–347, 352
anaerobic, 351
High-molecular-weight dissolved organic carbon (HMWDOC), 25, 46, 49, 54–55, 56
diurnal variation of, 50–51
seasonal distribution of, 49–50
High-molecular-weight dissolved organic phosphorus (HMWDOP), 127
High-performance liquid chromatography (HPLC), 34, 37, 39, 40, 52
High-temperature catalytic oxidation (HTCO), 27, 28–29
HindIII restriction endonuclease, 279, 280
HMWDOC, *see* High-molecular-weight dissolved organic carbon
HMWDOP, *see* High-molecular-weight dissolved organic phosphorus
HNF, *see* Heterotrophic nanoflagellates
Holomictic lakes, 314
Homothermy, 6, 105, 131, 135
HPLC, *see* High-performance liquid chromatography
HSB model, 24–25
HTCO, *see* High-temperature catalytic oxidation
Humic acids, 188
Humic matter, 119
dissolved, 25, 35, 49, 188
Hydrogen sulfate, 289
Hydrogen sulfide, 4, 287, 288, 293–294, 304, 305–308, 308, 311, 312, 317, 360
photosynthetic pigments and, 313, 314
redox potential and, 295
secondary primary production and, 316
Hydrolysis, 121, 163
acid phosphatase, 32
alkaline, 32
by APase, 127–128, 130, 131–132, 136, 138–139
of DOP, 127
gas-phase, 30
GLCase and, 149, 150
peptide, 153, 154, 159–160, 161
by 5PNase, 143–144
of polysaccharides, 144, 145
protein, 30, 152–162
Hydrolytic enzymes, 120
Hypereutrophic lakes, 246, 311
Hyphomicrobium sp., 186
Hypolimnion
APase in, 134, 135, 136
bacterial adaptation strategies in, 360
biopolymers in, 47
C:N ratio in, 32
DFAA in, 40
DFCHO in, 40
dissolved polyphenols in, 36
DON in, 30
DOP in, 31
DTAA in, 34
DTCHO in, 34
EOC in, 83
fish fauna in, 17, 18
GLCase in, 149
grazing in, 254, 256–257, 260, 261, 262, 263, 264, 267
inorganic nitrogen in, 8
iron bacteria in, 327, 331
MOD in, 345–346
N:P ratio in, 33
oligotrophic bacteria in, 179
phosphate in, 21
phytoplankton primary production in, 69, 83
5PNase in, 138, 139, 143
saprophytic bacteria in, 179
secondary production in, 105–106
SRP in, 6
sulfuretum organisms in, 287, 288, 290, 292, 293, 294, 296, 304, 305, 306, 307t, 308, 310, 312, 315, 316, 319
zooplankton in, 14, 15–17

I

5′-IMP, 137
Inorganic carbon, 222, 225
Inorganic nitrogen, 6–8, 97
Inorganic nutrients, 296–297
Inorganic orthophosphate, 124
Inorganic phosphate, 225

Inorganic phosphorus, 6, 31
  dissolved, *see* Dissolved inorganic phosphorus
  microbial enzymatic degradation and, 124–144
  oligotrophic bacteria and, 179
  saprophytic bacteria and, 179
  secondary production and, 97
In situ approach, 358
Iron, 295, 296, 304
Iron bacteria, 304, 326–333
  manganese-precipitating, 328, 330, 332–333
Iron-limited cyanobacteria, 82
Iron oxide, 326, 327, 331, 333
Iron sulfide, 296, 304
Isoline, 293, 295
Isotope dilution, 52, 53, 54, 97, 103

K

katF gene, 362
Kellicottia longispina, 13, 14
Keratella cochlearis, 14, 235
Keratella cochlearis var. hispida, 13
Keratella cochlearis var. tecta, 13
Keratella hiemalis, 14, 15f
Keratella quadrata, 235
Kiel Bight, 273
Kinetics
  Michaelis-Menten, *see* Michaelis-Menten kinetics
  of microbial enzymatic degradation, 121–123
  in sulfuretum, 309–311
Klebsiella sp., 187, 230, 235, 238
Kolksee, 311

L

Labile dissolved organic carbon pool (LDOCP), 25, 39, 40–44, 51, 55–56
  heterotrophic activities and, 52, 54–55
Lactate, 304, 305, 310
Lactic acid bacteria, 278, 283
Lactobacillus sp., 181
Lake Alexander, 186
Lake Biwa, 256
Lake Constance, 259, 262
Lake Glebokie, 124
Lake Kalandsvannet, 274
Lake Kinneret, 199, 272
Lake Michigan, 101, 139
Lake Mogilnoye, 315
Lake Oglethorpe, 258
Lake Vechten, 256, 315
Lake Windermere, 71, 75
Lamprocystis roseopersicina, 299f, 306, 308
Lamprocystis sp., 315
Land-water interface, 26
Lauterborn cycles, 13
Lauterborniola minima, 297
LDOCP, *see* Labile dissolved organic carbon pool
Lead, 154
Leptothrix echinata, 326, 327, 330
Leptothrix lopholea, 326
Leucaspius delineatus, 19
Leucine, 46, 159, 163
[^{3}H]Leucine, 94, 95–96, 108
  conversion factors for, 96, 98–103
  radiolabel concentration in, 97, 98
Leucine-aminopeptidase (AMPase), 46, 121–122, 153, 163
  activity in lake water, 159–160
  depth distribution of activity of, 157–159
  in euphotic zone, 154–157, 159, 160, 161
  kinetic data analysis of, 123
  protein hydrolysis by, 153–162
  secondary production and, 111, 157, 160
L-Leucine-4-methyl-coumarinylamide hydrochloride, 121–122
Leucyl-leucine, 160
Light-inhibited photosynthesis, 70, 73, 75
Light microscopy, 327
Light penetration, 65–66
Light-proportional photosynthesis, 71–72, 73
Light-saturated photosynthesis, 72, 73, 75
Light transmission, 291
Lignins, 26, 49
Lignocelluloses, 49
Limnothrix redekei, 9, 10, 12, 181, 217–227
  DOC and, 28
  in long-term cultures, 225–226
  lysis by, 220, 222–225, 227
  nitrogen compounds released by, 217, 218, 219–220, 222, 224, 226, 227

photosynthesis and, 70, 71, 72, 226
in sulfuretum, 306
trichomes in, 220–222
Limnothrix rosea, 217
Limnothrix sp., 69, 85
Linear regression, 99
Lineweaver-Burk reciprocal plots, 198, 343, 349
Lipids, 49, 119, 145, 363
Littoral zone
DOP in, 126
fish fauna in, 17, 18, 19
glucose uptake in, 199
grazing in, 233
sulfuretum organisms in, 318–319
LMWDOC, *see* Low-molecular-weight dissolved organic carbon
Low-molecular-weight carbon, 87
Low-molecular-weight dissolved organic carbon (LMWDOC), 25, 55
Lysis
by bacteriophages, *see* Lytic bacteriophages
by *Limnothrix redekei,* 220, 222–225, 227
by viruses, 359
Lysogenic bacteria, 278, 282, 283
Lytic bacteriophages, 120, 143, 270–278, 282, 283, 357

M

Macrophytes, 17, 18, 41, 277
Macrozooplankton, *see* Zooplankton
Magnesium, 4, 8, 137, 154
D,L-Malate, 317
Malate dehydrogenase, 205, 213
Mallomonas caudata, 10
Manganese, 137, 154, 295
Manganese oxide, 326, 327, 331, 333
Manganese-oxidizing bacteria, 331
Manganese-precipitating iron bacteria, 328, 330, 332–333
Mannose, 44
Mannose-6-phosphate, 128
Marine sediments, 24
Marine waters, 25, 29, 39, 40, 92
Marion Lake, 199
Mercury, 137, 154
Meromictic conditions, 314, 346
Mesembryanthemaceae, 202
Mesembryanthemum crystallinum, 202
Mesocosm approach, 358
Mesoeutrophic lakes, 258
Mesotrophic lakes, 65, 68, 208
Metalimnion
AMPase in, 157
APase in, 134, 135
dissolved polyphenols in, 36
EOC in, 83
GLCase in, 149
grazing in, 233, 234t, 242, 244, 254, 255–256, 257, 259–260, 261, 262, 263, 264
phytoplankton in, 10
phytoplankton primary production in, 66, 69, 75, 83
secondary production in, 105, 106
sulfuretum organisms in, 287, 289, 292, 297, 306, 309, 310
zooplankton in, 14
Metallogenium sp., 304, 305, 308, 326–327
morphology and fine structure of, 328–331
Metazoans, 92, 112
Methane, 337–353
dissolved, 337
GC determination of, 337–338
precursors of, 349–351
Methane budget, 351–353
Methane cycle, 337
Methane ebullition, 339, 351
Methane oxidation, 337, 339–344
anaerobic, 340
stoichiometry of, 339, 344
Methane oxidation demand (MOD), 344–347
Methane-oxidizing bacteria, 5, 339, 340–341
strain M 102, 338–339, 341–344
in sulfuretum, 292
Methane oxidizing potential (MOP), 345, 347
Methane oxygen demand (MOD), 338, 352
Methangenic bacteria, 316
Methanogenesis, 337, 338, 339, 347–351
Methanogens, 348–349
Methanol, 182, 349
Methanotrophs, *see* Methane-oxidizing bacteria

Methanotrophy, *see* Methane oxidation
Methylamines, 349
Methylene blue, 4
Methylocystis, 342
[^{3}H-Methyl]thymidine ([^{3}H]TdR) analysis
  grazing activity measured with, 252, 254
  secondary production measured with, 94, 95–96, 97–103, 108, 123
Methyltrophs, 341
4-Methylumbelliferone (MUF), 121
4-Methylumbelliferyl-β-D-glucopyranoside, 121
4-Methylumbelliferyl phosphate, 121, 138
Metopus sp., 265f, 267
MHE, *see* Microbial hydrolytic enzymes
Michaelis constants
  of AMPase, 157
  of APase, 129, 131, 134, 137
  of GLCase, 146, 149, 151–152
  of 5PNase, 140
Michaelis-Menten kinetics
  of biopolymers, 46
  of glucose uptake, 193, 195–196, 198, 290
  of methanogenesis, 349
  of microbial enzymatic degradation, 123
Microbial enzymatic degradation, 118–167
  assays in, 121–123
  kinetic data analysis in, 121–123
  of nucleotides, 136–144, 163
  of organophosphoric compounds, 121, 124–144
  of polysaccharides, 120, 121, 144–152, 163
  of proteinaceous compounds, 152–162
  significance of in ecosystem, 165–167
Microbial hydrolytic enzymes (MHE), 25, 51, 54–55
Microbial loop, 81, 118
  bacteriophages and, 277
  DOM and, 26, 56, 251–252
  grazing and, 242, 251–252, 263–264
  microbial enzymatic degradation and, 165–166, 167
  nutrient regeneration in, 263–264
  secondary production and, 112
Microcyclus aquaticus, 304
Microcystis, 9
Microflagellates, 126
Microheterotrophs, 26, 39, 45
Micromonas pusilla, 274
Microplankton
  AMPase and, 154
  APase and, 131, 132, 134, 135–136
  dephosphorylation and, 141
  DON and, 29
  SRP and, 5–6
Microplankton biomass
  APase and, 131, 135
  DOP and, 124, 126–127
  secondary production and, 106
Microspira desulfuricans, 304
Microzooplankton, 229, 242
Mixed model, 24
Mixotrophic nanoflagellate grazing, 251–264
MOD, *see* Methane oxygen demand
Moina sp., 246
Molluscs, 145
Monomeric substrates, 51–54
5′-Mononucleotides, 127
5′-Monophosphates, 136
Monosaccharides, 104
MOP, *see* Methane oxidizing potential
Morphometry, 1–3
mRNA, 202
MUF, *see* 4-Methylumbelliferone
Multiple regression analysis, 97
Mutations, 278
Myobacter sp., 186

N

NAD, 137, 312
NADH, 137, 205, 206, 213
Nanoflagellates, 10
  autofluorescent, *see* Autofluorescent nanoflagellate grazing
  heterotrophic, *see* Heterotrophic nanoflagellates
  mixotrophic, 251–264
Nanoplankton, 193
Nauplii, 244
NBT, *see* Nitroblue tetrazolium salt
Nickel, 137
Nitrate, 4, 6–8
  Limnothrix redekei and, 225
  methanotrophy and, 340
  in sulfuretum, 289, 296–297
Nitrite, 4, 219, 289

Nitroblue tetrazolium salt (NBT), 281
Nitrogen, 4, 26
bacteriophages and, 274
ectoenzymes and, 120, 164
inorganic, 6–8, 97
Limnothrix redekei and, 217, 218, 219–220, 222, 224, 225, 226, 227
methane-oxidizing bacteria and, 344
microbial loop and, 252
organic, *see* Organic nitrogen
particulate, 4
proteins as source of, 153
in sulfuretum, 296
Nitrogen fixation, 198
Nitrogen:phosphorus ratio, 32–33
Nitzschia sp., 235
Nocardia sp., 186
Nonaxenic cultures, 217–218, 219, 221t, 225
Nonlinear regression, 99
North Sea, 273
Nucleic acids
of bacteriophages, 278, 279, 284
microbial enzymatic degradation of, 119, 120, 163
5′-Nucleotidase (5PNase), 136–144
activity in lake water, 138–139
APase and, 137–138
DOM and, 121, 122, 123
DOP and, 127, 142
in phosphorus mineralization, 142–144
Nucleotide pyrophosphatase, 137, 139
Nucleotides, 136–144, 163
Nutrients
inorganic, 296–297
in microbial loop, 263–264

O

Okenone, 313
Oligocarbophilic bacteria, 198
Oligotrophic bacteria, 175–188, 360
abundance of, 176–178
functions of, 187–188
heterotrophic potential of, 193
taxonomy of, 181–185, 186
vertical stratification of, 178–181
Oligotrophic waters, 84, 274
OMR, *see* Organic matter reactivity
Ontario lakes, 124
Oochrobium sp., 332
Organic acids, 119
dissolved, 33, 39, 42
Organic carbon, 118, 123
APase and, 136
dissolved, *see* Dissolved organic carbon
extracellular released, *see* Extracellularly released organic carbon
labile dissolved, *see* Labile dissolved organic carbon pool
particulate, *see* Particulate organic carbon
phytoplankton primary production and, 68
5PNase and, 144
recalcitrant dissolved, *see* Recalcitrant dissolved organic carbon pool
in sulfuretum, 316
total, 165–166
Organic matter, 92–94, 104, 110; *see also* Dissolved organic matter; Particulate organic matter; Polymeric organic matter
Organic matter reactivity (OMR), 24, 25
Organic nitrogen, 144, 227
dissolved, 4, 29–31, 32, 225
Organic phosphorus, 124–127, 225; *see also* Dissolved organic phosphorus
Organic substrates, 289–290
Organophosphoric compounds, 121, 124–144
Organophosphoric esters, 120
Orthophosphate, 122, 123
dephosphorylation and, 140, 141
inorganic, 124
5PNase and, 139, 142–143
Oscillatoria agardhii, 179
Oscillatoria limnetica, 317
Oscillatoria sp., 235, 240
Osmotrophic microorganisms, 26
Oxaloacetate, 206
Oxycline, 14, 15, 254
Oxygen
dissolved, 4–5, 254, 257
fish fauna and, 17, 18
glycolic acid and, 42
grazing and, 256–257, 260, 262
iron bacteria, 327
methane and, 339, 340, 345–347, 352
methane-oxidizing bacteria and, 343
phytoplankton and, 10

Oxygen (*cont.*)
in sulfuretum, 288, 292–293, 294, 295, 305, 306, 308, 313, 316, 317
zooplankton and, 15, 16
Oxygen boundary, 14

P

PAGE, *see* Polyacrylamide gel electrophoresis
PAGIF, *see* Polyacrylamide gel isoelectric focusing
Pandorina, 235
PAR, *see* Photosynthetically active radiation
Paraphysomonas, 9
Particulate nitrogen, 4
Particulate organic carbon (POC), 25–26, 27, 40, 256, 263
Particulate organic matter (POM), 119, 120, 144, 362
ectoenzyme solubilization of, 162–163, 164, 165, 166–167
proteins in, 153
secondary production and, 93
Particulate phosphorus, 4
Pascherina, 10
Pectin, 144
Pediastrum sp., 235
PEG model, 13, 16, 85
Pelagic bacteria, 252
Pelagic ciliate grazing, 264–267
Pelagic zone
chemistry of, 4
DOM in, 24
DOP in, 124
fish fauna in, 19
grazing in, 263
microbial enzymatic degradation in, 121
secondary production in, 95
sulfuretum organisms in, 318, 319
zooplankton in, 14, 15
Pellicon cassette system, 231
Pelochromatium roseum, 297, 306, 308, 309f
Pelochromatium sp., 312
Pelodictyon aggregatum, 297
Pelodictyon clathratiforme, 306
Pelodictyon luteum, 315
Pelodictyon phaeum, 306, 315
Pelonema pseudovacuolatum, 299f
Pelonema sp., 302, 303t, 306, 308
Pelonema spirale, 299f
Pelonemataceae, 302–303, 312
Pelonema tenue, 299f
Peloplaca sp., 306
Peloploca fibrata, 308
Peloploca pulchra, 299f, 306
Peloploca sp., 301f, 302, 303t, 308
Peloploca taeniata, 306
Peloploca undulata, 299f, 306
PEP, *see* Phosphoenolpyruvate
PEP carboxykinase, *see* Phosphoenol-pyruvate carboxykinase
PEP carboxylase, *see* Phosphoenol-pyruvate carboxylase
Peptidases, 159
Peptide hydrolysis, 153, 154, 159–160, 161
Peptidoglycans, 145
Peptone, 349
Perca fluviatilis, 18
Perch, 18, 19
Peridinium sp., 12, 71, 78, 85
Peridinopsis, 85
PFUs, *see* Plaque-forming units
pH
DOC and, 36–37
DOM and, 38
EOC and, 84
GLCase and, 146
iron bacteria and, 327
of lake water, 4, 5
Limnothrix redekei and, 218, 224, 226
methane-oxidizing bacteria and, 344
in sulfuretum, 288, 291, 294, 295
Phacotus sp., 75, 76f
Phaeopigments, 63, 64, 65, 66, 73
Phages, *see* Bacteriophages
Phagotrophic microorganisms, 118
Phagotrophic phytoflagellates, 81
PhAR, *see* Photosynthetically available radiation
PHB, *see* Poly-β-hydroxybutyrate
PhDOC, *see* Photosynthetically fixed carbon
Phenylalanine, 220
Phormidium hypolimneticum, 317
Phosphate, 21, 225, 289, 296, 362
dissolved reactive, 4

enzymatically hydrolyzable, 127, 131, 136
inorganic, 225
Phosphoenolpyruvate (PEP), 206
Phosphoenolpyruvate carboxykinase (PEP carboxykinase), 207–208
Phosphoenolpyruvate carboxylase (PEP carboxylase), 201–214, 360
ecosystem level control of, 208–214
enzyme effector regulation of, 202, 205–207
Phospholipids, 120
Phosphoric esters, 119
Phosphorus, 19, 26
APase in metabolism of, 135–136
bacteriophages and, 274
dissolved total, 4
ectoenzymes and, 120, 164
heterotrophic potential and, 198
inorganic, *see* Inorganic phosphorus
Limnothrix redekei and, 222, 227
microbial loop and, 252
organic, *see* Organic phosphorus
particulate, 4
phytoplankton primary production and, 68
5PNase in mineralization of, 142–144
sestonic, 124–125, 130
soluble reactive, 5–6, 225
in sulfuretum, 296
surplus, 124
Photic zone, 101, 106
Photoautotrophic carbon dioxide fixation, 251
Photoautotrophs, 25
Photorespiration, 82, 87
Photosynthesis, 63–65, 78–79, 145
AMPase and, 159
DFCHO and, 42
EOC and, 82, 83, 87, 180, 359
GLCase and, 147
grazing and, 260–261, 262
in situ characteristics of, 69–76
light-inhibited, 70, 73, 75
light-proportional, 71–72, 73
light-saturated, 72, 73, 75
Limnothrix redekei and, 70, 71, 72, 226
oligotrophic bacteria and, 179, 180
PEP carboxylase and, 204
quantum yields of, 69–73
saprophytic bacteria and, 179, 180
seasonal variation in, 67, 78
secondary production and, 94, 107
in sulfuretum, 294, 306, 310, 315–316
Photosynthetically active radiation (PAR), 64, 65–66
Photosynthetically available radiation (PhAR), 260, 261
Photosynthetically fixed carbon (PhDOC), 24, 43, 50, 81
Photosynthetic carbon dioxide assimilation, 67
Photosynthetic pigments, 312–315
Phototrophic bacteria, 196, 327
Phototrophic sulfur bacteria, 287, 297–302
photosynthetic pigments and, 312–315
Phragmites, 18
Phycobilin, 252
Phycosphere, 359
Phytoflagellates, 81
Phytoplankton, 9–12, 150, 252, 357
AMPase and, 159, 160
APase and, 128
bacteriophages and, 277
carbon flux and, 107–108
DCCHO and, 50
DOC and, 27–28, 51
DOM and, 55
DON and, 30
EOC of, *see* Extracellularly released organic carbon
glycolic acid secreted by, 25, 42–44
grazing and, 257, 258, 260, 261, 262, 264, 267
inorganic nitrogen and, 6
LDOCP and, 40, 41
metabolic coupling between bacteria and, 361
oligotrophic bacteria and, 176, 178–181, 182, 188
particulate production by, 84
photosynthesis and, 4, 257, 292
5PNase and, 143
saprophytic bacteria and, 176, 178–181, 182
seasonal variation in, 9–10
secondary production and, 107–108
in sulfuretum, 290–297, 314, 319
vertical stratification of, 10
Phytoplankton assemblages, 9, 193

Phytoplankton biomass, 10–12, 20–21
  APase and, 136
  chlorophyll$_a$ and, 76
  DOC and, 28
  DTCHO and, 34
  EOC and, 85
  grazing and, 13, 16, 263
  heterotrophic potential and, 198
  secondary production and, 97, 106, 111
Phytoplankton bloom, 165
  AMPase and, 154, 157
  APase and, 128, 129, 135, 139
  chlorophyll$_a$ and, 66–67
  DIP and, 127
  DOP and, 124
  EOC and, 84
  GLCase and, 146–147, 150, 152
  grazing and, 257, 259, 260, 263, 264, 267
  oligotrophic bacteria and, 178, 179, 181
  photosynthesis and, 75
  5PNase and, 138
  saprophytic bacteria and, 178, 179, 181
  SRP and, 6
  in sulfuretum, 291, 310
  zooplankton and, 13
Phytoplankton maxima, 13
Phytoplankton primary production, 21, 63, 66–69, 81, 123–124, 165–166, 356, 359
  chlorophyll$_a$ and, 63–69, 70, 71, 72–75, 76–78, 79
  EOC and, 83, 85–86, 87
  heterotrophic potential and, 193, 195, 196–198
  photosynthesis and, 63–65, 67, 69–76, 78–79
  seasonal variation in, 66–69
  secondary primary production compared with, 315
  secondary production and, 94, 95, 97, 110
  in sulfuretum, 289
  vertical stratification of, 66
Picoplankton, 84, 147
Pike, 17, 19
Piscivorous fish, 18, 81
Planktivorous fish, 81
Plankton, 31, 108, 358; *see also* Bacterioplankton; Microplankton; Phytoplankton; Zooplankton
Planktonic algae, 9
Planktothrix agardhii, 9, 10, 217
Planktothrix rubescens, 9, 10
Planktothrix sp., 69
Plaque assays, 271, 273–274
Plaque-forming units (PFUs), 272–273
Plasmid DNA, 278, 283
5PNase, *see* 5′-Nucleotidase
POC, *see* Particulate organic carbon
Polyacrylamide gel electrophoresis (PAGE), 46, 48–49
Polyacrylamide gel isoelectric focusing (PAGIF), 46
Polyamide-6, 36, 38
Polyarthra sp., 13
Polyarthra vulgaris, 13, 14
Polyclonal antibodies, 271
Poly-β-hydroxybutyrate (PHB), 342, 363
Polymeric carbohydrates, 144, 150
Polymeric organic matter
  bacterial secondary production and, 93
  microbial enzymatic degradation of, 119–120, 163–165
Polymeric substrates, 54–55
Polymers, 45, 153, 161, 180, 363; *see also* Biopolymers
Polynucleotides, 49
Polyphenols, 25, 37, 49
  dissolved, 34–36
  dissolved combined, 44–45
  dissolved free, 38–39, 44–45
  dissolved free monomeric, 45
  dissolved total, 36, 39, 44–45
Polyphosphates, 363
Polysaccharides, 49, 51, 361
  microbial enzymatic degradation of, 120, 121, 144–152, 163
  oligotrophic bacteria and, 188
Polyvinylpyrrolidone (PVP), 36, 37, 38
POM, *see* Particulate organic matter
Postglacial period, 9
Precipitation, 1
Primary production, 3, 251
  bacteriophages and, 278
  biopolymers and, 47–48
  oligotrophic bacteria and, 180
  saprophytic bacteria and, 180
  see also Phytoplankton primary production

Profundal zone
  AMPase in, 157, 159, 160, 161, 162
  secondary production in, 96, 100, 102, 105, 109, 111, 112
Proline, 220
Proteinaceous compounds, 152–162
Protein hydrolysis, 30, 152–162
Proteins, 49, 119, 120, 121, 145, 163
  fibrous, 153
  globular, 153
  Limnothrix redekei and, 218, 220, 222
  oligotrophic bacteria and, 188
  single-cell, 342, 344
  stress, 202
  substrate-binding, 362
  in sulfuretum, 296
Proteolytic enzymes, 153
Protozoans, 81, 162
  APase in, 128
  bacteria biomass and, 109, 111
  grazing by, 167, 229, 242, 261, 357
  5PNase and, 143
  secondary production and, 92
Pseudanabaena hypolimnetica, 317
Pseudomonas putida, 230, 240, 242
Pseudomonas sp., 187, 198
Pulsed-field gel electrophoresis, 280
Purines, 143, 144
Purple nonsulfur bacteria, 287, 288, 314, 317–319
Purple sulfur bacteria, 287, 297, 314, 317, 318, 319
PVP, *see* Polyvinylpyrrolidone
Pyrimidine, 143, 144

R

R3 bacteria, 304
Radiation
  global, *see* Global radiation
  photosynthetically active, 64, 65–66
  photosynthetically available, 260, 261
Radiation maxima, 77–78
RDOCP, *see* Recalcitrant dissolved organic carbon pool
Recalcitrant dissolved organic carbon pool (RDOCP), 25, 39, 44–45, 56
  heterotrophic activities and, 52
  oligotrophic bacteria and, 187, 188
Redox potential, 295, 306
Red sulfur bacteria, 311, 312, 314
Refractory dissolved organic carbon, *see* Recalcitrant dissolved organic carbon pool
Reproduction
  bacterial, 278
  phytoplankton, 9
  zooplankton, 245
Respiration
  dark, 65, 72
  photo, 82, 87
Rhodomicrobium fulvum, 318
Rhodomicrobium sp., 317
Rhodomicrobium tenue, 318
Rhodomicrobium vannielii, 318
Rhodomonas minuta, 10
Rhodomonas sp., 257, 259, 264, 266t, 291
Rhodopinal, 313
Rhodopseudomonas capsulata, 318
Rhodopseudomonas gelatinosa, 317–318, 319
Rhodopseudomonas palustris, 317–318, 319
Rhodopseudomonas sp., 317
Rhodopseudomonas viridis, 317–318, 319
Rhodospirillaceae, 287, 314t, 317–319
Rhodospirillales, 308
Rhodospirillum rubrum, 318
Rhodospirillum sp., 317
Ribulose-1,5-bisphosphate, 42, 87
Ribulose-bisphosphate carboxylase, 82
RNA, 49, 137, 143–144
RNase A, 281
Roach, 18, 19
Rotifers, 12, 13, 14, 16, 229, 235, 242
Rottensee, 64, 77
rpoS gene, 362
rRNA, 341
Rudd, 18, 19
Rutilus rutilus, 18

S

Saprophytic bacteria, 175–188, 360
  abundance of, 176–178
  taxonomy of, 181–185
  vertical stratification of, 178–181
Sargasso Sea, 274
Scanning electron microscopy, 327
Scardinius erythrophthalmus, 18, 19

Scenedesmus quadricauda, 204, 232, 245, 317
Schmidlea luteola, 297
Schöhsee, 84, 180
SCP, *see* Single-cell protein
Seasonal variation
of HMWDOC, 49–50
of oligotrophic bacteria, 176–178
of photosynthesis, 67, 78
of phytoplankton, 9–10
of phytoplankton primary production, 66–69
of saprophytic bacteria, 176–178
of zooplankton, 13–14
see also Autumn; Spring; Summer; Winter
Secchi disk visibility, 65
chlorophyll$_a$ and, 77
grazing and, 257, 260
microbial enzymatic degradation and, 121
in sulfuretum, 288, 291
Secondary primary production, 315–316
Secondary production, 92–112, 123, 165, 357
AMPase and, 111, 157, 160
grazing and, 92, 94, 263
[^{3}H]leucine measurements of, *see* [^{3}H]Leucine
[^{3}H]TdR measurements of, *see under* [^{3}H-Methyl]thymidine analysis
regulators of, 108–111
Sediment/hypolimnion interface, 39
Sediments
DOM in, 24
fish fauna in, 17
glucose uptake in, 198–199
marine, 24
methanogenesis in, 339
secondary production in, 92
SRP and, 6
Sediment stagnant pool, 352
Sediment-water interface, 8, 47, 292
Sephadex G15, 38–39, 187
Sephadex G25, 206, 209
Sephadex G75, 46–48
Serine, 42, 160, 220, 342
Sestonic phosphorus, 124–125, 130
Siderocapsa geminata, 326, 327, 328t
morphology of, 331–333
Signal hypothesis, 360
Silent genes, 278
Silicate, 8
Silicon, 8
Silicon dioxide, 4, 8
Single-cell protein (SCP), 342, 344
Sloppy feeding, 51, 82, 85, 259
Sodium chloride, 202
Solar Lake, 294, 315
Soluble reactive phosphorus (SRP), 5–6, 225
Spectrofluorometry, 121, 124
Sphaerocystis sp., 70, 71, 75
Spined loach, 19
Spirilloxanthin, 313
Spirostoma minus, 265f
Spore-forming bacteria, 175
Spring
algae in, 19
AMPase in, 154
APase in, 128, 129, 131, 135, 136
C:N ratio in, 32
DCCHO in, 49
dephosphorylation in, 141
DFAA in, 40, 52
DFCHO in, 40, 53, 54
DIP in, 127
dissolved oxygen in, 4
dissolved polyphenols in, 36
DOC in, 28
DON in, 30
DOP in, 31, 124
DTAA in, 34
EOC in, 84
GLCase in, 146, 150
grazing in, 257–260, 263, 264, 267
inorganic nitrogen in, 6
iron bacteria in, 327
photosynthesis in, 70, 75
phytoplankton in, 9–10
phytoplankton primary production in, 66–67, 69, 70, 75
5PNase in, 138
secondary production in, 105
SRP in, 6
sulfuretum in, 290, 291
zooplankton in, 13, 14–16, 17
SRP, *see* Soluble reactive phosphorus
Stable-isotope analysis, 26
Starch, 120, 144, 163

Starvation, 53, 245, 253, 361–363
Staurastrum, 10, 235
Steemann-Nielsen technique, 83
Stephanodiscus alpinus, 9, 10
Stephanodiscus hantzschii, 9, 193, 264
Stephanodiscus minutulus, 9
Stephanodiscus sp., 10, 257, 266t, 291, 308
Stizostedion lucioperca, 19
Stratification
  chemical, 306
  thermal, *see* Thermal stratification
  vertical, *see* Vertical stratification
Stress proteins, 202
Substrate-binding proteins, 362
Substrates
  bacterial adaptation strategies and, 360–361, 362–363
  methanogenesis influenced by, 339
  monomeric, 51–54
  organic, 289–290
  polymeric, 54–55
Sugar phosphates, 163
Sulfate, 289, 297, 305, 340
Sulfate-reducing bacteria, 287, 297–302, 304–305, 310, 311, 313, 316
Sulfide, 288–289, 297, 312, 315
Sulfur, 26
sulfuretum, 287–319
  development and colonization of, 305–308
  microorganisms of, 297–305
  organic substrates in, 289–290
  physical and chemical analysis of, 288–289
  physical and chemical properties of, 290–297
  secondary primary production in, 315–316
  uptake kinetics in, 309–311
Sulfur-reducing bacteria, 313
Summer, 3
  algae in, 19
  AMPase in, 154, 157
  APase in, 128, 130, 131, 135, 139
  chlorophyll$_a$ in, 77
  C:N ratio in, 32
  DCCHO in, 49
  DFAA in, 40
  DFCHO in, 40
  DIP in, 127
  dissolved polyphenols in, 36
  DOC in, 27
  DON in, 30
  DOP in, 31
  DTAA in, 34
  EOC in, 85, 86–87
  grazing in, 260–261
  methane in, 352
  phytoplankton in, 10
  phytoplankton primary production in, 68, 75, 77
  5PNase in, 138
  secondary production in, 101, 105
  sulfuretum in, 291
  zooplankton in, 13, 16
Summer stagnation
  DOP and, 31
  grazing and, 264
  in sulfuretum, 288, 294, 296, 315
Summer stratification, 4, 63, 165
  AMPase and, 157
  APase and, 129, 132, 135, 136
  dissolved oxygen and, 5
  DOP and, 124, 126
  grazing and, 260, 266
  inorganic nitrogen and, 6–8
  iron bacteria and, 327
  oligotrophic bacteria and, 177
  photosynthesis and, 78
  phytoplankton primary production and, 78
  saprophytic bacteria and, 177
  SRP and, 6
  in sulfuretum, 291, 304
Surface waters
  APase in, 134
  biopolymers in, 47–48
  C:P ratio in, 32
  DCAA in, 51
  DFAA in, 42
  DFCHO in, 53, 54
  dissolved polyphenols in, 36
  DOC in, 39
  DOP in, 31
  LDOCP in, 43–44
  N:P ratio in, 33
Surplus phosphorus, 124
Synchaeta pectinata, 13, 14
Synchaeta sp., 244
Synechococcus sp., 10, 75, 274

T

T2 prokaryotes, 303–304, 306
T3 prokaryotes, 303–304, 306
T4 prokaryotes, 304, 306, 308
T5 prokaryotes, 304, 306, 308, 309f
Taxonomy
  of oligotrophic bacteria, 181–185, 186
  of saprophytic bacteria, 181–185
TCA cycle, *see* Tricarboxylic acid cycle
TDAN, *see* Total dissolved adenosine
[^{3}H]TdR, *see* [^{3}H-Methyl]thymidine analysis
TEM, *see* Transmission electron microscopy
Temperate bacteriophages, 278–284
Temperature, *see* Water temperature
Tench, 19
Thermal stratification, 1–3, 20, 360
  grazing and, 254, 257
  phytoplankton and, 10
  phytoplankton primary production and, 66, 70, 72, 73
  secondary production and, 105
  in sulfuretum, 290, 306
  summer, *see* Summer stratification
  winter, 290–291
Thermochemocline, 308
Thermocline, 1, 20
  APase in, 134
  biopolymers in, 47, 48
  dissolved polyphenols in, 36
  DOC in, 39
  DTCHO in, 34
  grazing in, 254, 256, 267
  methane in, 352
  oligotrophic bacteria in, 179
  phytoplankton primary production in, 66
  saprophytic bacteria in, 179
  sulfuretum organisms in, 287, 290, 315, 316
  zooplankton in, 16
Thin-layer chromatography (TLC), 38, 211–214
Thiobacillus denitrificans, 297
Thiobacillus sp., 294
Thiobiocoenosis, 288, 306, 311–312
Thiobiotope, 288
Thiocystis violacea, 306
Thiopedia rosea, 306
Third Sister Lake, 139
Threonine, 220
Tinca tinca, 19
Tintinnidium pusillum, 259, 264, 265f, 266f, 267
TLC, *see* Thin-layer chromatography
Top-down regulation, 108, 111, 112, 245
Total dissolved adenosine (TDAN), 142, 143
Total extracellular released organic carbon (EOC_t), 84, 86
Total organic carbon, 165–166
Transmission electron microscopy (TEM)
  bacteriophages examined with, 273–274
  iron bacteria identified with, 327
  phytoplankton examined with, 9
  sulfuretum organisms examined with, 297, 304
Tricarboxylic acid (TCA) cycle, 202, 205, 206, 208
Trichomes
  in *Limnothrix redekei,* 220–222
  in sulfuretum organisms, 302, 308
Trophic cascade, 112
Trophic levels, 167
  bacteria biomass and, 107, 111–112
  food chains in, 118, 119, 166
  secondary production and, 108
Trophogenic zone, 31, 32, 291
Tryptophan, 30
Typha, 18
Tyrosine, 220

U

UDOM, *see* Utilizable dissolved organic matter
Utilizable dissolved organic matter (UDOM), 93, 119–120, 164, 165
  allosteric enzyme regulation and, 201
  grazing and, 258

V

Vertical stratification
  of oligotrophic bacteria, 178–181
  of phytoplankton, 10
  of phytoplankton primary production, 66
  of saprophytic bacteria, 178–181
  of zooplankton, 14–15
Vibrio sp., 186
Viral loop, 56

Viruses, 56, 270–284, 357, 358, 359; *see also* Bacteriophages

W

Water column
- AMPase in, 157, 161
- APase in, 132–135
- biopolymers in, 47, 48
- DFAA in, 40
- DFCHO in, 40
- dissolved oxygen in, 4
- DOC in, 27, 36
- DOM in, 26, 33
- DON in, 30
- DOP in, 31
- DTAA in, 34
- GLCase in, 149
- glucose uptake in, 193–198, 199
- glycolic acid in, 43
- inorganic nitrogen in, 8
- iron bacteria in, 327
- methane in, 347, 348
- oligotrophic bacteria in, 178–179
- phytoplankton in, 10
- saprophytic bacteria in, 178–179
- secondary production in, 92, 94, 95, 101, 104–107, 108, 109, 111
- SRP in, 6
- in sulfuretum, 293, 310
- zooplankton in, 14, 15, 16–17

Water temperature, 65–66
- AMPase and, 154, 157
- dissolved oxygen and, 4
- EOC and, 86, 87
- GLCase and, 145–146
- grazing and, 240, 241t, 256, 257
- inorganic nitrogen and, 8
- phytoplankton primary production and, 75
- secondary production and, 101, 102
- in sulfuretum, 288, 290–291, 306
- zooplankton and, 14

WCO, *see* Wet chemical oxidation

Wet chemical oxidation (WCO), 27–28, 29, 30

Winter, 3
- APase in, 128, 131
- C:N ratio in, 32
- dissolved oxygen in, 5
- DOP in, 31, 124
- grazing in, 254–257, 263
- inorganic nitrogen in, 8
- methane in, 352
- oligotrophic bacteria in, 176, 184
- photosynthesis in, 70, 72
- phytoplankton in, 10
- phytoplankton primary production in, 67, 68, 69, 70, 72
- saprophytic bacteria in, 176, 184
- SRP in, 6
- sulfuretum in, 290–291, 293, 295

Winter stagnation, 293

Winter stratification, 290–291

Wood-Werkman reaction, 84

X

X-phosphate, *see* 5-Bromo-4-chloro-3-indolyl phosphate

Y

Yeast extract
- grazing and, 253
- PEP carboxylase and, 209, 211, 212t, 213, 214

Z

Zinc, 137

Zooplankton, 12–17, 18, 82, 358
- APase and, 128
- bacteria biomass and, 109, 111, 112
- EOC and, 81, 82, 83, 86
- herbivorous, 13, 82, 85, 126, 356
- heterotrophic, 19
- phytoplankton biomass and, 85
- 5PNase and, 143
- seasonal variation in, 13–14
- secondary production and, 92
- vertical stratification of, 14–15

Zooplankton assemblage, 12–13

Zooplankton biomass, 13, 21
- dynamics of, 15–17
- grazing and, 240–242

Zooplankton biomass maxima, 16, 17

Zooplankton grazing, 13, 16, 51, 83, 120, 167, 251–267, 356, 357, 359, 361
- DOP and, 126
- field experiments in, 233, 234f
- food quality/concentration and, 238–240
- *in vitro* experiments on, 232–233, 245

Zooplankton grazing (*cont.*)
in laboratory strains, 232, 235, 242
LDOCP and, 41
phytoplankton primary production and, 81
secondary production and, 92, 94, 263
species effect on, 235–238
in sulfuretum, 317
see also Ciliate grazing; Heterotrophic nanoflagellate grazing
Zooplankton grazing pressure, 179, 242, 244–245, 259
Zymogenous bacteria, 175

Ecological Studies

Volume 68
Stable Isotopes in Ecological Research (1989)
P.W. Rundel, J.R. Ehleringer, and K.A. Nagy (Eds.)

Volume 69
Vertebrates in Complex Tropical Systems (1989)
M.L. Harmelin-Vivien and F. Bourlière (Eds.)

Volume 70
The Northern Forest Border in Canada and Alaska (1989)
J.A. Larsen

Volume 71
Tidal Flat Estuaries: Simulation and Analysis of the Ems Estuary (1988)
J. Baretta And P. Ruardij (Eds.)

Volume 72
Acidic Deposition and Forest Soils (1989)
D. Binkley, C.T. Driscoll, H.L. Allen, P. Schoeneberger, and D. McAvoy

Volume 73
Toxic Organic Chemicals in Porous Media (1989)
Z. Gerstl, Y. Chen, U. Mingelgrin, and B. Yaron (Eds.)

Volume 74
Inorganic Contaminants in the Vadose Zone (1989)
B. Bar-Yosef, N.J. Barrow, and J. Goldshmid (Eds.)

Volume 75
The Grazing Land Ecosystems of the African Sahel (1989)
H.N. Le Houérou

Volume 76
Vascular Plants as Epiphytes: Evolution and Ecophysiology (1989)
U. Lüttge (Ed.)

Volume 77
Air Pollution and Forest Decline: A Study of Spruce (*Picea abies*) on Acid Soils (1989)
E.-D. Schulze, O.L. Lange, and R. Oren (Eds.)

Volume 78
Agroecology: Researching the Ecological Basis for Sustainable Agriculture (1990)
S.R. Gliessman (Ed.)

Volume 79
Remote Sensing of Biosphere Functioning (1990)
R.J. Hobbs and H.A. Mooney (Eds.)

Volume 80
Plant Biology of the Basin and Range (1990)
C.B. Osmond, G.M. Hidy, and L.F. Pitelka (Eds.)

Volume 81
Nitrogen in Terrestrial Ecosystem: Questions of Productivity, Vegetational Changes, and Ecosystem Stability (1990)
C.O. Tamm

Volume 82
Quantitative Methods in Landscape Ecology: The Analysis and Interpretation of Landscape Heterogeneity (1990)
M.G. Turner and R.H. Gardner (Eds.)

Volume 83
The Rivers of Florida (1990)
R.J. Livingston (Ed.)

Volume 84
Fire in the Tropical Biota: Ecosystem Processes and Global Challenges (1990)
J.G. Goldammer (Ed.)

Volume 85
The Mosaic-Cycle Concept of Ecosystems (1991)
H. Remmert (Ed.)

Volume 86
Ecological Heterogeneity (1991)
J. Kolasa and S.T.A. Pickett (Eds.)

Volume 87
Horses and Grasses: The Nutritional Ecology of Equids and Their Impact on the Camargue (1992)
P. Duncan

Volume 88
Pinnipeds and El Niño: Responses to Environmental Stress (1992)
F. Trillmich and K.A. Ono (Eds.)

Ecological Studies

Volume 89
Plantago: A Multidisciplinary Study (1992)
P.J.C. Kuiper and M. Bos (Eds.)

Volume 90
Biogeochemistry of a Subalpine Ecosystem: Loch Vale Watershed (1992)
J. Baron (Ed.)

Volume 91
Atmospheric Deposition and Forest Nutrient Cycling (1992)
D.W. Johnson and S.E. Lindberg (Eds.)

Volume 92
Landscape Boundaries: Consequences for Biotic Diversity and Ecological Flows (1992)
A.J. Hansen and F. di Castri (Eds.)

Volume 93
Fire in South African Mountain Fynbos: Ecosystem, Community, and Species Response at Swartboskloof (1992)
B.W. van Wilgen et al. (Eds.)

Volume 94
The Ecology of Aquatic Hyphomycetes (1992)
F. Bärlocher (Ed.)

Volume 95
Palms in Forest-Ecosystems of Amazonia (1992)
F. Kahn and J.-J. DeGranville

Volume 96
Ecology and Decline of Red Spruce in the Eastern United States (1992)
C. Eagar and M.B. Adams (Eds.)

Volume 97
The Response of Western Forests to Air Pollution (1992)
R.K. Olson, D. Binkley, and M. Böhm (Eds.)

Volume 98
Plankton Regulation Dynamics (1993)
N. Walz (Ed.)

Volume 99
Biodiversity and Ecosystem Function (1993)
E.-D. Schulze and H.A. Mooney (Eds.)

Volume 100
Ecophysiology of Photosynthesis (1994)
E.-D. Schulze and M.M. Caldwell (Eds.)

Volume 101
Effects of Land-Use Change on Atmospheric CO_2 Concentrations: South and South East Asia as a Case Study (1993)
V.H. Dale (Ed.)

Volume 102
Coral Reef Ecology (1993)
Y.I. Sorokin (Ed.)

Volume 103
Rocky Shores: Exploitation in Chile and South Africa (1993)
W.R. Siegfried (Ed.)

Volume 104
Long-Term Experiments With Acid Rain in Norwegian Forest Ecosystems (1993)
G. Abrahamsen et al. (Eds.)

Volume 105
Microbial Ecology of Lake Plußsee (1993)
J. Overbeck and R.J. Chrost (Eds.)

Volume 106
Minimum Animal Populations (1994)
H. Remmert (Ed.)